ADVANCES IN SUPERCONDUCTIVITY

New Materials, Critical Currents and Devices

Inauguration of ASMCCD '96 by Prof. V. Singh (Director, T.I.F.R.) in Homi Bhabha Auditorium, T.I.F.R., September 17, 1996 (inset).

Seated from left to right: Prof. S.K. Malik and Prof. R. Pinto (Chairpersons, ASMCCD '96), Prof. M. Murakami (S.R.L., ISTEC, Japan), Prof. J.R. Clem (Ames Laboratory, Iowa, USA and Editor, High T. Update) and Prof. S.S. Jha (present Director, T.I.F.R.).

ADVANCES IN SUPERCONDUCTIVITY
NEW MATERIALS, CRITICAL CURRENTS AND DEVICES

Editors

R. PINTO • S.K. MALIK
A.K. GROVER • P. AYYUB

Tata Institute of Fundamental Research, Mumbai, India

PUBLISHING FOR ONE WORLD

NEW AGE INTERNATIONAL (P) LIMITED, PUBLISHERS

New Delhi • Bangalore • Chennai • Cochin • Guwahati • Hyderabad
Jalandhar • Kolkata • Lucknow • Mumbai • Ranchi
Visit us at www.newagepublishers.com

Published by New Age International (P) Ltd., Publishers
First Edition: 1997
Reprint: 2009

Branches:

- 36, Malikarjuna Temple Street, Opp. ICWA, Basavanagudi, **Bangalore**. ✆ (080) 26677815
- 26, Damodaran Street, T. Nagar, **Chennai**. ✆ (044) 24353401
- CE-39/1016, Carrier Station Road, Near Paulson Park Hotel, Ernakulam South, **Cochin.** ✆ (0484) 4051303
- Hemsen Complex, Mohd. Shah Road, Paltan Bazar, Near Starline Hotel, **Guwahati**. ✆ (0361) 2543669
- No. 105, 1st Floor, Madhiray Kaveri Tower, 3-2-19, Azam Jahi Road, Nimboliadda, **Hyderabad**. ✆ (040) 24652456
- RDB Chambers (Formerly Lotus Cinema) 106A, 1st Floor, S.N. Banerjee Road, **Kolkata**. ✆ (033) 22275247
- 16-A, Jopling Road, **Lucknow**. ✆ (0522) 2209578
- 142C, Victor House, Ground Floor, N.M. Joshi Marg, Lower Parel, **Mumbai**. ✆ (022) 24927869
- 22, Golden House, Daryaganj, **New Delhi**. ✆ (011) 23262370, 23262368

ISBN : 978-81-224-1125-6

Rs. 600.00

C-09-06-3718

Printed in India at Glorious Printers, Delhi.

PUBLISHING FOR ONE WORLD
NEW AGE INTERNATIONAL (P) LIMITED, PUBLISHERS
4835/24, Ansari Road, Daryaganj, New Delhi-110002
Visit us at **www.newagepublishers.com**

ADVISORY COMMITTEE

NATIONAL

B.M. Arora (Mumbai)
R. Budhani (Kanpur)
P. Chaddah (Indore)
K.L. Chopra (Kharagpur)
A.D. Damodaran (Trivandrum)
B.A. Dasannacharya (Indore)
B. Ghosh (Calcutta)
A.K. Gupta (Delhi)
M.S. Hegde (Bangalore)
S.S. Jha (Mumbai)
P.K. Kaw (Ahmedabad)
N. Kumar (Bangalore)
R.A. Mashelkar (Delhi)
V.K. Misra (Delhi)
A.V. Narlikar (Delhi)
S.B. Ogale (Pune)
B.D. Padalia (Mumbai)
T.S. Radhakrishnan (Kalpakkam)
T.V. Ramakrishnan (Bangalore)
V.S. Ramamurthy (Delhi)
Ram Prasad (Mumbai)
C.N.R. Rao (Bangalore)
B.K. Sarkar (Mumbai)
K.N. Shankara (Ahmedabad)
R. Srinivasan (Indore)
G.P. Srivastava (Delhi)
O.N. Srivastava (Varanasi)
G.V. Subba Rao (Karaikudi)
R. Vijayaraghavan (Mumbai)

INTERNATIONAL

A. Barone (Italy)
J.G. Bednorz (Switzerland)
A.M. Campbell (U.K.)
P. Chaudhari (U.S.A.)
C.W. Chu (U.S.A.)
G.W. Crabtree (U.S.A.)
F. de La Cruz (Argentina)
G. Donaldson (U.K.)
R.B. Goldfarb (U.S.A.)
C.E. Gough (U.K.)
R.P. Griessen (Holland)
H. Hayakawa (Japan)
C. Heiden (Germany)
K. Kitazawa (Japan)
D.C. Larbalestier (U.S.A.)
K.H. Müller (Australia)
R. Nicolsky (Brazil)
Y. Nishihara (Japan)
N.F. Pedersen (Denmark)
K.V. Rao (Sweden)
B. Raveau (France)
J.M. Rowell (U.S.A)
T. Venkatesan (U.S.A)
H.W. Weber (Austria)
Y. Yeshurun (Israel)

ORGANIZING COMMITTEE

Chairpersons : R. Pinto and S.K. Malik

P.R. Apte
R.S. Chaughule
S. Ramakrishnan
S.K. Roy
E.V. Sampathkumaran
P. Ayyub
A.K. Grover
M. Randeria
S. Saha
V.V. Srinivasu

SCIENTIFIC PROGRAMME COMMITTEE

P. Ayyub
S. K. Malik
M. Randeria
S. Saha
A.K. Grover
R. Pinto
S.K. Roy

LOCAL ARRANGEMENTS COMMITTEE

P.R. Apte
S. Bhattacharya
C.M. Borde
Soma Chattopadhyay
O.B. Fernandes
K.V. Gopalakrishnan
S.S. Gurav
R. Kumar
V. Mahadevan
S.K. Paghdar
V.R. Palkar
R. Patil
S.C. Purandare
S.K. Roy
R.S. Sanabhatti
S. Srinivas
U. Vaidya
S. S. Banerjee
Hema Bhola
S.S. Chandvankar
R.S. Chaughule
S. Ghosh
A.K. Grover
J. Jesudasan
S.G. Lokhre
C. Mitra
S.P. Pai
N. Patil
S.M. Pattalwar
P. Raychaudhuri
S. Saha
S. Saran
V.V. Srinivasu
A. Venugopal

PREFACE

During the middle of 1995 some of us in the condensed matter group of the Tata Institute of Fundamental Research thought of holding an international meeting as a part of Golden Jubilee Celebrations of the TIFR. For many of us who have worked on superconductivity, there would not have been any better topic than that area. Further, since the year 1996 happened to be the tenth anniversary of the discovery of superconductivity in cuprates by Bednorz and Müller, an international meeting on Advances in Superconductivity was thought to be both timely and appropriate. So, what started as a mere idea grew into an International Symposium on Advances in Superconductivity.

Undoubtedly, the discovery of superconductivity in the cuprates by J.G. Bednorz and K.A. Müller produced an explosive growth in research, driven by the quest for higher and higher superconducting transition temperatures. In the initial stages, the excitement was tremendous both in the physical sciences and in engineering. However, the complexity of the new materials on the one hand, and the absence of a viable theory on the other, have made further developments much more difficult. It is to be expected therefore, that the early excitement and the subsequent rapid advances have paved the way for more systematic and detailed studies of all aspects of superconductivity. This, of course, is highly desirable.

Research in superconductivity commenced in the TIFR about three decades ago. Therefore, when high temperature superconductivity was discovered, because of the existing infrastructure and expertise, the TIFR scientists could immediately initiate studies on such systems. The united effort of material scientists, low temperature physicists, theoreticians and thin film device experts nucleated a major programme in the area of High Temperature Superconductivity at the TIFR.

The contributions to the field of superconductivity by the TIFR group are diverse. In the area of materials, Pr and Ca problem in $YBa_2Cu_3O_{7-\delta}$ (Y-123) and Lu-123 cuprates have been extensively studied. The Lu-123 system which is unstable in ceramic bulk form due to the small ionic radius of Lu, has been stabilized with partial substitution of Tb and Pr for Lu. The valence state of Pr and Tb in Lu-123 has also been studied using x-ray photoelectron spectroscopy (XPS). The most significant contribution in the area of new materials was the discovery of superconductivity in borocarbides. These new materials may be extremely useful for the understanding of coexistence of magnetism and superconductivity. There have also been significant contributions in the physics of cuprate superconductors. The most important among them are: magnetization and nuclear quadrupole resonance (NQR) studies in some rare earth cuprates, specific heat measurements, magnetic phase diagram of irreversible type I and type II superconductors, and finite size effects in nanoparticle based cuprates.

Due to the paucity of high quality single crystals, epitaxial thin films of cuprates have made tremendous progress, and the TIFR group has made very significant contributions, the most important among them being the beneficial effects of Ag- and Au- doping in Y-123 thin films leading to improved microstructural and transport properties. Other notable contributions include the realization of metastable Lu-123 in thin film form by pulsed laser ablation deposition, study of self-induced flux creep and flow in cuprate thin films, development of a new technique for the measurement of metal-superconductor thin film contact conductance, and microwave transmission studies in cuprate thin films using microstrip resonator technique.

This International Symposium was intended to provide a forum to review the progress in selected areas in superconductivity. The emphasis was on experimental and theoretical studies of the new superconductors, advances in the theoretical understanding, progress in studies of flux pinning and vortex dynamics which affect critical currents, and development of novel material synthesis methods. Recent developments in the twin areas of thin films and devices was extensively discussed. The Symposium program comprised plenary lectures, invited talks, contributed oral and poster presentations.

About 200 scientists and research students including about thirty from abroad, participated in the four day long meeting. The ASMCCD '96 Symposium was inaugurated by Professor Virendra Singh, Director, TIFR. In the Opening Session, two eminent scientists, Professor J.R. Clem of Iowa State University, U.S.A. and Professor M. Murakami of Superconductivity Research Laboratory, Japan delivered stimulating plenary lectures. Professor T. Venkatesan of Centre for Superconductivity Research, University of Maryland, U.S.A. gave a very interesting colloquium. There were 37 invited talks, 8 oral presentations and 88 poster papers. At the end, Professor R. Srinivasan, Raman Research Institute, Bangalore, India beautifully summarized the Symposium. We are very glad that the ASMCCD '96 Symposium was one of the highly successful meetings held in the TIFR which gave an opportunity for Indian participants to interact with distinguished scientists from abroad.

We are indebted to the Advisory Committee for their valuable suggestions and for their personal interest in attracting some of the leading experts as invited speakers. We take this opportunity to thank the plenary and invited speakers for their interesting lectures, the contributors to Poster Sessions and all the other participants in the Symposium. The response from the participants was overwhelming. We also thank all who have contributed manuscripts to this Proceedings.

Finally, we take this opportunity to thank the Tata Institute of Fundamental Research and the Department of Science and Technology, Government of India for providing financial support for holding the ASMCCD '96 Symposium. We would also like to express our appreciation to various departments of the TIFR, the administration and others, for providing support during the Symposium. Finally, we thank the members of various local organizing subcommittees for their dedicated work and valuable contributions in making the ASMCCD '96 Symposium a grand success.

R. Pinto
S.K. Malik
A.K. Grover
P. Ayyub

CONTENTS

CONTRIBUTED PAPERS

THIN FILMS AND DEVICES

INVITED LECTURES

CONTRIBUTED PAPERS

VORTEX STRUCTURE, PINNING AND DYNAMICS

INVITED LECTURES

CONTRIBUTED PAPERS

THEORETICAL ASPECTS OF HIGH-T_C SUPERCONDUCTORS

INVITED LECTURES

CONTRIBUTED PAPERS

STUDIES ON THE OXYGEN STOICHIOMETRY AND COPPER VALENCY IN SUPERCONDUCTING CUPRATES BY MEANS OF CHEMICAL ANALYSIS

M. Karppinen[1,2], H. Yamauchi[2] and L. Niinistö[1]

[1]*Laboratory of Inorganic and Analytical Chemistry, Helsinki University of Technology, FIN-02150 Espoo, Finland*
[2]*Materials and Structures Laboratory, Tokyo Institute of Technology, Yokohama 226, Japan*

Several methods of chemical analysis, *i.e.* various redox titrations and thermogravimetric approaches, have been developed and/or adapted to establish the oxygen content and to determine the higher oxidation states of copper and other mixed-valent metals in various superconducting and related cuprates. The present contribution discusses the potential and applicability of each method and summarizes the experimental results obtained.

I. INTRODUCTION

Like many other perovskites, high T_c oxide superconductors show variable oxygen contents depending on the conditions during the processing. Superconductivity occurs only in a narrow doping range, controlled both by the oxygen adjustment and aliovalent cation substitutions. Without reliable knowledge of the oxygen stoichiometry the properties of the prepared sample can not be interpreted. Furthermore, the distribution of the excess charge between the superconducting CuO_2 planes and non-superconducting blocks is an important question.

The number of holes or electrons is commonly associated with the oxidation state of copper. However, although the changes occurring in the carrier concentration are discussed here in terms of metal valences, it is recognized that the excess positive or negative charge is likely to be delocalized over the frameworks formed by the metals and the surrounding oxygen atoms. Especially in the case of hole-doped CuO_2 planes, it is generally agreed that there exist no localized Cu^{3+} ions but the conduction band is derived from the Cu $3d(x^2\text{-}y^2)$ and O 2p states. Fortunately, trivalent copper and peroxide ion behave identically in the utilized wet chemical redox experiments, and the actual quantity being measured is the sum of holes belonging to either copper or oxygen, *i.e.* the concentration of $[CuO]^+$. In the case of the alkaline earth or most rare earth cuprates the ratio of the cations to oxygen content determines the degree of mixed-valency of copper. In Bi-, Pb-, Tl- and Hg-based superconductors, on the other hand, the analytical difficulties arise from the fact that besides copper also bismuth, lead, thallium and mercury may exhibit mixed oxidation states, since all of these metals possess a 6s or 6p band close to the Cu $3d(x^2\text{-}y^2)$ - O 2p band[1]. This means that calculating the valence values for copper from the oxygen content results the corresponding valences of the other multi-valent metals have to be determined.

II. DETERMINATION OF THE OXYGEN CONTENT

Several methods of chemical analysis are available to establish the oxygen contents in superconducting cuprates[2]. In the following the applicability of oxidation-reduction titrations based on iodide/iodine or Cu(+I)/Cu(+II) chemistry as well as thermogravimetric (TG) analyses measuring the weight loss accompanying either total or partial reduction of the cuprate are discussed.

Iodometry

Iodometric titration is probably the most widely used method for analyzing the oxygen content of rare earth cuprates[3]. The sample is dissolved in an acidic solution of KI resulting in the reduction of all copper to solid Cu(+I) iodide according to Eqs. 1 and 2. Back-titration of the liberated iodine by thiosulfate (Eq. 3) gives the average oxidation state of copper and thereby the oxygen content of the sample. The equivalent point of the titration is in most cases determined visually using starch as indicator, but potentiometric detection is possible as well[4].

$$Cu^{2+} + 2\,I^- \rightarrow CuI + 0.5\,I_2 \quad (1)$$
$$[CuO]^+ + 3\,I^- + 2\,H^+ \rightarrow CuI + I_2 + H_2O \quad (2)$$
$$I_2 + 2\,S^2O_3^{2-} \rightarrow 2\,I^- + S_4O_6^{2-} \quad (3)$$

High-valent bismuth, lead and thallium oxidize iodide as well, enabling in principle the determination of the total oxidation power of the Bi-, Pb- and Tl-based superconductors by iodometry. In the case of bismuth and lead cuprates, however, the end point detection may be somewhat inaccurate because of the formation of strongly coloured Bi(+III) and Pb(+II) iodide complexes. In the case of Tl-based superconductors, the formation of a passive surface coating during the dissolution in a mixture of KI and HCl prevents further dissolution of bulk material, and as a consequence a slightly modified titration route, proceeding *via* HBr dissolution and a subsequent iodometric determination of the liberated bromine, has to be applied[5]. Oxygen contents in mercury cuprates have been analyzed iodometrically by assuming a valence value of +2 for mercury[6].

Reduction with Cu(+I)

The oxygen contents in superconducting rare earth[7], bismuth[8], lead[9,10] and thallium[10] cuprates can also be analyzed by a redox process utilizing the reduction capability of Cu(+I). The total oxidation power corresponding to the oxidation states of +III, +V, +IV and +III for Cu, Bi, Pb and Tl, respectively, is determined according to Eqs. 4-7 by dissolving the sample in HCl solution containing CuCl. The excess Cu(+I) is then titrated by electrochemical oxidation as indicated in Eq. 8. In the case of RE-, Bi- or Pb-based materials it is most convenient to use coulometric back-titration of the excess Cu(+I), while for the Tl cuprates potentiostatic electrolysis has to be applied.

$$[CuO]^+ + Cu^+ + 2\,H^+ \rightarrow 2\,Cu^{2+} + H_2O \quad (4)$$
$$BiO_3^- + 2\,Cu^+ + 4\,H^+ \rightarrow BiO^+ + 2\,Cu^{2+} + 2\,H_2O \quad (5)$$
$$Pb^{4+} + 2\,Cu^+ \rightarrow Pb^{2+} + 2\,Cu^{2+} \quad (6)$$
$$Tl^{3+} + 2\,Cu^+ \rightarrow Tl^+ + 2\,Cu^{2+} \quad (7)$$
$$Cu^+\ (\text{excess}) \rightarrow Cu^{2+} + e^- \quad (\text{coulometry/electrolysis}) \quad (8)$$

The Cu(+I)/Cu(+II) reduction method can be applied to the characterization of oxygen-deficient samples as well. This is important *e.g.* for the lead cuprates which, although being hole-type superconductors, exhibit average Cu valence values below +II[12].

Thermogravimetric reduction

In the thermogravimetric reduction method the sample is heated to high temperatures in a reducing atmosphere, and from the total weight loss during the thermal treatment the amount of oxygen in the starting material can be calculated if the reduction products are known. Hydrogen is the most commonly used reductant for the total reduction of copper to metallic state. For example, heating the $YBa_2Cu_3O_{7-\delta}$ compound under a H_2/Ar atmosphere results at 950 ^{0}C in a selective reduction of copper to metallic state, while yttrium and barium are maintained as oxides (Eq. 9). Like yttrium, all the other rare earth elements remain as trivalent oxides[13]. The H_2/TG reduction method can not be recommended for the Bi-, Pb-, Tl- or Hg-based materials due to the easy volatilization of these metals.

$$2\,YBa_2Cu_3O_{7-\delta} + (7-2\delta)\,H_2 \rightarrow Y_2O_3 + 4\,BaO + 6\,Cu + (7-2\delta)\,H_2O \qquad (9)$$

Recently, a large number of new homologous series of superconducting cuprates and related phases has been discovered utilizing very high oxygen pressures for the syntheses[14]. The mass of the samples synthesized in the existing high-pressure apparatus is characteristically too small for allowing efficient oxygen content determination by means of conventional chemical analysis. In the high-pressure synthesized samples, *e.g.* in the $(Cu,Cu)Ba_2Ca_{n-1}Cu_nOz$[15] and $LaCuO_{3-\delta}$[16] systems, TG measurements under controlled O_2/Ar atmosphere have been the only technique so far used for the estimation of the oxygen stoichiometry.

Comparison between the various methods

In Table 1 the applicability of the above described analysis methods for the oxygen content determination in various cuprate systems is summarized. Standard laboratory apparatus are sufficient in each of the techniques. The wet chemical analyses, however, have to be carried out under an inert atmosphere in order to prevent air-oxidation of the reductants. Additionally, the titration methods have to be standardized frequently: the $Na_2S_2O_3$ titrant *e.g.* against a known CuO sample and the CuCl reductant for the Cu(+I) content by blank experiments. Iodometric titration is the fastest to perform in a routine-type use, but the visual end point detection may cause difficulties for an unaccustomed analyst. The Cu(+I)/Cu(+II) redox method is applicable to a wide range of different cuprates and enables also the analysis of samples which contain monovalent copper.

*Table 1. Applicability of the different analysis methods for the determination of the oxygen content in various superconducting and related cuprates: *** most reliable, ** applicable, * possible but not recommended, ( ) predicted but not yet applied.*

	I(-I) / I (0)	H_2 / TG	Ar / TG	Cu (+I) / Cu (+II)	
				Coul.	Pot.
RE-cuprates	***	**	*	***	**
Bi-cuprates	*	*	-	***	**
Pb-cuprates	-	-	*	***	**
Tl-cuprates	*	-	-	-	**
Hg-cuprates	**	-	-	(**)	(*)
High-p. phases	-	(**)	**	-	-

The precisions of the different analytical methods are on the same level showing reproducibilities of about ± 0.01 for the oxygen content values. The results obtained for the same sample by different wet chemical redox analyses are usually within limits of reproducibility, but, for example, in the case of $YBa_2Cu_3O_{7-\delta}$ the thermogravimetric reduction method tends to give somewhat higher oxygen contents[17]. This discrepancy has been explained by the presence of unreacted $BaCO_3$ which decomposes in the thermobalance increasing the observed weight loss considerably even in small amounts. Other impurity phases than the carbonates do not usually have drastic influence on the accuracy of the results. For example, the presence of small amounts of $YBa_2Cu_3O_{7-\delta}$ and CuO does not affect the observed oxygen content value of the $YBa_2Cu_4O_{8-\delta}$ phase at all[4].

III. DETERMINATION OF THE DISTINCT METAL VALENCES

In order to distinguish the high oxidation states of copper and bismuth in bismuth cuprates, a permanganate titration method, based on a selective reaction between Bi(+V) and Mn(+II) in the presence of Cu(+III), and a subsequent reduction of the liberated MnO_4^- ions by an excess of Fe(+II), has sometimes been adopted[8,16]. The present authors prefer, however, a direct reduction of Bi(+V) by Fe(+II) for the separation of the Cu and Bi valences[9]. Furthermore, this method has been found to be applicable to the analyses of lead containing materials as well[9,11]. Finally, a possible route, based on different reactivities with water, has been proposed to distinguish the valences of copper and thallium in thallium cuprates[18-21].

Reduction with Fe(+II)

In the Fe(+II) reduction method the sample is dissolved in HCl solution containing $FeCl_2$ $4H_2O$[8,11]. Pentavalent bismuth and tetravalent lead react with Fe(+II) according to Eqs. 11 and 12, while for Cu(+III) the reaction with water (Eq. 10) is the preferred route. Any Cu(+I) present in the sample reacts with the formed Fe(+III) according to Eq. 13. The valences of bismuth and lead are obtained from the coulometric titration data of the excess Fe(+II) (Eq. 14).

$$2\,[CuO]^+ + 2\,H^+ \rightarrow Cu^{2+} + H_2O + 0.5O_2 \qquad (10)$$
$$BiO_3^- + 2\,Fe^{2+} + 4\,H^+ \rightarrow BiO^+ + 2\,Fe^{3+} + 2H_2O \qquad (11)$$
$$Pb^{4+} + 2\,Fe^{2+} \rightarrow Pb^{2+} + 2\,Fe^{3+} \qquad (12)$$
$$Cu^+ + Fe^{3+} \rightarrow Cu^{2+} + Fe^{2+} \qquad (13)$$
$$Fe^{2+}\ (\text{excess}) \rightarrow Fe^{3+} + e^- \quad (\text{coulometry}) \qquad (14)$$

The quantitativities of the processes 10-12 have been confirmed using $YBa_2Cu_3O_{6.87}$, $BiNa_{3-\delta}$, PbO_2 and Pb_3O_4 as reference materials for Cu(+III), Bi(+V) and Pb(+IV), respectively[9]. By adding extra Bi^{3+}, Pb^{2+} and Cu^{2+} ions to the studied reference solutions it has also been verified that under the experimental conditions concerned Cu(+III) does not oxidize Bi(+III) or Pb(+II) and Cu(+II) is not oxidized by Bi(+V) or Pb(+IV). When the studied sample contains both bismuth and lead the distinct Bi and Pb valences can not be detected by this method but only an average (Bi,Pb) oxidation number is observed. However, since bismuth and lead typically occupy same lattice sites in superconducting cuprates, the distribution of the holes between separate layers can be analyzed after all[12].

Utilizing different stabilities in aqueous solutions

When separating the Cu and Tl valences the thallium cuprate sample is dissolved in 0.5 M HCl in which only the trivalent copper reacts with water (Eq. 10). After the reduction of Cu(+III) is completed and the resulting oxygen gas is released from the solution, enough CuCl is added to reduce the more stable Tl(+III) according to Eq. 15. The Tl valence is then determined by analyzing the amount of excess Cu(+I) with constant potential electrolysis (Eq. 8).

$$2\,[CuO]^+ + 2\,H^+ \rightarrow Cu^{2+} + H_2O + 0.5\,O_2 \qquad (10)$$
$$Tl^{3+} + 2\,Cu^+ \rightarrow Tl^+ + 2\,Cu^{2+} \qquad (15)$$
$$Cu^+\ (\text{excess}) \rightarrow Cu^{2+} + e^- \qquad (\text{electrolysis}) \quad (8)$$

The fast and complete reaction between Cu(+III) and H_2Ohas been corroborated by means of the $YBa_2Cu_3O_{6.87}$ reference material. By adding extra Tl^+ ions to the studied $YBa_2Cu_3O_{6.87}$ solutions it has been further confirmed that trivalent copper does not oxidize monovalent thallium under the experimental conditions concerned. The proposed method has also been standardized against known Tl_2O_3 samples. The stability of Tl(+III) in 0.5 M HCl and at temperatures between 15-25 ^{0}C was found to be sufficient (> 95 %) to permit the successful separation of the Cu and Tl valences[18,19].

IV. STUDIES ON VARIOUS SUPERCONDUCTING AND RELATED CUPRATES

The analytical possibilities of the methods discussed above are best demonstrated with experimental results of oxygen content and metal valence studies on various superconducting and non-superconducting cuprate phases. Examples of such studies will be briefly discussed in the following.

Rare earth cuprates

In the $YBa_2Cu_3O_{7-\delta}$ (Cu-1212) system the oxygen stoichiometry can vary within a wide range ($0 < \delta < 1$) depending on the synthesis conditions. In material with δ equal to zero the square planar coordination of the Cu(1) atoms in the Cu-O chains refers to the +III oxidation state, while in the Jahn-Teller distorted CuO_5 polyhedrons the Cu(2) atoms are basically divalent. The Cu(1) atoms of the fully oxygen depleted $YBa_2Cu_3O_6$, having only two neighbouring oxygen atoms, are considered to be monovalent. The structure of $YBa_2Cu_4O_{8-\delta}$,(Cu-2212) differs from that of the Cu-1212 phase in having a double Cu-O chain. The double Cu-O chain structure, in which each oxygen is bonded to three copper atoms, is considered to be a rigid system with a rather constant oxygen content, and the oxygen stoichiometry seems to depend only very weakly on the oxygen pressure applied during the synthesis: values slightly below the stoichiometric number ($\delta < 0.05$) have been obtained by iodometric and Cu(+I)/Cu(+II) coulometric titrations for samples prepared under atmospheric O_2 pressure[4]. Although the average Cu valence in the stoichiometric RE-124 (+2.25) is lower than in $YBa_2Cu_3O_7$,(+2.33), nothing can be concluded about the effective valence of the conducting CuO_2 planes since distribution of the charge between the Cu(1) and Cu(2) sites may be different in these two phases.

In the Cu-1212 and Cu-2212 systems, even though the determination of the average valence of copper is straightforward, the distribution of the holes between the superconducting CuO_2 planes and the non-superconducting Cu-O chains cannot be evaluated from the data of chemical analyses. Bond valence calculations (based on structure data from refs. 22,23) give, however, useful information on the charge distribution. In both phases the positive charge is shared by the two copper lattice sites. In Cu-1212 the valence of Cu(1) decreases linearly from +2.38 to +1.30 as δ varies from 0 to 1, while the valence of Cu(2) changes in two steps from +2.21 to +2.14. In stoichiometric Cu-2212 the valences of Cu(1) and Cu(2) are +2.35 and 2.22, respectively.

Bismuth cuprates

For the $Bi_2Sr_2Ca_{n-1}Cu_nO_z$ series control over the hole concentration is best achieved in the Bi-2212 (n = 2) phase. Through aliovalent substitutions of ions located in between the superconducting CuO_2 planes it is possible to tune the charge balances and thereby the critical temperatures. This kind of effects have been demonstrated with the $Bi_2Sr_2Ca_{1-x}Y_xCu_2O_{8+\delta}$ system studied by the Cu(+I)/Cu(+II) and Fe(+II)/Fe(+III) coulometric titrations[12]. A continuous increase in the oxygen content has been observed upon substituting calcium by yttrium in the fully-oxidized material. The extra oxygen taken into the structure only partially counteracts the trivalent to divalent substitution, and consequently the valences of both copper and bismuth were found to be depressed. The highest T_c value of 86 K at x = 0.2 corresponds to the +2.11 valence of copper, proposing that the non-doped, air-annealed Bi-2212 with the Cu valence of +2.18 is in an over-doped state. Furthermore, it is evident that the hole concentration of 0.06 per copper atom observed for the non-superconducting x = 0.6 sample represents a lower limit for the disappearance of superconductivity in this system. On the other hand, when bismuth is partly replaced by lead in the N_2 -annealed Bi-2212 the oxygen content decreases gradually indicating that lead in principle enters the structure as a divalent ion. With increasing Pb content the Cu valence (+2.10) and the critical temperature of superconductivity (84 K) remain constant.

The average valence of +2.08 observed for copper in a $Bi_{1.85}Pb_{0.22}Sr_{1.92}Ca_{2.09}Cu_{3.10}O_{10.40}$ (Bi-2223; n = 3) sample (annealed in 7.6 % O_2 atmosphere) by a combination of the Cu(+I)/Cu(+II) and Fe(+II)/Fe(+III) coulometric titrations seems to be quite low if compared to the value of +2.18 in the air-oxidized Bi-2212 sample[12]. It is noteworthy, however, that in the Bi-2212 structure Cu atoms occupy only one crystallographic site, but in the case of Bi-2223 the charge may be inhomogeneously distributed between the five- and four-fold coordinated copper.

Thallium cuprates

Superconductivity in the $Tl_2Ba_2Ca_{n-1}Cu_nO_z$ series is supposed to be promoted by the supply of electrons from the CuO_2 planes to the empty 6s orbitals of thallium. Also the existence of thallium vacancies and/or intrinsic metal substitutions may contribute to the available charge of copper. The results of band structure calculations[24] and XPS experiments[25], showing an overlap between the Tl 6s and Cu 3d orbitals and valence values below +3.00 for thallium, are consistent with the above self-doping assumption. On the other hand, bond valence calculations performed by Otto *et al.*[26] and ^{205}Tl NMR measurement results of Jol *et al.*[27] and Winzek *et al.*[28] support the +III oxidation state of thallium. Wet chemical analyses based on different stabilities of Cu(+III) and Tl(+III) in aqueous solution for various Tl double layer cuprates ($Tl_{1.75}Ba_2Cu_{1.2}O_{5.98}$, $Tl_{1.63}Ba_2Ca_{1.3}Cu_2O_{7.94}$, $Tl_{1.62}Ba_2Ca_{2.3}Cu_3O_{9.92}$ and $Tl_{1.60}Ba_2Ca_{3.3}Cu_3O_{11.89}$) revealed Tl valence values very close to +3.00[20,21]. The average valence of copper was found to decrease from +2.26 to +2.09 with increasing number of the consecutively stacked CuO_2 planes (n going from 1 to 4). It is, of course, realized that the valence values exhibited by Cu and Tl in the solid state may change upon dissolving the material. However, the observed Cu valence values are sufficient for high T_c superconductivity, thus making the +III oxidation state of Tl possible in the samples studied.

Lead cuprates

The structure of $Pb_2(Sr,Ba)_2(RE,Ca)Cu_3O_8$ differs from the corresponding Bi- and Tl-2212 phases in respect to the PbO-Cu-PbO block layer. The Cu(1) atoms between the PbO layers have a linear coordination to the nearest oxygen atoms and are, according to charge balance considerations, monovalent (*cf.* Cu(1) atoms in $YBa_2Cu_3O_6$). According to coulometric Cu(+I)/Cu(+II) titrations the oxygen excess δ of Ar-annealed $Pb_2(Sr,Ba)_2RECu_3O_{8+\delta}$ material is around 0.10[12]. When the oxygen deficient layer between the pyramidal CuO_2 planes is totally composed of rare earth elements, the material is semiconducting but by substituting trivalent RE partly by divalent Ca holes are introduced in the superconducting CuO_2 planes. In the Ca-doped samples two types of copper atoms with oxidation states below and over +II coexist in the same structure. In an Ar-annealed $Pb_2Ba_{0.6}Y_{0.6}Ca_{0.8}Cu_3O_{8.10}$ sample 0.92 Cu^+ ions per formula unit have been detected by Cu(+I)/Cu(+II) coulometric titrations performed for the sample as such (without extra Cu(+I) or Fe(+II) addition)[12]. By assuming that Cu(+III) and Pb(+IV) do not react at all with Cu(+I) originating from the same sample but are completely reduced by H_2O a valence value of +1.08 for the Cu(1) atoms could be calculated. A further assumption that the partial replacement of yttrium by calcium has an effect only on the Cu(2) valence results in the valence values of +2.20 and +2.06 for Cu(2) and Pb, respectively. The valence values estimated above show a good agreement with the structurally (bond valences) calculated values by Marezio *et al.*[29].

The $Pb_2(Sr,Ba)_2(RE,Ca)Cu_3O_8$ lattice easily accommodates more oxygen on the empty sites of the Cu(1) layers, resulting in the non-superconducting composition of $Pb_2(Sr,Ba)_2(RE,Ca)Cu_3O_{10}$ as an extreme case. The amount of the incorporated oxygen has been conveniently controlled and determined by performing the O_2-annealing in a thermobalance[30]. Oxygen loading up to δ = 1.8 is possible without destroying the structure. The extra oxygen above δ = 0 raises most probably the valences of lead and Cu(1) atoms, since the oxygen atoms are accommodated in the empty sites on the Cu(1) layer, close to the PbO layer. For the O_2-annealed $Pb_2Ba_2YCu_3O_{9.72}$ sample the results obtained by the coulometric Fe(+II) titration method suggest that the valence of copper (observed Cu valence +2.01) can not be oxidized above the average oxidation state of +II by accommodating extra oxygen to the structure, but the so produced holes are trapped on lead (observed Pb valence +3.21)[12]. This result is consistent with the fact that the $Pb_2Ba_2YCu_3O_{9.72}$ sample is not superconducting.

V. CONCLUSIONS

For the determination of the oxygen contents a variety of different redox chemical methods are available yielding results which are in good internal agreement and have acceptable accuracy. Although being both accurate and precise, the simple chemical analyses involving the dissolution or heating up of the whole sample detect only "average bulk properties". Therefore, the results obtained by chemical methods for the oxygen content should be compared with the more "phase specific" values evaluated *e.g.* from the occupation factors of crystal structure refinements based on diffraction experiments. Diffraction analyses do not give highly accurate results but, on the other hand, impurity phases do not lead to systematic errors. Furthermore, not only the amount but also the exact location of oxygen in the superconductor lattices has an influence on the properties of these materials.

Various wet chemical routes have been developed for the distinct metal valence studies in bismuth, lead and thallium cuprates. A critical question for all the solution based studies is how well the solid state properties, such as metal valences, are maintained upon dissolving the material. In that respect, the importance of various comparative characterization methods as well as extensive studies on sample series with gradually changing properties should be emphasized. Especially bond valence calculations based on neutron diffraction data can provide interesting reference data as the Cu valence values are reflected directly in the Cu-O distances. At best the chemical analyses are of great significance for understanding the relations between the oxidation state of copper and superconductivity.

REFERENCES

1. A.W. Sleight, Acc. Chem. Res. **28,** 103 (1995).
2. M. Karppinen, A. Fukuoka, L. Niinistö and H. Yamauchi, Supercond. Sci. Technol. **9,** 121 (1996).
3. E.H. Appelman, L.R. Morss, A.M. Kini, U. Geiser, A. Umezawa, G.W. Crabtree and K.D. Carlson, Inorg. Chem. **26,** 3237 (1987).
4. A. Fukuoka, M. Karppinen, N. Seiji, J. Valo, A. Kareiva, L. Niinistö, M. Leskelä, N. Koshizuka and H. Yamauchi, Supercond. Sci. Technol. **8,** 673 (1995).
5. M. Paranthaman, A. Manthiram and J.B. Goodenough, J. Solid State Chem. **87,** 479 (1990).
6. A. Fukuoka, A. Tokiwa-Yamamoto, M. Itoh, R. Usami and H. Yamauchi, unpublished.
7. G. Kawamura and M. Hiratani, J. Electrochem. Soc. **134,** 3211 (1987).
8. K. Kurusu, H. Takami and K. Shintomi, Analyst **114,** 1341 (1989).
9. M. Karppinen, H. Yamauchi and S. Tanaka, J. Solid State Chem. **104,** 276 (1993).
10. M. Wakata, S. Takano, J. Wang, M. Karppinen and H. Yamauchi, *Advances in Superconductivity V,* eds. Y. Bando and H. Yamauchi, Springer-Verlag, Tokyo, p. 423 (1993).
11. M. Karppinen, A. Fukuoka, T. Kaneko and H. Yamauchi, Supercond. Sci. Technol. **6,** 265 (1993).
12. M. Karppinen, A. Fukuoka, J. Wang, S. Takano, M. Wakata, T. Ikemachi and H. Yamauchi, Physica C **208,** 130 (1993).
13. M. Karppinen, O. Antson, P. Baulés, T. Karlemo, T. Katila, J. Lindén, M. Lippmaa, L. Niinistö, C. Roucau, I. Tittonen and K. Ullakko, Supercond. Sci. Technol. **5,** 476 (1992).
14. H. Yamauchi and M. Karppinen, Superlatt. Microstructr., in press.
15. H. Yamauchi, M. Karppinen, T. Ito, H. Suematsu and O. Fukunaga, Physica C **266,** 345 (1996).
16. M. Karppinen, H. Yamauchi, H. Suematsu, K. Isawa, M. Nagano, R. Itti and O. Fukunaga, J. Solid State Chem., unpublished.
17. M. Karppinen and L. Niinistö, Supercond. Sci. Technol. **4,** 334 (1991).
18. M. Karppinen, Ann. Acad. Sci. Fenn., Ser. A II. Chem. **249,** 51 (1993).
19. M. Karppinen, J. ISTEC **6/2,** 15 (1993).
20. A. Fukuoka, M. Karppinen, M. Ito, K. Hamada and H. Yamauchi, *Advances in Superconductivity*, eds. Y. Bando and H. Yamauchi, Springer-Verlag, Tokyo, p. 283 (1993).
21. A. Fukuoka, M. Karppinen, M. Itoh, K. Hamada and H. Yamauchi, Phys. Rev. B **51,** 12759 (1995).
22. R.J. Cava, A.W. Hewat, E.A. Hewat, B. Batlogg, M. Marezio, K.M. Rabe, J.J. Krajewski, W.F. Peck Jr. and L.W. Rupp Jr., Physica C **165,** 419 (1990).

23. P. Lightfoot, S. Pei, J.D. Jorgensen, Y. Yamada, T. Matsumoto, F. Izumi and Y. Kodama, Acta Cryst. C **47,** 1143 (1991).
24. D. Jung, M.-H. Whangbo, N. Herron and C.C. Torardi, Physica C **160,** 381 (1989).
25. T. Suzuki, M. Nagoshi, Y. Fukuda, S. Nakajima, M. Kikuchi, Y. Syono and M. Tachiki, Physica C **162,** 1387 (1989).
26. H.H. Otto, R. Baltrusch and H.-J. Brandt, Physica C **215,** 205 (1993).
27. J.C. Jol, D. Reefman, H.B. Brom, T. Zetterer, D. Hahn, H.H. Otto and K.F. Renk, Physica C **175,** 12 (1991).
28. N. Winzek, Hj. Mattausch, S.G. Eriksson, R.K. Kremer, A. Simon and M. Mehring, Physica C **205,** 45 (1993).
29. M. Marezio, A. Santoro, J.J. Capponi, E.A. Hewat, R.J. Cava and F. Beech, Physica C **169,** 401 (1990).
30. M. Karppinen, J. Lindén, A. Fukuoka, L. Niinistö and H. Yamauchi, *Advances in Superconductivity VI*, eds. H. Yamafuji and T. Morishita, Springer-Verlag, Tokyo, p. 273 (1995).

HARMONIC GENERATION EFFECT IN HIGH-Tc FILMS

Neeraj Khare, S.K. Shrivastava, V.P.N. Padmanabhan, Sangeeta Khare and A.K. Gupta
National Physical Laboratory, New Delhi-110012, India

Harmonic generation in thick BPSCCO and thin YBCO films are reported. The application of an ac field ($H_{ac}>H_{c1}$) of frequency f causes the generation of odd harmonics of frequency (2n+1)f. The application of dc field in addition to the ac field causes the appearance of even harmonics also in the BPSCCO film. However, the appearance of even harmonics is not observed in YBCO film with high J_c~$1.6x10^6$ A/cm^2 and appearance of second harmonic with small magnitude is observed in YBCO film with low J_c~$2x10^3$ A/cm^2. The variation of amplitudes of these harmonics are studied as a function of magnitude of ac and dc field and the results are explained in the framework of critical state model. A high-T_c film magnetometer based on the measurement of the amplitude of second harmonic has been developed whose field sensitivity is ~$1.5x10^{-8}$T.

I. INTRODUCTION

Harmonic generation in high-T_c bulk superconductor has been reported by several workers[1-4]. Application of an ac field ($H_{ac}>H_{c1}$) of frequency f to a small piece of high-T_c bulk superconductor leads to the generation of higher odd harmonics of frequency (2n+1)f. The application of a dc field besides an ac field causes the generation of even harmonics of frequency 2nf in addition to odd harmonics. The amplitudes of these harmonics are found to modulate with dc field. The grain boundaries present in the bulk sample acting as pinning centers are responsible for this harmonic generation effect. The application of $H_{ac}>H_{c1}$ causes the vortices to sweep in and out of the material during each ac cycle. The field inside the superconductor lags behind the applied ac field in a non linear fashion. This non linearity causes the harmonic generation. Bean critical state model[5] which assumes critical current density to be independent of local magnetic field predicts the generation of odd harmonics even in the presence of dc field. Unlike the Bean model, the critical state model of Kim et al [6] assumes a critical current density J_c that decreases with the increasing local field H_i. Several workers have used the Kim type critical state model to explain the harmonic generation and dc field modulation of the harmonic amplitudes [2,3]. Few reports on the observation of harmonic generation in YBCO thin film has also been published [7-11]. Yamamoto et al[7] reported the observation of third harmonic in YBCO thin film prepared by evaporation technique. Shaulov et al [8] showed the use of third harmonic to investigate the multiphase nature of YBCO thin film. Revenaz and Dumas [9] reported the harmonic generation in YBCO thin film due to magnetically modulated microwave absorption, Palenque et al[10] reported the study of third harmonic near T_c and Khare et al[11] reported study of harmonic generation in laser ablated YBCO thin film with and without dc field. In this paper, we report harmonic generation effects in high-T_c BPSCCO thick film and in YBCO thin film of different critical current densities for understanding the relation between harmonic generation effect and J_c of the films. The effect of dc field on the amplitudes of harmonics has been studied and a thick film magnetometer has been developed using this harmonic generation phenomena.

II. EXPERIMENTAL

BPSCCO thick film has been prepared on single crystal MgO(100) substrate by screen printing technique. The starting composition of the powders for printing was taken in the ratio of Bi:Pb:Sr:Ca:Cu=1.7:0.4:2:2:3. The details of thick film preparation are described elsewhere[12]. The resultant film has T_c (R=0) of 107K and J_c~20 A/cm^2 at 77K. YBCO thin film was deposited in-situ on single crystal $SrTiO_3$ (100) substrate by dc magnetron sputtering from a superconducting single phase stoichiometric target. Details of the deposition conditions are described elsewhere[13]. The films (A and B) used were deposited at two different substrate temperatures 720°C and 750°C respectively. Film A has T_c (R=0) ~ 86K and J_c(at 77K) ~ $2x10^3$ A/cm^2 and film B has T_c (R=0) ~ 89K and J_c(at

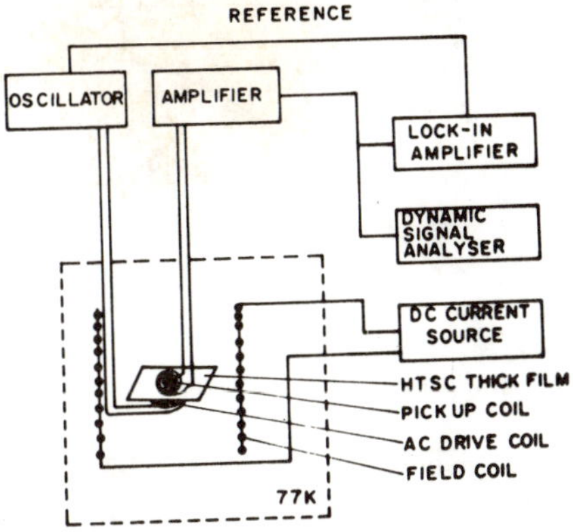

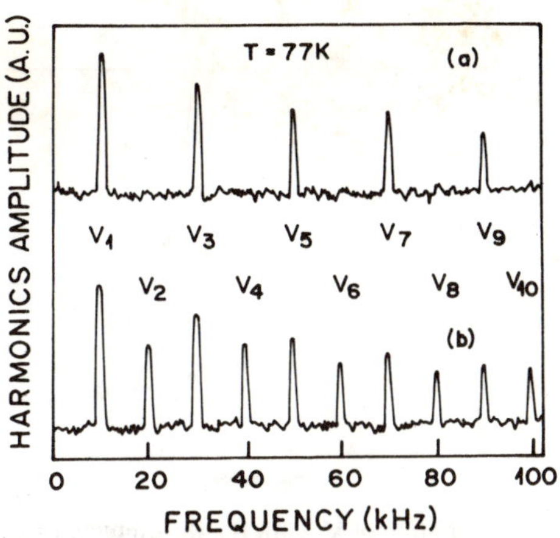

Fig.1. (Left) Schematic of the experimental setup used for observing harmonic generation effect.
Fig.2. (Right) Frequency spectrum of the signal at pick-up coil for BPSCCO thick film: (a)H_{ac}=4.3G, f_{ac}=10kHz, H_{dc}=0, T=77K (b) H_{ac}=4.3G, f_{ac}=10kHz, H_{dc}=0.45G, T=77K

77K)~ 1.5×10^6 A/cm^2 . Fig. 1 shows the schematic of the experimental setup used for observing harmonic generation effect in high-T_c films. A twenty five turn coil of diameter 6mm was glued on the back side of substrate for applying an ac field. A flat spiral coil of twenty five turn was glued on the surface of the film. This coil was used to pickup signal of the film. A solenoid of diameter 2.8 cm surrounding the film was used to apply dc field. Both ac and dc fields were applied perpendicular to the surface of the film. The primary coil was connected to the oscillator for producing an ac field while the pickup coil was connected to a low noise amplifier. The output of the amplified signal was connected to a dynamic signal analyser for observing the frequency spectrum of the pickup signal. A lock-in-amplifier was also used for measuring amplitude of second harmonic.

Results and Discussion:

The application of H_{ac} greater than a critical value causes the generation of odd harmonics in all three films. In BPSCCO film, the application of dc field in addition to the ac field causes the generation of even harmonics (fig. 2). In YBCO film A, very small magnitude of V_2 was observed and in YBCO film B, no even harmonics appeared in the presence of dc field. Fig. 3 shows the variation amplitude of the third harmonic (V_3) with H_{ac} for BPSCCO film and YBCO film A. For BPSCCO

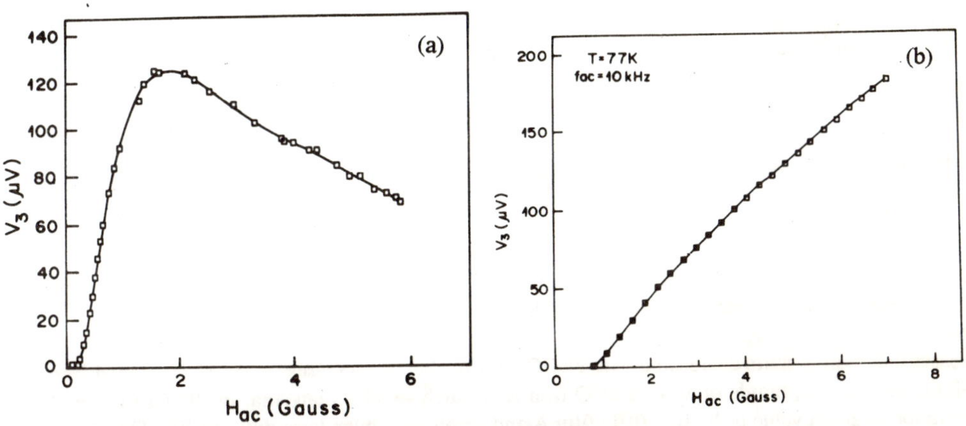

Fig.3. Variation of V_3 with H_{ac} (f_{ac}=10kHz) at 77K for (a) BPSCCO thick and (b) YBCO thin films.

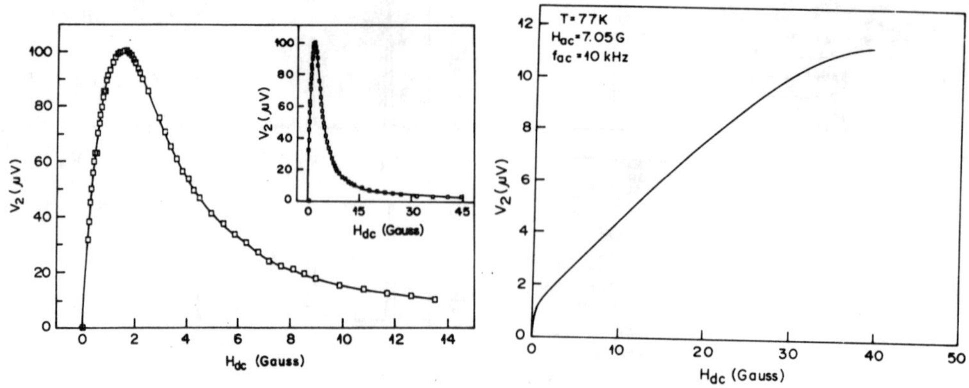

Fig.4. Variation of V_2 with H_{dc} for BPSCCO thick film (left) and YBCO thin film A (right).

film, initially V_3 is zero and it starts growing when $H_{ac} > 0.1$ Gauss. The growth of V_3 is fast in the range of 0.1-1.6 Gauss and when H_{ac} exceeds 2.1 Gauss, V_3 starts decreasing. For YBCO film A, V_3 starts growing when $H_{ac}>0.8$ Gauss. It shows no sign of saturation even at $H_{ac} \sim 7$ Gauss. For YBCO film B, V_3 has been found to grow when $H_{ac}>2.1$ Gauss. It has also been observed that the magnitude of V_3 is larger in YBCO film A. These observations can be understood in the framework of critical state model. Higher harmonics will start growing when flux starts penetrating into the films. Thus in the beginning, when H_{ac} is very small, third harmonic will not observed and it will start growing when H_{ac} is larger than a critical H_{c1} value which depends upon critical transport current of the film. Transport critical current for YBCO film B is higher ($J_c \sim 1.5 \times 10^6$ A/cm^2) and for BPSCCO film, it is lower ($J_c \sim 20$ A/cm^2). Thus, the value of H_{ac} at which V_3 starts growing is expected to be lower for BPSCCO film and highest for YBCO film B. Similar trends have been observed experimentally. Fig. 4 shows the variation of second harmonic amplitude (V_2) with H_{dc} in BPSCCO film and YBCO film. For BPSCCO film, H_{ac} and f_{ac} was kept fixed at 4.3 Gauss and 10kHz. In this film the value of V_2 first

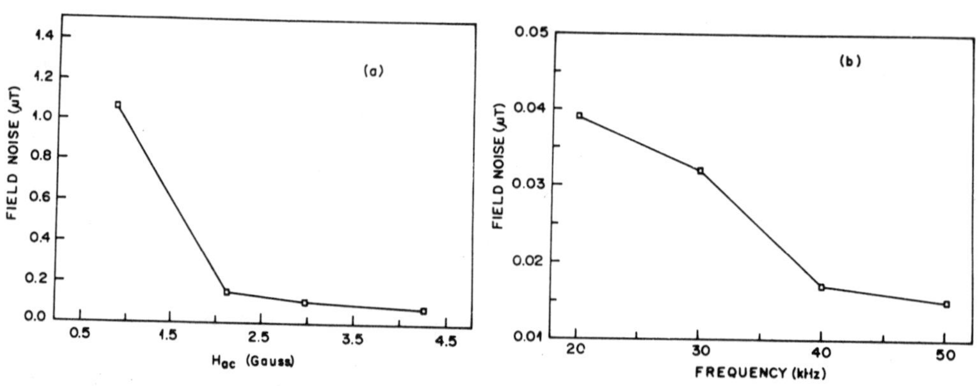

Fig.5. Variation of the field noise of BPSCCO high-T_c thick film magnetometer with (a) ac driving field H_{ac}, and (b) frequency of the ac field.

increases and then it starts decreasing . In YBCO film A, the V_2 increases when H_{dc} is varied from 0 to 40 Gauss. The magnitude of V_2 in YBCO film A is much smaller than that in BPSCCO film. The observation of small value of V_2 in YBCO film A and absence of even harmonics in YBCO film B can also be understood in light of Bean and Kim critical state models [5,6]. The Kim model[6] for the critical state, which assumes the dependence of J_c on local field, explains the appearance of even harmonics on application of a dc field. However if J_c is independent of the local field (Bean model[5]), then even harmonics will not appear. In BPSCCO film critical current is small, so it can be assumed that grain boundary currents are local field dependent and thus Kim model is applicable. In YBCO thin film B, J_c is very large so Bean model is expected to be applicable. Thus, no even harmonics will be generated even in the presence of dc field. For YBCO film A, J_c is lower than that in film B but higher than BPSCCO film, thus, it is reasonable to assume that in film A, for most of the grain boundaries J_c will be independent of local field and only for a few grain boundaries J_c will depend on the local field. In such a case, application of dc field will result in the appearance of small value of V_2 as has been observed experimentally.

For BPSCCO film, it has been observed that in the presence of small varying dc field (in the range of 0-0.6 Gauss), the amplitude of even harmonics varies linearly. A magnetometer has been developed based on the measurement of the amplitude V_2 of the second harmonic. The field noise ($\sqrt{S_{BA}}$) of this magnetometer has been calculated using the formula: $\sqrt{S_B} = [\Delta V/(\delta V/\delta H_{dc})]$, where ΔV is the voltage noise at the output of the lock-in-amplifier and $(\delta V/\delta H_{dc})$ is the transfer function of the magnetometer. $\sqrt{S_B}$ represents the smallest field change which could be detected by the magnetometer. Fig. 5 (a) shows the variation of field noise of the high-T_c thick film magnetometer when H_{ac} was varied from 0.8 to 4.3 Gauss and the driving frequency was kept fixed at 10kHz. Fig. 5(b) shows the variation of the $\sqrt{S_B}$ with frequency of the ac field when H_{ac} is kept constant at 4.3 Gauss. The highest sensitivity of the magnetometer is found to be ~ 1.5×10^{-8} T when the H_{ac} =4.3 Gauss and f_{ac}=50 kHz. It is to be noted that in the present case, the noise is dominated by the pickup coil and the instrument. Thus, by proper design of the pickup coil and detection electronics, the sensitivity of the magnetometer can be improved.

ACKNOWLEDGMENT

The authors are grateful to Prof. E.S.R. Gopal, Director NPL for his keen interest and encouragement. One of the author (SKS) is grateful to University Grants Commission, New Delhi for the award of Research Fellowship.

REFERENCES

1. N. Khare, A.K. Gupta and G.B. Donaldson in *Studies in High Temperature Superconductors* Ed. A. Narlikar, Pub., Nova Science Publishers (N.Y.) **14**, 421 (1995) and references therein.
2. K.H. Muller, J.C. Macfarlane and R. Driver, Physica C **158**, 366 (1989).
3. L. Ji, R.H. Sohn, G.C. Spalding, C.J. Lobb and M. Tinkham, Phys. Rev. B **40**, 10936 (1989).
4. S.B. Roy, S. Kumar, A.K. Pradhan, P. Chaddah, R. Prasad, N.C.Soni and K. Adhikary, Pramana-J. Phys. **41**, 51 (1993).
5. C.P. Bean, Rev. Mod. Phys. **36**, 31 (1964).
6. Y.B. Kim, C.F. Hempstead and A.R. Strnad Phys. Rev. **129**, 528 (1963) P.W. Panderson and Y.B. Kim Rev. of Mod. Phys. **36**, 39 (1964).
7. K. Yamamoto, H. Mazaki, H. Yasuoka, K. Hirata, T. Terashima, K. Iijima and Y. Bando, Jpn. J. Appl. Phys. **28**, L1568 (1989).
8. A. Shaulov, R. Bhargava and S. Shatz, J. Appl. Phys. **69**, 6738 (1991).
9. S. Revenaz and J. Dumas, Physica C **219**, 450 (1994).
10. E.R. Palenque, N.J. Appleyard, T.J. Jackson and S.B. Palmer, Supercond. Sci. Technol. **8**, 279 (1995).
11. N. Khare, J.R. Buckley, R.M. Bowman, G.B. Donaldson and C.M. Pegrum, Pramana J. Phys. **46**, 283 (1996).
12. V.P.N. Padmanabhan and A.K. Gupta, Indian J. Crygenics (in press).
13. S.K Shrivastava, Vijay Kumar and A.K. Gupta, Indian J. Cryogenics **20**, 6 (1995).

ELECTRICAL TRANSPORT AND PERCOLATION BEHAVIOUR OF SUPERCONDUCTOR-INSULATOR COMPOSITE SYSTEM

J. Koshy, Asha M. John, R. Jose, J. Kurian and K.S. Kumar
Regional Research Laboratory (CSIR)
Trivandrum 695 019, India

The electrical transport properties and percolation behaviour of superconducting Bi(2223)-$DyBa_2SnO_{5.5}$ and YBCO-$DyBa_2SnO_{5.5}$ composite systems have been studied by X-ray diffraction and temperature resistivity measurements. In both the superconductor-insulator systems, the normal state percolation threshold values are found to be ~ 20 vol% of the superconductor in the composite. However, the superconducting percolation threshold value is found to be 20 vol% of Bi(2223) in the Bi(2223)-$DyBa_2SnO_{5.5}$ system and 30 vol% of YBCO in the YBCO-$DyBa_2SnO_{5.5}$ system. The values obtained for the critical exponents describing the normal state transport behaviour of the system matched with the theoretically expected values for an idealized metal-insulator composite system. No detectable chemical reaction was observed between the superconductor and the ceramic insulator $DyBa_2SnO_{5.5}$ (within the precision of XRD technique), even under severe heat treatment at processing temperature. The implications are discussed.

I. INTRODUCTION

The discovery of superconducting transition at elevated temperatures in ceramic materials has received tremendous attention because of its scientific and practical potential. The study of superconducting small aggregates, clusters or particles are important from both fundamental and technological stand points[1]. Due to the granular nature of these materials with short coherence length and large penetration depth[2,3], it is interesting to study the percolation and fractal properties, quantum size effects, thermal fluctuations and size effects on superconductivity. Transport properties of normal-metal insulator percolation system have been extensively studied over the years[4]. Recently there were few studies on the percolation behaviour of supercoductor-normal metal composites based on electrical transport and magnetisation properties[5-8]. Since the resistivity ratio ρ_s/ρ_n (where ρ_s the resistivity of the superconductor and ρ_n the resistivity of the normal metal at room temperature) is very small for the superconductor-normal metal composite system, the percolation model which describes the normal state transport behaviour of the metal-insulator composite syster cannot be applied strictly to such a system. It is difficult to have a continuous network for ceramic superconductor system to exist in another ceramic system without deteriorating the superconductivity properties. The YBCO phase requires high temperature annealing in an oxygen atmosphere in order to retain its superconducting properties. Under such adverse condition, most metals and insulators will react with YBCO compound and degrade the superconductivity[1]. In the case of Bi(2223)-insulator composite systems, the synthesis of a single phase Bi(2223) was found to be difficult and addition of most of the insulators in Bi(2223) produces the intergrowth of an appreciable amount of Bi(2212) phase with Bi(2223) in the composite during the annealing. After an extensive search, recently we synthesised a new insulating ceramic material $DyBa_2SnO_{5.5}$ (DBNO)[9,10] which is found to be chemically compatible with Bi(2223) and YBCO superconductors even after severe heat treatment, making percolation studies in Bi(2223)-$DyBa_2SnO_{5.5}$ and YBCO-$DyBa_2SnO_{5.5}$ composite systems possible. In the present paper, we report on the electrical transport and percolation behaviour of Bi(2223)-$DyBa_2SnO_{5.5}$ and YBCO-$DyBa_2SnO_{5.5}$ composite systems based on electrical measurements and X-ray diffraction studies.

II. EXPERIMENTAL DETAILS

Phase pure Bi(2223), (Tc(o) = 110K), with nominal stoichiometry $Bi_{1.5}Pb_{0.5}Sr_2Ca_2Cu_3O_x$, was prepared from high purity (99.9%) Bi_2O_3, PbO, $SrCO_3$, $CaCO_3$ and CuO by solid state reaction method and details of the preparation of Bi(2223) are given in our earlier publication[11]. The YBCO was prepared from high purity Y_2O_3, $BaCO_3$ and CuO by the conventional solid state reaction method. The compound $DyBa_2SnO_{5.5}$ (DBSO) was also synthesised by solid state reaction method and the details of the preparation and characterisation are reported in our previous publications[9,10]. The Bi(2223)-DBSO and YBCO-DBSO composites were prepared by mixing different vol% of Bi(2223), YBCO, and DBSO using the theoretical densities of 6.31 g/cm^3 for Bi(2223), 6.4 gm/cm^3 for YBCO and 6.92 gm/cm^3 for DBSO. The composite samples were pressed in the form of pellets of 10 mm dia and 2 mm thickness and sintered at 850 °C for 20 h in air for Bi(2223)-DBSO composite and at 950-1000 °C for for 12 h in air in the case of YBCO-DBSO composites. The samples were then cooled slowly at a rate of 1°C/min to room temperature in air. The chemical reactivity between the superconductors and DBSO were studied by X-ray diffraction method throughout the composition range using nickel filtered Cu Kα radiation from a Rigaku X-ray diffractometer. The electrical resistivity measurements of the composites were carried out by four probe method. For superconducting samples, resistance measurements were made using a Keithley nanovoltmeter (model 181) and a Keithley current source (model 220). For higher resistance ranges, resistance measurements were made by Keithley electrometer (model 602) at room temperature. Temperatures of the samples were measured by a copper-constantan thermocouple, calibrated with a RF-800 rhodium-iron resistance sensor with an accurancy of ±0.2K.

III. RESULTS AND DISCUSSION

The X-ray diffraction (XRD) pattern of sintered DBSO for 2θ values between 5 and 90 degrees is shown in Fig. 1. The ceramic insulator DBNO is found to be isostructural with other complex cubic perovskites with general formula $A_2BB'O_6$ such as $ErBa_2SbO_6$, YBa_2NbO_6, $DyBa_2NbO_6$, etc., in JCPDS file in which doubling of the basic perovskite unit cell is observed. The XRD peaks including the minor peaks of $DyBa_2SnO_{5.5}$ are indexed for a complex cubic perovskite structure with the lattice constant a=8.52 Å. The DTA experiments showed that there is no phase transition in DBSO material upto a temperature of 1300 °C. The room temperature resistivity of the sintered DBSO pellets were found to be 10^{10} Ω cm. The dielectric constant (ε) and loss factor (tan δ) are found to be ~11 and 2 x 10^{-3} respectively at 13 MHz.

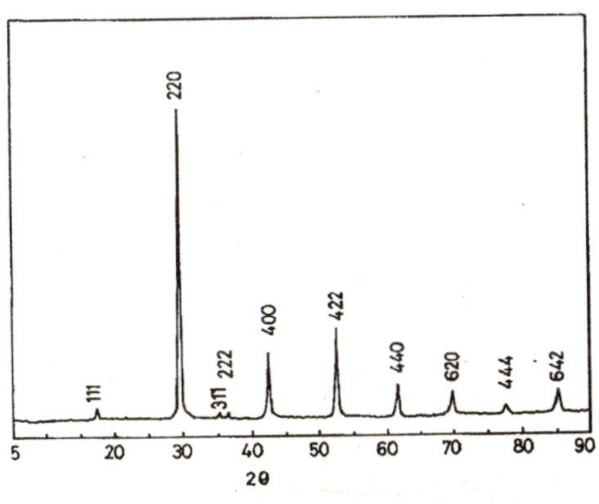

Fig. 1 Powder XRD pattern of sintered DBSO

(a) Bi(2223)-DBSO composite system

The XRD patterns of Bi(2223)-DBSO composites for 0 to 100 vol% of DBSO in the system are shown in Fig.2. In the XRD patterns of the composites, two distinct sets of peaks, one corresponding to Bi(2223) and the other corresponding to DBSO are clearly visible. There was no evidence of any additional peak other than those of Bi(2223) and DBSO in the XRD patterns indicating that there is no detectable chemical reaction (within the precision of XRD) between the two components even after severe heat treatment at 850 °C for 20 h.

The resitivities of the Bi(2223)-DBSO composites with different vol% of DBSO were

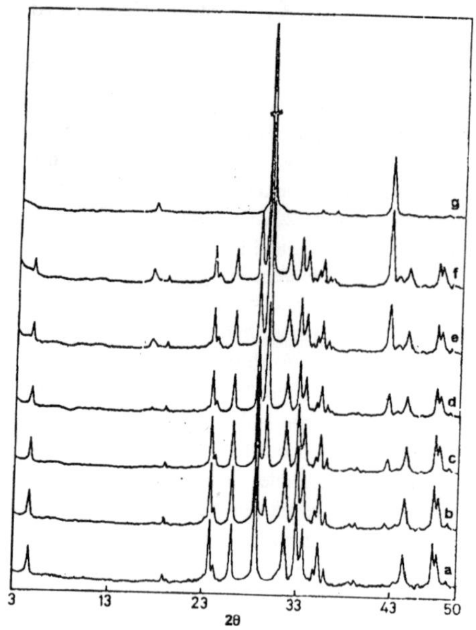

Fig. 2 X-ray diffraction for different vol % of DBSO in the Bi(2223)-DBSO composites; (a) 0%, (b) 10%, (c) 30%, (d) 50%, (e) 70%, (f) 80% and (g) 100%.

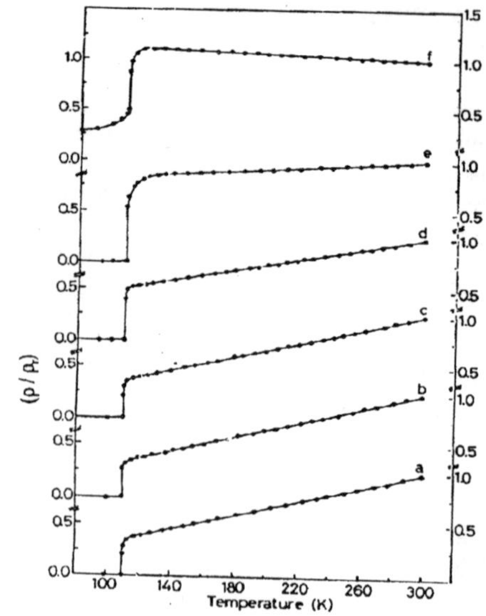

Fig.3 Variation of normalised resistivity with temperature for different vol % of DBSO in the Bi(2223)-DBSO composites; (a) 0%, (b) 10%, (c)30%, (d) 50%, (e) 70% and (f) 80%

measured in the temperature range 300-77K. The temperature dependence of the normalised resistivities with different vol% of DBSO is shown in Fig.3. Composites upto 70 vol% of DBSO showed metallic behaviour and gave zero resistivity superconducting transition temperature of 110K. For 80 vol% of DBSO, the composites showed semiconducting behaviour with a superconducting onset transition of 110K but no zero resistivity transition was obtained upto 77K. Above 80 vol% of DBSO, the Bi(2223)-DBSO composites showed semiconducting behaviour with no superconducting onset transition. Thus the superconducting percolation threshold for Bi(2223)-DBSO composites lies around 20 vol% of Bi(2223) in the composite.

Now we discuss the electrical transport properties of these composites in the normal state. The resistivity and temperature coefficient of resistivity $\alpha\, [= (1/\rho)(d\rho/dT)]$ at room temperature are shown in Fig.4 as a function of vol% of Bi(2223) (V_s) in the composite. In the normal state Bi(2223) is a metallic conductor with resistivity ~1 mΩcm, nearly 12 orders of magnitude lower than that of DBSO. The normal state resistivity of Bi(2223)-DBSO composite samples is dominated by Bi(2223) with a significant drop of ρ occurring near V_s ~20 vol%. The behaviour of ρ correlates with that of α which increases sharply towards that of Bi(2223) starting from V_s ~20 vol%. Therefore the percolation threshold value V_c for the normal state transport properties of the composite is ~20 vol% of Bi(2223). Thus the superconducting percolation threshold and normal state percolation threshold values of Bi(2223)-DBSO composites lie in the same range.

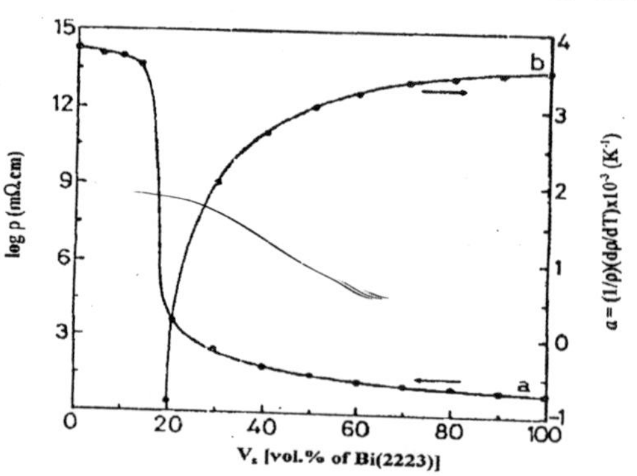

Fig. 4 Variation of (a) ρ and (b) α at room temperature for different vol% of Bi(2223) (Vs).

The normal state transport properties of the superconductor-insulator system can be described by a set of equations[12,13]

$$\rho = \rho_o (V_s - V_c)^{-t} \quad \text{for } V_s > V_c \qquad (1)$$

$$\rho' = \rho_o' (V_c - V_s)^{u} \quad \text{for } V_c > V_s \qquad (2)$$

where ρ_o and ρ_o' are constants, V_c is the critical volume fraction at which transport properties change drastically called percolation threshold, V_s is the vol% of Bi(2223) in the composite and t, u are critical exponents describing the transport properties of the composite system. The values of ρ_o, ρ_o', t and u are found from the log-log plot of ρ vs $(V_s - V_c)$ and ρ' vs $(V_c - V_s)$. For an idealised metal-insulator system $V_c \sim 17\%$, $t = 1.7$ and $u = 0.7$. The value of Vc is taken such that a log-log plot of equations (1) and (2) gives a straight line. Figs.5(a) and (b) show the log-log plot of ρ vs $(V_s - V_c)$ and ρ' vs $(V_c - V_s)$

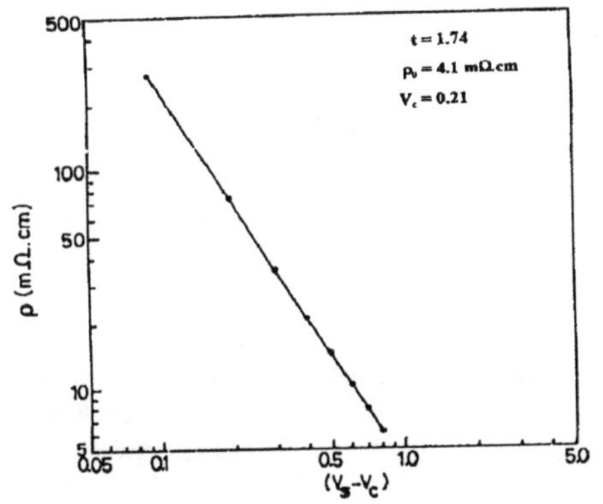

Fig. 5a. Log-log plot of ρ vs (Vs-Vc) where Vs is the vol% of Bi(2223) in the composite and Vc is the percolation threshold.

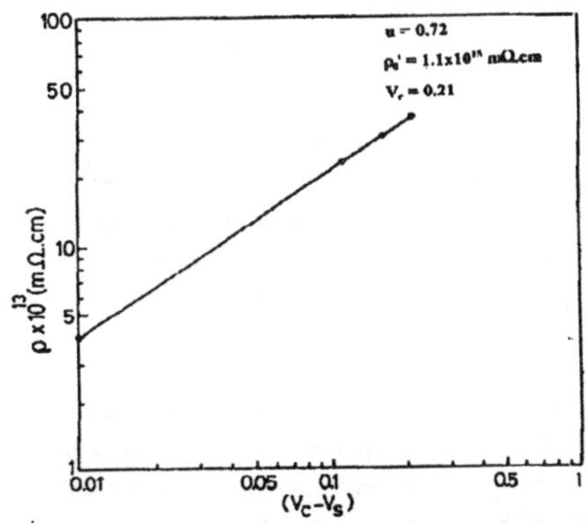

Fig. 5b. Log-log plot of ρ' vs (Vc-Vs) where Vc is the percolation threshold and Vs is the vol% of Bi(2223) in the composite.

respectively. The least square fits were performed to determine the slope of the plots which gave the exponents t and u as 1.74 and 0.72 respectively with a percolation threshold $V_c = 0.21$ and the values for ρ_o and ρ_o' are 4.1 and 1.1×10^{15} mΩ cm respectively.

(b) YBCO-DBSO composite system

Figure 6 shows the XRD patterns of YBCO-DBSO composites for 0 to 100 vol% of DBSO in the system. In the XRD patterns of the composites having upto 75 vol% DBSO, two distinct sets of peaks, corresponding to YBCO having an orthorhombic structure and DBSO are clearly visible. There was no additional peak other than those of YBCO and DBSO in the XRD patterns indicating that there is no detectable chemical reaction (within the precision of XRD) between the two components even after severe heat treatment above 950 °C.

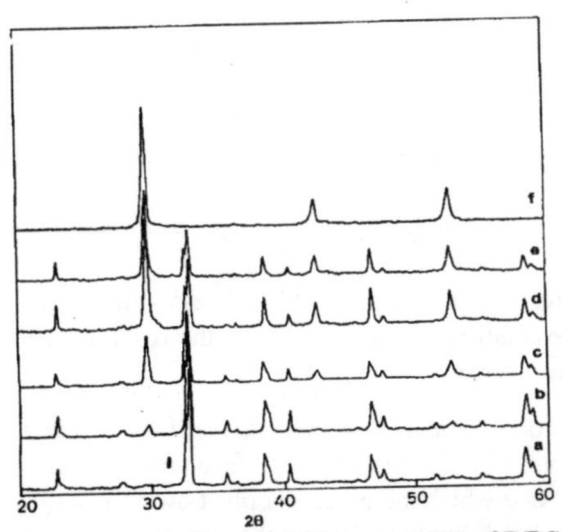

Fig. 6. X-ray diffraction for different vol% of DBSO in YBCO-DBSO composites; (a) 0%, (b) 10%, (c) 30%, (d) 50%, (e) 70% and (f) 100%.

The resistivities of the YBCO-DBSO composites with different vol% of DBSO were measured in the temperature range 300-77K. Figure 7 shows the temperature dependence of the normalised resistivities with different vol% of DBSO. Composites upto 30 vol% of DBSO showed metallic behaviour and gave zero resistivity superconducting transition temperature, above liquid N_2 temperature. The YBCO-DBSO composite with 75 vol% of

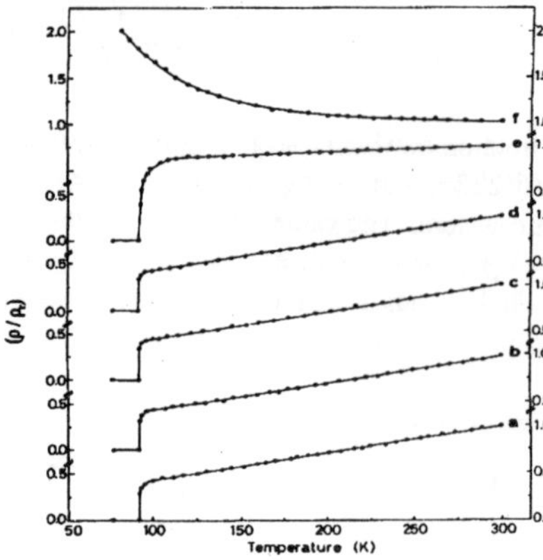

Fig. 7. Variation of normalised resistivity with temperature for different vol% of DBSO in YBCO-DBSO composites; (a) 0%, (b) 10%, (c) 30%, (d) 50%, (e) 70% and (f) 75%.

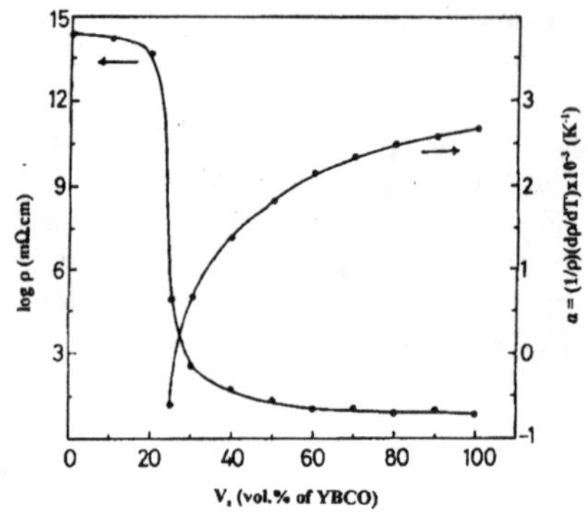

Fig. 8. Variation of ρ and (b) α at room temperature for different vol% of YBCO

DBSO showed semiconducting behaviour and did not show any superconducting transition. This may be due to the absence of a superconducting network through the matrix of the composite, because of the low vol% of YBCO. Thus the superconducting percolation threshold for YBCO-DBSO composites lies between 25 and 30 vol% of YBCO in the composite.

The resistivity and temperature coefficient of resistivity α [= $(1/\rho)(d\rho/dT)$] at room temperature are shown in Fig.8 as a function of vol% YBCO (V_s) in the composite. In the normal state, YBCO is a metallic

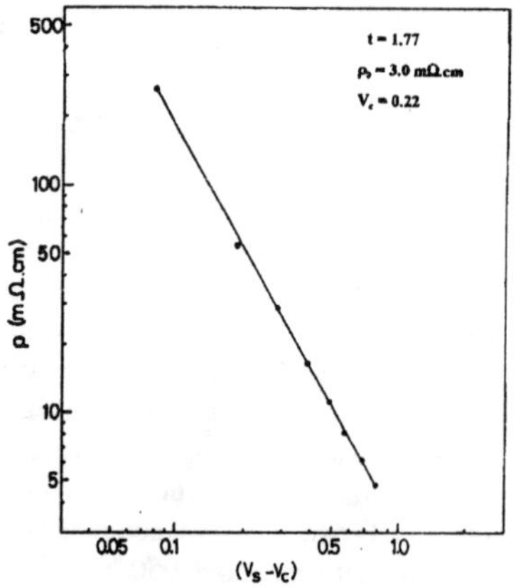

Fig. 9a. Log-log plot of ρ vs (Vs-Vc) where Vs is the vol% of YBCO in the composite and Vc is the percolation threshold.

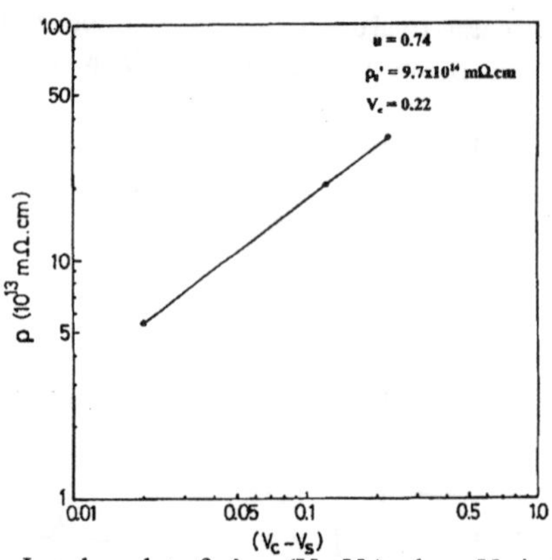

Fig. 9b. Log-log plot of ρ' vs (Vc-Vs) where Vc is the percolation threshold and Vs is the vol% of YBCO in the composite.

conductor with resistivity $\rho \sim 1$ mΩ cm, nearly 12 orders of magnitude lower than that of DBSO. The normal state resistivity YBCO-DBSO composite samples is dominated by YBCO with a significant drop of ρ occurring near $V_s \sim 20$ vol%. The behaviour of ρ correlates with that of α which increases sharply t oward that of YBCO starting from $V_s \sim 20$ vol%. Therefore the percolation threshold value V_c for the normal state transport properties of the composite is ~ 20 vol% of YBCO. Figs.9(a) and (b) show the log-log plots of ρ vs $(V_s - V_c)$ and ρ' vs $(V_c - V_s)$ respectively. The least-square-fits gave the exponents t and u as 1.77 and 0.74, respectively with a percolation threshold $V_c = 0.22$, and the values for ρ_0 and ρ_0' are 3.0 and 9.7×10^{14} mΩ cm respectively.

The values of the critical exponents obtained for the YBCO-DBSO composites also agree reasonably well with the theoretical values for an idealised metal-insulator percolation system.

In addition to the normal percolation threshold there exists a different threshold for the superconducting matrix in the case of YBCO-DBSO composite system. As shown in Fig.7 the superconducting transition in YBCO-DBSO composites is essentially undistributed by the addition of DBSO as long as the volume fraction of YBCO is greater than 30%. As we could not get a zero resistivity superconducting transition in samples having less than 30 vol% of YBCO, 30 vol% can be considered as the superconducting percolation threshold. For a random continuous two component medium with black and white symmetry, the percolation threshold for both components should be around 17 %. In the present YBCO-DBSO system, the fact that the percolation threshold differs is indicative of a loss of black and white symmetry because the presence of large number of insulating grains of DBSO, which effectively reduces the number of superconducting path and thus results in a higher superconducting percolation threshold, as observed in the present case.

CONCLUSION

The results obtained from the XRD and resistivity studies showed that in Bi(2223)-DBSO and YBCO-DBSO composites, the superconductors are found to remain as two separate phases with their own characteristics even under severe heat treatment. The normal state percolation and superconducting percolation threshold values are found to be almost same, i.e. ~20 vol% of the superconductor in both Bi(2223)-DBSO and YBCO-DBSO composites. However in YBCO-DBSO composite, superconducting threshold is ~30% of YBCO in the composite. The critical exponents describing the transport behaviour of the composite system match fairly well with those for an idealised metal-insulator composite system. The observation of high temperature superconductivity with the transition temperature of 110K in Bi(2223)-DBSO composites having 80 vol% of insulating material (DBSO) may be of fundamental importance in view of the very short coherence length (~15 A°) of Bi(2223) superconductors. The chemical non-reactivity of Bi(2223) and YBCO with non-metal ceramic, DBSO, under severe heat treatment makes DBSO a suitable substrate material for both Bi(2223) and YBCO films. The suitability of DBSO as substrate for Bi(2223) and YBCO superconductor was confirmed by developing a superconducting Bi(2223) thick film with Tc(0) = 110K and YBCO thick film with Tc (0)= 92K on polycrystalline DBSO substrates and these results are reported elsewhere[9,10].

REFERENCES

1. G. Xiao, F.H. Streitz, M.Z. Cieplak, A. Bakhshai, A. Govrin and C.L. Chien, Phys. Rev. **B38**, 776 (1988).
2. T.K. Worthington, W.J. Gallagher and T.R. Dinger, Phys. Rev. Lett. 59, 1160 (1987).
3. D.R. Harshman, G. Aeppli, E.J. Ansaldo, B. Batlogg, J.H. Brewer, J.F. Arolan, R.J. Cava, M. Celio, A.C.D. Chaklader, W.N. Hardy, S.R. Kreitzman, G.M. Luke, D.R. Noakes and M. Sneba, Phys. Rev. B **36,** 2386 (1987).
4. Proceedings of the Second International Conference of Electrical Transport and Optical Properties of the Inhomogeneous Media, ed. J. Lafait and D.B. Tanner, Physica A **152**, (1989).
5. P. Umadevi Muralidharan, K.V. Paulose, J. Koshy and A.D. Damodaran, J. Am. Ceram. Soc. **74,** 2679 (1991).
6. B. Ropers, R. Canet, F. Carmona and S. Flandrois, Solid State Commun. **75**, 791 (1990).
7. L. Ganapath, A. Kumar and J. Narayanan, J. Appl. Phys. **66**, 5935 (1989).
8. D. Payuna, H. Berger, J.L. Tholence, M. Affronte, R. Sanjines, A. Dubas, P. Bugnon and F. Vasey, Physica C **153**, 1339 (1988).
9. J.Koshy, K.S.Kumar, J.Kurian, Y.P.Yadava and A.D.Damodaran, Appl. Phys. Lett. **65,** 2857 (1994).
10. J. Koshy, K.S. Kumar, J. Kurian, Y.P. Yadava and A.D. Damodaran, Physica C **234**, 211 (1994).
11. J. Koshy, K.S. Kumar, Y.P. Yadava and A.D. Damodaran, J. Mater. Sci. Lett. **13**, 554 (1994).
12. D. Stauffer, Phys. Rep. **54,** 1 (1979).
13. J.P. Straley in *Percolation structures and processes,* ed. G. Deutscher, R. Zalen and J. Adler, Annals of the Israel Physical Society, Vol. 5 p. 353 (AIP, New York, 1984).

MECHANICAL PROPERTIES OF OCMG-PROCESSED Nd-Ba-Cu-0 BULK SUPERCONDUCTORS

T. Murayama[1], N. Sakai[1], S.I. Yoo[1,2] and M. Murakami[1]

[1]Superconductivity Reseach Laboratory-ISTEC, 1 - 16-25 Shibaura, Minatoku, Tokyo 105, Japan
[2]Railway Technical Research Institute, 2-8-38 Kokubuji-shi, Tokyo 185, Japan

The fracture toughness and hardness of Nd-Ba-Cu-O (NdBCO) bulk superconductors fabricated by the oxygen-controlled-melt-growth (OCMG) process were measured using an indentation method on the (001) plane and the ac plane perpendicular to the (001) plane. The fracture toughness on the (001) plane of NdBCO was nearly the same as that of YBCO. The effects of the second phases such as $Nd_4Ba_2Cu_2O_y$ (Nd422) and silver particles on the fracture toughness were also investigated, and it was found that finely dispersed Nd422 particles and silver particles are effective in increasing the fracture toughness.

I. INTRODUCTION

When NdBCO superconductors were melt-processed in air, they showed a low T_c and a broad superconducting transition, which was due to the presence of a partial substitution of Nd on Ba site, thus forming a $Nd_{1+x}Ba_{2-x}Cu_3O_y$ type solid solution. Recently, we have developed the OCMG process, using low oxygen partial pressures of 10^{-2}~10^{-3} atm during the melt growth, and have succeeded in achieving high T_c for $NdBa_2Cu_3O_y$ (Nd123) superconductors. It was also interesting to note that the anomalous secondary peaks were commonly observed in the M-H loops at 77 K, leading to large J_c values in a high field region, which were larger than that of high quality MPMG-processed YBCO superconductor. Hence, NdBCO superconductors have a high potential for bulk-type applications. For most applications, however, a large electromagnetic force will act on the superconductors, and therefore the improvement of mechanical properties are crucial. Fracture toughness and hardness value of YBCO using the indentation technique were reported by many researchers[1-7]. All of these data indicate that the fracture toughness of YBCO is extremely low.

It is common to add the second phase to improve the fracture toughness. In the NdBCO samples, Nd422 particles are trapped in the Nd123 matrix, and these particles may increase the fracture toughness when their distribution is well controlled. It is also known that silver doping is effective in toughening YBCO superconductors without decreasing superconducting properties[8].

In this paper, we report the fracture toughness and hardness values for the OCMG-processed NdBCO, determined by using an indentation technique, and discuss the effect of the second phase such as Nd422 and silver on the fracture toughness.

II. EXPERIMENTAL

Sample preparation

Nd-Ba-Cu-0 samples were prepared by using powders of as-sintered Nd123 and Nd422. The sample A was prepared from Nd123 + 10%mol Nd422 + 0.5wt% Pt, and the sample B was prepared from Nd123 + 10%mol Nd422 + 0.5wt % Pt + 10wt% Ag_2O.

Precursors were pressed into pellets of 30 mm in diameter using a cold-isostatic press. The pellets were then melt-processed in a gas mixture of 1%O_2 in Ar using a tube furnace. For the melt growth, thc furnace was ramped up to 1080°C for the sample A and 1060°C for the sample B in 1%O_2 in Ar for 3 h, held for 30 min, cooled to 1030°C for the sample A and 1015°C for the sample B at a rate of 10°C/min, slowly cooled to 960°C for sample A and 940°C for sample B at a rate of 1°C/min, and then furnace-cooled. The flow rate of the mixture gas was 500 ml/min. All melt-processed samples were annealed in flowing pure O_2 gas at a rate of 300 ml/min from 500°C to 300°C for 300 h. Melt-grown samples consisted of multiple textured domains.

Samples were first cleaved along the (001) basal plane, and then cut perpendicularly to the cleavage surface. Both the cleavage surface and the surface perpendicular to the cleavage surface were mounted in epoxy resin and polished. The final samples were rectangular shape of about 1 to 2 mm in their size. Microstructures of the samples were observed using an optical microscope.

Hardness and Fracture toughness measurements

Indentations on the polished surfaces were made with a Matsuzawa hardness tester, using a Vickers Diamond Pyramid indenter at room temperature. Contact loads were 0.245, 0.49, and 0.98N for a peak load contact time of 15 s. Indentation was made at least 10 times. Orientation of the diagonals of square impression of indentations was parallel to the [100] and [010] directions on the cleavage plane (001), and was parallel and perpendicular to the [001] direction on the ac plane perpendicular to the (001) plane.

For the indentations made on the (001) cleavage surface, the diagonals of impression were aligned in the [100] and the [010] directions. This is relatively easy to accomplish since the traces of the twin planes (110)/(110) on the (001) cleavage plane are exactly at 45° to the [100] and the [010] directions. For indentations made on the ac plane perpendicular to the (001) plane, the diagonals of impression were aligned in the directions coincident with [001] and the [100] or [010]. The diagonals were parallel or perpendicular to the cleavage plane. We measured the width of diagonals of indentation and crack length from the center of indentation to the crack tip using a scanning laser microscope.

Elastic modulus measurements

In order to estimate the fracture toughness, the value of Young's modulus, E, should be known beforehand. Here we measured Young's modulus with a Fischer dynamic hardness tester H100V, using knoop indenter on the (001) plane and the ac plane. We measured load-indentation depth during loading and unloading. Load-indentation depth curve during loading contains depth due to both the elastic and plastic effects. On the other hand, load-indentation depth curve during unloading after loading was directly related to the elasticity. Then, we calculated E from load-indentation depth curve which is a linear part during unloading. Indentations were made on the Nd123 matrix such that the effect of the second phase on the elastic modulus could be avoided. The initial part of the unloading curve was linear, and fit a tangent to the unloading curve at maximum load. Here we assumed Poisson's ratio v is 0.28 for the calculation of E. The loads were 0.4mN~100mN. Details for this method were reported elsewhere[9].

III. RESULTS AND DISCUSSION

Hardness and fracture toughness of NdBCO

The critical stress intensity factor (Kc) and hardness (H) were calculated using the following equations[5], $H=1.854P/(2a)^2$ and $Kc=0.016(E/H)^{1/2}\ P/C^{3/2}$, where P is the applied indentation load, 2a is the indentation diagonal, E is the elastic constant, and C is the radial crack length measured from the center of the indent to the crack tip. Figure 1 shows optical micrograph of the impression diagonal and radial cracks on the (001) plane and the ac plane. Cracks propagated along [100] or [010] directions on the (001) plane, while the cracks propagated parallel to the basal plane than along [001] direction for the ac plane. In this case, we measured the crack length along basal plane. These results show that the (001) plane fractures more easily than the (100) or the (010) planes. Indentation was made with each load; however, in the case of indentation load with 0.25 N, we could not observe well-developed cracks, while with the load of 0.98 N almost all indentations were crashed. Therefore, we measured indentation diagonals and crack lengths with the indentation load of 0.49 N. For sample B with silver doping, since the size of the silver particles was relatively large, we could not obtain reliable Kc values.

Table 1 shows calculated hardness and fracture toughness and measured Young's modulus. In the case of highly aligned YBCO, Goyal et al.[4] reported the presence of a large anisotropy of hardness between (001) and perpendicular (001) plane. However, in the case of NdBCO, the value of hardness on the (001) plane was almost the same with that of the plane perpendicular to the (001) plane. The value of Kc on the (001) plane was the same as that reported for YBCO[4]. And cleavage fracture toughness Kc [001] was about 20% lower than that on the (001) plane. However, the anisotropy of crack length along [100] and [010] directions on the (001) plane was observed. Here we neglected it for calculating Kc, and used the average crack length. It is necessary to distinguish the crack length for calculating absolute Kc in each direction. The data obtained can be regarded as the fracture toughness of the Nd123 pseudo-crystals because the indentation was made on the Nd123 matrix.

Elastic modulus of NdBCO

The modulus values show the anisotropy between (001) and perpendicular (001) plane. This anisotropy can be understood in terms of the layered crystal structure of Nd123. Weak coupling between layers will result in lower modulus along the c-direction. It is not peculiar that difference in the moduli along these two directions was detected since these samples are highly aligned and behave like pseudo-crystals.

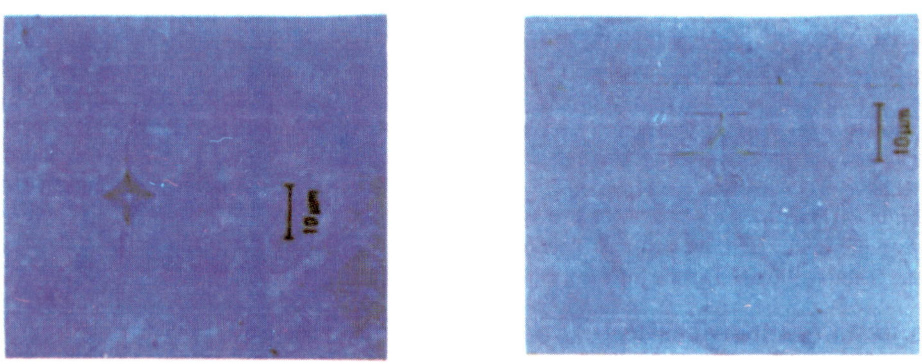

(a) **(b)**

Fig. 1. Optical micrographs of indentation mark and cracks of sample A for the (001) plane in (a) and the ac plane in (b).

Table 1. Data of mechanical properties

NdBCO

Indented plane	Elastic Modulus (GPa)
(001)	158
(100)/(010)	191

Sample	Plane	H (Gpa)	E/H	Kc($MPa^{1/2}$)
NdBCO	(001)	7.79	24.5	0.703
NdBCO	(100)/(010)	7.71	20.49	0.566
YBCO[4]	(001)	6.74	27	0.67

Second phase effect on fracture toughness of NdBCO

Figure 2 shows the crack propagation in sample A and sample B. For sample A, it is evident that cracks are distorted when they interact with the Nd422 particles. When interacting with large Nd422 particles, the cracks propagate through them. However, in the case of small Nd422 particles, the cracks propagate by passing around their surface, which suggests that smaller particles are more effective in toughening the Nd123 matrix. For sample B, cracks stopped at silver particles, which implies that silver can absorb crack propagation energy with its elasticity. This clearly shows that silver particles are effective in toughening the Nd123 phase.

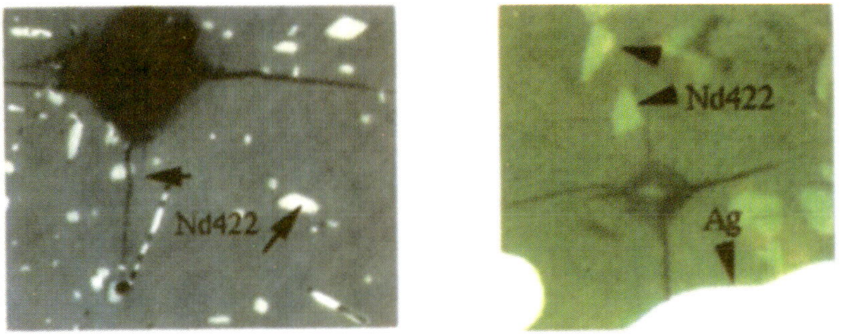

(a) (b)

Fig. 2. Laser micrograph of sample (A) in (a) and optical micrograph of sample (B) in (b) for indentation and cracks.

IV. SUMMARY

Fracture toughness and hardness of OCMG-processed NdBCO were measured by using the indentation method on the plane (001) and the ac plane perpendicular to the (001) plane. For both planes, while the anisotropy in the hardness was not observed in contrast with YBCO, the anisotropy in the fracture toughness was commonly observed. The Kc values were calculated for both planes. The Kc value on the (001) plane of NdBCO was equivalent to that of YBCO. The effects of the second phase such as Nd422 and silver on the fracture toughness were also investigated. It was found that small Nd422 particles and silver particles are effective in toughening NdBCO superconductors.

ACKNOWLEDGMENT

This work is partially supported by New Energy and Industrial Technology Development Organization (NEDO) for the R&D Industrial Science and Technology Frontier Program.

REFERENCES

1. R. F. Cook, T. R. Dinger and D. R. Clarke, Appl. Phys. Lett. **51**, 454 (1987).
2. A. S. Raynes, S. W. Freiman, F. W. Gayle, and D. K. Kaiser, J. Appl. Phys. **70**, 5254 (1991).
3. S. Ochiai, K. Osamura and T. Takayama, Jpn. J. Appl. Phys. **27**, L 1101 (1988).
4. A. Goyal, P. D. Funkenbusch, D. M. Kroeger and S. J. Burns, J. Appl. Phys. **71**, 2363 (1992).
5. H. Fujimoto, M. Murakami, T. Oyama, Y. Shiohara, N. Koshizuka and S. Tanaka, Jpn. J. Appl. Phys. **29**, L1793 (1990).
6. A. Goyal, W. C. Oliver, P. D. Funkenbusch, D. M. Kroeger and S. J. Burns, Physica C **183**, 221 (1991).
7. H. Fujimoto, M. Murakami and N. Koshizuka, Physica C **203**,103 (1992).
8. M. Murakami, T. Oyama, H. Fujimoto, T. Taguchi, S. Gotoh, Y. Shiohara, N. Koshizuka and S. Tanaka, Jpn. J. Appl. Phys. **29**, L1991 (1990).
9. J. L. Loubet, J. M. Georges and G. Meille, *Microindentation Techniques in Materials Science and Engineering,* eds. P. J. Blau and B. R. Lawn, Am. Soc. for Testing and Materials, Philadelphia, p. 72 (1986).

Pr IN RE 123 SUPERCONDUCTORS - TEN YEARS AFTER

R. Suryanarayanan[a], A. Das[b] and I. Zelenay[c]
[a]*Laboratoire de Chimie des Solides, UA 446, CNRS, Université Paris-sud, 91405 Orsay, France*
[b]*Laboratoire de Physique des Soldies, CNRS, 92195 Meudon, France*
[c]*Physics Department, Faculty of Electrical Engineering, Ilkovica 8, 81219 Bratislava, Slovakia*

Soon after the discovery of superconductivity at 92K in $YBa_2Cu_3O_7$, it was shown that Pr suppressed superconductivity in $Y_{1-x}Pr_xBa_2Cu_3O_7$ when the concentration of Pr exceeded a critical value x_c of 0.55. Further, the value of x_c depended on the ionic radius of the rare earth (at the Y site), x_c being as low as 0.3 for Nd. Several models have been proposed to account for these data but without much success. In the first part of the talk, we will review the experimental data obtained since 1987. In the second part of the talk, recent results obtained in the last year on the effect of isovalent Sr substitution at the Ba site on x_c will be presented. In particular, It was shown that x_c could be increased from 0.3 to 0.45 when 50% of Ba was substituted by Sr in $Nd_{1-x}Pr_xBa_2Cu_3O_7$. Further, heating the samples in argon prior to oxygen annealing increased the T_c by 6 to 15K depending on x. In addition, annealing in high pressure oxygen induced superconductivity in certain insulating samples. The normal state properties including Seebeck coefficient will be reported. Neutron diffraction studies indicated a relative slow variation of apical oxygen distance as a function of x(Pr) in Sr substituted systems. In the third part of the talk, these data will be discussed in the light of certain theoretical models invoking the effect of Prf-CuO_2 hybridisation. Our data show that Sr retards this hybridisation and in addition stress the important role played by oxygen reordering along the chains in the Pr substituted Ln123 compounds.

I. INTRODUCTION

The superconducting properties of $Y_{1-x}Pr_xBa_2Cu_3O_7$ have been an enigma in the field of high temperature superconductivity[1-4]. As early as 1987, it was shown that the superconducting critical temperature T_c of this compound decreased progressively as a function of Pr(x) concentration and finally when the concentration exceeded a critical value $x_c \geq 0.55$, the compound lost the superconductivity. For x=l, the compound is an antiferromagnetic insulator with T_N=17 K, Further, the value of x_c depends[5,6] on the specific rare earth R in $R_{1-x}Pr_xBa_2Cu_30_7$. Thus, the value of x_c is 0.3 for R=Nd. To interpret some of these results, mechanisms involving hole filling and /or pair breaking have been proposed. In the hole filling model[7,8], Pr adopts a valence of 4+ and the extra electron missing from the Y site may transfer to the Cu-0 planes to decrease the hole density and hence the critical temperature T_c. However, the valence of Pr was found to be close to 3+(for all x > 0) from $M_{4,5}$ X-ray absorption edge spectra[9], valence-band resonant photoemission[10], and coulometric experiments[11]. On the other hand, both the ^{141}Pr Mössbauer isomer shift[12] and the neutron diffraction studies[13] indicated a valence of 3.4(1) for $PrBa_2Cu_30_7$. According to the pair-breaking model[14], the decrease in T_c as a function of x(Pr) results from the interaction of extended Pr^{3+} orbitals with the charge carriers in the Cu-0 planes. On the theoretical front, several models[15-22] have been proposed to account for these observations. Of these, our attention is drawn in particular to the model proposed by Fehrenbacher and Rice[17]. These authors propose the existence of a strong hybridisation of f-states of Pr with p-orbitals of 8 surrounding oxygens with possible localisation of holes due to it. The resulting wave function has the following form $|\Psi\rangle = \alpha|4f^1p^6\rangle + \beta|4f^2p^5\rangle$. When the weight of β is large, many local properties of Pr as seen by high energy spectroscopy would be similar to those of Pr^{+3}. Further, it was pointed out that the extreme sensitivity of the 1D chains and the Pr^{III} - Pr^{IV} mixed valent state to 0 vacancies and other defects may play a role.

Lately there has been a renewed interest around a central theme, whether a change in the ion size[23-27] at the Ba site would influence the superconducting properties of $R_{1-x}Pr_xBa_2Cu_3O_7$? In this work, we will present the current situation describing the effect of Sr substitution on several properties. In particular, we discuss the effect of heat treatment on the structural, transport and thermoelectric power (TEP) data on $Y_{1-x}Pr_xSrBaCu_3O_7$, $Nd_{1-x}Pr_xBa_{2-k}Sr_kCu_3O_7$ and $Y_{1-x}Pr_xSr_2Cu_{2.8}Re_{0.2}O_7$. Let us note that the last compound contains Re substituted at the Cu chain sites[28,29] in order to stabilise the perovskite phase of the Sr analog of YBCO. Our data indicate that the Sr substitution and the heat treatment result in a slow charge transfer and reordering of oxygen along the chains which retard the Pr 4f hybridization leading to a remarkable increase in x_c.

II. EXPERIMENTAL TECHNIQUES

The polycrystalline samples have been prepared by solid state reaction of the respective oxides or carbonates and further sintering . Y_2O_3, Nd_2O_3, Pr_6O_{11}, $SrCO_3$, $BaCO_3$, CuO, and ReO_2 were thoroughly mixed in required proportions and calcined at 950 °C in air for a period of 18 hours. The resulting product was ground, mixed pelletized (two pellets of 750 mg each) and heated in air at 980 °C for a period of 24 hours. This was repeated twice. The samples containing no Ba ($Y_{1-x}Pr_xSr_2Cu_{2.8}Re_{0.2}O_7$ compounds) were annealed under oxygen at 1000 °C for a period of 24h, then furnace cooled. As for the other series, for each composition, two heat treatments[24] were carried out : i) annealing in oxygen at 450 °C for about 72h [O]; ii) heating in argon at 850 °C for about 24h furnace cooling to 20 °C followed by oxygen annealing at 450 °C for about 72h [AO]. X-ray diffraction (XRD) data of the samples were collected using an automatic Philips diffractometer fitted with a secondary beam graphite monochromator and using Cu K_α (40KV/20mA) radiation. The oxygen content of the samples was determined by the iodometry titration method. The diamagnetic onset in ac susceptibility measurements was taken as T_c. The resistivity was measured in the Van der Pauw configuration. The TEP (S) was measured by a slow dynamical method as a function of temperature employing a 10nV resolution digital voltmeter.

III. RESULTS AND DISCUSSION

A. $Y_{1-x}Pr_xSrBaCu_3O_7$

X-ray analyses of the samples studied here have been reported earlier[24]. In the case of $Y_{1-x}Pr_xSrBaCu_3O_7$ there was a gradual increase in the lattice parameters as the concentration of Pr was increased. This could be attributed to the larger ionic radius of Pr. As x increased the ratio c/a and the orthorhombicity defined as (b-a)/(b+a) decreased. An orthorhombic to tetragonal transformation was observed (Fig.1) for the sample x=0.5 [O]. The samples [O] with x=0.6 and x=0.65 showed tetragonal symmetry and were superconducting, while the samples [O] x>0.65 were insulators. After the [AO] annealing both the ratios c/a and the orthorhombicity increased, and there was a net decrease in the volume. The system $Y_{1-x}Pr_xSrBaCu_3O_7$ [AO] remained superconducting till x=0.7 (Fig.1) in an orthorhombic symmetry and then for x>0.7 the samples [AO] were insulating and showed tetragonal symmetry. Though there was a slight decrease in oxygen content as x increased, (6.98 for x=0 to 6.84 for x=0.7) we observed no change of oxygen content after the [AO] treatment.

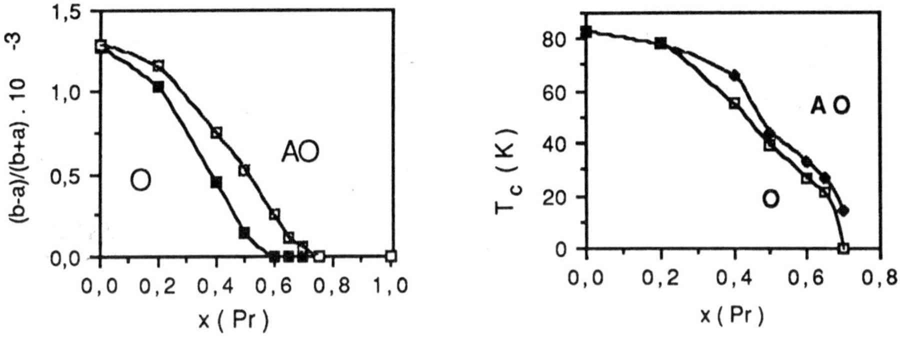

Fig.1. Orthorhombicity (left) and T_c (right) of $Y_{1-x}Pr_xSrBaCu_3O_7$.

Let us look at the transport and TEP data (Fig. 2). For x=0, there was practically no change in ρ as a result of heat treatment. As x increased, ρ at 300K increased gradually and T_c decreased but for a given x, the heat treatment [AO] not only decreased the value of ρ at 300K but also increased the T_c. For x=0.6, the linearity in ρ extended to lower temperatures after the [AO] annealing. The residual resistivity ρ_0, for example, for x=0.6 reduced from 1.041 to 0.75mΩcm after the [AO] treatment indicating a reduction in scattering by defects(oxygen).The effect of the heat treatment on TEP data is much more revealing. For x=0 [O], the value of S at 300 K (S_{rt}) is negative indicating an overdoped state following Obertelli et al[30]. This value was still more negative for [AO] indicating that this treatment increased the doping. For x=0.1[O], S_{rt} was positive as a result of a decrease in doping but again the value became negative for the [AO] sample pointing to an increase in doping. This is much more apparent for x=0.2 (Fig.3). The decrease in S_{rt} for the [AO] samples continues till x ≤ 0.7 indicating the doping cannot be increased further for x>0.7. However, it is interesting to compare the S_{rt} [O] as a function of x (Fig.3) with that obtained on the Ba2 system[7]. The value of S_{rt} increased rapidly for x≥0.4 for the Ba2 system whereas for the SrBa system, the increase is gradual up to x=0.7 indicating a retarded localization of holes. If this is due to changes in structural parameters which retard the Pr f-Cu-O2 hybridization, then it should be more apparent if one could replace completely Ba by Sr. The results are presented in the next section.

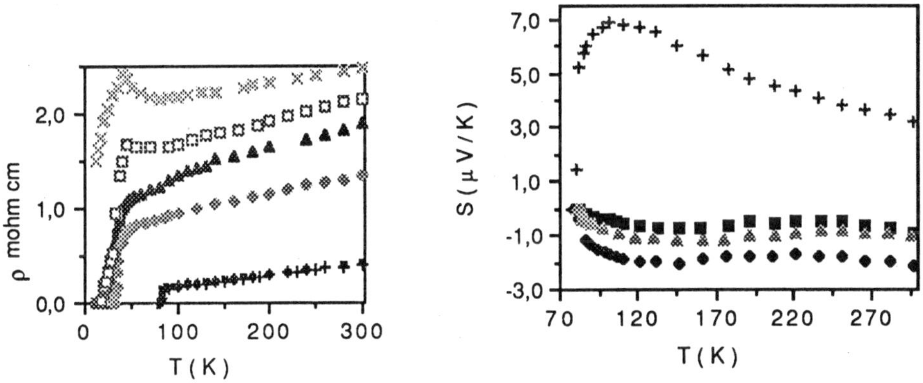

Fig.2 Resistivity and TEP of $Y_{1-x}Pr_xSrBaCu_3O_7$ as a function of temperature. (left) from bottom x=0 [O] and [AO]; x=0.6 [O] and [AO]; x=0.7 [O] and [AO]; (right) from top, x=0.1[O]; x=0[O]; x=0.1[AO]; x=0 [AO]

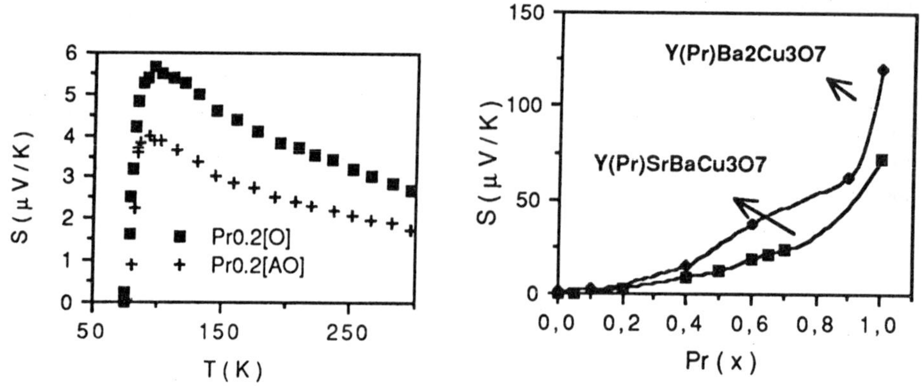

Fig.3. (left) TEP of $Y_{1-x}Pr_xSrBaCu_3O_7$ as a function of temperature. (right) TEP at 300 K for the SrBa and the Ba2 systems.

B. $Y_{1-x}Pr_xSr_2Cu_{2.8}Re_{0.2}O_7$

The XRD analyses of $Y_{1-x}Pr_xSr_2Cu_{2.8}Re_{0.2}O_7$ revealed[31] a tetragonal symmetry for all values of x. Figure 4 shows T_c and S_{rt} as a function of x. Indeed, the x_c increased from 0.55 (for the Ba2 system) to 0.75. Further, it is interesting to note that in spite of the relatively low doping level of this system for x=0 (T_c=43 K) as revealed by the S_{rt} value (15µV/K) the doping decreased rather gradually as a function of Pr concentration indicating again that in this system the hybridization effects are retarded. Three recent results are worth noting in this respect. (i)The rate of decrease[32] in the apical distance was smaller in the Sr2 system compared to the Ba2 system; (ii) The T_c of the sample x=0.75 could be increased[33] from 0 to 14 K as a result of annealing under 200 bar of oxygen; (iii) In $Pr_{0.9}Sr_{0.1}Sr_2Cu_{2.7}Mo_{0.3}O_7$ superconductivity was observed[34] at 17K. These data further show the strong influence of structural parameters and oxygen defects (along chain) on x_c and T_c. In the next section, we discuss the effect of Sr on T_c in Nd substituted compound.

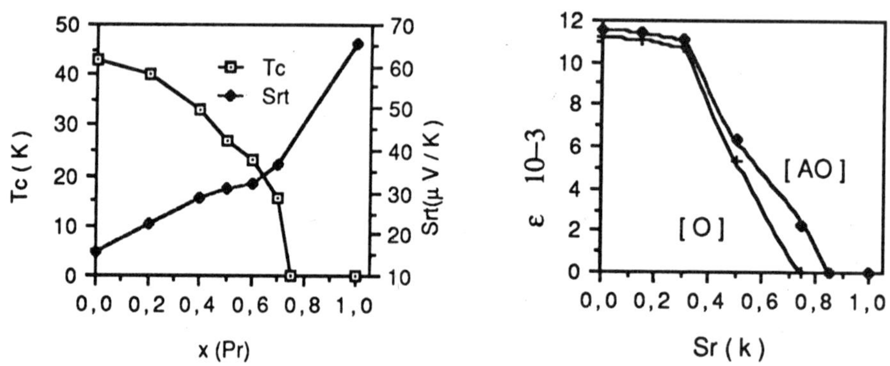

Fig.4. (left) T_c and S_{rt} of $Y_{1-x}Pr_xSr_2Cu_{2.8}Re_{0.2}O_7$. (right) Orthorhombicity of $Nd_{0.7}Pr_{0.3}Ba_{2-k}Sr_kCu_3O_7$ as a function of Sr(k).

C. $Nd_{0.7}Pr_{0.3}Ba_{2-k}Sr_kCu_3O_7$

An orthorhombic to tetragonal transition was observed at k=0.7 which was strongly influenced by the [AO] treatment (Fig.4, right). As noted earlier, there was no change in the oxygen content after the [AO] treatment. Superconductivity was restored in the insulator with Pr(0.3) for $k \geq 0.15$ and again the T_c of the [AO] samples was higher by 6 to 14 K depending on k (Fig. 5). From the S_{rt} values, we calculated the hole density p per Cu-02 plane following Obertelli et a1[30] (Fig.5 right). This is an approximate estimation. However, the data show clearly that the isovalent Sr substitution could delocalise holes and superconductivity is restored with 30% of Pr. It is important to note that the increase in Sr content did not result in an increase in the oxygen content but probably increased the chain fragments.

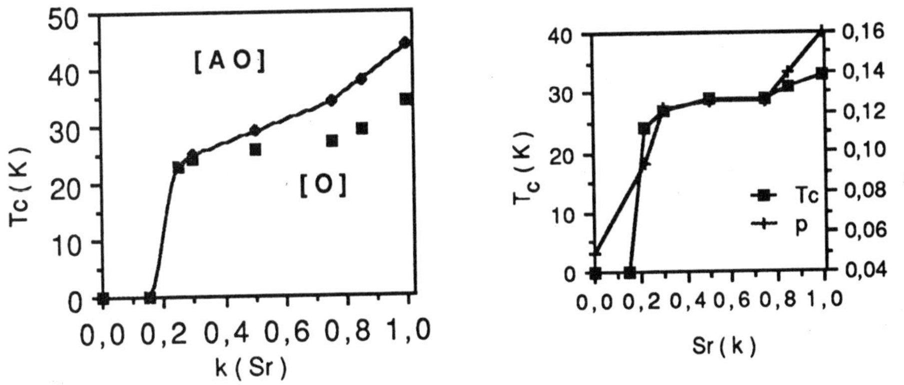

Fig.5. (left) T_c of $Nd_{0.7}Pr_{0.3}Ba_{2-k}Sr_kCu_3O_7$ as a function of k(Sr). (right) T_c and hole density per Cu-O2 plane as a function of k(Sr) derived from TEP data.

IV. CONCLUSIONS

In order to examine the effect of crystal chemistry on the T_c suppression behaviour of Pr in RE-123 system, we have studied two series:(A) $RE_{1-x}Pr_xBa_{2-k}Sr_kCu_3O_7$ (RE=Y,Nd), and (B) $Y_{1-x}Pr_xSr_2Cu_{2.8}Re_{0.2}O_7$. In the case of (A), Sr substitution increased x_c considerably. The [AO] treatment further reduced the ρ_{rt} and ρ_o indicating a substantial reduction in scattering by defects (oxygen disorder). In the case of the system B, which had a lower T_c (43K) for x=0 is underdoped as revealed by the value of S_{rt}. In spite of this, the effect of Pr was much less drastic as shown by a smaller slope dT_c/dx and a relatively higher value of x_c=0.7. It is interesting to note that this value is quite comparable to that observed[35] in the double chain compound $Y(Pr)Ba_2Cu_4O_8$. It is possible that the value of x_c will increase further in a double chain compound with Sr partially/fully replacing Ba. Our S_{rt} data as a function of Pr concentration indicate that the hole depletion is retarded considerably due to Sr substitution. However, to understand the slow charge transfer of holes and the effect of [AO] treatment, one has to take into account of the role played by Cu-O chains and oxygen defects in the basal plane. We suggest that efforts should be directed to determine the optimizing conditions for preparation such as annealing procedures and high pressure oxygen annealing. To fully substantiate the arguments developed here and to further examine the model suggested by Fehrenbacher and Rice[17], detailed microscopic experiments are needed to determine the role of the chains and how they affect the Pr(f) hybridisation in the CuO2 planes. In this respect, the recent data obtained by Takata et al[36] are quite revealing. A distinct difference in the charge density of YBCO and PBCO was observed that was taken as an experimental evidence of the hybridization between Pr 4f-O 2pπ orbitals. Such studies in the Sr substituted systems should be interesting.

REFERENCES

1. L. Soderholm, K.Zhang, D.G.Hinks, M.A.Beno, J.D.Jorgenson, C.U.Segre and I.K.Schuller, Nature **328**, 604 (1987).
2. J.K Liang, et al., Z. Phys. B **69**, 137 (1987).
3. Y. Dalichaouch, M.S. Torikachrili, E.A. Early, B.W.Lee, C.L.Seaman, K.N.Yang, H.Zhou and M.B.Maple, Solid State Commun. **65**, 1001 (1988).
4. For a general review and references see H.B Radousky, J.Mater.Res, **7**, 1917 (1992).
5. W. Guan, Y.Xu, S.R. Sheen, Y.C. Chen, J.Y.T. Wei, H.F. Lai, M.K. Wu, and J.C. Ho, Phys.Rev.B **49** 15993 (1993); Y. Xu and W. Guan, Solid State Comm. **80**, 105 (1991).
6. S.K.Malik, C.V.Tomy and Parag Bhargava, Phys. Rev. B **44**,7042 (1991).
7. A.P.Goncalves, et al., Phys. Rev. B **37**, 7476 (1988).
8. A. Matsuda, et al., Phys. Rev. B **38**, 2910 (1988).
9. U.Neukirch, C.T. Simmons, P.Sladeczek, C.Laubschat, O.Strebel, G.Kaindl and D. D.Sarma, Europhys. Lett. **5**, 567 (1988).
10. J.S.Kang, J.W.Allen, Z.X.Shen, P.W.Ellis, J.J.Yeh, B.W.Lee, M.B.Maple, W.E.Spicer and I.Lindau, J. Less Common Met. **148**, 121 (1989).
11. M.Kato, et al. Physica C **235**, 815 (1990).
12. A. A. Moolenaar, P. C. M. Gubbens, J. J. van Loef, M. J. V. Menken and A. A. Menkovsky, Hyperfine Interactions **93**, 1717 (1994).
13. M. Guillaume, et al., Phys. B **90**, 13 (1993).
14. J.J. Neumeier, M.B. Maple and M.S. Torikachvilli , Physica C **156**, 574 (1988); A.Kebede, et al., Phys. Rev. B **40**, 4453 (1989).
15. G. Y. Guo and W. M. Tammerman, Phys. Rev. B **41**, 6372 (1990); Pan Wei and Hu Wen Ying, Physcia C **209**, 400 (1993).
16. D. Khomskii, J.Supercond. **6**, 69 (1993).
17. R. Fehrenbacher and T.M.Rice, Phys. Rev. Lett. **70**, 3471 (1993).
18. D. J. Singh, Phys. Rev. B **50**, 4106 (1994).
19. M. Muroi and R. Street, Physica C **253**, 205 (1995).
20. H. A. Blackstead and John D. Dow, Phys. Lett. A **206**, 11 (1995).
21. Pan Wei and Hu Wen Ying, Physcia C **209**, 400 (1993).
22. M. Biagini, C. Calandra and Stefano Ossicini, Phys. Rev. B **52**, 104668 (1995).
23. G. Cao et al., Phys. Lett. A **196**, 263 (1994).
24. A.Das and R.Suryanarayanan, J.Phys. I France **5**, 623 (1995).
25. G. Cao, Y. Qian, X. Li, Z. Chen, C.Wang, K. Ruant, Y. Qiut, L. Cao, Y. Ce and Y.Zhang, J.Phys. Condens. Matter. **7**, 287 (1995).
26. G. Cao, Y.Qian, X.Li, H.Wu., Z.Chen. and Y.Zhang, Physica C **248**, 92 (1995).
27. A. Das and R. Suryanarayanan, Mater. Res. Bull. **30**,1363 (1995).
28. Tohru Den and Tamaki Kobayashi, Physica C **196**,141 (1992).
29. R.L.Harlow, et al. Physica C **257**,125 (1996).
30. S.D. Obertelli, J.R. Cooper and J.L. Tallon., Phys. Rev. B **46**,14928 (1992).
31. A.Das et al., J. Solid State Chem. **118**, 215 (1995).
32. C.R.Taylor et al., Physica C **261**, 135 (1996).
33. A. Das, R. Suryanarayanan and B.Dabrowski, Solid State Commun. **99**, 413 (1996).
34. H. L. Tsay, Y. C. Chen, F. L. Juang, W. M. Wang and H. D. Yang, Phys. Rev. B **53**, 11340 (1996).
35. K. Koyama, S. Taga and S. Noguchi, Physica C **185**, 809 (1991).
36. M. Takata, et al., Physica C, **263**, 340 (1996).

SUPERCONDUCTING PROPERTIES OF THE MERCURY AND Cu/C PHASES

J.L.Tholence[1], P.Toulemonde[1], A.Sulpice, C. Acha, M. Nunez-Regueiro, J.J. Capponi[2], C.Chaillout[2], M.Alario-Franco[2], E. Gautier[2], S.Le Floch[2], M.Marezio[2], P.Bordet[2], S.M.Loureiro[2], S.N.Putilin[2,3], E.V.Antipov[2,3]

Centre de Recherches sur les Très Basses Tempèratures ([1] LEPES) and [2]Laboratoire de Cristallographie Laboratoires associès à l'Universitè Joseph Fourier, CNRS, BP 166, F38042 Grenoble-Cèdex 09, France [3]Department of Chemistry, Moscow State University, Moscow 119899, Russia

In Hg-cuprates high critical temperatures T_c are reached with a maximum value of 135K for n = 3 in $HgBa_2Ca_{n-1}Cu_nO_{8+\delta}$. T_c goes through a maximum as a function of the oxygen content for n = 1,2,3 and increases with pressure for 1 <n<5. It has been found to reach 160K under pressure in Hg-1223. Single crystals of good quality are now available. The critical current densities, irreversibility line and anisotropy compare favourably with other (Bi, Tl, ...) compounds. Phases with two mercury planes have been studied too. Many substitutions have been tried, they often form new superconducting phases with one or two mercury planes, or reach T_c larger than 140K, up to now. In the $C_yCu_{1-y}Ba_2Ca_{n-1}Cu_nO_x$ cuprates, prepared under pressure, the overdoped n = 3 phase has reached a T_c of 120K after annealing in Ar, the as prepared n = 4 phase is very stable with an optimal T_c of 117K and a (still unknown) phase with T_c = 127K has been observed. Progress is made to obtain phases and single crystals. Most of the obtained phases have been studied under pressure. The pressure dependence of T_c is often an indication of the doping (under or overdoped) state; however, even in optimally doped Hg phases, T_c can increase, which indicates that the optimal doping depends on the pressure.

I. INTRODUCTION

Since the discovery of superconductivity at high temperature in copper mixed oxides by Bednorz and Muller (1986) a number of new compounds and of families of compounds has been discovered. Besides $YBa_2Cu_3O_7$ with a T_c of 94K, the highest T_c values have been obtained for n=3 in the Bi and Tl families of general formulas: $Bi_2Sr_2Ca_{n-1}Cu_nO_{2n+2+\delta}$ and $Tl_2Ba_2Ca_{n-1}Cu_nO_{2n+2+\delta}$ with T_c =108K for Bi-2223 and T_c =125K for Tl-2223. The structures of these compounds have some common features. They are made of blocks of CuO_2-Ca-CuO_2- layers having the perovskite structure which alternate with blocks made of BaO-TlO-TlO-BaO or SrO-BiO-BiO-SrO having the rock-salt structure. The perovskite blocks are at the origin of superconductivity when the Cu valence is higher than 2 and in the range of +2.1 to +2.2. For a given compound, T_c is strongly dependent on the valence of copper in the CuO_2 planes, which itself strongly depends on the doping or on the oxygen content in the rock-salt blocks called charge reservoirs. The study of the fundamental properties of these HTSC compounds has given increasing evidence in favor of a d-wave like superconductivity, which does not exclude a coexistence of s and d-wave characters depending on the direction of observation. At the same time, the possibilities of applications of all these materials have increased by preparing better textured ceramics and by increasing the pinning of vortices with appropriate defects (columnar defects in the Bi compounds). The search of new superconducting cuprates with higher T_c s becomes more difficult, but the discovery of the mercury, the (Cu,C), the infinite layer, and of the collapsed superconducting phases is promising, as well as the possibility of their applications.

II. THE MERCURY FAMILY

The $HgBa_2CuO_{4+\delta}$ compound has been obtained by substituting one Cu (the Cu of the chains) of $YBa_2Cu_3O_7$ by mercury, and reducing the number of CuO_2 planes from 2 to 1. The critical temperature, T_c =97K, of this n=l compound was among the highest of all cuprates with one CuO_2 layer (except for $Tl_2Ba_2CuO_{6+\delta}$ which has a T_c of up to 100K)[1]. Therefore, record high T_c's have been obtained with n=2 (T_c=127K) and n=3 (T_c=135K) in the mercury family $HgBa_2Ca_{n-1}Cu_nO_{2n+2+\delta}$ [2,3]. Then, T_c decreases for n larger than 3, like in other known families[4-8]. Values of n up to 9 have been observed in some grains, by electron microscopy, and T_c reaches 114K, 100K and 88K respectively for n=5, n=6, n=7[4-8]. It is possible to prepare the samples from the precursors by a classical technique or under pressure (around 20kbar) for n=1-4, and only under pressure for n>3. This last technique, besides its safety, allows to obtain samples with a large oxygen content (overdoped for n<3). Further annealing can be made to vary the oxygen content, and to obtain the maximum of the curves of T_c versus oxygen content (or copper valence), as for instance in Hg-1212[9]. However, for n>3 the maximum of T_c cannot be attained even after annealing in a strong oxygen pressure, as if the necessary oxygen stoichiometry could not be reached.

Surprisingly, T_c has been found to strongly depend on the applied pressure, even when the maximum of T_c versus oxygen content was reached for various elements of the mercury family[10-12] (this was not observed in other families). In particular, values as high as 164K (onset T_c) have been mentioned for Hg- 1223[11]. This increase of T_c is associated with changes in the structure and in the charge transfer. In particular a decrease of the lattice parameters, and more precisely of the distance of the apical oxygen from the CuO_2 planes is observed[13-14]. This suggests that it could be possible to reach such high T_c values if similar pressure effects were obtained by chemical substitutions. Several substitutions have been tried on the Ba and Hg sites. Many of them give superconducting materials[15-20], but without improvement of T_c (except may be for a partial substitution of Hg by Tl or by Pb which would respectively give a T_c of 138K[21] or of 143K[22]). The Ba could be fully replaced by Sr in compounds where simultaneous substitutions can also be made on the Hg or Hg and Ca sites[18,20]. In particular, compounds with Re or Mo on the mercury site remain superconducting. Moreover, the mercury site, or $[HgO_2]^{-2}$ dumbell, can be occupied by a carbonate[23] $[CO_3]^{-2}$ or by a sulfate[24] $[SO_4]^{-2}$ which form new phases. However, their superconducting properties deteriorate rapidly with increasing substitution.

The most interesting substitutions on the mercury site have been realized with Cu (in fact Cu/C) to form superconducting copper oxycarbonates as discussed later[25-27], and with gold[28]. The gold substitution is not interesting, a priori, since T_c decreases. The decrease of T_c with the gold content corresponds to a decrease of the a and c lattice parameters,, which means that the expected increase of T_c with chemical pressure effect is not attained[29]. However, the decrease of T_c with gold content in Hg- 1223 is mainly due to an overdoping of the CuO_2 layers due to the extra oxygen anions bonded to the Au^{3+} cations. This overdoping is confirmed by the increase of T_c from 121 to 125K after annealing in an Ar atmosphere a Hg-1223 sample with 30% of Au on the Hg sites. Such an overdoping could never be obtained in pure Hg-1223 and Hg compounds with n>3, which may be due to the formation of a mercury-oxygen valence band. The gold substitution has been successfully used to directly obtain a T_c of 110K in the (Hg,Au)-1245 phase with 30% of Au[28]. A similar T_c could only be reached previously in pure Hg-1245 by annealing in 1300 bars of oxygen for 10 hours at 280°C[8]. This way of increasing the oxygen stoichiometry appears as promising to reach the maximum of T_c in otherwise underdoped compounds.

Finally, superconducting compounds with two mercury layers (analogous to the Tl_2- and Bi_2- families) have been synthesized. They are very difficult to prepare as pure phases and their superconducting properties did not appear very interesting (T_c = 45-70K)[30,31]. However, a variety of new phases can be prepared, and recent resistivity measurements under pressure show that their superconducting properties can be strongly improved [32]. Eventually, and like for the Bi and Tl families, the Hg elements with two mercury planes could have T_c's larger than the elements with only one mercury plane.

III. SUPERCONDUCTING PROPERTIES OF THE MERCURY COMPOUNDS

The properties of HTSC cuprates are strongly anisotropic since their conduction is much better along the CuO_2 planes in the normal state. The superconducting properties, too, are anisotropic and it is very important to measure this anisotropy, the critical current densities (in various directions), and the irreversibility line, to appreciate the interest of a new phase for applications when compared with the existing superconductors. Some of these properties can be estimated as soon as a pure phase is synthesized in the form of ceramics[33]. For instance the resistivity, magnetoresistance, susceptibility in a dc field, magnetization and specific heat have been obtained on good ceramics[33]. From the critical current density, irreversibility line, and anisotropy of Hg-1223, associated with its large T_c, it was evident that this new material had good potential for applications.

The irreversibility line $H_{irr}(T)$, can be taken at the point where the magnetization M(H) becomes reversible, and DM=0. The irreversibility line of Hg-1223 is above that of YBCO for T>70 K, and well above those of the Bi_2- and Tl_2- families. A similar behavior is observed for the irreversibility lines obtained from magnetoresistance measurements (in this case the irreversibility lines have been taken at 1/100 of the normal state resistance)[33] and shown in Fig. 1.

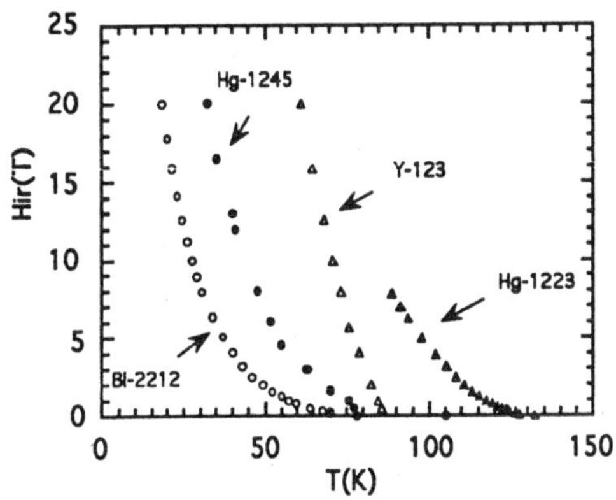

Fig. 1. Irreversibility lines, deduced from magnetoresistance measurements for some mercury phases are compared with other HTSC.

The growth of crystals or films, needed for more detailed and accurate studies, is now possible[34-36]. Several groups have already reported results on various crystals. Transport properties on single crystals of Hg-1223 were among the first to be reported[37]. They have been followed by magnetization[38-39], Specific heat [40-41] and NMR[42-43] measurements.

The anisotropy, γ, of the high T_c superconductors is inherent to the bidimensional character of their structure. It influences the normal state (resistivity: $\gamma = \rho_c/\rho_{a,b..}$), the superconducting parameters ($\xi_{a,b}/\xi_c$, $\lambda_c/\lambda_{a,b..}$) and the properties associated with the pinning of the vortex: critical current density, J_c, irreversibility line H_{irr} (T). It can be obtained from the resistivity in the normal state, the magnetoresistance, the reversible magnetization, torque measurements, etc. Some general trends are observed: γ increases with the number, m = 1 or 2, of layers of the rock-salt block, and tends to decrease with the number, n, of CuO_2 layers [see for instance ref. 38,39]. The anisotropy is small in the case of $YBa_2Cu_3O_7$ with γ = 4-6, larger in Tl-1212 (γ = 50) and much larger in $Bi_2Sr_2Ca_{n-1}Cu_nO_{2n+2+d}$ and $Tl_2Ba_2Ca_{n-1}Cu_nO_{2n+2+d}$, with γ values of the order of 100. It is quite remarkable that for Hg-1223 it is rather small with a value between 16[33] and 50[38]. Larger values are obtained for Hg- 1234 or Hg- 1245, but these phases are not so pure. This is in agreement with the relatively large values of $H_{irr}(T)$ for Hg-1223, which, with a high T_c and large critical current densities, must be a good candidate for the applications. This has been confirmed by magnetic measurements, which show the same general features as in most HTSC crystals: the fishtail effect, and a strong dependence of J_c versus the annealing conditions[39]. Like in other materials (Y, Bi, Tl families) the critical current density must improve very much in the Hg family with the quality of the crystals and the presence of columnar defects.

The Hg-1223 crystals prepared up to now are underdoped which results into a small T_c and downward curvatures in the resistivity versus temperature curves. Specific heat measurements have been made on single crystals. They give evidence of strong thermal fluctuations, which give a quasi symmetric shape for the anomaly, very similar to what is observed in other quasi-two-dimensional HTSC like the Bi and Tl family, and in contrast with the behavior observed in YBCO. A quasi-logarithmic divergence of the specific heat is observed at T_c in zero field[40,41]. In a large magnetic field the specific heat anomaly is strongly reduced, with only a weak shift in temperature. NMR measurements on ^{63}Cu have shown the opening of a spin-gap in Hg-1223 below about 230K, the highest temperature observed in HTSC, together with large energy characteristic of spin-fluctuations[42-43].

As expected, the anisotropy is larger in systems where the charge reservoir block contains two planes (Tl_2-, Bi_2-, families) than in $YBa_2Cu_3O_7$ or the Tl1- and Hg1- families. A strong anisotropy is found in other superconducting properties, in particular the critical current density is the highest when the current flows along the CuO_2 layers. It remains constant in large fields when the magnetic field is parallel to these planes. Therefore, for the applications it is important to realize these conditions: for instance, coils made of ribbons with the CuO_2 layers in the plane of the ribbon (the highest field being then parallel to CuO_2 planes), or to use HTSC with a small anisotropy.

Another very important problem is to make wires or ribbons from materials which naturally exist in the form of ceramics. It is then important to obtain the growth of crystals as large as possible which have good superconducting contacts with the next crystals, to make wires with large critical current densities. Obviously, it is easier to make short superconducting connections (a few cms long) as superconducting leads for superconducting coils. It is also easy to make cylinders for magnetic screening or to trap a magnetic field, as well as to realize a superconducting cavity. But the best performances will be obtained only if the number of grains is limited and the contact between grains optimized (in quality and size). The addition of Ag (a few per cent) helps to increase the size of the grains and the contact between them (not for the Hg-family!). Much work has been carried out to realize rods or wires of $YBa_2Cu_3O_7$ and ribbons of $Bi_2Sr_2Ca_{n-1}Cu_nO_{2n+2+d}$ and $Tl_2Ba_2Ca_{n-1}Cu_{2n+2+d}$ sheathed with silver and some superconducting coils have already been made. Such coils have apparently weak performances: 5-6 T, but (since they have larger critical fields) they can be put inside classical superconducting coils (giving up to 20T) and add their field to this 20T field.

Films of all the main superconducting phases, have very good performances in terms of critical current densities. Films of Hg-1212 have been successfully prepared. Very large critical current densities have been measured (10^7 A/cm^2 at 77K in the best conditions, with J and H in the a-b plane, and J perpendicular to H) in various experimental conditions[44].

IV. THE COPPER-CARBONATE FAMILY

One of the successful substitutions on the mercury site has been obtained by trying to put silver instead of mercury[25-27]. In fact, silver does not enter in the structure but copper is found on one Hg site over two (50% of Cu on the mercury sites), the other Hg site being occupied by carbonate $(CO_3)^{2-}$ ions. Moreover, and contrary to the mercury family, this copper family (of general formula $Cu_{1-x}C_xBa_2Ca_{n-1}Cu_nO_y$) needs to be prepared under pressure (70kbar). The presence of carbonates comes naturally from the carbon dioxide of the air which reacts with the starting precursor. In the copper family, n is found to vary only from n=3 to n=6 (observed by electron microscopy). In the as-prepared samples, T_c is equal to 70K for n=3 and 117K for n=4, it decreases for larger n. It would be the only family where T_c would be maximal for n=4 instead of n=3. However, the n=3 member is obtained overdoped. Its T_c increases up to 120K after heat treatment in Ar, and the (Cu,C) family presents the highest T_c for the third member as other superconducting families.

For every synthesis performed at ≈ 800 °C and under 8GPa besides the usual transition at 117.6K corresponding to the n=4 member, often mixed with other members of the family with lower critical temperatures, we also observed a well defined transition at ≈ 126.5K (Fig. 1). This temperature is very close to the highest observed T_c within the thallium family of cuprates, and only second to the best in the mercury based system at atmospheric pressure (T_c=133K for the n=3 member of this family with optimal doping). The amount of superconducting phase with T_c = 126.5K is very small[45]. It can be estimated to be ≈ 1 % as compared with about 60% for the n=4 member of the cupro/carbonate family observed in the same synthesis, from magnetization measurements on ceramics. It could even be much less as indicated by ac susceptibility measurements on powders. Although we could not identify the material responsible for this high T_c, one can reasonably assume that it is either an optimised member of the cupro/carbonate family, or that, under the high - pressure used during the synthesis, silver has entered the structure in the Charge Reservoir Block (CRB), improving the superconducting properties. Obviously, more work is needed in order to increase the amount of this phase in the sample so as to properly identify it.

V. RESISTIVITY UNDER PRESSURE

Although for the reasons indicated above we have not been able as yet to prepare in pure form a single member of the cupro/carbonate family, we have been able to follow the pressure dependence of T_c for the three different phases. Selecting chips from the sample that showed Meissner expulsion below 126.5K we measured their electrical resistance as a function of temperature in the diamond anvil apparatus used previously[45]. As shown in Fig. 2 it is easy to determine three different transition temperatures (T_{c1}, T_{c2}, T_{c3}) that correspond to different superconducting phases.

We attribute T_{c2} ≈ 117K to the n=4 member of the family of cupro/carbonates[25,45]. As shown in Fig. 3, T_{c2} has a small pressure dependence; it goes through a maximum around 10GPa. As it is usually assumed that, due to a most subtle compromise between oxygen content and interatomic distances, dT_c/dP is small in optimaly doped samples, we can infer that the n=4 member of the family is nearly optimised under the present synthesis conditions (see also ref. 45). T_{c1} ≈ 80K should correspond to the n=3 member of the cupro/carbonate family. It decreases strongly with pressure indicating clearly that overdoping in oxygen exists in this phase, confirming previous observations [45] (see Fig. 3). The most interesting result is the marked increase of the highest T_c, namely T_{c3} ≈ 126.5K, observed in samples of this system. At the highest pressure used, 21GPa, T_{c3} attains ≈ 136K. It might still increase if higher pressures were applied. This is the second highest T_c ever observed for a superconducting sample under pressure, the first being that of the n=3 member (Hg- 1223) of the mercury based HTSC cuprates.

Undoubtedly, more work has to be done to synthesize monophase samples of these high pressure phases, to fully characterize their superconducting and other properties, and identify the unknown phase with the highest T_c. Yet the present results show that, in the absence of toxic thallium or mercury, very high critical temperatures can be achieved in this cupro/carbonate family.

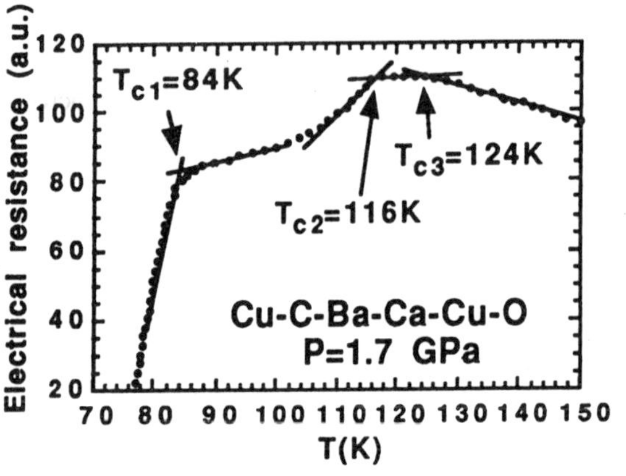

Fig. 2. Existence of three different phases in a (Cu,C) compound prepared under pressure.

The structural refinement based on neutron diffraction powder data reported by Shimakawa et al.[45] confirmed the structure envisaged by high resolution electron microscopy. However, it is not very conclusive about the local arrangement of the (Cu,C) CRB, which should reveal the doping mechanism and eventually allow one to optimize the superconducting properties.

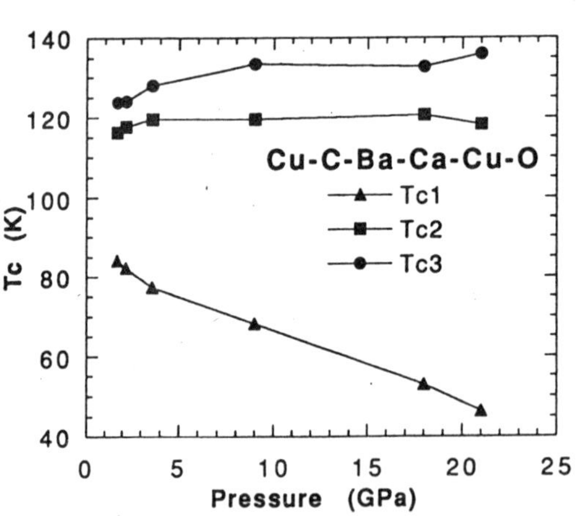

Fig. 3. Effect of pressure on T_c of various (CuC) phases.

Chaillout et. al.[46] also collected neutron powder data on a sample containing 60% of (Cu,C)-1234, 19% of (Cu,C)-1223, and other impurities, some of which have not been identified. This refinement is not conclusive either, but together with the results of other experiments allows us to make some conjectures about the structure of the CRB.

These cupro/carbonates seem to be insensitive to heat-treatments under oxygen pressure. This would indicate that the CRB of these compounds does not contain mobile oxygen atoms as the Hg counterparts. In this case the Cu cations of the CRB would have the dumbbell coordination and the only oxygen atom of this layer would be that bonded to C. This conjecture is corroborated by the results of the structural refinement of ref.45. The formula deduced from preliminary results would be $Cu_{0.22}C_{0.78}Ba_2Ca_3Cu_4O_{11}$, which corresponds to V_{Cu} = 2.16 v.u. for the Cu cations of the CuO_2 layers. This value is very close to the critical Cu valence corresponding to the optimal carrier concentration for superconductivity. The same reasoning for the n=3 member yields V_{Cu} =2.22 v.u. which would correspond, indeed to an overdoped state as the high pressure experiments indicate. As stated before, successive annealings in Ar atmosphere result into larger T_c for (Cu,C)- 1223 which tends to saturate to 120K (Fig. 4)[46].

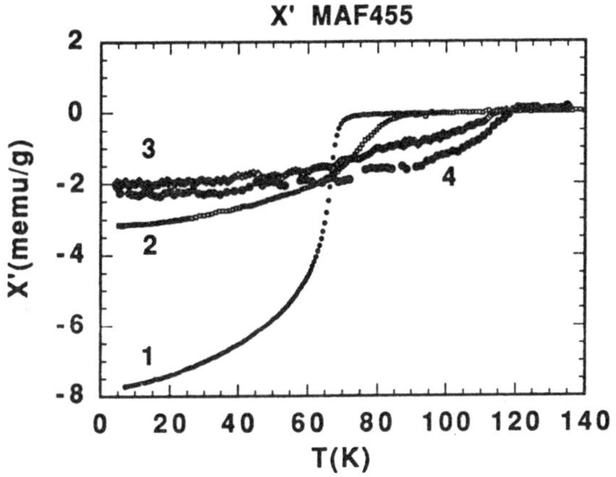

Fig. 4. The effect of successive annealings at 400 °C in Ar atmosphere (1: as prepared, 2:2h, 3:4h, 4: 4.5h) increases the T_c of (Cu,C)-1223.

As revealed by high resolution electron microscopy, some crystallites of the cupro/ carbonates contain sequences corresponding to the n=5, 6, or 7 members. The existence of the higher members of the series, shared with the Hg-based cuprates, would indicate that the CuO_2 layers are flat as in the Hg counterparts. The Hg and (Cu,C) are good families to check if T_c could increase with the number n of CuO_2 planes if the charge transfer with the charge reservoir was sufficient. Besides the synthesis under pressure, a layer by layer deposition technique could be successful to obtain members of the Hg and (Cu,C) families with large n values.

V. OTHER SUPERCONDUCTING FAMILIES

New compounds with high critical temperature can be prepared with success, by the application of shearing mechanisms to cuprates and oxycarbonates[17]. The observation of collapsed phases by High Resolution Electron Microscopy is often at the origin of the synthesis of a new phase. Many combinations are possible; however, may be due to the ondulations of the CuO_2 planes, the highest observed T_c remains below 100K.

Many attempts have been made to realize an infinite layer compound, where an infinite number of CuO_2 planes (non superconducting when they are only separated by Ca or Sr layers) would become superconducting if they were correctly charged. However, when some superconductivity is found in these materials, it appears that it is due to other phases, for instance with 2 CuO_2 layers. Moreover, the superconducting properties obtained are not excellent and well reproducible (maximum T_c = 110K and 30% superconducting volume). The deposition, layer by layer, is another way to obtain new compounds and structures which would not exist naturally, even after synthesis under pressure.

VI. SUMMARY

In summary, the discovery of new HTSC phases, which already show excellent potential for applications, is quite promising, even though T_c has not yet reached values of 160K observed under pressure. The efforts towards the applications are also fruitfull, and there is no doubt that these materials will have an increasing importance in the future.

REFERENCES

1. S.N. Putilin, E.V. Antipov, O. Chmaissen and M. Marezio, Nature, **362**, 226 (1993).
2. S.N. Putilin, E.V. Antipov and M. Marezio, Physica C**212**, 266 (1993).
3. A. Schilling, M. Cantoni, J.D. Guo and H.R. Ott. Nature, **363**, 56 (1993).
4. S.M.Loureiro et al., Physica C **235**, 905 (1994). See also Physica B **197**, 570 (1995).
5. E.V.Antipov et al., Physica C **215** 1 (1993).
6. B.A.Scott et al.,Physica C**230** 239 (1994).
7. E.M.Kopnin et al., Physica C**243**, 222 (1994).
8. J.J.Capponi et al., Physica C **256**, 1 (1996).
9. E.V. Antipov et al., Physica C **218**, 348 (1995). See also S.M. Loureiro et al., Physica C **217** 253 (1993). E.V.Antipov et al., Physica C **235**, 21 (1994).
10. M.Nunez-Regueiro et al., Science **262**, 97 (1993).
11. C.W.Chu, L.Gao, F.Chen, ZJ.Huang, R.L.Meng and Y.Y. Xue, Nature **365**, 323 (1993). L.Gao, Y.Y.Xue, F.Chen, Q.Xiong, R.L.Meng, D.Ramirez, C.W.Chu, J.H.Eggert and H.K.Mao, Phys.Rev.B **50**, 4260 (1994), and Physica C **235** 1493 (1994).
12. D.Tristan Jover et al., Phys. Rev. B**54**, 4265 (1996) and Physica C **235**, 893 (1994). See also M.Marezio and J. Chenavas, J. Solid State Chem. **121**, 24 (1996), and H.Ihara et al., Jpn.J.Appl.Phys. **32**, L1732 (1993); and H.Takahashi and N-.Mori in *Studies of High Temperature Superconductors*, vol 16/17, ed. A.V.Narlikar, Nova Science Publishers, New-York, USA.
13. A.R. Armstrong et al., Phys. Rev. B**52**, 15551 (1995).
14. V.L. Aksenov et al., *High pressure research*, vol. 14, 127, OPA,Gordon and Breach Publishers SA (1995).
15. Z.H.He, Q.M. Lin, L.Gao, Y.Y.Sun, Y.Y.Xue and C.W.Chu, Physica C**241**, 211 (1995).
16. A. Maignan et al., Physica C**243**, (1995) 233. See also Physica C**243**, 214 (1995).
17. B. Raveau, C. Nfichel and M. Hervieu, in *Coherence in high temperature Superconductors,* World Scientific, eds. G. Deutcher and A. Revcolevschi, p.42 (1996).
18. J.Shimoyama, S.Hahakura, K.Kitazawa, K.Yamafuji and K.Kishio, Physica C **224**, 1 (1994).
19. K.K. Singh, V. Kirtikar, A.P.B.Sinha and D.E.Morris, Physica C**243**, 233 (1995).
20. S.Hahakura et al., Physica C**235**, 915 (1994).

21. G.F.Sun, K.W.Wong, B.R.Xu, Y.Xin and D.F.Lu, Phys. Lett. A, **192**, 122 (1994). P.Dai, B.C.Chakoumakos, G.F.Sun, K.W.Wong, Y.Xin and D.F.Lu. Physica C **243**, 201(1995).
22. H.M. Shao et al., Physica C**246**, 207 (1995).
23. S.M.Loureiro, E.V.Antipov, J.J.Capponi and M.Marezio, Mat. Res. Bull. **30**, 1463 (1995). See also E.M.Kopnin et al., Physica C **243**, 222 (1996).
24. S.M.Loureiro, P.G.Radaelli, E.V.Antipov, J.J.Capponi, B.Souletie, M.Brunner and M.Marezio, J.Solid State Phys. **121**, 66 (1996).
25. M.A.Alario-Franco, C.Chaillout, J.J.Capponi, J.L. Tholence and B.Souletie, Physica C **222**, 52 (1994). For a partial replacement of Hg by Cu see: E.T.Alexandre et al., Physica C **245**, 207 (1995).
26. H.Ihara et al., Jpn. J. Appl. Phys. **33**, L503 (1994).
27. C.-Q.Jin, S.Adachi, X.J.Wu, H.Yamauchi, and S.Tanaka, Physica C**223**, 238 (1994). See also X.J.WU et al., in Physica C **223**, 243 (1994).
28. P.Bordet et al., Physica C (1996).
29. C.Acha and M.Nunez-Regueiro, unpublished.
30. P.G. Radaelli et al., Science **265**, 380 (1994).
31. P.G. Radaelli et al., J.Phys.Chem.Solids **56**, 1471 (1995).
32. C. Acha et al., unpublished.
33. O.Laborde, J.L.Tholence, J.Voiron, M.Saint-Paul, B.Souletie and JJ.Capponi, Physica C **235**, 2717 (1994). See also: M.Couach et al., Phys. Lett. A**188**, 85 (1995); R.Calemczuck et al., Physica C **235**, 1743 (1994); O. Jeandupeux, A.Schilling, H.R. Ott, Physica C **216**, **17** (1993); O.Laborde et al., Solid State Commun. **90**, 443 (1994).
34. D.Colson, A.Bertinoti, J.Harnmann, J.F.Marucco and A.Pinatel, Physica C **233**, 231 (1994). See also A.Bertinotti et al., Physica C **250**, 213 (1995).
35. L.W.Finger, R.M.Hazen, R.T.Downs, R.L.Meng and C.H.Chu, Physica C **226**, 216 (1994).
36. H.Schwer et al., Physica C **254**, 7 (1995).
37. A.Carrington et al., Physica C **234**, 1 (1994).
38. V.Vulcanescu et al., Physica C **259**, 131 (1996).
39. A.Maignan, S.N.Putilin, V.Hardy, Ch.Simon and B.Raveau, Physica C **266**, 173 (1996), and references therein. For a comparison with Tl- and Tl-Pb, see V.Hardy et al., Physica C **232**, 347 (1994), and A.Wahl et al., Phys. Rev. B **51** 9123 (1995).
40. A.Carrington, D.Colson, Y.Dumont, C.Ayache, A.Bertinotti and J.F.Marucco, Physica C **234**, 1 (1994). C.Marcenat, R.Calemczuk and A. Carrington, in *Coherence in High Temperature Superconductors*, eds. G.Deutcher and A.Revcolevschi, World Scientific, p. 101 (1996).
41. A.Carrington et al., Proceedings of LT21 (Prague, 1996).
42. M.-H.Julien et al., Phys. Rev. Lett. **76**, 4238 (1996).
43. C.C.Tsuei, A.Gupta, G.Trafas, and D.Nfitzi, Science **265**, 1259 (1995). See also A.Gupta, J.Z.Sun and C.C.Tsuei, Science **265**, 1075 (1995). L.Krusin-Elbaum, C.C.Tsuei and A.Gupta, Nature **373**, 679 (1995).
44. M. Jaime, M.Nunez-Regueiro, M.A.Alario-Franco, C.Chaillout, JJ.Capponi, A.Sulpice, S. de Brion, P.Bordet, M.Marezio, J.Chenavas and B.Souletie, Solid State Commun. **97**, 131 (1996).
45. Y. Shimakawa, J.D. Jorgensen, D.G. Hinks, H. Shaked, R.L. Hitterman, F. Izumi, T. Kawashima, E. Takayama-Muromachi and T. Kamiywna, Phys. Rev. B **50**, 16008 (1995).
46. C.Chaillout, S.LeFloch, E.Gautier, P.Bordet, C.Acha, Y.Feng, A.Sulpice, J.L.Tholence and M.Marezio, Physica C **266**, 215 (1996).

NMR STUDIES IN SUPERCONDUCTORS

R. Vijayaraghavan
Society for Applied Microwave Electronics Engineering and Research
IIT Campus, Powai, Mumbai 400 076, INDIA.

Resonance techniques have played an important role in the study of superconductors. In early works, the resonance shift had been effectively used to identify the type of pairing model. Subsequently the relaxation studies gave an insight to the nature of interactions in the superconductors. In some favorable cases. NQR frequency (nuclear quadrupole resonance) reflected the types of dissimilar sites and the doping effects in superconductors. In strongly correlated electron systems like heavy fermions and high temperature superconductors the deviations from simple BCS model were seen. Coexistence of magnetism and superconductivity could be inferred from NMR and NQR studies. In the recently discovered borocarbide superconductors, NMR investigations have provided interesting information about the nature of carrier band. In this talk, it is proposed to review the work done in this field by various workers.

I. INTRODUCTION

Nuclear Magnetic Resonance (NMR) studies probe the local magnetic field at the nucleus. The NMR field at the nucleus for a metallic sample in fixed external frequency depends on several factors; significant among them are (1) spin paramagnetism of conduction electron due to direct contact interaction (2) core polarisation (3) orbital Van Vlack paramagnetism (4) effects of spin orbital coupling in crystalline lattice and (5) spin reversal at surfaces and impurities. These contributions get modified in superconductors. Even in superconductors, they would greatly depend upon the types of superconductivity (arising out of s-wave, p-wave or d-wave). In this review article, it is proposed to select few examples of NMR measurements in superconductors and indicate the power of this technique in understanding the nature of superconductivity.

In simple monovalant cubic metals, the polarisation of the conduction electron spins becomes zero at the absolute zero temperature in agreement with the BCS prediction[1]. The large reduction in the Knight Shift (K.S.) in the superconducting state in metals like Al, Sn etc. below the onset of superconducting T_c greatly supported the singlet s state of Cooper pairs. Soon, it became clear that in transition metals and alloys (with impurities) the effects of orbital para magnetism and spin orbital scattering at imperfections, impurities and surfaces are important in determining the values of Knight Shift in the superconducting state [2-5].

NMR has been increasingly used in the study of strongly correlated electron systems like heavy fermions and oxide superconductors to understand the nature of magnetism and superconductivity in these materials. The quantity measured in NMR experiments are the shift (K.S.), relaxation times T_1 and T_2, nuclear quadrupole resonance frequencies, internal fields, and line width as a function of temperature.

II. THEORY

The interaction between the electron spin $\vec{S}$ and nuclear spin $\vec{I}$ can be written as fermi contact interaction,

$$H_s = \vec{I} \bullet \vec{A} \bullet \vec{S} \tag{1}$$

where $\vec{A}$ is the hyperfine coupling tensor. H_s can be related to the hyperfine field $\vec{H}_{hf}$ per Bohr magneton of electron spin as

$$H_s = \gamma_n \hbar \vec{I} \cdot \vec{H}_{hf}^{s} \qquad (2)$$

γ_n is gyromagnetic ratio of nucleus. A similar relation can be written for the hyperfine field due to electron orbital momentum. The Knight Shift which is the additional field due to electron polarisation is a sum of K_s and K_{orbit}.

The nuclear electric quadrupole moment eQ and the electronic field gradient due to electron charge distribution produce an electrostatic interaction which is written as

$$\frac{e^2qQ}{4I(2I-1)}\left[3I_z^2 - I(I+1) + \eta(I_x^2 - I_y^2\right] \qquad (3)$$

where $\eta = \left(\frac{d^2v}{dx^2} - \frac{dv^2}{dy^2}\right) \Big/ \frac{d^2v}{dz^2}$

and x, y, z are the principal axis of the field gradient tensor. This interaction gives rise to pure quadrupole resonance in some systems and the resonance can be observed without the application of external field.

The nuclear spin lattice relaxation time T_1 is related to the perpendicular component of dynamical susceptibility (imaginary part) x″ (q, ω) and the Korringa relation connects K.S. and T_1 as

$$T_1 T K^2 = \text{constant}$$

This is true where s electrons are primarily responsible for the shift and relaxation and has been verified experimentally in many cases.

In the BCS superconducting state, with singlet Cooper pairing, due to the uniform energy gap, the Knight Shift and the spin susceptibility χ_s decrease exponentially with temperature. At $T \rightarrow 0$, K.S. and χ_s become zero. In heavy elements, with strong spin orbit scattering, Knight Shift does not become zero at $T \rightarrow 0$ and remains finite. When the Cooper pair is in odd parity state with parallel spin pairing, spin susceptibility remains same below T_c. So the Knight Shift does not change with temperature as seen in some heavy fermion superconductors[6]. The copper oxides in their insulating state are antiferromagnetic before they are doped. The electron spin associated with one copper atom is antiparallel to all of its nearest neighbours[7]. After doping, though the long range order disappears, the spins exhibit short-range antiferromagnetic fluctuations, suggestive of spin glass state or short range order in very short time scale. The theories which predict that such fluctuations are responsible for superconductivity are based on dx^2-dy^2 wave pairing. In such cases, the knight shift decreases below superconducting temperature due to spin singlet.

The scattering of excited normal electrons provide the relaxation of nucleus in the superconducting state. The density of states N(E) at temperatures close to T_c is very large and gives rise to a divergence in the relaxation rate. Hebel and Slichter[8] observed such a peak in $1/T_1$ just below T_c in aluminum metal. At lower temperatures, it decreases exponentially if the energy gap is uniform as in the BCS state. The gap anisotropy brings down the enhancement of the relaxation rate, which can also be caused by impurities of magnetic or nonmagnetic origin. This peak is not seen in many heavy fermion superconductors, copper oxide superconductors and the recently discovered borocarbide superconductors[9]. At lower temperatures, the relaxation rate seem to follow T^3 law. In rare earth based copper oxide superconductors, the relaxation rate below T_c is greatly influenced by the magnetic moment of the rare earth ion.[10] In the resonating valence bond scheme of Anderson[11] a spinon state is made of one unpaired spin in the background of resonating singlets. Also a charge carrying excitation can be produced by removing spins at two sites. The holons are empty sites with resonating singlets in the background and are charged excitations. Spinons are neutral excitations carrying no charge. Baskaran has attempted to explain the relaxation rate in the copper oxide superconductors in the normal state using the above model[12], which is an order of magnitude more than that in ordinary metals and band theory estimates.

III. EXPERIMENTS AND DISCUSSION

Rare earth 123 Compounds (RE $Ba_2Cu_3O_7$ and O_6 compounds)

Nuclear magnetic resonance[13] (also pure quadrupole resonance PQR) has been extensively used to understand the nature of superconductivity and magnetic properties of these compounds. O_7 compounds are superconducting around 90K while, O_6 compounds are insulators. There are two types of copper atoms viz plane copper atoms and chain copper atoms. From the NMR relaxation rates of Cu isotopes in $YBa_2Cu_3O_7$ two energy gaps are shown to open up in the superconducting state[14]. The quadrupole resonance of Cu(1) and Cu(2) isotopes are observed close to 20MHz and 30 MHz respectively in O_7 compounds while in O_6 compounds Cu(1) isotope is observed around 30MHz. Cu(2) isotope experiences an internal field due to the antiferromagnetism in O_6 compounds. Table I gives T_1 values at 4.2K of copper isotopes in $REBa_2Cu_3O_7$ and O_6 compounds[15].

Table I (T_1 in msec at 4.2K)

Rare earth atoms	O_7		O_6
	Cu(1)	Cu(2)	Cu(1)
Y	35.9	21.50	418
Eu	---	2500	44.3
Nd	1.54	0.16	---
Sm	10.0	0.51	13.5
Gd	1.0	0.04	0.93
Ho	---	0.05 (10K)	1.50
Er	---	0.05	2.25

It is clear that the relaxation rates are affected by the moments on the rare earth ion. It is found that in O_6 systems, the relaxation times are independent between 4K and 77K. In O_7 systems interaction between rare earth moment and conduction electrons is suppressed due to formation of Cooper pairs. The quadrupole resonance frequency ν_a of Cu(1) nuclei does not depend upon the rare earth ion in comparison to O_7 oxides where it decreases with the decrease of the size of the rare earth ion[15].

NMR in $TlBa_{2-x}La_xCuO_{5-y}$ Superconductors

Unlike 123 superconductors, where quadrupolar interaction of copper nucleus was studied to understand the nature of normal and superconductivity of high T_c materials, thallium nucleus with spin I= 1/2 is devoid of such interaction. In layered Tl compounds, the nuclear relaxation time in the normal state of Tl isotopes in $Tl_2Ba_2\ CaCu_2O_8$ and $Tl_2Ba_2Ca_2Cu_3O_{10}$ does not follow the Korringa relation and in the superconducting state it also deviates from the prediction of the simple BCS model. In some Tl-O bilayer compounds, deviation which occurs from normal behaviour is seen well above T_c. Rajarajan et al[16] studied extensively the Tl based superconductors having Tl-O monolayers, with La substituting partially Ba, in $TlBa_2CuO_{5-y}$. In $TlBa_{2-x}La_xCu_3O_{5-y}$ (x = 0.2 and 0.5), the superconductivity was observed at 45K and 32 K respectively.

For x = 0.2, the line width $\Delta\omega$ of ^{205}Tl resonance at 300K in a field of 12.6 kilogauss is 60 KHz and is weakly temperature dependent in the normal state upto T_c. The width starts increasing rapidly below T_c and is close to 250 KHz at 4.2K. The increase in the linewidth is due to the formation of a flux line lattice in the superconducting state resulting generally in inhomogeneous field distribution at the nuclear site. The Knight shift decreases from the normal state to 0.11% and 0.16% in x = 0.2, and 0.5 respectively as in normal superconductors.

The relaxation times in the two cases follow Korringa relation from 300 K $\leq$ T $\leq$ T^* where T^* is 60K and 90K, respectively for the above materials. T* is considerably higher than T_c (45K and 32K respectively). At temperature below T*, T_1^{-1} (T) deviates from the Korringa relation and there is no anomaly in the relaxation rate at T_c. The relaxation rate below T_c does not follow the exponential relation. $T_s^{-1} \sim \exp(-Eg / K_BT)$, where Eg is the energy gap in the superconducting state. No Hebel- Slichter peak is seen close to T_c.

Slichter et al[17] interpret the additional relaxation rate to fluctuations of Cu_3d spin in $YBa_2Cu_3O_7$ type superconductors. The antiferromagnetic fluctuations are seen to persist at temperatures far higher than T_c and contribute to nuclear relaxation rate below T^*. However, the linewidth increases markedly only below T_c and not at T^*, which means static line broadening occurs only in the superconducting state. This feature is similar to that observed in some superconductors.

Borocarbide Superconductors

Discovery of superconductivity in quaternary rare earth transition metal borocarbide superconductors (REM_2B_2C) and subsequently in REPdBC alloys at a high 23 K transition temperature[18] laid down the foundation of new area of research in superconductivity of multi-elements system. We will discuss here NMR studies in two superconductors viz. 1) YNi_2B_2C[19] and $Y_{0.98}Er_{0.02}Ni_2B_2C$ and 2) $LaPt_{1.5}Au_{0.6}B_2C$[20], ^{89}Y, ^{11}B and ^{195}Pt are nuclei which are amenable for NMR studies.

YNi_2B_2C

Since boron atoms in YNi_2B_2C have a uniaxial local symmetry electric quadrupole interaction with asymmetry parameter η = 0 gives rise to the quadrupole frequency ν_q to be 0.7MHz ($T_c \approx$ 15.5K) (see Fig.1 of ref. 19). In the superconducting state, two ^{11}B resonance are seen though crystallographically there is only one site for boron atom. One of the signals moves to higher fields as temperature is lowered through T_c, while the other is nearly constant with the change of temperature. We attribute the two signals arising from boron atoms in the superconducting and normal metal regions, respectively. The normal metal region is where the occupancy of boron and carbon atoms varies at a microscopic level. A similar thing has been seen[21] in the single phase sample of $YBa_2Cu_3O_{7-x}$. The relaxation rate of 11B and 89Y decreases drastically below T_c, with no Hebel-Slichter peak at T_c. The Knight Shift of ^{11}B decreases from + .06% to +0.04% as temperature is decreased in the superconducting state.

The temperature dependence of relaxation rates may now be discussed in some detail. According to the selfconsistent renormalisation (SCR) theory[22] for itinerant antiferromagnetic spin fluctuation (AF), relaxation rate T_1^{-1} is proportional to $(T/T - T_N)$ and $T/(T-T_N)^{1/2}$ for 2D and 3D itinerant antiferromagnets respectively (T_N = Neel temperature). NMR results of Kohara et al[19] (See Fig. 3), the relaxation rates seem to fit the picture of 2D and 3D antiferromagnets. It may be inferred that Ni atoms may carry moments and YNi_2B_2C can be regarded as itinerant antiferromagnetic spin fluctuating material where 3D electronic correlation plays a significant role.

In order to understand the role of local moments in the relaxation times of ^{11}B, two new superconductors were studied[20], 1) $Y_{0.98}Er_{0.02}Ni_2B_2C$ and 2) $LaPt_{1.5}Au_{0.6}B_2C$[20], ^{89}Y, Er and Ni have moments and in the latter there are no moment bearing atoms.

The ^{11}B NMR results on $LaPt_{0.5}Lu_{0.6}B_2C$ show that the Knight Shift is small and independent of temperature T. (See Fig. 2 of ref. 20). The relaxation time follows Korringa relation which is in contrast to the behaviour in YNi_2B_2C[19]. These two results would mean that there are no magnetic fluctuation in the La system where the d-band of Pt is presumably filled. This is confirmed from the fact the K.S of ^{195}Pt is small (+0.13%), positive and temperature independent, $(T/T)^{-1}$ is also independent of T indicating normal Fermi liquid behaviour.

In the case of $Y_{0.98}Er_{0.02}$, the relaxation rate of ^{11}B is an order of magnitude more than that in YNi_2B_2C $[(T_1T)]^{-1}$ = 0.8 (sec K)$^{-1}$ and $(T_1T)^{-1}$ = 0.1 (sec K)$^{-1}$, respectively at 4.2K; This is clearly due to the influence of Er moments whose fluctuations strongly affect the relaxation rate of ^{11}B.

IV. SUMMARY

It has been shown that NMR technique is a powerful probe to understand the mechanism and the nature of superconductivity at a microscopic level. In simple systems, it served to verify the predictions of BCS theory. In favorable cases, it is able to distinguish the type of superconductivity (due to BCS like s-wave or p-wave or d-wave or modified d-wave). In oxide superconductors, the nuclear quadrupole resonance of Cu isotopes gives lot of insights into the nature of plane and chain copper atoms. The relaxation studies are powerful enough to give clues about the type of excitations both in the normal and superconducting state. Combined with neutron studies (which is not discussed here), NMR studies have given a wealth of information about the mechanism of superconductivity in various types of superconductors.

ACKNOWLEDGMENT

The author expresses his gratitude to Prof. L.C. Gupta for helpful discussions and the members of the Solid State Physics Group of Tata Institute of Fundamental Research for their cooperation and assistance during the author's stay at the Institute.

REFERENCES

1. K. Yosial, Phys. Rev **110**, 769 (1958).
2. A.M. Clogston, V. Jaccarino and Y. Yafet, Phys. Rev. **134 A**, 650 (1964).
3. R.J. Noer and W.D. Knight, Rev. Mod. Phys. **36**, 177 (1964).
4. P.W. Anderson, Phys. Rev. Lett. **3**, 325 (1959).

5. M.A. Abrikosov and L.P. Gorkov, JETP **15**, 752 (1962).
6. H. Ton et al., Phys. Rev. Lett. **77**, 1374 (1996).
7. J. Rossat-Mignod et al., *Selected topics in superconductivity*, Vol. 1 ed. L.C. Gupta and M.S. Multani, World Scientific **265-347** (1992) and references therein.
8. L.C. Hebel and C.P. Slichter, Phys. Rev. **113**, 1504 (1959).
9. T. Kohara et al., Phys. Rev. B **51**, 3985 (1995).
10. M. Itoh et al., Physica C **160**, 177 (1989).
11. P.W. Anderson, Science **235**, 1196 (1987).
12. G. Baskaran *Selected topics in superconductivity*, Vol. 1 ed. L.C. Gupta and M.S. Multani, World Scientific, **107-137** (1992) and references therein.
13. K. Asayama et al., Physica B **171**, 226 (1991) and references therein.
14. W.W. Warren et al., Phys. Rev. Lett. **59**, 1860 (1987).
15. A.K. Rajarajan, L.C. Gupta and R. Vijayaraghavan, Physica C **193**, 413 (1992) and references therein.
16. A.K. Rajarajan et al., J. Phys. Condensed Matter **4**, 6971 (1992).
17. C.P. Slichter et al., *Physical Properties of High Temperature Superconductors* ed. D.M. Ginsburg, Singapore, World Scientific, p. 269 (1990).
18. Zakir Hossain et al., Solid State Commun. **92**, 341 (1994) and references therein.
19. T. Kohara et al., Phys. Rev. B **51**, 3985 (1995).
20. K. Elankumaran et al., Physica B **223 & 224**, 90 (1996) and references therein.
21. H. Yasuoka, *Mechanism of High temperature superconductivity,* Springer Verlag, Berlin, p. 156 (1988).
22. T. Moriya et al., J. Phys. Soc. Jpn. **59**, 2905 (1990).

HOMOLOGOUS SERIES OF LAYERED CUPRATES AND SUPERCONDUCTIVITY

H. Yamauchi[1] and M. Karppinen[1,2]

[1]*Materials and Structures Laboratory, Tokyo Institute of Technology, Nagatsuta, Yokohama 226, Japan*
[2]*Laboratory of Inorganic and Analytical Chemistry, Helsinki University of Technology, FIN-02150 Espoo, Finland*

The structures of the existing high T_c superconductors are described as members of different homologous series with a general expressio of $M_mA_2Q_{n-1}Cu_nO_{m+2+2n\pm\delta}$ or M-*m2(n-1)n*. In this designation the $M_mO_{m\pm\delta}$ unit refers to the charge-reservoir block between two adjacent $Q_{n-1}Cu_nO_{2n}$ infinite-layer blocks. The same representation system is extended to cover also the structures containing fluorite blocks. In that case the structure is expressed by $M_mA_{2k}B_sCu_{1+k}O_{m+4k+4}$ or M-*m(2k)s (1+k)*, where the B metal atoms form the B_sO_s fluorite block unit. A generalized view on the functions of each structural block in terms of charge balance and then hole doping is discussed based on the above structural classification scheme and bond-valence considerations. Moreover, a concept of "confined net holes" in the innermost CuO_2 plane(s) is introduced for $n \geq 3$ phases, as the main parameter to control the value of T_c.

1. INTRODUCTION

The versatility of the chemistry of copper exhibiting various coordination numbers and valence values extending even up to +IV has enabled the creation of numerous mixed copper oxides with interesting electronic, magnetic and superconducting properties. Stabilization of the highest oxidation states requires, however, high oxygen pressures and/or presence of basic cations such as alkaline or alkaline earth metals [1]. For p-type high-T_C superconductivity pyramidally coordinated mixed (+II/+III)-valent copper in CuO_2 planes has been considered essential, while the structures of n-type cuprates are based on square-planar CuO_2 planes without the apical oxygen or other anions. In both cases the adjustment of the carrier concentration in the superconducting CuO_2 planes is achieved by controlling either the oxygen content or the size and valence of the cations in the various bare metal/metal oxide spacing layers. Classification of the existing superconducting structures enables us to see more clearly the functions of each structural unit in respect to charge balance and superconductivity. In the present contribution the known superconducting cuprates are classified as members of homologous series and the doping mechanisms and charge transfer conditions between the different structure blocks are discussed.

2. CLASSIFICATION OF THE STRUCTURES

The structures of superconducting cuprates consist of three different types of building blocks: the perovskite blocks including the so-called infinite-layer blocks, the rock-salt blocks and the fluorite blocks. The rock-salt blocks possessing positive charge are considered to act as hole donors for the infinite-layer blocks, while the fluorite blocks with their negative charge are electron donors. Based on the above defined structural blocks the existing superconducting cuprate phases can be categorized into two main groups [2]. The structures of the first compound group contain only infinite-layer and rock-salt blocks, while in the second group the structures may be built up by all the three possible block types.

To further simplify the denotation, the various stoichiometries and structures of high T_c cuprates and related phases can be described by viewing the unit cells as collections of sheets rather than blocks. For this purpose the four-integer naming scheme as described by Shaked *et al.* [3] has been widely used, although it fails to describe structures containing fluorite blocks or other anions than oxygen. The first entry in this number series tells the number of consequtively stacked layers in the so-called charge-reservoir block between the adjacent infinite-layer blocks. These layers are usually built up with various heavy metals or transition metals and oxygen. The second entry tells the number of spacing alkaline earth (A) or rare earth (RE) oxide layers which are located between the charge-reservoir and infinite-layer blocks. The fourth and third entries, respectively, correspond to the number of conducting CuO_2 planes and the number of bare alkaline earth and/or rare earth layers separating these CuO_2 planes in one infinite-layer block. If the fourth figure is indicated by n, the third figure is n-1 in all the existing superconducting cuprates except in the so-called infinite-layer compounds and in those structures containing a fluorite block. Crystal structures fulfilling the above given criteria will be designated as M-$m2(n-1)n$, where M is the chemical name of the cations in the charge-reservoir block. In the case of m = 0, however, in order to include not only the superconducting $02(n-1)n$ series but also the "not-yet-superconducterized" $01(n-1)n$ series, the member with n = 1 being the conventional perovskite structure, the constant "2" in $02(n-1)n$ should be rather considered as a variable having two possible values of 1 and 2.

In Table 1 a possible extension of the four-integer naming system to cover also the infinite-layer structure and the structures containing the fluorite block is summarized. The structure of the infinite-layer compounds consists of an infinite alternating stacking of bare alkaline earth elements and CuO_2 planes, and it should be denoted by 0011, as the $n = \infty$ limit of $mj(n-1)n$ (where m = finite and j = 1 or 2). This 0011 structure can be considered as a parent phase for the second structure group. Furthermore, all the existing second structure group phases can be expressed by a general formula of $M_mA_{2k}B_sCu_{1+k}O_{m+4k+2s}$ or M-$m(2k)s(1+k)$, where the B metal atoms form the B_sO_s fluorite block unit (k = 0 or 1, $s \geq 2$) [4]. In the case of k = 0, only two phases with m = 0 and s = 1 (0011 or infinite-layer structure) and s = 2 (0021 structure, which is nothing else but the T' structure) exist. But for k = 1, m can vary between 0 and 3 and s ≥ 2, as far as the data presently available are concerned.

Table 1. Four-integer naming scheme proposed for cuprate structures containing a fluorite block.

Name	Chemical formula
0011	$(Sr,Nd)CuO_2$
0021	$(Nd,Ce)_2CuO_4$
0222	$(Nd,Sr,Ce)_2CuO_4$
(Pb,Cu)-1222	$(Pb,Cu)(Sr,Eu)_2(Eu,Ce)_2Cu_2O_9$
Cu-1222	$Cu(Ba,Eu)_2(Eu,Ce)_2Cu_2O_9$
Bi-2222	$Bi_2Sr_2(Gd,Ce)_2Cu_2O_{10}$
(Pb,Cu)-3222	$(Pb_2Cu)Sr_2(Eu,Ce)_2Cu_2O_{11}$
(Pb,Cu)-1232	$(Pb,Cu)Sr_2(Ho,Ce)_3Cu_2O_{11}$

3. HOMOLOGOUS SERIES

The concept of a homologous series in the case of cuprate superconductors can be defined as "a family of layered copper oxides that differ from one another only by the number of pairs (n-1) of CuO_2 and bare cation planes in the infinite-layer block". Each member of one homologous series thus has the same charge-reservoir block (M-m is fixed) but the number of CuO_2 planes (n) is different. Furthermore, it is feasible to speak about a homologous series only if the members at least up to n = 3 have been experimentally found.

By the end of the year 1992 only three homologous series fulfilling the above definition, *i.e.* Bi-22(*n*-1)*n*, Tl-22(*n*-1)*n* and Tl-12(*n*-1)*n*, had been verified. In January 1993, the $n = 2 - 5$ members of the 02(*n*-1)*n* series were synthesized in the Sr-Ca-Cu-O system [5,6], and since then many other series, most of them being stabilized only under very high oxygen pressures (~ 2 - 8 GPa), have been discovered. Replacing Sr by Ba in the Sr-Ca-Cu-O system did not result in the 02(*n*-1)*n* structures but in the discovery of a new homologous series of Cu-12(n-1)nP with single $CuO_{1-\delta}$ planes as a perovskite-type (P) charge-reservoir block [7-9]. Part/half of the $CuO_{1-\delta}$ units in this system can also be replaced by carbonate groups [10].

The importance of categorizing the known structures and observing various periodical trends or empirical rules was clearly realized when the "Eka-Silicon's" in the Pb-Sr-(RE,Ca)-Cu-O system, *i.e.* the Pb-1212 and Pb-1223 phases, were first predicted, then designed and finally successfully synthesized [11]. In that case the empirical rule, stating that the average valence of the Q cations in the bare cation layer sandwiched by two adjacent CuO_2 planes must be equal to or less than +2 in an infinite-layer block containing more than two CuO_2 planes [2], was obeyed and the trivalent RE totally replaced by divalent Ca. On the other hand, when building up new types of series the connection criteria between the adjacent blocks become important. That is, (1) sharing of a common cation sheet with/without oxygen by the rock-salt/fluorite and the infinite-layer blocks and (2) lattice constant matching between all the blocks.

In order to find new superconducting structures the fact worthy of note is that the parent 01(*n*-1)*n* series of structures has not been superconducterized yet. Although the $LaCuO_{3-\delta}$ system with the 0101 structure ($n = 1$) has been studied by several groups [12-14], less interest has been paid on the second member in that family, *i.e.* the 0112 structure, which has been so far stabilized in the pyramidal $BaY[Cu(+II)_{0.5}M(+III)_{0.5}]_2O_5$ system [15] only with a fixed oxygen content and half-occupation of the Cu-site by trivalent transition metal ions. Finally, the discovery of the [(Cu,C)Ba]-32(*n*-1)*n*P series [16] demonstrates that it is in principle possible to increase the number of superconducting structures, besides by increasing the *n* value, also by making thicker (increase *m*) charge-reservoir blocks.

4. CHARGE BALANCE AND HOLE-DOPING MECHANISMS

The net nominal charge of the charge-reservoir block varies in the range of 0 and +2 in all the existing homologous series. Since the rock-salt-type spacing layers between the $M_mO_{m\pm\delta}$ charge-reservoir and the $Q_{n-1}Cu_nO_{2n}$ infinite-layer blocks are always formed by formally neutral A(+II)O units, the available overall charge of the infinite-layer block depends on the charge of the charge-reservoir block: more positively charged charge-reservoir leads to less doped infinite-layer block, and *vice versa*. If a "non-doping charge-reservoir" is defined to exhibit a net charge of +2, a simple calculation implies that the infinite-layer block could exhibit p-type superconductivity without doping from the charge-reservoir only in the case of the valence of the metal Q (V_Q) in the bare metal spacing layers between the CuO_2 planes being less than +II. Decreasing the net charge of the $M_mO_{m\pm\delta}$ unit below +2 by changing either the metal composition M or the oxygen stoichiometry δ, enables us to increase the V_Q value above +II without destroying superconductivity. In practice most of the charge-reservoir blocks, *e.g.* Tl(+III)O, $Tl(+III)_{2-x}O_2$, $Bi(+III)_2O_{2+\delta}$, $Pb(+II)_2Cu(+I)O_2$, $Hg(+II)O_{1-\delta}$ and $CuO_{1-\delta}$ [varying from Cu(+III)O to Cu(+I)], exhibit a net nominal charge close to +1. The $[CO]^{2+}$ group is, at least formally, the most reducing charge-reservoir constituent, and when partly substituting the $[CuO]^+$ units in the (Cu,C)-12(*n*-1)*n*P system is assumed to suppress the hole concentration strongly [17]. The structures of the 02(*n*-1)*n* series, on the other hand, are built up without any charge-reservoir blocks, *i.e.* the net charge of the charge-reservoir can be considered to be zero. Consequently, if we suppose a nearly full occupancy of the apical oxygen site in the A_2O_2 rock-salt layers as well as nearly perfect CuO_2 planes, these materials should be much more oxidized in terms of (nominal) copper valence than the other high T_c cuprate phases. However, recent neutron diffraction data for the 0201 phase suggest that the CuO_2 plane is strongly oxygen-deficient [18,19]. Another possibility to maintain the low enough Cu-valence in the 02(n-1)n series is to replace the neutral A(+II)O units by $(A_{1-y}RE_y)O^{y+}$ in the rock-salt layer, as is the case in the $(La,Sr)_2CuO_4$ phase (0201 structure).

The above general consideration gives only a rough estimation of the available charge in the infinite-layer block, usually imagined as "nominal average copper valence". The actual hole concentration in the CuO_2 planes is determined as a delicate balance within the exact oxygen content of the charge-reservoir block and the nature (valence and size) of the cations in the charge-reservoir block and in the rock-salt or bare-cation spacing layers. For example, the nominal net charge of the $[CuO_{1-\delta}]^+$ chain-block in $CuBa_2YCu_2O_{7-\delta}$ does not depend on the oxygen deficiency δ, since the valence of the Cu atoms in the charge-reservoir block increases with increasing δ. The fully oxidized $[Cu(+III)O]^+$ chains, however, dope more holes into the CuO_2 planes than the oxygen-depleted $[Cu(+I)]^+$ chains, as evidenced by the well-established T_c - δ relation in the $CuBa_2YCu_2O_{7-\delta}$ system.

The concept of bond-valence developed by Brown and Altermatt [20,21] provides us a more quantitative means for discussing the inter- and intra-block charge transfer balances. The bond-valence s_{ij} of the bond between the species i and j (atoms or ions) in a solid matrix is calculated from the bond length R_{ij} using the following equation: $s_{ij} = \exp[(R_0-R_{ij})/B]$, where R_0 and B are empirically determined parameters, R_0 being specific for each different type of i-j pairs. The bond-valence-sum around a species i is then described by $V_i = \pm\Sigma_j s_{ij}$ (+ for a cation and - for an anion). Since the holes responsible for superconductivity may reside either on copper or on oxygen in the CuO_2 planes, a parameter obtained by summing the excess positive charge above +II for Cu (V_{Cu} - 2) and -II for O_{pl} (V_{Opl} + 2) in one CuO_2 unit is supposed to correlate better than the V_{Cu} value with the effective or "net" hole concentration, and further with the superconducting properties [22,23]. In the resulting $p(CuO_2)$ value the contributions from the planar Cu-O_{pl} bonds are actually counteracted, *i.e.* the positive Cu-O_{pl} bond-valences from V_{Cu} exactly cancel the corresponding negative values from V_{Opl}, and only the vertical bonds from Cu to the apical oxygen (O_{api}) and from O_{pl} to the Q and A cations have net effects on the hole concentration. Furthermore, based on this observation it is plausible to distinguish three different ways of doping the CuO_2 planes, *i.e.* (1) shortening the Cu-O_{api} bond and (2) lengthening the effective (R_0-R_{ij}) O_{pl}-A or (3) O_{pl}-Q bond [24]. Whether the above discussed different doping mechanisms result in different hole distributions in the Cu-O_{pl} bond is not known but certainly worth interesting consideration as well.

Mechanisms (1) and (3) are possible *e.g.* in the $CuBa_2(RE,Ca)Cu_2O_{7-\delta}$ system: incorporation of oxygen into the $Cu(1)O_{1-\delta}$ block is seen as a gradual movement of O_{api} towards the Cu(2)-Opl planes as the oxidation proceeds, while the partial substitution of trivalent RE by divalent Ca introduces the holes into the CuO_2 planes by increasing the R_0-R_{ij} value of the O_{pl}-(RE,Ca) bond. In the $HgBa_2Ca_{n-1}Cu_nO_{3+2n-\delta}$ system, on the other hand, the oxygen deficiency d in the $[Hg(+II)O_{1-\delta}]^{2\delta+}$ block is usually far above the value of 0.5, and consequently the Cu_{py}-O_{api} distance is long. The excess oxygen (O_{cr}) in the $HgO_{1-\delta}$ charge-reservoir block is located quite far away from Hg but right above Ba (A cation), and therefore mainly shortens the O_{cr}-Ba bond and lengthens the O_{pl}-Ba bond according to mechanism (2).

In the phases with $n \geq 3$ the holes may be inhomogeneously distributed between the pyramidal (py) and square-planar (sp) CuO_2 planes in the infinite-layer block, and the calculation of the distinct $p(CuO_2)_{py}$ and $p(CuO_2)_{sp}$ values from the crystallographic data published for various $n \geq 3$ phases in fact suggests that the net holes are concentrated in the innermost planar CuO_2 planes (Table 2). This observation is really surprising since pyramidally coordinated copper has been considered essential for the occurrence of p-type high-T_c superconductivity. The $p(CuO_2)_{sp}$ value, which varies between *ca.* +0.15 and +0.30, seems to correlate with the critical temperature (Table 2). Neither over-doped region nor even saturation with increasing $p(CuO_2)_{sp}$ is observed for the phases analyzed. In each homologous series found so far, T_c shows a maximum at n = 3 or 4, and then slowly decreases with higher values of n. Unfortunately for the $n \geq 4$ phases the availability of reliable crystallographic data is very limited, but the results obtained for the Hg-12(n-1)n system (Table 2) support this experimentally observed T_c - *versus* - n trend. For comparison, the $p(CuO_2)_{sp}$'s for infinite-layer compounds of the $(Sr,Ca)CuO_2$ system are very close to zero, while for the n-type $(Sr_{0.9}La_{0.1})CuO_2$ infinite-layer compound a negative $p(CuO_2)_{sp}$ value of -0.132 was obtained.

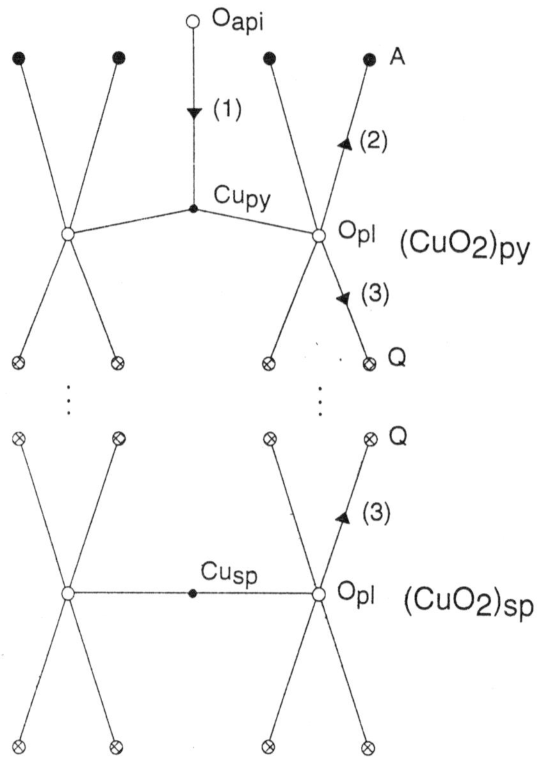

Fig. 1. Different ways of hole-doping the $(CuO_2)_{py}$ and $(CuO_2)_{sp}$ planes.

Finally, in order to find "the universal maximum T_c" the hole concentration, $p(CuO_2)_{sp}$, in the innermost CuO_2 planes should be increased, and moreover the hole distribution in the Cu-O_{pl} bond should be considered. Understanding of the role of hole distribution might also help us to explain why the thin film sample of the infinite-layer $Ca_{1-x}Li_xCuO_2$ ($0.15 \leq x \leq 0.45$) phase [25], even though successfully p-type doped has not shown superconductivity.

Table 2. Calculated $p(CuO_2)_{py}$ and $p(CuO_2)_{sp}$ values for Hg-12(n-1)n phases with n = 3 - 5. The T_c's and bond lengths are from refs. [26-30].

phase	Hg-1223	Hg-1234	Hg-1245	Tl-2223	(Cu,C)-1234
$p(CuO_2)_{py}$	-0.251	-0.274	-0.274	-0.097	+0.021
$p(CuO_2)_{pl1}$	+0.281	+0.296	+0.060	+0.286	+0.222
$p(CuO_2)_{pl2}$	-	-	+0.208	-	-
T_c [K]	133	125	101	124	117

5. CONCLUSIONS

All the existing high T_c superconducting cuprates can be named and classified by two types of four-integer scheme. Especially the scheme written as M-$m2(n-1)n$ has led us to the concept of "homologous series". Till now, nearly 20 different homologous series have been synthesized, clearly showing that in each series T_c increases as n increases from 2 to 3. The concept of "net holes" in individual CuO_2 planes in $n \geq 3$ phases as estimated from the existing crystallographic data via a bond-valence-sum calculation, revealed that the net holes are concentrated and/or confined in the innermost CuO_2 plane(s), predominantly controlling the value of T_c as compared with those in the other CuO_2 planes Finally, the efficiency of the different doping mechanisms and their effects on the charge distribution in the Cu-O_{pl} bond are interesting questions.

ACKNOWLEDGEMENT

This work was partially supported by a Grant-in-Aid for Scientific Research under a contract No. 08455299 from The Ministry of Education, Science and Culture of Japan.

REFERENCES

1. S. Darracq, S.G. Kang, J.H. Choy and G. Demazeau, J. Solid State Chem. **114,** 88 (1995).
2. T. Wada, A. Ichinose, H. Yamauchi and S. Tanaka, J. Cer Soc. Jpn. Int. Ed. **99,** 420 (1991).
3. H. Shaked, P.M. Keane, J.C. Rodriguez, F.F. Owen, R.L. Hitterman and J.D. Jorgensen, *Crystal Structures of the High T_c Superconducting Copper Oxides*, Elsevier, Amsterdam 1994.
4. H. Yamauchi, New Ceramics **3,** 47 (1990) [in Japanese).
5. S. Adachi, H. Yamauchi, S. Tanaka and N. Mori, Physica C **208,** 226 (1993).
6. S. Adachi, H. Yamauchi, S. Tanaka and N. Mori, Physica C **212,** 164 (1993).
7. C.-Q. Jin, S. Adachi, X. -J. Wu, H. Yamauchi and S. Tanaka, Physica C **223** 238 (1994).
8. M.A. Alario-Franco, C. Chaillout, J.J. Capponi, J.-L. Tholence and B. Souleite, Physica C **222,** 52 (1994).
9. H. Ihara, K. Tokiwa, H. Ozawa, M. Hirabayashi, A. Negishi, H. Matsuhata and U.S. Song, Jpn. J. Appl. Phys. **33,** 503 (1994).
10. T. Kawashima, Y. Matsui and E. Takayama-Muromachi, Physica C **224,** 69 (1994).
11. H. Yamauchi, T. Tamura, X. -J. Wu, S. Adachi, and S. Tanaka, Jpn. J. Appl. Phys. **34,** 349 (1995).
12. G. Demazeau, C. Parent, M. Pouchard and P. Hagenumuller, Mater. Res. Bull. **7,** 913 (1972).
13. Z. Hiroi and M. Takano, Nature **377,** 41 (1995).
14. M. Karppinen, H. Yamauchi, H. Suematsu and O. Fukunaga, Physica C **264,** 268 (1996).
15. L. Er-Rakho, C. Michel, Ph. Lacorre and B. Raveau, J. Solid State Chem. **73,** 531 (1988).
16. T. Kawashima, Y. Matsui and E. Takayama-Muromachi, Physica C **233,** 143 (1994).
17. H. Yamauchi, M. Karppinen,T. Ito, H. Suematsu and O. Fukunaga, Physica C **266,** 345 (1996).
18. Y. Shimakawa, J.D. Jorgensen, J.F. Mitchell, B.A. Hunter, H. Shaked, D.G. Hinks, R.L. Hitterman, Z. Hiroi and M. Takano, Physica C **228,** 73 (1994).
19. J.P. Hodges, P.R. Slater, P.P. Edwards, C. Greaves, M. Slaski, G. Van Tendeloo and S. Amelinckx, Physica C **260,** 249 (1996).
20. I.D. Brown and D. Altermatt, Acta Cryst. **B 41,** 244 (1985).

21. I.D. Brown, J. Solid State Chem. **82**, 122 (1989).
22. J.L. Tallon, Physica C **168**, 85 (1990).
23. I.D. Brown, Physica C **169**, 105 (1990).
24. M. Karppinen and H. Yamauchi, to be published.
25. K. Kubo, M. Ichikawa, N. Sugii, K. Yamamoto and H. Yamauchi, Phys. Rev. B **49**, 6919 (1994).
26. O. Chmaissem, Q. Huang, E.V. Antipov, S.N. Putilin, M. Marezio, S.M. Loureiro, J.J. Capponi, J.L. Tholence and A. Santoro, Physica C **217**, 265 (1993).
27. M. Paranthaman and B.C. Chakoumakos, J. Solid State Chem. (in press).
28. Q. Huang, O. Chmaissem, J.J. Capponi, C. Chaillout, M. Marezio, J.L. Tholence, and A. Santro, Physica C **227**, 1 (1994).
29. D.C. Sinclair, M.A.G. Aranda, P. Attfield and J. Rodriguez-Carvajal, Physica C **225,** 307 (1994).
30. Y. Shimakawa, J.D. Jorgensen, D.G. inks, H. Shaked, R.L. Hitterman, F. Izumi, T. Kawashima E. Takayama-Muromachi and T. Kamiyama, Phys. Rev. B **50** 16008 (1994).

ANOMALOUS NON-RESONANT MICROWAVE ABSORPTION IN YBCO/INSULATOR, Nb/INSULATOR AND YBCO/Nb MIXTURES

A. Anand, Amit Rastogi, M.S. Hegde* and S.V. Bhat
Department of Physics,
** Solid State and Structural Chemistry Unit,*
Indian Institute of Science, Bangalore 560012, India

Non resonant microwave absorption of YBCO/CuO, YBCO/PrBCO, Nb/CuO and YBCO/Nb powder mixtures are studied as functions of temperature and low magnetic fields. Anomalous absorption of microwave radiation is observed in composites of ceramic YBCO. A study of powder mixtures of Nb/Insulators did not show any anomaly. Peculiar temperature dependence of absorption has been observed in YBCO/Nb mixtures. These observations are explained by extending the RSJ model and incorporating phase difference across the junctions lying between $\pi/2$ and $3\pi/2$. Junctions with such a phase difference have important consequences to the symmetry of the order parameter and paramagnetic Meissner effect. Simulations based on the extended RSJ model consistently reproduce the observed effects.

1. INTRODUCTION

The microwave and rf absorption in ceramic samples of copper oxide based high temperature superconductors (HTSC) is a very sensitive function of externally applied low DC magnetic fields [1]. This is due to field induced 'diffraction reduction' of the critcal current density J_c of the intergranular contacts acting as Josephson junctions abundantly present in the ceramic samples. Normally the absorption in these junctions is an increasing function of the field with a minimum at low applied fields. However, there is a number of cases where it is 'anomalous' , in that it decreases with increasing fields at low applied fields [2]. This anomalous behaviour has important implications for the symmetry of the order parameters well as exotic phenomena like paramagnetic Meissner effect (PME). In this paper, we study the nature of the non resonant microwave absorption (NRMA) in various superconductor insulator composites. Composites of YBCO/Insulator and YBCO/Nb of various relative fractions of YBCO, Nb and insulator are subjected to non resonant microwave absorption (NRMA).Two insulators were considered, CuO and PrBCO. The NRMA of monophasic YBCO and monophasic Nb has a monitonic field dependence, i.e., the field dependent absorption is a minimum at zero field and it increases with the increase in the static field. However, anomalous absorption is observed in composites of YBCO at low magnetic fields. The absorption in monophasic YBCO, Nb and Nb/CuO mixtures does not show anomalous absorption at zero field. We explain the anomalous absorption at low fields by extending the RSJ model and incorporating junctions with phases such that the phase values lies between $\pi/2$ and $3\pi/2$. The mixtures of YBCO/Nb shows peculiar temperature dependence of absorption, below an above the T_c of Nb. This peculiar dependence is explained by the extended RSJ model.

2. EXPERIMENTAL

Ceramic pellets of $YBa_2Cu_3O_7$ were prepared by the standard solid state technique. Some of these pellets were powdered and mixed in different proportions by weight with Nb powder and pelletised to a standard size (1mm thick and 4mm diameter) at a pressure of 250psi. Different proportions of YBCO/Nb and YBCO/insulator were chosen for the study. Along with this, a sample with 50% Nb and 50% CuO was also studied with the technique. A standard X-band EPR spectrometer (Bruker ER200D) equipped with an Oxford Instruments continuous flow He-cryostat was used. The NRMA responses vs DC magnetic field at different temperatures were recorded for pure YBCO, Nb and powder mixtures of the above samples. Magnetic field modulation at 100kHz and phase sensitive detection are used to record the derivative of the field dependent absorption.

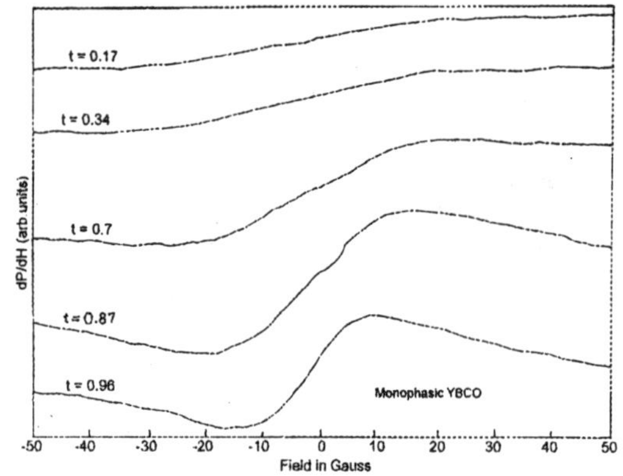

Fig 1: NRMA of monophasic YBCO pellet.

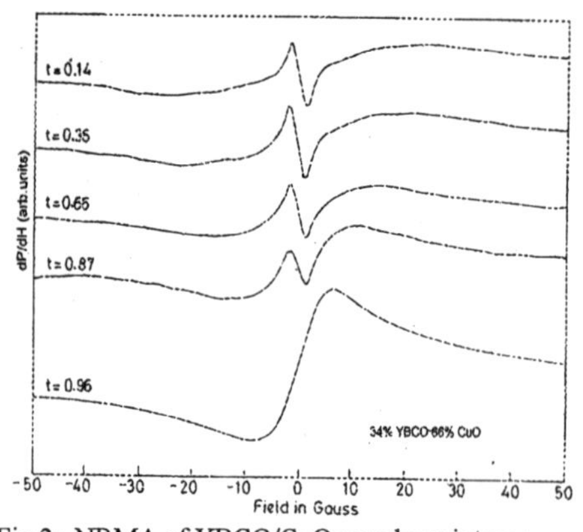

Fig 2: NRMA of YBCO/CuO powder mixtures

3. OBSERVATIONS AND RESULTS

Figure 1 shows the NRMA of monophasic YBCO. Figure 2,3 and 4 shows the NRMA of YBCO/CuO, YBCO/PrBCO and YBCO/Nb mixtures respectively. Figure 5 shows NRMA of Nb/CuO powder mixtures. Figure 6.a 6.b and 6.c are simulations based on the extend RSJ model for a given proportion of YBCO/Nb mixture at 7.5K, close to T_c of Nb and above T_c of Nb respectively.

The magnitude of the anomalous phase is dependent on the concentration of Nb below the critical temperature of Nb. Higher the concentration of Nb, lower is the magnitude of anomalous absorption. Close to the transition temperature of Nb the anomalous component in the mixture almost disappears and it reappears with a larger amplitude beyond the critical temperature of Nb. The absorption after the transition temperature of Nb is similar to absorption in composites of YBCO/Insulator and YBCO/metal [3]. The anomalous absorption is seen in composites of YBCO only. Other composites like Nb/CuO did not show any anomalous absorption.

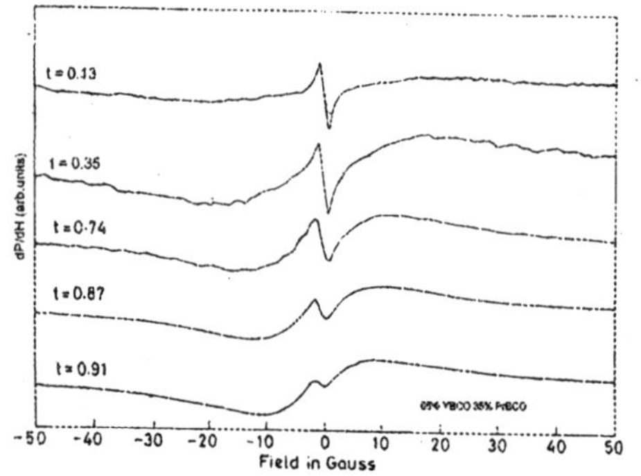

Fig 3 : NRMA of YBCO/PrBCO powder mixtures

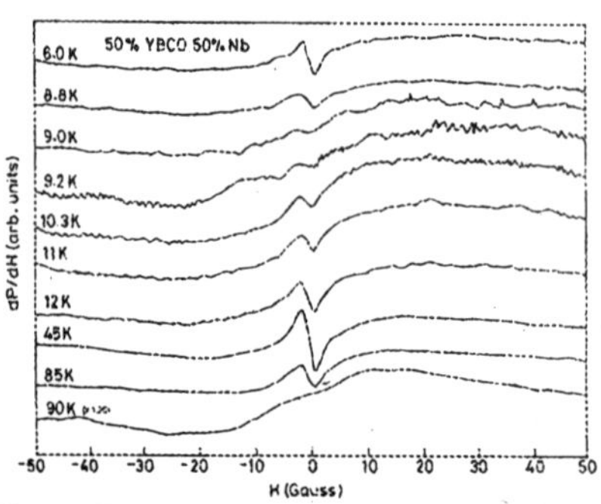

Fig 4 : NRMA of YBCO/Nb powder mixtures

4. ANALYSIS

The field dependent dissipation in granular superconductors is mostly due to the presence of junctions, since the junction critical current decreases steeply with the applied field.

The system is made of the following types of junctions: YBCO insulator (CuO/PrBCO) YBCO, Nb/YBCO/Nb and YBCO/Nb/YBCO. Above the critical temperature of Nb we do not have Nb-YBCO-Nb junctions. The Nb pieces form a metal matrix in YBCO after the critical temperature of Nb is crossed. Since the composite samples consist of superconducting regions of YBCO separated by Nb or Insulator grains, they can be considered to constitute an assembly of Josephson junctions of varying coupling strengths and orientations. Considering any one such junction close to the surface of the pellet, the applied magnetic field

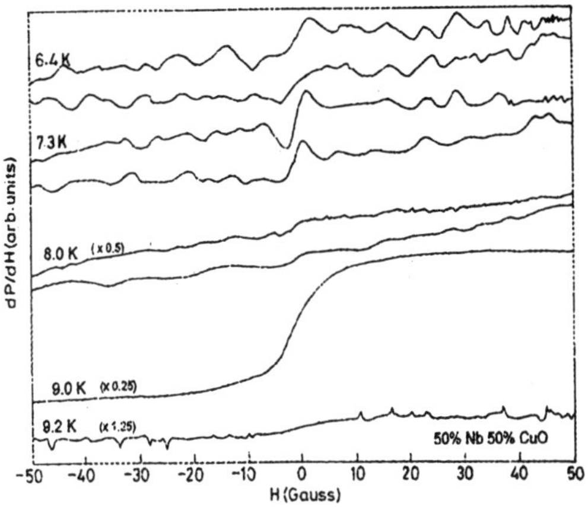

Fig 5 : NRMA of Nb/CuO powder mixtures

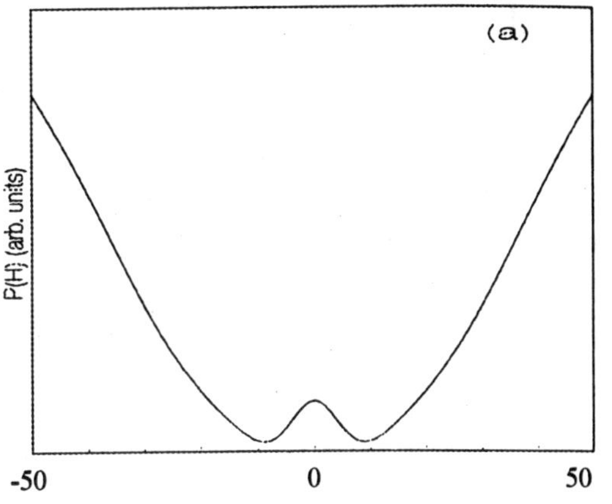

Figure 6a : Simulated P(H) for YBCO/Nb mixtures at7.5K

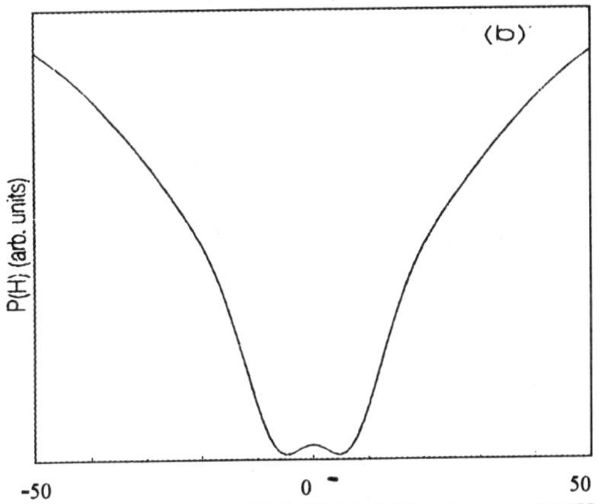

Figure 6b : Simulated P(H) for YBCO/Nb mixtures at8.8K

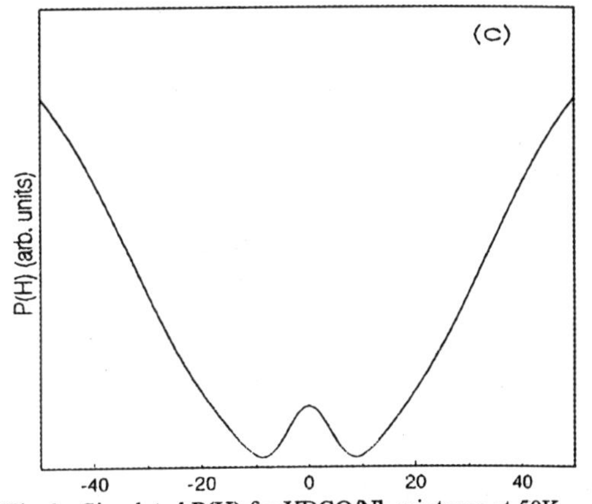

Fig 6a :Simulated P(H) for YBCO/Nb mixtures at 50K

induces a boundary current I_o across it, which determines the phase difference across the junction through $I_c \sin\psi_o = I_o$, where I_c is the critical current of the junction. The field dependent microwave absorption at low fields in such a junction is commonly explained using the RSJ model. Following the model and neglecting the capacitive term, the phase difference $\psi = \psi_o + \psi_{mw}$, across the junction satisfies the equation

$$\frac{d\psi}{d\tau} + \frac{2\pi I_c R_n}{\Phi_o}\sin\psi = \frac{2\pi I_c R_n}{\Phi_o} + \frac{2\pi I_{mw} R_n}{\Phi_o}\cos\omega_{mw}\tau \qquad (1)$$

where R_n is the normal state resistance of the junction, $I_{mw}\cos\omega_{mw}\tau$ is the microwave field induced current and Φ_o is the flux quantum. At t=0, the static field is zero therefore $I_c \sin\psi_o = I_o = 0$ thus the initial condition is given by

$$\left.\frac{d\psi}{d\tau}\right|_{\tau=0} = \frac{2\pi I_c R_n \cos\omega_{mw}\tau}{\Phi_o \omega_{mw}} \qquad (2)$$

The power absorbed $P_{mw}(H)$ is given by the time average of the product of the voltage $V = \frac{\Phi_o}{2\pi}\frac{d\psi_{mw}}{d\tau}$ and the microwave current $I_{mw}\cos\omega_{mw}\tau$. The power absorbed $P_{mw}(H) = \langle VI \rangle$ is given by the equation :

$$\frac{1}{2} I_{mw}^2 R_n \left[\frac{1}{(1+\eta^2(H))} + \frac{\eta^3(H)}{\pi}\left(\frac{(1-\exp(-2\pi\eta(H))}{(1+\eta^2(H))^2}\right)\right] \tag{3}$$

In Figs. [6.a,b,c] we present the results of the simulations using (3) along with the experimentally observed field dependent microwave absorption. For the simulation, we model the overall absorption to be a resultant of two contributions: first, coming from the `normal' junctions (with positive $\cos\psi_o$) and second, from the `anomalous' junctions (with negative $\cos\psi_o$). By writing $P_{mw}(H,t) = (1-x)P1_{mw}(H,t) + xP_{mw}(H,t)$ where the first term corresponds to $0 < \psi_o < \pi/2$ and the second to $\pi/2 < \psi_o < 3\pi/2$ and `x' is the fraction of latter type of junctions. The above expression with different temperature dependences are used for SNS, SIS and weak junctions of YBCO and Nb. The critical currents are different for different types of junctions. We assume temperature dependence of critical currents of SNS junctions as $(1-t)^2$[4] and Ambegaokar-Baratoff relations for SIS junctions.

To the first approximation we can account for the absorption of microwave radiation to be sum of independent contributions of absorption due to absorption in YBCO-Nb-YBCO and weak links of YBCO and Nb. The absorption in weak links and SNS junctions can have anomalous field dependence of absorption since the phase of critical currents are sensitive to the strength of the junctions. These junctions have low critical current densities and they are sensitive to magnetic field and temperature.

The disappearance of the anomalous phase close to the transition temperature of Nb (as well as YBCO) is due to complete penetration of the magnetic field in the superconductor. As a consequence, the normal field dependent absorption dominates over the anomalous absorption . An approximate calculation of absorption fractions from Nb and YBCO close to the T_c of Nb shows that the absorption in Nb and its junctions are 10^4 more than that in YBCO and its junctions as per the RSJ model. The absorption of electromagnetic radiation is entirely dominated by Nb granules where as absorption in YBCO is confined to a few angstroms of its surface close to transition temperature of Nb. This explains the reduction of anomalous absorption close to the transition temperature of Nb. We also see that the anomalous absorption disappears close to the transition temperature of YBCO, at which the weak links are broken down and absorption in these junctions will be independent of field. The residual field dependent dissipation very close to T_c, are therefore, attributable to the damped motion of fluxons in the bulk of superconductors.

5. ACKNOWLEDGEMENTS

Anand.A and Amit Rastogi thank the Council of Scientific and Industrial Research, India for fellowships.

REFERENCES

1. S.V. Bhat, P. Ganguly, T.V. Ramakrishnan and C.N.R. Rao, J. Phys. **C 20**, 559 (1987).
2. S.V. Bhat, V.V. Srinivasu and N.Kumar, Phys. Rev. **B 44**, 10121 (1991).
3. A. Anand, Amit Rastogi, M.S. Hegde and S.V. Bhat, Solid. State. Comm. **99**, 665 (1996).
4. B. Nebendahl, C. Kessler, D.N. Peligrad and M. Mehring, Physica **C 209**, 362 (1993).

SUPERCONDUCTING $Pb_{0.5}Sr_{2.5}Y_{0.5}Ca_{0.5}Cu_2O_y$ SINGLE CRYSTALS AND ROLE OF SILVER ADDITION

D.K. Aswal[#], S.N. Narang[#], S.K. Gupta[#], S.C. Purandre[$] and S.C. Sabharwal[#]

[#]*Technical Physics and Prototype Engineering Division,*
Bhabha Atomic Research Centre, Trombay, Mumbai 400 085, India
[$]*Tata Institute of Fundamental Research,*
Homi Bhabha Road, Colaba, Mumbai 400 005, India.

Single crystals of $Pb_{0.5}Sr_{2.5}Y_{0.5}Ca_{0.5}Cu_2O_y$ have been grown using CuO as a flux and adding 10% (by weight) of $AgNO_3$ to the starting charge. The morphological, structural and superconducting properties of the grown crystals have been investigated. The as grown crystals are found to be superconducting at 50 K. The role of Ag doping in high T_c superconductor is briefly discussed.

I. INTRODUCTION

Single crystals of $YBa_2Cu_3O_x$ (123) require a prolonged annealing treatment at the elevated temperatures and under flowing oxygen to superconduct at 90 K.[1] Recently, it has been shown that the addition of $AgNO_3/Ag_2O$ to the starting charge for growth facilitates the growth of superconducting crystals with T_c of 90 K[2,3] X-ray fluorescence measurements performed on crystals grown from charges containing 10% by weight of $AgNO_3$ showed that the amount of silver, if present, in the crystal lattice was much lower than 0.1 wt%. This study raised an interesting question: whether the effect of the presence of Ag is specific to 123 lattice? The $Pb_{0.5}Sr_{2.5}Y_{0.5}Ca_{0.5}Cu_2O_y$ (1212) superconductor has structural resemblance with 123. The polycrystalline material of 1212 phase prepared is known to exhibit a superconducting transition at 53 K.[4,5] However, the growth of superconducting single crystals of this phase has not been reported so far. In the light of the results obtained for 123 crystals grown from charges containing silver, we have recently studied the crystal growth of 1212 material from charges containing 10 wt% (by wt.) of $AgNO_3$. The as grown crystals have been found to undergo transition at 50K. In this paper, we present some results on growth and characterization of 1212 crystals and briefly discuss the role of $AgNO_3$ addition to the starting charge on the superconducting properties of both 1212 and 123 phase crystals.

II. EXPERIMENTAL

The single crystal growth was effected in a covered alumina crucible using CuO as a flux.[6,7] The starting material used for crystal growth was prepared by mixing a well formed polycrystalline $Pb_{0.5}Sr_{2.5}Y_{0.5}Ca_{0.5}Cu_2O_y$ material and CuO powder in the molar ratio of 70 and 30 respectively. A few crystal growth runs were also given after adding 10% (by wt.) of $AgNO_3$ to this starting charge. The crucible, containing about 30 gm of charge, was placed inside a vertical resistance furnace. The crucible temperature was raised to 1120°C and held there for 2 h. Subsequently, it was cooled down to 1050°C at the rate of 30°C /h and then to 900°C at the rate of 1°C/h. Finally, it was brought to room temperature at the rate of 30°C/h.

The grown crystals were characterized for morphological, structural and superconducting properties. X-ray diffraction (XRD), energy dispersive X-ray analysis (EDAX) measurements for composition, scanning electron microscopy(SEM) for surface morphology and four-probe resistivity measurements for the superconducting properties were performed to characterize the crystals.

III. RESULTS AND DISCUSSION

The crystals were found to have grown over the entire volume of the solidified melt. They were removed mechanically from the crucible. The cleaved crystals were of typical dimensions 1.5 x 1 x 0.5 mm^3. The growth features and smoothness of the grown crystal face as revealed by the scanning electron microphotograph is shown in Fig. 1.

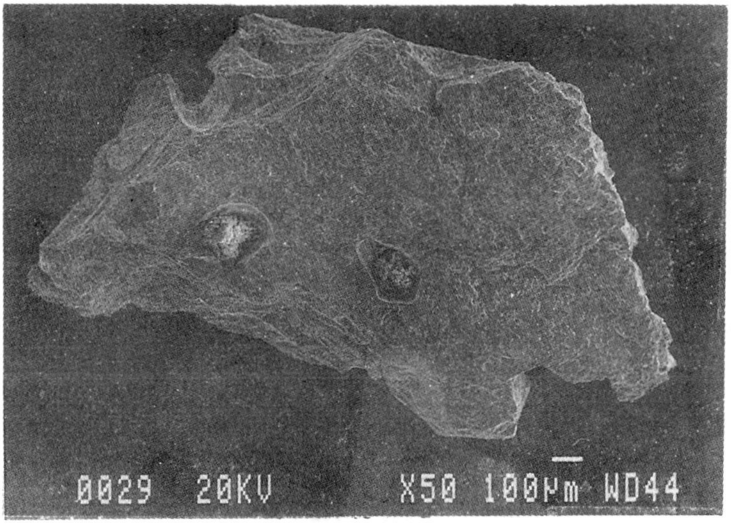

Fig.1. SEM photograph of a $Pb_{0.5}Sr_{2.5}Y_{0.5}Ca_{0.5}Cu_2O_y$ single crystal.

The crystal composition was found to be $Pb_{0.27}Sr_{2.67}Y_{0.5}Ca_{0.51}Cu_{2.07}O_y$, which showed a partial substitution of Sr for Pb. The results of crystal composition measurements mentioned here are quite close to that reported for polycrystalline sample.[4,5] It may be mentioned that the composition of crystals reported by others is highly deficient in Sr.[6,7] That is why the superconductivity could not be observed in these crystals. Like in the case of 123 crystals grown from charges containing $AgNO_3$, the concentration of silver in this case is also found to be below the detection limit of EDAX measurement.

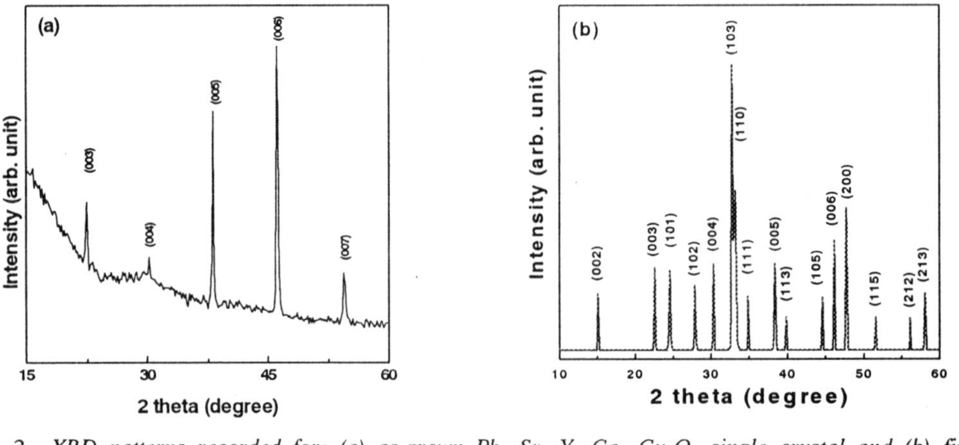

Fig. 2. XRD patterns recorded for: (a) as-grown $Pb_{0.5}Sr_{2.5}Y_{0.5}Ca_{0.5}Cu_2O_y$ single crystal and (b) fine powder of $Pb_{0.5}Sr_{2.5}Y_{0.5}Ca_{0.5}Cu_2O_y$ crystals.

In the XRD pattern recorded and reproduced in Fig. 2 (a), the presence of (00*l*) reflections shows the preferred c-axis orientation for the growth. An analysis of the XRD pattern obtained after crushing the crystals into fine powder and reproduced as plot-b in Fig. 2, shows that (013) reflection is stronger than (110) reflection, implying a tetragonal crystal structure.[3] The XRD pattern also reveals the mono phasic character of the grown crystal. The lattice parameters have been computed from the XRD data as a = 3.813(3) Å and c = 11.785(7) Å. The c lattice parameter is observed to be slightly smaller than that reported for the polycrystalline material.[4,5]

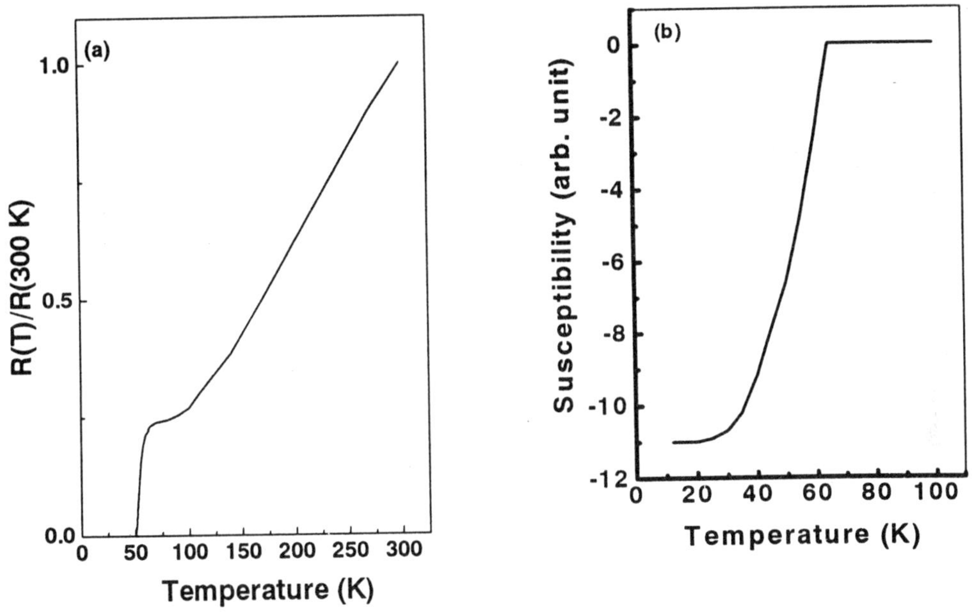

Fig. 3 *(a) Normalized resistance versus temperature plot for a $Pb_{0.5}Sr_{2.5}Y_{0.5}Ca_{0.5}Cu_2O_y$ single crystal grown using charges containing 10 wt.% $AgNO_3$. (b) The temperature dependence of AC susceptibility for the same crystal.*

The crystals grown from charges free of $AgNO_3$ did not undergo a transition from normal to superconducting state on cooling down to 15K. For crystals grown from charges containing $AgNO_3$, the results obtained are shown in Fig. 3(a). In the resistance versus temperature plot, the presence of two slopes just above the transition temperature is noted. The onset of superconducting transition (T_c^{on}) for the crystals was found to occur around 65K and the zero resistance state was reached at 50K. The onset of superconducting state at 65K was also revealed by the results of AC susceptibility measurements shown in Fig. 3(b). The presence of superconductivity in as-grown crystals shows that the addition of $AgNO_3$ in the starting charge for growth facilitates their oxygenation.

The results on crystal characterization presented here and those reported for 123 elsewhere[3] show that the addition of silver in the starting charge assists in the oxygenation of both these crystals during growth. Although, we have been unable to detect the presence of silver in the grown crystals, minute amount of this dopant in the crystal lattice could not be ruled out. The results of our measurements on the oxygen diffusion in undoped and Ag-doped crystals show that the diffusion coefficient is significantly reduced in the later case.[8] The implications of this result are that Ag-doped crystals could not pickup enough oxygen from the ambient during cooling from the growth temperature to room temperature to yield superconductivity in as-grown crystals. It appears that the presence of this dopant helps in stabilizing the superconducting phase at the growth temperature itself. This view is consistent with the result on silver doped thin films which when quenched from 750°C have been found to be superconducting.[9] Further investigations are needed to understand as to how the orthorhombic phase is stabilized by the presence of silver.

REFERENCES

1. W. Assmus and W. Schmidbauer, Supercond. Sci. Technol. **6**, 555 (1993).
2. Y. Haung, Y.F. Lin, M.J. Wang, M.K. Wu and K.J. Ma, J. Appl. Phys. **73,** 3580 (1993).
3. D.K. Aswal, S.K. Gupta, S.C. Gadkari, K.D.S. Mudher, L.C. Gupta, S.C. Sabharwal and M.K. Gupta, Supercond. Sci Technol. **8**, 710 (1995).
4. T. Rouillon, J. Provost, M. Hervieu, D. Groult, C. Michel and B. Raveau, Physica C **159**, 201 (1989).
5. T. Rouillon, J. Provost, M. Hervieu, D. Groult, C. Michel and B. Raveau, J. Solid State Chem **84**, 375 (1990).
6. H. Jin, N.L. Wang, Y. Chong, M. Ding, L.Z. Cao, Z.J. Chen, G.E. Zhou and Z.W. Mao, J. Mater. Res. **10,** 2211 (1995).
7. H. Jin, Z.J. Chen, L.Z. Cao, G.E. Zhou, and Z.W. Mao, J. Crystal Growth **148**, 106 (1995).
8. D.K. Aswal, S.K. Gupta, S.N. Narang, S.C. Sabharwal and V.C. Sahni, Solid State Physics Symposiam, D.A.E., India, 1996.
9. R. Pinto, D. Kumar, S.P. Pai, A.G. Chourey and P.R. Apte, Supercond. Sci. Technol. **7**, 95 (1994).

STRUCTURAL ASPECTS AND SUPERCONDUCTIVITY OF $CaLa_{1-x}Pr_xBaCu_3O_7$ SYSTEM

V.P.S. Awana[1], S.K. Malik[2], W.B. Yelon[3], Rajvir Singh[4], D.A. Landinez Tellez[1], J. Albino Aguiar[1] and A.V. Narlikar[4]

[1]*Depto de Fisica, Univ. Federal de Pernambuco, 50670-901. PE, Brasil*
[2]*Tata Institute of Fundamental Research, Colaba, Mumbai - 400 005, India*
[3]*University of Missouri Research Reactor Facility, Columbia, Mo 65211, U.S.A.*
[4]*National Physical Laboratory, K.S. Krishnan Marg, New Delhi - 110 012, India*

The result of superconducting, magnetic and neutron diffraction studies on tetragonal $CaLa_{1-x}Pr_xBaCu_3O_7$ system with x = 0.0, 0.3, 0.7 and 1.0 are presented. The compound $CaLaBaCu_3O_7$ is superconducting with transition temperature T_c of 80K. AC susceptibility measurements show that the T_c of the $CaLa_{1-x}Pr_xBaCu_3O_7$ samples decreases with increasing Pr concentration; the samples with x = 0.7 and 1.0 being not superconducting down to 2K. In the normal state, the (dc) susceptibility largely follows Curie-Weiss (CW) behaviour with effective paramagnetic moment of Pr intermediate between that of the free Pr^{3+} and Pr^{4+} ions. For the non superconducting samples with x = 0.70 and 1.0, a low temperature deviation from CW behaviour is observed at 4K and 8K, respectively, which may be due to antiferromagnetic ordering of the Pr moments in these compounds.

I. INTRODUCTION

Substitution of Pr at the rare earth (R) site in $RBa_2Cu_3O_{7-\delta}$ (R:123) type compounds has attracted considerable interest due to the peculiar and unique behaviour of Pr in these compounds (see, for instance, ref. 1,2 and references therein). The Pr substitution at the R-site has been widely reported in other oxide systems also, e.g. R:124[3], R:247[4] and $RSr_2Cu_{3-x}M_xO_7$ (M = Nb, Ti, Re, V, W, Mo etc.) [5,6] type compounds. Recently, detailed studies have been carried out on $PrBa_2Fe_3O_8$, where once again the behaviour of Pr is entirely different than that of the other $RBa_2Fe_3O_8$ compounds in terms of structural and magnetic properties [7,8].

There are several explanations for the peculiar behaviour of Pr-in R:123 and related compounds. These are based on either the tetravalent state of the Pr ions or the strong hybridization of Pr-4f electrons with the Cu(2)-O(2) conduction band[1,2].

In the present work, we have studied the effect of Pr substitution in a new R:123 type compound, namely, $CaLaBaCu_3O_7$. This compound crystallizes in the tetragonal structure, similar to that of the oxygen deficient R:123 compounds. However, in the present system the oxygen content remains close to 7.0 [9,10]. In

$RBa_2Cu_3O_7$ compounds, the oxygen content of ~7 results in an orthorhombic structure with nearly complete occupancy of the O(1) (0,0.5,0) site and very small or no occupancy of the O(5) (0.5,0,0) site. In the present system, both the O(1) and O(5) sites are equally occupied which results in the tetragonal structure with oxygen content of nearly 7. The Pr substitutional studies on this interesting system have been carried out and the results are reported here.

II. EXPERIMENTAL

The samples of $CaLa_{1-x}Pr_xBa_2Cu_3O_7$ series with $x = 0.0$, 0.30, 0.70 and 1.0 were synthesized through a solid state reaction route. In brief, the constituents Pr_6O_{11}, La_2O_3, $BaCO_3$, $CaCo_3$ and CuO were mixed in stoichiometric proportion and fired at 975°C to 995°C depending on the x values for several hour. Repeated powdering of the resulting product and sintering were carried out at least three times. In the final step the samples were annealed in oxygen at the same temperature for 12 hours, with an additional sintering in the same atmosphere at 575°C for 24 hours. Samples were characterized using powder X-ray diffraction technique. AC and dc magnetic susceptibility measurements were carried out using a SQUID magnetometer, in an applied r.m.s field of 0.10 gauss and 31 Hz for ac measurements and in a dc field of 5kOe for dc measurements. Room temperature neutron diffraction studies were carried out at the Research Reactor Center of University of Missouri, Columbia.

III. RESULTS AND DISCUSSION

Figure 1 shows the room temperature x-ray diffraction patterns of the samples $CaLa_{1-x}Pr_xBa_2Cu_3O_7$ ($x = 0.0$, 0.30, 0.70 and 1.0). All the samples show sharp x-ray lines with maximum impurity phases of <8% in x=1.0 sample. The lattice parameters of all these samples, obtained from the observed d values, are given in Table 1. A clear trend of the variation of c lattice parameter suggests substitution of Pr at La site in the host system. Neutron diffraction studies confirm the structure. Detailed analysis of the neutron diffraction work will be reported elsewhere.

Table 1: Lattice parameters (a and c), superconducting transition temperature (T_c), paramagnetic Curie temperature (θ_p), temperature independent susceptibility (χ_0) and effective magnetic moment/Pr ion in $CaLa_{1-x}Pr_xBaCu_3O_7$ system

x	a [Å]	c [Å]	T_c(dia) [K]	θ_p [K]	$\chi_0(10^{-4})$ [emu/mole]	μ_{eff} [μ_B]
0.0	3.869	11.645	69	–	–	–
0.30	3.866	11.632	40	−14.4	3.9	3.51
0.70	3.872	11.621	-	−17.8	9.6	3.07
1.00	3.875	11.603	-	−27.4	4.1	3.40

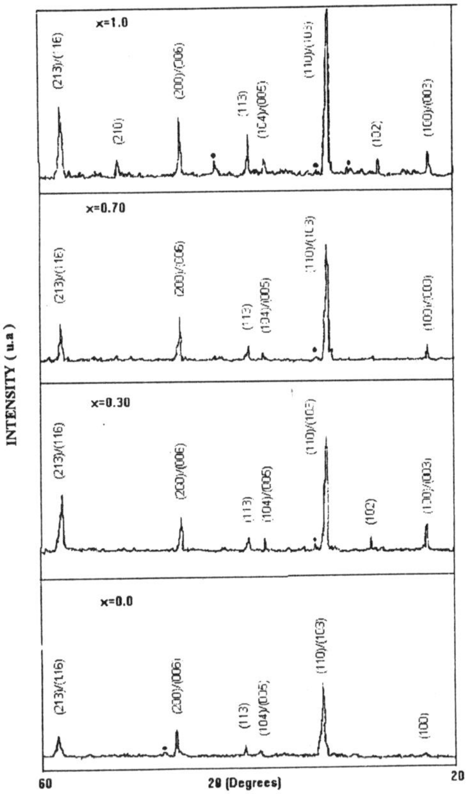

Fig. 1. X-ray diffraction patterns of $CaLa_{1-x}Pr_xBaCu_3O_7$ (x=0.0, 0.3, 0.7, 1.0)

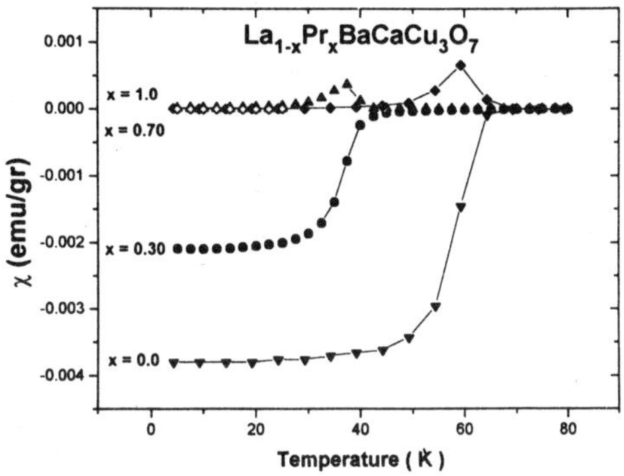

Fig. 2. AC susceptibility vs. temperature for $CaLa_{1-x}Pr_xBaCu_3O_7$ (x=0.0, 0.3, 0.7, 1.0)

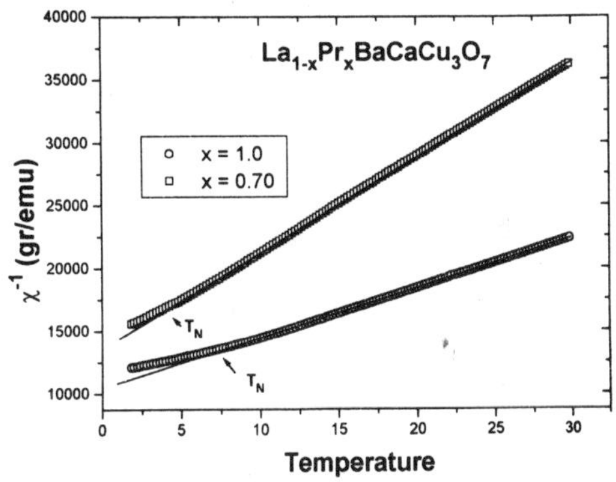

Fig. 3. χ^{-1} vs temperature for $CaLa_{1-x}Pr_xBaCu_3O_7$ (x=0.70 and 1.0), between 2–30K, marking deviation from Curie-Weiss behaviour.

Figure 2 shows ac susceptibility versus temperature for $CaLa_{1-x}Pr_xBa_2Cu_3O_7$ samples. The superconducting transition temperature, T_c, decreases with increasing Pr concentration; the samples with x=0.7 and 1.0 being not superconducting down to 2K. The Tc suppression due to Pr substitution appears to be weaker in $CaLa_{1-x}Pr_xBaCu_3O_7$ system compared to that in $R_{1-x}Pr_xBa_2Cu_3O_7$ with R=light rare earth, say Nd (it may be noted that, based on R-ionic size considerations [11], the suppression of T_c with R=La is expected to be even larger).

The normal state high field dc molar susceptibility (χ) versus temperature for $x = 0.30$, 0.70 and 1.0 samples can be fitted to the modified Curie-Weiss (CW) behaviour, $\chi = \chi_0 + C/(T - \theta_p)$, where χ_0 is the temperature independent susceptibility, C is the Curie-constant and θ_p is the paramagnetic Curie temperature.

The values of effective paramagnetic moment, μ_{eff} and θ_p are given in Table 1 along with the lattice parameters and the T_c values. Figure 3 shows χ^{-1} in the low temperature (2K to 30K) range for $La_{1-x}Pr_xBaCaCu_3O_7$ samples with x=0.7 and 1.0. The low temperature deviations from CW behaviour at 4K and 8K may be due to antiferromagnetic ordering of Pr moments in these compounds as is the case in Pr:123 also. The T_N values for these samples are much lower than that corresponding values in the $R_{1-x}Pr_xBa_2Cu_3O_7$ system [11]. The lower T_N of Pr in the present system suggests weaker hybridization of Pr with the conduction of Cu-O band than in Pr:123. This may be responsible for the weaker T_c suppression in the present system. However, antiferromagnetic ordering in the present compounds need to be confirmed by other measurements.

ACKNOWLEDGMENT

This work was financed by the Brazilian Science Agencies CNPq and FINEP. V.P.S. Awana thanks CNPq for a visiting post doctoral fellowhip. The work at MURR was supported by the U.S. Department of Energy grant No. DE-FE02-90ER45427 through the Midwest Superconductivity Consortium.

REFERENCES

1. H.B. Radousky, J. Mater. Res. 7, 1917 (1992).
2. S.K. Malik and C.V. Tomy, in *Physical and Material Properties of High T_c Superconductors*, edited by S.K. Malik and S.S. Shah, NOVA Science Publishers, New York, USA, p. 283, (1994).
3. S.K. Agarwal, R. Lal. V.P.S. Awana, S.P. Pandey and A.N. Narlikar, Phys. Rev. B 50, 10265 (1994).
4. J. Herrmann, U.C. Boehnke, M. Krötzch, B. Lippold and F. Schlenkrich, Physica C 221, 76 (1994).
5. G. Hilscher, E. Holland-Moritz, T. Holubar, H.D. Jostarndt, V. Nekvasil, G. Schaudy, U. Walter and G. Fillion, Phys. Rev. B 49, 535 (1994).
6. V.P.S. Awana, Latika Menon and S.K. Malik, Physica C 262, 266 (1996).
7. V.P.S. Awana, Latika Menon, Apurva Mehta, W.B. Yelon and S.K. Malik, Physica B 223 & 224, 558 (1996).
8. I. Felner, G. Hilscher, M. Michor, B. Wortzmann, J. Dumschat, V.P.S. Awana and S.K. Malik, Phys. Rev. B 54, 11999 (1996).
9. D.M. de Leeuw, C.A.H.A. Mutsaers, H.A.M. Van Hal, H. Verwij, A.H. Carim and H.C.A. Smoorenburg, Physica C 156, 126 (1988).
10. T. Yagi, M. Domon, Y. Okajima and K. Yamaya, Physica C 173, 453 (1991)
11. Yunhui Xu and Weiyan Guan, Phys. Rev. B 45, 3176 (1992).

STRUCTURE, SITE OCCUPANCIES AND SUPERCONDUCTIVITY IN $NdBa_{2-x}Pr_xCu_3O_{7+\delta}$ SYSTEM

D.K. Falzgraf[1], M. Park[1], M.J. Kramer[1], R.W. McCallum[1], S.K. Malik[2] and W.B. Yelon[3]

[1]*Ames Laboratory, Iowa State University, Ames, IA 5001, U.S.A.*
[2]*Tata Institute of Fundamental Research, Colaba, Mumbai - 400 005, India*
[3]*University of Missouri Research Reactor Facility, Columbia, Mo 65211, U.S.A.*

Structural and superconducting studies have been carried out on the system $NdBa_{2-x}Pr_xCu_3O_{7+\delta}$ ($0 \le x \le 0.6$). The system transforms from orthorhombic to tetragonal for $x \ge 0.1$ and the oxygen content progressively increases beyond 7 to charge balance the replacement of Ba^{2+} by Pr^{3+}. The superconducting transition temperature, T_c, decreases faster in this system compared to that in $RBa_{2-x}R_xCu_3O_7$ system with R = Nd, Sm, Eu and Gd. Neutron diffraction studies suggest that Pr has a preference for the Ba site in $NdBa_{2-x}Pr_xCu_3O_{7+\delta}$, but there is some occupancy of Pr at the Nd site also. It is known that Pr suppresses T_c when substituted at the R site in $RBa_2Cu_3O_{7-\delta}$ system. Thus in the $NdBa_{2-x}Pr_xCu_3O_{7+\delta}$ system, T_c suppression has two contributions: one from Pr at the Nd site and the other from Pr/Nd at the Ba site. This may be responsible for the faster decrease in T_c in the present system.

I. INTRODUCTION

The compound $PrBa_2Cu_3O_{7-\delta}$ (Pr123) is an antiferromagnetic semiconductor although most of the other isostructural R123 compounds exhibit superconductivity with a transition temperature, T_c, in the 90 K range. The absence of superconductivity in Pr123 has been explained on the basis of various models, such as hole filling, hole localization, hybridization of Pr-4f electrons with the conduction electrons, Pr occupancy at the Ba site and the pair breaking effect due to these Pr ions, etc., etc. [1-3].

The ionic radii of light rare earths (LR) approach that of Ba allowing for a large substitution of LR^{3+} at Ba^{2+} site. Thus, compounds of the type $RBa_{2-x}R'_xCu_3O_{7+\delta}$ can be prepared. The extent of such solid solution of rare earth at the Ba site depends on the sizes of the rare earths R and R′. Depending on the processing parameters, single phase compounds can be made up to $x \sim 1.0$ for La decreasing to 0 for ionic radii greater than Gd [4]. When R and R′ are Pr or Nd, x is typically 0.6–0.8. The preference of the R^{3+} ion to occupy the Ba^{2+} site increases with increasing ionic size of the rare earth ion. Thus if the above compounds are made with two different rare earths such as R=Nd and R′=Pr, it is expected that Pr will preferentially occupy the Ba site. In order to unravel the role of Pr in suppressing superconductivity, we have utilized the ability of LR^{3+} to substitute for Ba^{2+} and have investigated the system with nominal composition $NdBa_{2-x}Pr_xCu_3O_{7+\delta}$ to study the structure, Ba site occupancy, oxygen content and their effect on the T_c of this system.

II. EXPERIMENTAL

The samples with the nominal composition $Nd_{1.05}Ba_{1.95-x}Pr_xCu_3O_{7+\delta}$ (to be referred to as $NdBa_{2-x}Pr_xCu_3O_{7+\delta}$), with $0 \leq x \leq 0.6$, were prepared by the usual ceramic technique using thoroughly dried Nd_2O_3, Pr_6O_{11}, $BaCO_3$ and CuO powders. The samples were examined by x-rays and differential thermal analysis and found to be single phase materials. The superconducting transition temperatures were measured by low field dc susceptibility measurements in 10 Oe applied field using a SQUID magnetometer. Neutron diffraction studies were carried out at room temperature at the University of Missouri Research Reactor Center.

III. RESULTS AND DISCUSSION

Both x-ray and neutron diffraction studies show that the $NdBa_{2-x}Pr_xCu_3O_{7+\delta}$ system changes from orthorhombic to tetragonal as Pr concentration increases beyond $x = 0.1$ [Fig. 1]. This is analogous to the Y123 system with decreasing oxygen, but in this case the anti-chain oxygen site, O5 ($\frac{1}{2}$,0,0) is also filled in order to charge balance the R^{3+} substitution for Ba^{2+}. Rietveld refinements of the neutron diffraction data show that there is indeed a distinct preference for Pr to occupy the Ba site as evidenced by the Pr/Nd ratio at the Ba site in comparison to the same at the Nd site. However, superconductivity persists in this system up to $x \sim 0.30$. This casts some doubts on the hypothesis that a small amount of Pr at the Ba site in pristine Pr123 is responsible for suppression of superconductivity in this compound

Earlier studies on $NdBa_{2-x}Nd_xCu_3O_{7+\delta}$ have revealed that superconducting transition temperature, T_c, decreases with increasing Nd concentration at the Ba site and superconductivity is suppressed at about x=0.35. In fact, in the $RBa_{2-x}R_xCu_3O_{7+\delta}$ systems, superconducting T_c decreases systematically with increasing x and these systems show nearly identical depression in T_c over the single phase regions when R is Nd, Sm, Eu and Gd. This is essentially due to hole localization when Ba^{2+} is replaced by R^{3+} and is independent of the magnetic state of R^{3+} ion. Figure 2 shows a plot of T_c versus x for $NdBa_{2-x}R_xCu_3O_{7+\delta}$ systems with R=Pr and, a normal representative rare earth, Nd. It is noted that the T_c vs x curve for $NdBa_{2-x}Pr_xCu_3O_{7+\delta}$ with Pr at the Ba site, and hence the rate of suppression of T_c, is relatively steeper in this system compared to that in the $NdBa_{2-x}Nd_xCu_3O_{7+\delta}$ system. This may be due to the fact that part of the Pr occupies Nd site and reduces the T_c of the system (even when Ba site is fully occupied by nothing else but Ba). The starting T_c value is no longer 90K in the above system when Pr partly goes to the Nd and partly to the Ba site. For instance, Rietveld analysis of the neutron diffraction pattern shows that the site occupancies in nominally $NdBa_{1.7}Pr_{0.3}Cu_3O_{7+\delta}$ are such that its formula may be

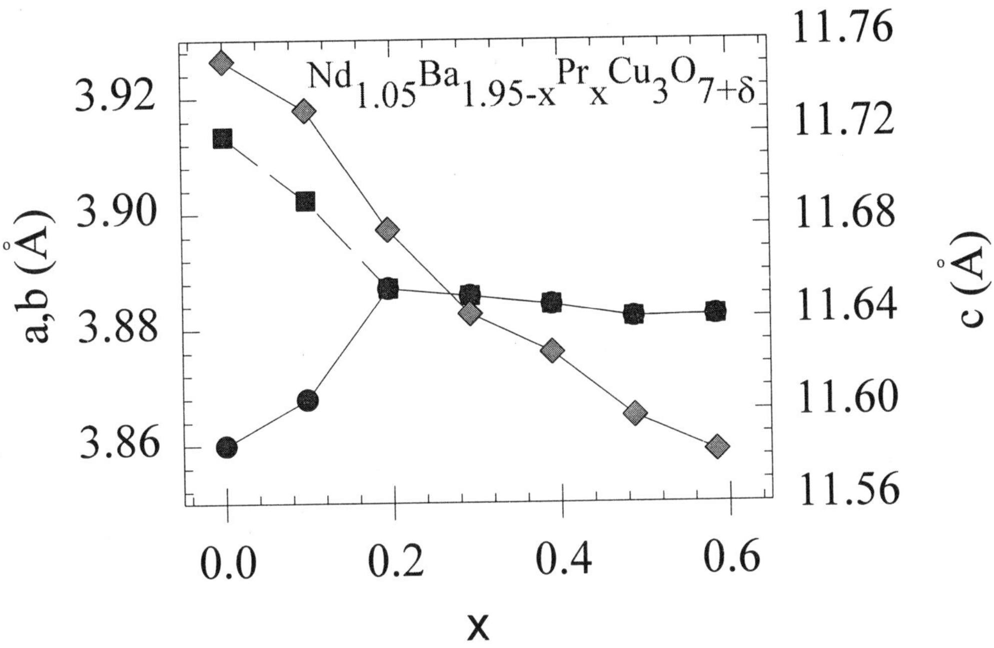

Figure 1. Lattice parameters for $NdBa_{2-x}Pr_xCu_3O_{7+\delta}$.

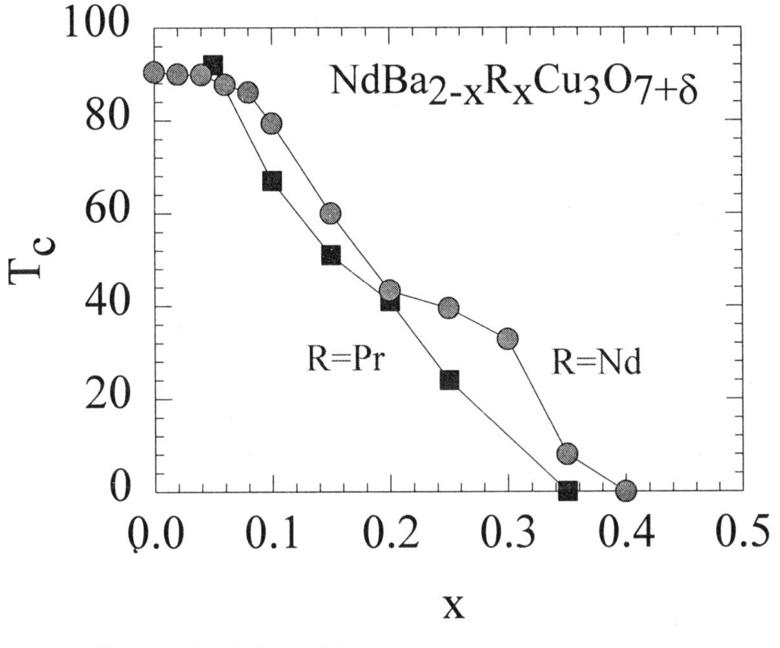

Figure 2. T_c vs x in $RBa_{2-x}R'_xCu_3O_{7+\delta}$ with R=Nd and R′=Pr and Nd. Samples with nominally Pr on the Ba site show a much faster depression in T_c.

represented as $[Nd_{0.84}Pr_{0.16}][Ba_{1.7}Pr_{0.14}Nd_{0.16}]Cu_3O_{7+\delta}$

Thus, the T_c suppression in $NdBa_{2-x}Pr_xCu_3O_{7+\delta}$ may be thought to have two contributions: (1) from Pr on the R site and (2) from Pr on the Ba site. As for other rare earths on Ba site (e.g., $RBa_{2-x}R_xCu_3O_{7+\delta}$), as mentioned above, the T_c suppression is due to hole localization and is independent of R [5,6]. If we assume the same for Pr on the Ba site, and also take into account the effect due to Pr on the R site, rough estimates suggests that the T_c suppression in $NdBa_{2-x}Pr_xCu_3O_{7+\delta}$ can be accounted for by combination of the above two effects. It would be interesting to study T_c suppression in other $RBa_{2-x}Pr_xCu_3O_{7+\delta}$ systems where the ionic sizes are such that relatively Pr has even lower or no preference to substitute at the R site so that the first contribution to T_c suppression is absent.

ACKNOWLEDGMENT

The work performed at Ames Laboratory was supported by the Director of Energy Research, U. S. Department of Energy under contract No. W-7405-ENG-82. The work at MURR was supported by the U.S. Department of Energy grant No. DE-FE02-90ER45427 through the Midwest Superconductivity Consortium.

REFERENCES

1. For a review see: H. B. Radousky, J. Mater. Res. 7, 1917 (1992); S.K. Malik and C.V. Tomy, in *Physical and Material Properties of High T_c Superconductors*, edited by S.K. Malik and S.S. Shah, NOVA Science Publishers, New York, USA, p. 283, (1994).

2. L. Soderholm, C.K. Loong, G.L. Goodman and B.D. Dabrowski, Phy. Rev. B 43, 7923 (1991).

3. H. D. Blackstead and J. D. Dow, Superlattices and Microstructures, 17, 473 (1995).

4. K. Osamura and W. Zhang, Z. Metailkd 84, 93 (1993); B. W. Veal and A. P. Paulikas, Physica C184, 321 (1991).

5. M. J. Kramer, S. I. Yoo, R. W. McCallum, W. B. Yelon, H. Xie, P. Allenspach, Physica C 219, 145–155 (1994).

6. M. J. Kramer, A. Karion, K. W. Dennis, M. Park, and R. W. McCallum, J. Elec. Mat. 24, 1931 (1995).

SEARCH FOR SUPERCONDUCTIVITY IN SOME CHALCOGENIDES

Padmakar Kichambare and Maheshwar Sharon
Department of Chemistry, Indian Institute of Technology, Powai, Bombay - 400 076, India.

The quaternary chalcogenides $SnBi_{4-x}Te_{3-y}Se_{4-z}$ and $PbBi_{4-x}Te_{3-y}Se_{4-z}$ have been prepared and found to possess orthorhombic structure. The chemical composition of materials is confirmed by EDX. $SnBi_{4-x}Te_{3-y}Se_{4-z}$ and $PbBi_{4-x}Te_{3-y}Se_{4-z}$ exhibit a low resistivity 0.012 ohm.cm and 0.133 ohm.cm at room temperature and carrier concentration 1.62 $\times 10^{16}$ and 2.34 $\times 10^{16}$ cm^{-3} respectively. The Hall measurements show n-type behaviour with carrier mobility 3036 and 1670 M^2/V, and band gaps by diffuse reflectance are found to be 4.01 eV and 3.99 eV for $SnBi_{4-x}Te_{3-y}Se_{4-z}$ and $PbBi_{4-x}Te_{3-y}Se_{4-z}$, respectively.

I. INTRODUCTION

A great deal of research in the physics and chemistry of binary and ternary chalcogenides during the past decade has led to a more complete understanding of these solid solutions. This progress has been made possible by significant advances in material preparation techniques. These chalcogenides are of much interest because of their rich structural chemistry[1] and unusual physical properties[2].

SbSI is a ferroelectric high resistance orthorhombic structured material[3]. It's structure has some similarity with Bi-based and Thallium based superconductor except that the number of lattice planes in one unit cell in the direction perpendicular to c-axis is not as high as those of high T_c materials. Can the structure of SbSI be modified by incorporating Bi, Pb, Se and Te such that the structure remains unaltered and the compounds exhibit unusal physical properties ? This question has been examined. Sb in SbSI is replaced with SnBi or PbBi and SI is replaced by SeTe to get quaternary chalcogenides of composition $SnBi_4Te_3Se_4$ and $PbBi_4Te_3Se_4$. This composition was fixed by assuming that each constituent of compound may be present in each possible valent state e.g. 3, 5 for bismuth and 2,4 for lead. However, single phase compound of such combination would be formed with only one of these valent states. Since we did not know which valent state would predominate, the compound during its heat treatments under vacuum could distill off unreacted (i.e. excess) either selenium or tellurium or both, giving perhaps a single phase stable compound. The present work, therefore, is carried out with the intention to examine whether we can induce unusal physical properties in compound of SbSI series by modifying its structure and constituents. In this paper, we describe the preparation, structure and electrical resistivity of $SnBi_{4-x}Te_{3-y}Se_{4-z}$ and $PbBi_{4-x}Te_{3-y}Se_{4-z}$.

II. EXPERIMENTAL

Synthesis of $SnBi_{4-x}Te_{3-y}Se_{4-z}$ and $PbBi_{4-x}Te_{3-y}Se_{4-z}$.

To synthesize the polycrystalline quaternary chalcogenide, $SnBi_{4-x}Te_{3-y}Se_{4-z}$ and $PbBi_{4-x}Te_{3-y}Se_{4-z}$, powders of the elements Sn (Alpha 99.8 %), Bi (E. Merck 99.99 %), Pb (BDH 99.98 %), Te (Fluka AG, 99.7 %) and Se (Nacalai Tesque 99.99 %) have been used. Sn, Bi, Se and Te in the elemental form were loaded in a quartz tube in the atomic ratio Sn/Pb: Bi: Te: Se: 1:4:3:4. The tube was evacuated to 10^{-3} Torr and then sealed. The tube was heated in a furnace at 900°C for 24 hours. The product synthesized gave black, shiny needles. The ingot of this treatment was regrounded and pressed into pellets for further studies. A semiquantitative elemental analysis by Energy Dispersive X-ray Analysis (EDX) revealed the compounds to have composition $SnBi_{3.94}Te_{2.87}Se_{3.96}$ and $PbBi_{3.92}Te_{2.89}Se_{3.91}$.

Physical Properties

The X-ray powder diffraction (XRD) patterns were recorded on Phillips PW 1820 X-ray diffractometer with CuK_α radiation. The diffuse reflectance spectra have been scanned on Shimadzu UV-260 double beam UV-Visible spectrophotometer. The nature of electric conduction was measured by a hot-probe technique. The electrical resistivity above and below the room temperature was measured by conventional two probe and four probe methods, respectively. Hall co-efficient at room temperature was measured by a Hall-effect assembly (Polytronic Electromagnet Type HEM-100). During the heat treatment, few small single crystals were also obtained. The result of single crystal XRD of these materials are being under analysis and will be published shortly.

III. RESULTS

X-ray Powder Data Analysis

The synthesis of quaternary compounds was carried out by maintaining stoichiometry of the elements corresponding to the chemical formula. X-ray diffraction patterns of the $SnBi_{4-x}Te_{3-y}Se_{4-z}$ and $PbBi_{4-x}Te_{3-y}Se_{4-z}$ prepared by heating mixture of elemental charge at 900°C for 24 hours were recorded. All the lines of diffraction pattern obtained with polycrystalline sample of these materials can be indexed in terms of orthorhombic symmetry as a single phase. The lattice parameters a, b, and c of $SnBi_{4-x}Te_{3-y}Se_{4-z}$ are 11.57, 27.54 and 7.62 Å; whereas for $PbBi_{4-x}Te_{3-y}Se_{4-z}$ are 4.59, 25.59, 7.37 Å. These lattice parameters were obtained by using Powder Diffraction Package PdP 1.1.[4].

Optical Properties

The diffuse reflectance spectra were scanned using Shimadzu UV-260 Spectrophotometer with integrating sphere between 200 to 900 nm to calculate the band gap of these chalcogenides. The reflectance spectra of $SnBi_{4-x}Te_{3-y}Se_{4-z}$ (prepared at 900°C for 24 hours) shows an intense absorption band around 309 nm. Around 260 nm, a small shoulder is observed. The shoulder around 260 nm in the spectrum of $SnBi_{4-x}Te_{3-y}Se_{4-z}$ can be assigned to the crystalline field transition. The position of this transition depends strongly on the crystalline field strength.

The diffuse reflectance spectrum of $PbBi_{4-x}Te_{3-y}Se_{4-z}$ shows a very strong absorption in the wavelength region from 300 to 320 nm as compared to the absorptivity of $PbBi_{4-x}Te_{3-y}Se_{4-z}$.

Hall Measurements

The type of conductivity and carrier concentration were calculated from the Hall co-efficient measured for the pressed pellets. The Hall co-efficient is negative. These chalcogenides are n-type in nature with carrier concentration 1.62 $\times 10^{16}$, and 2.34 $\times 10^{16}$ cm^{-3} and carrier mobility 3036 and 1670 M^2/V for $SnBi_{4-x}Te_{3-y}Se_{4-z}$ and $PbBi_{4-x}Te_{3-y}Se_{4-z}$ chalcogenides, respectively. It is observed from these studies that though carrier concentration values of these chalcogenides are low but Hall mobility values at room temperature are rather high.

Electrical Properties

The electrical resistivity of $SnBi_{4-x}Te_{3-y}Se_{4-z}$ and $PbBi_{4-x}Te_{3-y}Se_{4-z}$, prepared at 900 °C for 24 hours, was measured in the range 298-450K. It is also observed that the resistivity of $SnBi_{4-x}Te_{3-y}Se_{4-z}$ and $PbBi_{4-x}Te_{3-y}Se_{4-z}$ chalcogenides gradually falls with increasing temperature thereby exhibiting a negative temperature co-efficient of resistance, which is a characteristic of semiconductor, with the activation energy for conduction 1.49 and 0.89 eV, respectively. The fact that these chalcogenides have high band gaps (4 eV) but smaller activation energy indicates that these materials are extrinsic in nature at temperature around room temperature to 380 K. Their Fermi level is very close to their conduction band. It is observed that these chalcogenides have a higher carrier concentration and lower resistivity. EDX results of samples (in pellet form) were measured at different positions and showed a small variation in tellurium and selenium composition, suggesting nonuniformity in the composition.

The normalized resistivity of these chalcogenides as a function of temperature (17 K to 298 K) has been studied. The resistivities at 298 K for $SnBi_{4-x}Te_{3-y}Se_{4-z}$ and $PbBi_{4-x}Te_{3-y}$ Se_{4-z} are 0.012 and 0.133 ohm-cm, respectively. The resistivity increases monotonically with decreasing temperature and approaches a residual value of σ_0. The resistivity ratio, $\sigma(300\ K)/\sigma_0$, varies from sample to sample.

IV. CONCLUSION

The structural, electrical and optical studies of quaternary $SnBi_{4-x}Te_{3-y}Se_{4-z}$ and $PbBi_{4-x}Te_{3-y}Se_{4-z}$ have been investigated. None of these materials could exhibit any trace of superconductivity. The Hall mobilities for $SnBi_{4-x}Te_{3-y}Se_{4-z}$ and $PbBi_{4-x}Te_{3-y}$ Se_{4-z} are 3036 and 1670 M^2/V, respectively. The donor level concentrations are in the range of 10^{16} cm^{-3}, which make these materals interesting to study for applications.

ACKNOWLEDGMENTS

One of the authors (PDK) wishes to acknowledge the CSIR, New Delhi for the award of Senior Research Associateship.

REFERENCES

1. A. Mar and J.A. Ibers, J. Chem. Soc. Dalton Trans. 639 (1991).
2. S. Nagata, S. Ebisu, T.Aochi, Y. Kinoshita, S. Chikazawa and K. Yamaya, J. Phys. Chem. Solids **52**, 761 (1991).
3. Landolt-Bornstein, III/16b, Spinger-Verlag, Berlin (1982).
4. M. Calligaris and S. Germia, Powder Diffraction Package PDP11, Universita 'di Trieste, Italy.

NEUTRON DIFFRACTION AND MAGNETIC STUDIES ON $PrBa_2Cu_3O_{7-\delta}$ WITH REDUCED ORTHORHOMBICITY

S.K. Malik[1], S. Srinivas[1] and W.B. Yelon[2]

[1]*Tata Institute of Fundamental Research, Mumbai - 400 005, India*

[2]*University of Missouri Research Reactor, Columbia, MO 65211, USA*

By varying the heat treatment, we have been able to make nearly single phase $PrBa_2Cu_3O_{7-\delta}$ (Pr123) which shows an oxygen content of ~7 but considerably less orthorhombicity. Rietveld analysis of the neutron diffraction pattern shows that the oxygen is redistributed with occupancy of 0.76 at the O (1) [0, ½, 0] site and 0.26 at the normally vacant O(5) [½, 0, 0] site. The structure-magnetic property relationship has been studied and it is observed that oxygen redistribution affects the antiferromagnetic ordering temperature of the Pr moments in this compound.

I. INTRODUCTION

It is well known that $PrBa_2Cu_3O_{7-\delta}$ (Pr123), which is isostructural to other 90 K superconducting $RBa_2Cu_3O_{7-\delta}$ (R123) compounds, is a semiconductor [1] in which Pr moments are antiferromagnetically (AFM) ordered [2] with a Néel temperature, T_N, of ~17K. The absence of superconductivity in Pr123 is of great interest and its origin is still not understood. Various models have been proposed to explain this. These include hole filling, hole localization, hybridization between the Pr-4f electrons and the conduction electrons, etc. [1–4]. In the R123 compounds, the O(1) $[0, \frac{1}{2}, 0]$ site is nearly fully occupied while the O(5) $[\frac{1}{2}, 0, 0]$ site is nearly empty. In the oxygen deficient R123 compounds, the O(1) and O(5) sites become equally occupied and the system transforms to a tetragonal structure with reduced oxygen content. We have been examining the possibility of making Pr123 samples with oxygen content ~7 but with reduced orthorhombicity or even tetragonal structure. This is expected to lead to redistribution of oxygen at the available sites, in particular, the O(5) site. In this paper, we present the results of neutron diffraction and magnetic studies on Pr123 with reduced orthorhombicity to see whether oxygen redistribution has any effect on the magnetic ordering temperature of the Pr moments.

II. EXPERIMENTAL

The sample of $PrBa_2Cu_3O_{7-\delta}$ was prepared by the standard solid state route using 4N pure Pr_6O_{11}, $BaCO_3$, CuO and La_2O_3 powders. Calcination was done at 900°C for 72 hrs with intermediate grinding. The resulting product was pulverized and reheated. This process was repeated at least three times and final oxygenation was carried out for 24 hrs at 930°C and for 48 hrs at 550°C. The sample was examined using both X-ray and neutron diffraction methods. Neutron diffraction (ND) patterns were obtained at room temperature on the powder diffractometer at the University of Missouri Research Reactor. Magnetic susceptibility measurements were carried out in the temperature range of 2-300K using a SQUID Magnetometer.

III. RESULTS AND DISCUSSION

(a) Neutron diffraction: The data were analyzed by the Rietveld refinement procedure using the generalized structural analysis system (GSAS) program. The ND pattern could be fitted on the basis of orthorhombic structure (space group *Pmmm*). Figure 1 shows the observed and fitted neutron diffraction pattern. The structural parameters are given in Table 1. The lattice constants are a = 3.87563(31)Å, b = 3.92523(32)Å and c = 11.71646(90)Å. The oxygen content is found to be close to 7 with occupancy of 0.758(24) at the O(1) site and 0.264(21) at the O(5) site. Thus it appears that the O(1) and O(5) site occupancies can be changed by heat treatment while retaining a total oxygen content which is close to 7 unlike that in oxygen deficient R123 compounds. The orthorhombicity $[(b-a)/a]$ of the present compound is (0.012) and is about 2/3 of that seen in fully oxygenated Pr123 with very little or no O(5) occupancy [4].

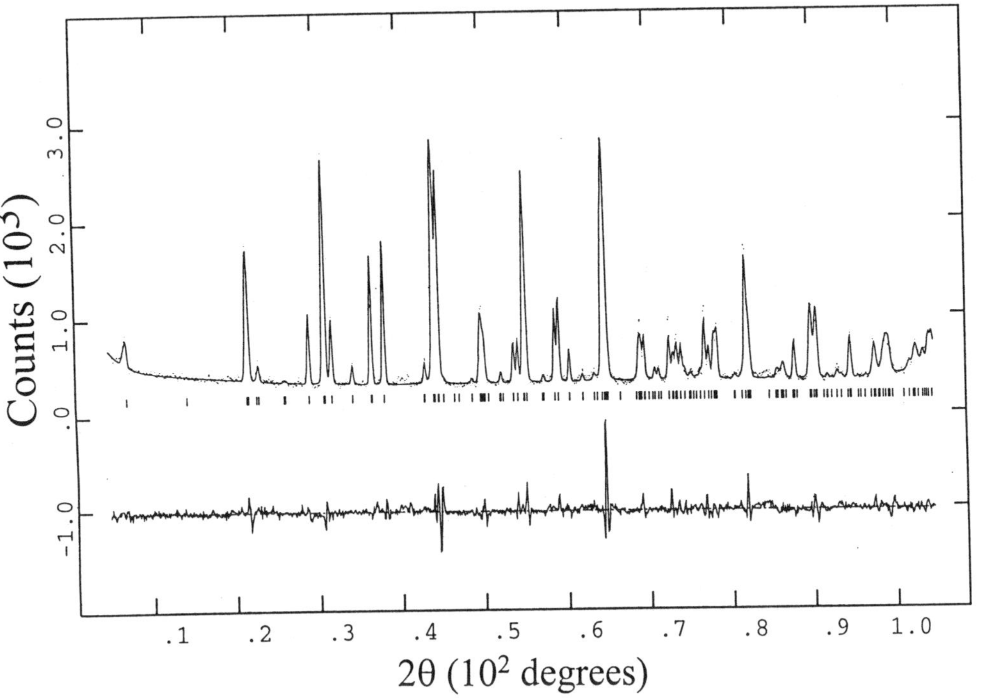

Fig. 1. Observed and fitted neutron diffraction pattern of $PrBa_2Cu_3O_{7-\delta}$

Table 1: Structural parameters of $PrBa_2Cu_3O_7$ used in the present investigation (space group *Pmmm*). The table gives atomic positions, (x, y, z), occupancy and isotropic thermal parameter (U_{iso}).

Atom	Space Group position	x	y	z	Occupancy	$100 * U_{iso}$ (Å^2)
Pr	1h	0.5	0.5	0.5	1	0.06(24)
Ba	2t	0.5	0.5	0.1817(5)	1	1.21 (13)
Cu(1)	1a	0	0	0	1	1.21 (31)
Cu(2)	2q	0	0	0.34894(30)	1	0.78(10)
O(1)	1e	0	0.5	0	0.758(24)	2.48(39)
O(2)	2s	0	0	0.37182(26)	1	0.94(10)
O(3)	2r	0	0	0.37182(26)	1	0.94(10)
O(4)	2q	0	0	0.15752(46)	1	1.67(65)
O(5)	1b	0.5	0	0	0.264(21)	2.48 (39)

$wRp = 0.08450$, Rp=0.0645 and $\chi^2 = 3.687$

(b) Magnetic measurements: Figure 2 shows a plot of dM/dT vs temperature, where M is the magnetization of the sample. A break in this plot occurs at T_N. The antiferromagnetic ordering of the Pr moments in the present compound is found to occur at about 16.2K. This is lower than the T_N value of about 17K for the same in normally prepared Pr123 with very little or no occupancy of O(5) sites. The T_N of the Pr moments is also observed to decrease in tetragonal Pr123 with decreased oxygent content.

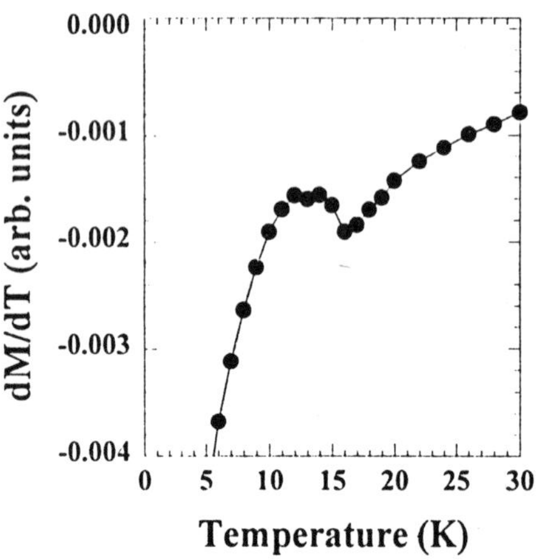

Fig. 2. dM/dT vs temperature for $PrBa_2Cu_3O_{7-\delta}$

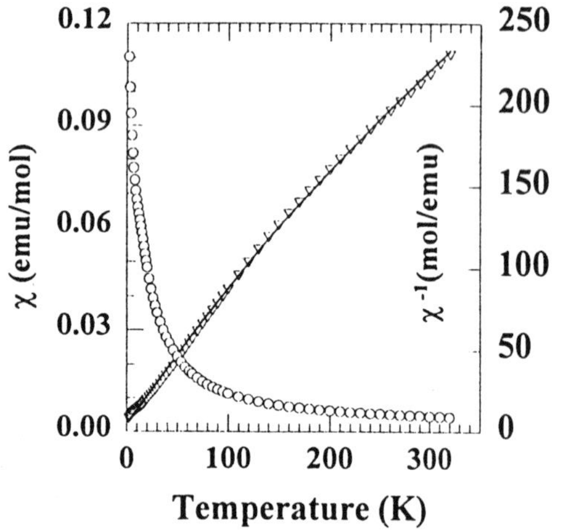

Fig. 3. χ and χ^{-1} vs temperature for $PrBa_2Cu_3O_{7-\delta}$.

Figure 3 shows the plot of magnetic susceptibility (χ) and its inverse vs temperature over the full temperature range 2–300 K for the present Pr123. The susceptibility follows Curie-Weiss behavior with effective paramagnetic moment μ_{eff}=2.975μ_B and a paramagnetic curie temperature $\theta_p = -4.4$K. The μ_{eff} value is intermediate between the free ion values for Pr^{3+} (2.56μ_B) and Pr^{4+} (3.56μ_B). The intermediate μ_{eff}, high magnetic ordering temperature of the Pr moments and large electronic specific heat coefficient of Pr123 may arrise due to strong hybridization of the Pr-4f states with the Cu-O states that also suppress the superconductivity in this system [1-4]. Low value of T_N obtained in the present Pr123 with redistributed oxygen is also consistent with the above model since such a redistribution will influence the hybridization of the Pr-4f states with the electronic states in the planar Cu-O network.

ACKNOWLEDGMENT

The work at MURR was supported by the U.S. Department of Energy grant No. DE-FE02-90ER45427 through the Midwest Superconductivity Consortium.

REFERENCES

1. For a review see: H. B. Radousky, J. Mater. Res. 7, 1917 (1992); S.K. Malik and C.V. Tomy, in *Physical and Material Properties of High T_c Superconductors*, edited by S.K. Malik and S.S. Shah, NOVA Science Publishers, New York, USA, p. 283, (1994).

2. A. Matsuda, K. Kinoshita, T. Ishii, H. Shibata, T. Watanabe and T. Yamada, Phy. Rev. B 38, 2910 (1988).

3. S. K. Malik, W.B. Yelon, J.J. Rhyne, W.J. James, Ram Prasad, K. Adhikary and N.C. Soni, Solid State Commun. 89, 383 (1994).

4. S.K. Malik, H. Jhans, W.B. Yelon and J.J. Rhyne, Solid State Commun. 85, 849 (1993).

UNIVERSAL RARE EARTH - SIZE DEPENDENCE OF T_c IN RE- METAL STABILIZED Sr-BASED 1-2-3 CUPRATES

M. Murugesan, T. Ishigaki and S. Ebisu
Department of Materials Science and Engineering,
Muroran Institute of Technology, Muroran - 050, Japan

Single phase $Tm_{1-x}RE_xSr_2Cu_{2.85}Re_{0.15}O_z$ hybrid compounds, where RE = Pr, Nd, Sm, Eu, Gd, Tb & Dy, have been characterized for their structural and superconducting properties. All the compounds crystallize in tetragonal (P4/mmm) structure. However, a small degree of orthorhombic distortion is found in all the superconducting samples. The T_c of hybrid compound [where the weighted average rare earth ionic radius (i.r.) is close to that of Ho (i.r. ≈ 1.015 Å, T_c = 41K) and Y (i.r. ≈ 1.019 Å, T_c = 32K)] is relatively higher as compared to the T_c value of individual members. The T_c value of 41K for Ho in the hybrid compounds is nearly equal to the T_c value of (Y/Ho/Er) $Sr_2Cu_{2.85}Re_{0.15}O_z$, which confirms the strong RE ionic size dependence of T_c. Magnetic moment of the RE ion is found to have no effect on structure and T_c.

I. INTRODUCTION

It is widely known that T_c is highly insensitive to the rare earth (RE) ionic size and magnetic moment in the orthorhombically-distorted layered perovskite $REBa_2Cu_3O_7$[1] (90-92K, RE = lanthanides except Ce, Tb and Pr). On the other hand, in the case of $REBa_2Cu_4O_8$ (RE=Sm, Eu, Gd, Dy, Ho, Er and Tm; Y)[2], T_c varies monotonically to a lesser extent with RE ionic size (69-80 K). However, it has recently been shown that Sr- based metal stabilized superconductors, $RESr_2Cu_{2.85}Re_{0.15}O_z$, T_c varies drastically but not monotonically with respect to RE ionic size[3]. T_c changes very interestingly that it is maximum for a optimum RE ionic size and is minimum on either side of this optimum size. Although such a T_c variation is interesting, to prove this we have mixed two REs (whose ionic radii (i.r.) lie far from the optimum ionic size such that the compounds neither form nor superconduct) to obtain the optimum ionic radius of Ho, Y and Er so that the T_c is maximum.

II. EXPERIMENTAL

Polycrystalline compounds with nominal composition $Tm_{1-x}RE_xSr_2Cu_{2.85}Re_{0.15}O_z$ were prepared from RE_2O_3 (RE=Y, Tm, Er, Ho, Dy, Gd, Eu, Sm and Nd), Pr_6O_{11}, Tb_4O_7, $SrCO_3$, CuO and ReO_3 as per the procedure described elsewhere[3]. Structural characterization was performed by X-ray diffraction (XRD). Superconducting properties were studied by both four probe resistivity, ac- and dc- magnetic susceptibility measurements.

III. RESULTS AND DISCUSSION

Room temperature powder XRD pattern of hybrid compounds reveals that these compounds form a near single phase with distorted tetragonal (p4/mmm) structure. A small amount of secondary phase belonging to the Sr_2CuO_3 & $Sr_{1.75}Cu_3O_{5.13}$ ternary oxide phases is also observed. Lattice parameters (for $Tm_{1-x}Pr_xSr_2Cu_{2.85}Re_{0.15}O_z$, a = 3.809 (1) Å ; c = 11.494(2) Å) were obtained by least square fit method by considering 8-10 higher angle Bragg reflections.

Published results[4,5] on the structural analysis of metal stabilized Sr-based 1-2-3 superconductors show that these compounds belong to tetragonal space group. However, we consistently observe the existence of small orthorhombic distortion in the XRD pattern (inferred from the splitting of (200) and (213) Bragg reflections around 2θ = 47-48 deg & 58-59 deg, respectively) in all the superconducting hybrid compounds. Similar kind of phenomenon has previously been observed by others in metal stabilized Sr-123 samples[3,6]. It is well known that orthorhombicity (e) is also one of the main criteria for realizing superconductivity in 1-2-3 cuprates. The importance of e and the relation between e and T_c were discussed in detail in our earlier report[3].

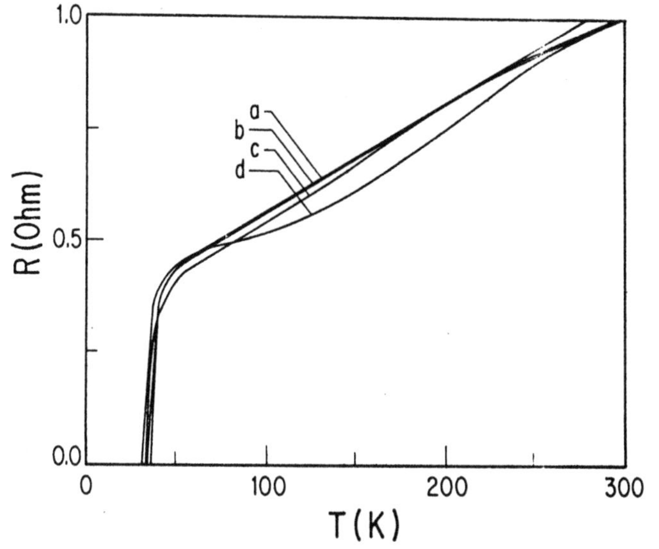

Fig.1. Normalized R-T plot of $Tm_{1-x}RE_xSr_2Cu_{2.85}Re_{0.15}O_z$ (a) RE = Dy, (b) RE = Tb, (c) RE = Gd and (d) RE = Eu.

T_c values were determined from resistivity measurements using conventional four probe technique. Shown in Fig. 1 is the normalized R-T plot of various hybrid compounds with Tm. The results show that samples in which average RE i.r. is close to that of Ho, superconduct around 36-41K. These samples showed a good metallic behavior between room temperature and T_c. The ΔT in these samples is approximately 5 to 6 K. The T_c value obtained inductively from ac-χ measurements is in close agreement with that of resistive measurements. The bulk superconductivity of the sample is inferred from the saturation of the χ' signal at temperatures well below T_c. It is clear from the earlier report[3] that there exists an optimum RE ionic size regime which gives highest T_c irrespective of the RE ion, viz. Ho, Er and Y, while for other REs whose i.r. fall either side of this plateau region the compound gives either lower T_c or only $T_{c\ onset}$ or the phase does not exist. [Tm by itself gives T_c only 23K; Pr, Nd or Sm do not form 123 phase; Eu by itself is semiconducting; Gd by itself shows only $T_{c\ onset}$; Tb superconducts at 22K; and Dy superconducts at 27K]. On the basis of the dependence of T_c or rare earth i.r., one would expect qualitatively higher T_c compounds than that of individual $RESr_2Cu_{2.85}Re_{0.15}O_z$. It is instructive to note that T_c of the above mixed composition is considerably high. It is 41-36K for Ho, 30-32K for Y and ≈ 22K T_c onset for Er. [see Table 1].

Table 1

	Dy	Tb	Gd	Eu	Sm	Nd	Pr
Ho ionic radius (1.015 Å)	x = 0.63 37 K	x = 0.45 35 K	x = 0.35 36 K	x = 0.29 36 K	x = 0.25 41 K	x = 0.18 41 K	x = 0.16 40 K
Y ionic radius (1.019 Å)	x = 0.76 32 K	x = 0.54 32 K	x = 0.42 30 K	x =0.34 32 K	----	----	----
Er ionic radius (1.004 Å)	x = 0.31 12 K	x = 0.23 22 K(on)	x = 0.18 18 K(on)	x =0.15 20 K(on)	-----	-----	----

We have observed higher T_c of 36 K in $Tm_{0.84}Pr_{0.16}Sr_2Cu_{2.85}Re_{0.15}O_z$ (see Fig. 2) as compared to the individual members ($TmSr_2Cu_{2.85}Re_{0.15}O_z$, T_c = 28K and $PrSr_2Cu_{2.85}Re_{0.15}O_z$ phase does not exist). This finding very well confirms the RE ionic size dependence of T_c in Sr-based 1-2-3 superconducting materials. In addition, it showed the bulk superconductivity with the superconducting volume fraction of 40%. It is interesting to note that the partial replacement of Tm by Pr does not show any detrimental effect on T_c. Indeed, it gave higher T_c value as one would expect. However, it is well studied that the partial substitution of Pr in the place of RE in Ba based 1-2-3 cuprates invariably reduces T_c. From the above results one may also conclude that Pr ion plays entirely different role in Sr-based 1-2-3 superconductors as compared to the Ba-analog.

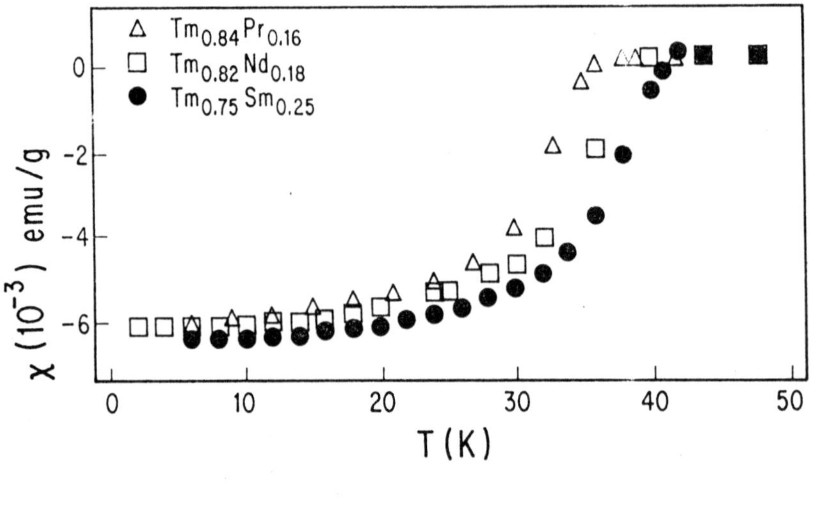

Fig.2. Plot of dc-χ vs T for $Tm_{1-x}RE_xSr_2Cu_{2.85}Re_{0.15}O_z$

Although all trivalent lanthanide ions have varying magnetic moments, the T_c of the hybrid compounds with RE = Pr, Nd, Eu, Gd, Tb, Dy is same ; 36K and 32K for i.r. equivalent to Ho and Y, respectively. Thus, the magnetic moment of the RE ion does not play any role in the above hybrid compounds and the observed variation is solely due to the variation of RE ionic size. Although the compounds, whose weighted average i.r. is close to that of Ho and Y, resulted in higher T_c as expected, we have observed semiconducting behavior at higher temperatures (just above the superconducting transition point). In the case of Er, there is a drop in resistance ($T_{c\ onset}$) around 25K but zero resistance is not observed down to 10K for all compounds (except for

$Tm_{0.69}Dy_{0.31}Sr_2Cu_{2.85}Re_{0.15}O_z$, whose T_c is around 12 K). Though the XRD pattern shows the existence of single phase, the lower T_c values may be attributed to the fact that Er i.r. lies just at the verge of plateau region (optimum i.r.).

IV. CONCLUSION

In summary, we have stabilized $Tm_{1-x}RE_xSr_2Cu_{2.85}Re_{0.15}O_z$ compounds with RE= Pr, Nd, Sm, Er, Ho, Dy, Tb, Gd, Eu in 1-2-3 structure. XRD results show the existence of small orthorhombic distortion in all superconducting phases. The T_c value is maximum for the hybrid compounds with average RE i.r. nearly equal to that of Ho or Y, and this clearly indicates that there exists a strong RE ionic size dependence on T_c of Sr-based 1-2-3 superconductors. The magnetic moment of the RE ions does not play any role. It is clear from the present and earlier reports[3] that one may observe similar nature of strong RE ionic size dependence of T_c in all other metal stabilized Sr-based 1-2-3 superconductors.

REFERENCES

1. P.H. Hor, L. Gao, R.L. Meng, Z.J. Huang, Y.Q. Wang, K. Foster, J. Vissitium, C.W. Chu, M.K. Wu J.R. Ashburn and C.J. Torng, Phys. Rev. Lett. **58**, 911 (1987).
2. D.E. Morris, J.H. Nickel, J.Y.T. Wei, N.G. Asmar, J.S. Scott, U.M. Scheven, C.T. Hultgren, A.G. Markelz, J.E. Post, P.J. Heaney, D.R. Veblen and R.M. Hazen, Phys. Rev. B **39**, 7347 (1989).
3. M. Murugesan, M.S.R. Rao, L.C. Gupta, R. Pinto, M. Sharon and R. Vijayaraghavan, Phys. Rev. B **53**, 8604 (1996).
4. A. Sequiera, H. Rajagopal, M. Murugesan, M.S.R. Rao, L.C. Gupta, R. Pinto, M. Sharon and R. Vijayaraghavan, Physica B **223**, 574 (1996).
5. R.L. Harlow, G.H. Kwei, R. Suryanarayanan, M.A. Subramanian, Physica C **257**, 125 (1996).
6. Y. Zhao, F. Shi, H.K. Liu, C. Andrikidis and S.X. Dou, Physica C **212**, 451 (1993).

MÖSSBAUER STUDIES ON $YSr_2Cu_{3-x}Fe_xO_z$ (X = 0.1, 0.2, 0.3 AND 0.4)

M. Murugesan, T. Yamada, T. Ishigaki and H. Kuwano
Department of Materials Science and Engineering,
Muroran Institute of Technology, Muroran - 050, Japan

Structural properties of perovskite type $YSr_2Cu_{3-x}Fe_xO_z$ materials have been characterized using X-ray diffraction and Mössbauer effect spectroscopy. Single phase 1-2-3 structure is obtained for x = 0.30. Room temperature Mössbauer spectra consist of three doublets for all the Fe concentrations examined. The possible relative site occupancies by Fe atom in the 1-2-3 lattice are discussed.

I. INTRODUCTION

Sr-based 1-2-3 compounds, $YSr_2Cu_{3-x}M_xO_z$, have earlier been stabilized by the partial replacement of Cu with various d-transition elements[1]. The Fe stabilized compounds are of particular interest in view of the possible occupation of both the Cu sites in the 123 unit cell. Whereas in compounds stabilized by other than Fe, the substituted metal atoms preferentially occupy the Cu (1) site. In order to find out the Fe occupancy, we have used ^{57}Fe Mössbauer spectroscopy (MES) as a probing tool and the results are discussed in what follows.

II. EXPERIMENTAL

Samples were prepared by solid state reaction method following the procedure described in elsewhere[2]. Phase purity was checked by X-ray diffraction (XRD) analysis. Room temperature (RT) MES were obtained using a ^{57}Co source of Rh matrix.

III. RESULTS AND DISCUSSION

The lattice structure and unit cell parameters were determined from the least square fitting analysis using 8 - 10 higher angle Bragg reflections of XRD [see Fig. 1]. Results show that $YSr_2Cu_{3-x}Fe_xO_z$ compounds crystallize in tetragonal structure (a=3.812(7) Å and c=11.419(4) Å for $YSr_2Cu_{2.7}Fe_{0.3}O_z$) for all the Fe concentrations studied. For lower concentration of Fe, there exists concomitant fomation of impurity phases namely, $Sr_{1.75}Cu_3O_{5.13}$ and Sr_2CuO_3. Also, the d-value of impurity peaks is matching for $SrFeO_{3-z}$ ternary oxide phase. Hence, we have separately prepared $SrFeO_{3-z}$ compound and studied for its MES. Having found that the hyperfine parameters (isomer shift relative to α-Fe (δ) and quadruple splitting (Δ)) of $SrFeO_{3-z}$ are largely different from δ and Δ of $YSr_2Cu_{3-x}Fe_xO_z$, we confirm that the impurity peaks are mainly due to $Sr_{1.75}Cu_3O_{5.13}$ and Sr_2CuO_3 ternary oxide phases.

MES spectra of $YSr_2Cu_{3-x}Fe_xO_z$ for various values of x are shown in Fig. 2. These spectra were analyzed by a least-square fitting routine using a Lorentzian molecular field approach allowing independent peak positions & amplitude. The results are tabulated in Table 1. A MES spectrum for a given Fe concentration, x, is the result of summation of all subspectra which correspond to the Fe ions

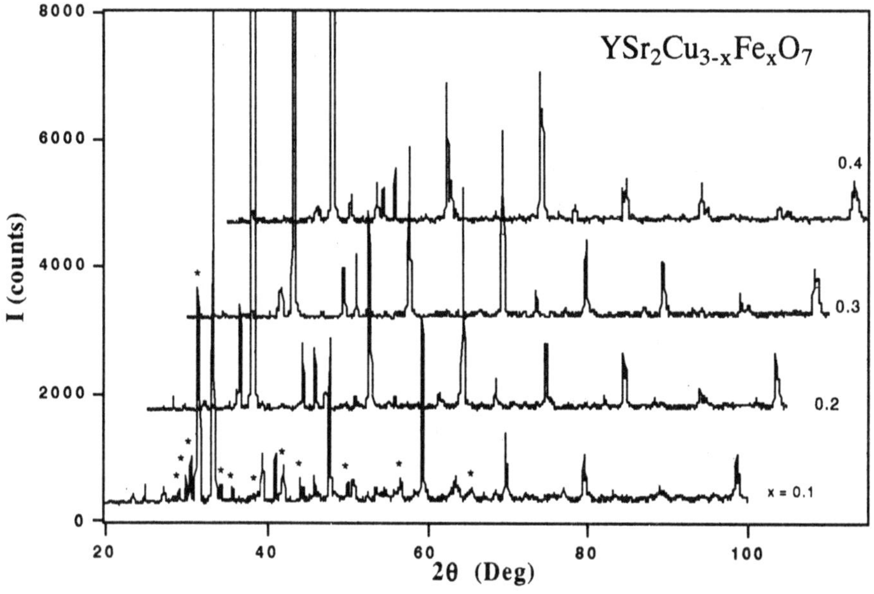

Fig. 1. XRD pattern of $YSr_2Cu_{3-x}Fe_xO_7$ (Impurity peaks are marked by asterisk)

present in the different local geometry and oxygen coordination. Three resolvable subspectra are found for all the concentrations of Fe. Assuming that Fe is substituting for copper nonpreferentially in the 123 lattice, one can correlate these subspectral area to the relative site occupancies. As can be seen in fig. 2, as the Fe concentration increases the intensity of the Fe(III) doublet increases while the Fe(I) doublet decreases progressively. On the other hand, the intensity of Fe(II) doublet is found to be x independent. Since the Fe(II) doublet has nearly 2/3 of the overall area, we assign this doublet to square-pyramidal co-ordination of the Cu(2) sites (as reported earlier[3]), which correspond to the two of the three Cu-sites [one site corresponds to the Cu(1) site and the other two sites belong to the Cu(2) site] available in the 123 unit cell. The combined area of Fe(I) and Fe(III) is little over the 1/3 of the total spectral area and thus well corroborate to the relative number of Cu atoms in the square planar (Cu(1)) site. It is evident from the above observation that Fe occupies both the Cu sites irrespective of the Fe concentration.

The δ value of the Fe(I) doublet increases with the increase in the Fe concentration which is ascribed to the decrease in the 4s e^- density at the Fe nucleus. While there is a marginal decrease in δ value of the Fe(III) doublet. The magnitude of δ and Δ for Fe(III) are typical of octahedrally coordinated

Fe^{3+}. The result that δ of the Fe(II) doublet is very close to zero (0.01 mm/sec) and independent of Fe concentration reveals that Fe could exist as either Fe^{3+} (ferric ion) or metallic iron. However, the extremely oxidizing synthesis condition of these materials makes the existence of metallic iron very unlikely. The point charge model predicts that a large Δ would appear at the square planar site with the magnitude approximately twice that of the square pyramidal site. Hence, the large Δ value of Fe(I) doublet must be due to the ferric ion in the square planar geometry. One can also infer that the fraction of

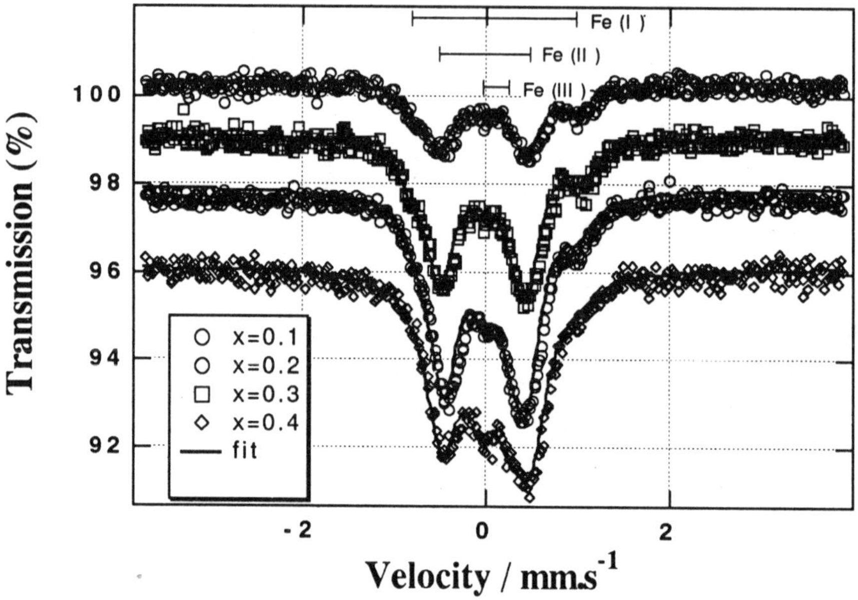

FIG. 2. Room temperature Mossbauer spectra of $YSr_2Cu_{3-x}Fe_xO_z$ (x = 0.1, 0.2, 0.3 & 0.4)

Fe^{3+} ion in the square planar geometry is small at the large x regime, because the relative spectral area of Fe(I) doublet is small for the higher value of x. Thus, the linear increase in the relative area of doublet (III) at the expense of Fe(I) with the Fe concentration may be an indicative of the preference of more number of ferric ions in octahedral co-ordination which is well supported by the earlier results on the $YBa_2(Cu_{1-x}Fe_x)_3O_z$ system[3].

IV. CONCLUSION

We have stabilized $YSr_2Cu_{3-x}Fe_xO_z$ phase at ambient pressure otherwise $YSr_2Cu_3O_z$ does not exist. XRD results show that this compound crystallizes in tetragonal structure. Based on the MES results, we have shown that Fe substitutes for both the Cu sites non-preferentially and Fe ions exist is trivalent (Fe^{3+}) state.

TABLE 1. Room temperature hyperfine parameters (δ & Δ in $mm.s^{-1}$) and relative areas (in %) for doublets I, II & III of $YSr_2Cu_{3-x}Fe_xO_z$

	x = 0.10			x = 0.20			x = 0.30			x = 0.40		
	δ	Δ	Area	δ	Δ	Area	δ	Δ	Area	δ	Δ	Area
I	0.12(8)	1.75(2)	24	0.14(4)	1.82(4)	16	0.16(2)	1.71(8)	11	0.18(9)	1.68(2)	6
II	-0.003	0.96(2)	60	-0.004	0.93(3)	67	0.01(1)	0.88(8)	67	0.03(7)	0.96(2)	58
III	0.14(6)	0.36(8)	16	0.14(0)	0.16(5)	17	0.13(1)	0.29(6)	22	0.12(4)	0.32(2)	36

REFERENCES

1. S.A. Sunshine, L.F. Schneemeyer, T. Siegrist, D.C. Douglass, J.V. Waszczak, R.J. Cava, E.M. Gyorgy and D.W. Murphy, Chem. Mater. **1**, 331 (1989).
2. T. Den and T. Kobayashi, Physica C **196**, 141 (1992).
3. X.Z. Zhou, A.H. Morrish, Q.A. Pankhurst and M. Raudsepp, J. Phys. C **8**, 2213 (1988) and references there in.

GROWTH AND SUPERCONDUCTING PROPERTIES OF TERNARY $REBa_2Cu_3O_{7-x}$ BY THE OXYGEN-CONTROLLED-MELT-GROWTH (OCMG) PROCESS

M. Muralidhar[1], H.S. Chauhan[1], T. Saitoh[2], K. Segawa[2], K. Kamada[2] and M. Murakami[1,3]

[1]*SRL-ISTEC, Morioka Lab., Iioka-Shinden, Morioka-Shi, Iwate 020, Japan*

[2]*Iwate Indust. Res. Inst., 3-35-2, Iioka-Shinden, Morioka-Shi, Iwate 020, Japan*

[3]*SRL-ISTEC, Tamachi Lab. 1-16-25 Shibaura, Minato-Ku, Tokyo, 105, Japan*

In this report we present our investigations on ternary RE123 (RE: Nd, Sm, Eu and Gd) systems prepared by the OCMG process which is the melt process in a reduced oxygen atmosphere. We could grow a relatively large domained (Sm-Eu-Gd) $Ba_2Cu_3O_{7-x}$ (SEG 123) crystal by combining MgO (100) seeding technique with the OCMG process. Microstructural investigations on the cross-section of the sample reveal that the domain size is very large. XRD and EPMA results indicate that the matrix is uniform 123 phase and the second phase has 211 composition. SQUID measurements on the oxygenated ternary RE 123 system showed that T_c (onset) was in the range of 92.6-94.1K.

INTRODUCTION

The oxygen-controlled-melt-growth (OCMG) process first developed for Nd123[1] system and later extended to other RE123 systems such as Sm123[2] has proven excellent technique for achieving enhanced T_c in the range of 94-96K with a sharp transition and large J_c values. Recently, it was found that binary RE123 systems also exhibit good T_c and large J_c values[3,4]. However, so far no detailed report has been appeared in the literature concerning the melt growth of ternary RE123 systems except one on (Nd-Sm-Eu)123[3]. In this presentation we report the growth and superconducting properties of ternary RE123 systems prepared by the OCMG process. By using top seeded growth we could fabricate the relatively large domained sample of ternary SEG123 (1% O_2 partial pressure) system by the OCMG process. We also report T_c measurements on ternary RE123 systems and microstructural investigations for SEG123.

EXPERIMENTAL

The first of all different rare-earth oxides, $BaCO_3$ and CuO were mixed to have nominal composition of $(A_{0.33}B_{0.33}C_{0.33})Ba_2Cu_3O_{7-x}$, where A, B, C= Nd, Sm, Eu and Gd. The three RE's were in equal ratio in all the ternaries RE123 made in this study. These powders were ground thoroughly and calcinated at 880 °C for 24 hours in air with intermediate grindings. Then pellets were made and sintered at 1020 °C for 48 h. After that DTA measurements were made on samples pre-treated at 950 °C in lower oxygen partial pressure (1% or 0.1% O_2) for 15 hours. The peritectic temperatures (T_p) were obtained from these DTA measurements. The peritectic temperature obtained were used for the determination of the heat treatment in the OCMG process of different ternary RE123 systems.

The precursor ternary RE123 powders were first pelletized in cylindrical shape 20mm in diameter and 15mm in thickness, which were then subjected to CIP (cold isostatic pressing) with the pressure of 2000 kg/cm^2. Such compact pellets were put in the furnace at room temperature after placing a small MgO(100) seed on the center of the pellet surface. These samples were OCMG processed in 1% and 0.1% partial pressure of O_2. In the OCMG process the furnace was heated to T1 (T_p + 40°C) from RT in four hours and held for 30 minutes, then cooled to T2 (T_p + 10°C) in 5 minutes, next temperature was lowered by 100 °C with slow cooling rate of 1 °C/hr. In the final stage sample was furnace cooled. The gas flow was kept about 300 ml /minute. Small samples with dimensions of $1.5x1.5x1.0\ mm^3$ were cut from the grown crystals. These samples were annealed in flowing O_2 gas following heat treatment- first furnace temperature was raised to 600 °C from RT in 2 h and hold for 1h. After that furnace was cooled to 500 °C in 12 h, then to 400 °C in 24 h. Finally, the temperature was lowered to 300 °C in 24 h and hold for 100h and after that sample was furnace cooled. We have made XRD, optical microstructural, SEM and EPMA measurements on the SEG123 sample. The oxygenated samples of ternaries RE123 were used for T_c measurements.

RESULTS AND DISCUSSION

Figure 1(a) shows the DTA measurements on ternaries RE123 in 1% O_2 partial pressure. Figure 1(b) summarizes the peritectic temperatures of ternaries RE123 in 1% and 0.1% O_2 partial pressures obtained by DTA and plotted against average ionic radii of RE's in each ternary. It is evident that the peritectic temperature shows linear dependence on average

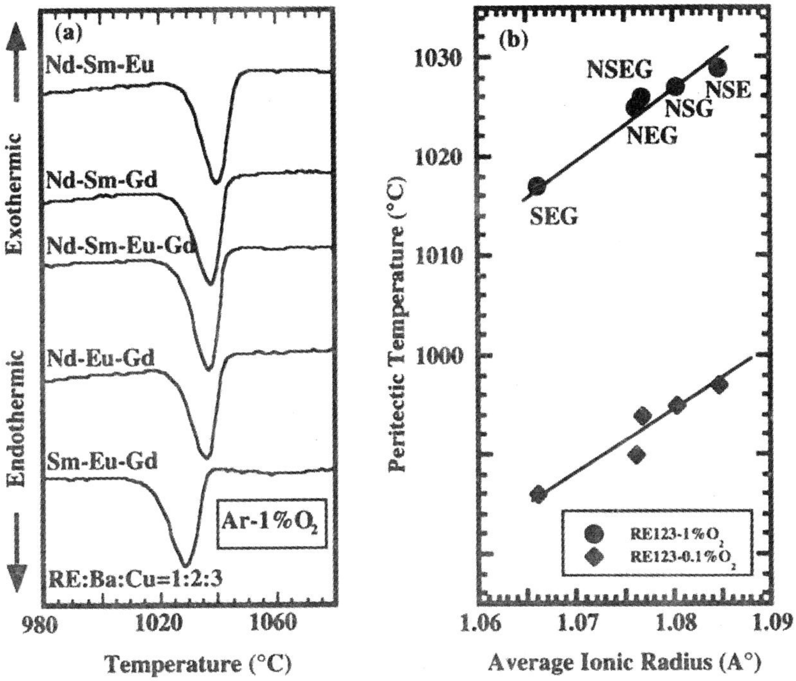

Figure 1 (a) Differential thermal analysis curves of ternary and quaternary RE123 systems for 1% oxygen partial pressure, (b) Peritectic temperature versus average ionic radii of RE's in ternary and quaternary RE123 systems.

ionic radii of RE's in ternary RE123 same as observed in case of RE123 systems. This also supports the fact that RE mixing is uniform in the formation of RE123 matrix. Further, it provides the way of controlling the peritectic temperature by varying the composition of ternaries and consequently the average ionic radii of RE's. The figure 2(left) and 2(right) display the top and polished cross-sectional views of the grown ternary SEG123 sample. In top view photograph there is a small MgO (100) seed crystal in the center of the pellet surface The closer inspection of the top view of the grown sample reveals that growth has taken place mostly from the seed. However, two other nucleations mainly from the circumference of the pellet have prevented the growth from the seed otherwise the growth of larger grain was possible. These nucleations from circumference may have taken place during growth at larger undercoolings in the OCMG process. One other possible reason of these nucleations may be associated with the temperature gradient. Also by looking at cross-sectional photograph one can clearly see that the large domained growth has taken place in this sample. Figure 3 show the comparison of XRD measurements on Sm123, Eu123, Gd123 and SEG123 which reveals that ternary system consists mainly of 123 phase with slight presence of second phase.

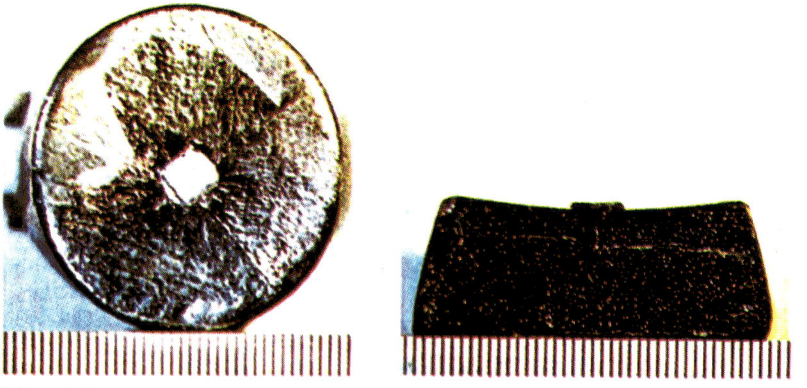

Figure 2 Top (left) and cross-sectional (right) view of top seeded grown ternary SEG123 sample.

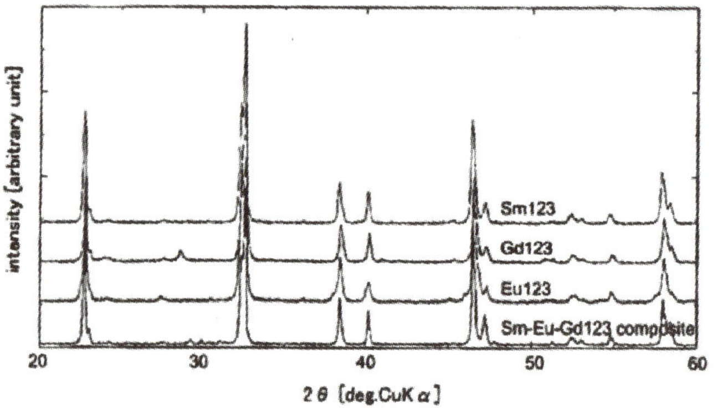

Figure 3 XRD pattern from Sm123, Eu123, Gdl23 and ternary SEG123.

Table 1. Atomic ratios for matrix RE123 and second phase SP in SEG123 sample measured by EPMA.

Element	Matrix 1	Matrix 2	Matrix 3	SP 1	SP 2	SP 3
Sm	0.334	0.318	0.322	0.701	0.604	0.680
Eu	0.302	0.307	0.297	0.585	0.572	0.780
Gd	0.324	0.301	0.319	0.680	0.647	0.670
Ba	1.982	1.960	1.961	0.980	0.960	1.000
Cu	2.999	2.999	2.999	0.999	0.999	0.999

Table 2. T_c (on set) data of ternary RE123 systems (1% and 0.1% Oxygen) measured by SQUID.

Sample	1% Oxygen	0.1% Oxygen
SEG123	93.1	92.6
NEG123	93.1	93.6
NSG123	94.1	92.6
NSE123	92.6	94.1
NSEG123	94.1	93.6

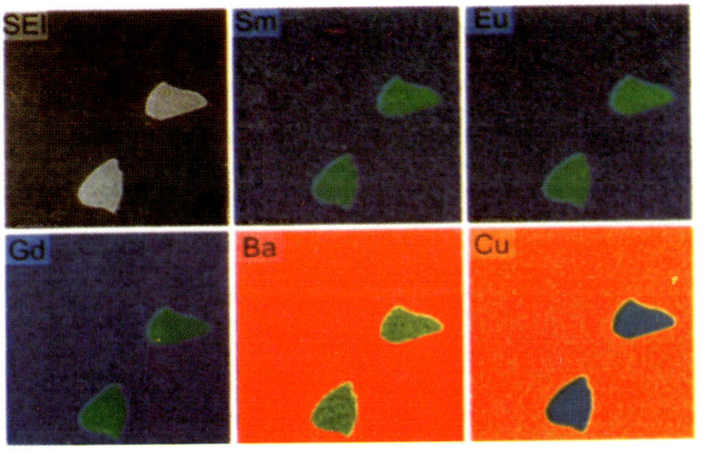

Figure 4 SEM and EPMA mapping for Sm_2O_3, Eu_2O_3, Gd_2O_3, BaO and CuO measured for SEG123.

The SEM and EPMA investigations on SEG123 are shown in figure 4. The SEM image shows that main phase is 123 in which entrapped 211 particles may be observed. Our optical microstructural investigations (not shown here) also support the same contention. EPMA mapping indicates that matrix composition is uniform 123 which is further supported by elemental analysis as shown in table 1. However, there is slight difference in 211 composition which may perhaps be attributed to the lower processing temperature used in the sample preparation. The table 2 enlists the T_c onset measured on post annealed samples of ternary RE123 systems and found to be in the range of 92.6–94.1K. The T_c value of SEG123 (1% O_2) was 93.1K. When comparing the T_c values for two cases in some samples processed at lower oxygen partial pressure show slightly lower values which perhaps arise due to unoptimised oxygenation treatment.

CONCLUSIONS

The linear dependence of peritectic temperatures on average ionic radii of RE's in ternary RE123 systems reveal their uniform mixing. The ternaries RE123 could be prepared successfully by the OCMG process. For SEG123 a large domained growth was achieved. In SEG123 and all other ternaries matrix is 123 and has very uniform distribution of RE's. The T_c values in these systems are comparable to RE123 systems.

ACKNOWLEDGMENTS

This work was partially supported by NEDO for the R&D of Indust. Sci. and Tech. Frontier Program.

REFERENCES

1. S.I. Yoo, N. Sakai, H. Takaichi, T. Higuchi and M. Murakami, Appl. Phys. Lett. 65, 633(1994).
2. M. Murakami, S.I. Yoo, T. Higuchi, N.Sakai, J. Weltz, N. Koshizuka and S. Tanaka, Jpn. J. Appl. Phys. 33, L715 (1994).
3. T. Saitoh, K. Segawa, K. Kamada, N. Sakai, S.I. Yoo and M.Murakami, in extended abstract, Intl. Wks. on Superconductivity, June 18-21, 1995, Hawaii, USA, p. 330.
4. T. Saitoh, K. Segawa, K. Kamada, N. Sakai, S.I. Yoo and M.Murakami, in extended abstract, Intl. Wks. on Superconductivity, June 24-27, 1996, Iwate, Japan, p. 185.

OPTIMUM SUPERCONDUCTING COMPOSITION OF $Nd_{2-x}Ce_xCuO_y$

Gunadhor S. Okram[1,2], B.D. Padalia[2], Om Prakash[3], S.K. Agarwal[4] and A. V. Narlikar[4]

[1]*Nuclear Science Centre, P.B. 10502, Aruna Asaf Ali Marg, New Delhi 110067, India*
[2]*Department of Physics, Indian Institute of Technology, Mumbai 400076, India*
[3]*Department of Metallurgical Engineering and Materials Science*
Indian Institute of Technology, Mumbai 400076, India
[4]*National Physical Laboratory, Dr. K.S. Krishnan Road, New Delhi 110 012, India*

It is confirmed that the x = 0.145 composition in $Nd_{2-x}Ce_xCuO_y$ exhibits almost identical T_c vs y behavior with the highest T_c (R = 0) ~ 17K comparing well with that of the x = 0.150 composition. The smooth and narrow ΔT (≤ 3K) in these two compositions are also similar. On the contrary, the T_c vs y of the x = 0.155 composition is distinctly different from that of x = 0.145. Further, the T_c values are also significantly lower and the superconducting transition regions are broader and are associated with double transition features for all the values of y.

Oxygen non-stoichiometry is a key parameter in cuprate superconductors. In the nominally electron-doped $Nd_{2-x}Ce_xCuO_y$ (NCCO) superconductor system, a range of Ce content (0.14 ≤ x ≤ 0.175) and then controlling the oxygen content y arc cssential for the material to become superconductor [1,2]. The T_c may thus be considered as a function of x and y. Practically, x could be fixed and vary y to obtain a spectrum of T_c [3]. Due probably to these material parameter intricacies, there were wide variation in the value of x at which this system exhibits maximum T_c but later on settled quite satisfactorily (ref. 1 and references therein). The importance of constructing an accurate superconducting phase diagram for other sub-systems has also been realized [4]. Thus, our aim in this work is to confirm the highest T_c composition in NCCO system in a still more precise way. This has been performed from the T_c vs y curves for a fixed x value around the controversial composition viz. x = 0.150. For this, we choose x = 0.145 and x = 0.155 as the case of x = 0.150 samples has already been reported earlier [3].

The samples of NCCO, x = 0.145 and x = 0.155, were synthesized using our standardized procedure [1, 3]. Six pellets each of a given x value were prepared. After the sintering step, each pellet of x = 0.145 was annealed at 1000° C in flowing high purity argon for 5h, 10h, 15h, 20h, 30h and 40h and subsequently quenched. Identical heat treatment was performed for the x = 0.155 samples.

XRD data analysis revealed that the samples were of single phase and possess T-type structure. However, for the samples, annealed longer than 20h, weak impurity peaks appeared. This observation supports the earlier reports [5]. The lattice parameters, calculated using least square fitting, for $Nd_{1.855}Ce_{0.145}CuO_y$ and $Nd_{1.845}Ce_{0.155}CuO_y$ samples for different values of y are given in Table 1. No significant change in the lattice parameters with the change of the oxygen content (y) is evident from the table. Within the experimental error (±0.01), the value of y decreases with increase in the duration of annealing in argon (Table 1). The overall value of y for different samples clearly indicates that variation of y is possible by changing annealing duration in argon atmosphere.

Fig. 1 shows the plots of normalized resistance, $R = R_T/R_{300}$, as a function of temperature (T) on $Nd_{1.855}Ce_{0.145}CuO_y$ samples for different values of y. For y = 4.02, the T_c onset is ~22K while T_c(R = 0) is ~13K, the normal state resistance is semiconducting-like, and transition from T_c onset to T_c(R = 0) is smooth. When y = 3.99, the T_c onset improves to ~24K, T_c(R = 0) to ~15.5K, transition width (ΔT) is narrower while the semiconducting-like behavior remains almost unchanged.

Table 1: Lattice parameters and oxygen content of the $Nd_{2-x}Ce_xCuO_y$, x=0.145 and 0.155 samples annealed for different periods of time in argon.

x	Annealing Time (h)	Lattice Parameters a(Å)	c(Å)	V(Å^3)	Oxygen Content y (±0.01)
0.145	5	3.9441(5)	12.079(2)	187.92	4.02
0.145	10	3.9475(6)	12.085(2)	188.33	3.99
0.145	15	3.9465(4)	12.083(1)	188.19	3.97
0.145	20	3.9465(5)	12.082(3)	188.19	3.96
0.145	30	3.9472(2)	12.084(2)	188.27	3.95
0.145	40	3.9492(5)	12.091(5)	188.58	3.95
0.155	5	3.9452(4)	12.072(3)	187.89	4.04
0.155	10	3.9464(4)	12.076(2)	188.08	4.03
0.155	15	3.9434(6)	12.072(3)	187.72	4.02
0.155	20	3.9442(8)	12.066(4)	187.70	4.02
0.155	30	3.9476(4)	12.076(2)	188.18	4.01
0.155	40	3.9478(2)	12.078(1)	188.23	4.00

Unlike y=3.99, at y=3.97, T_C(R=0) attains ~17.5K while T_C onset is ~24K, whereby the transition width (ΔT) decreases further (to <3K). When y is further decreased, T_C(R=0) degrades; the R=0 is not observed. The resistivity increases after it falls down to a minimum value below T_C onset. These results indicate that T_C(R=0) is substantially dependent on y.

The plots of R_T/R_{300} vs T on $Nd_{1.845}Ce_{0.155}CuO_y$ samples for different values of y are shown in Fig.2. The figure indicates that for y=4.04, normal-state R is semiconducting-like, T_C onset is ~22K and a second T_C onset-like feature appears at ~21K, below this temperature R falls steep to a minimum at ~19K but does not attain zero value. In some samples, there is an increase in R with lowering of temperature. This may be due to the presence of non-SC regions in these samples. When y=4.03, T_C onset increases to ~22.5K, and the minimum of R decreases further. However, the width of transition region (or ΔT), between second T_C onset and minimum of R, increases while other features almost remain unchanged. Interestingly, improvement in the normal state and superconducting properties is seen in the sample annealed for 15h in argon. The sample shows R=0 at ~12K. Further, when the sample was annealed for 20h in argon, the R versus T_C curve exhibits R=0 at ~12K, T_C onset at ~24K and narrower ΔT; while other features remain the same. Further improvement of T_C(R=0)

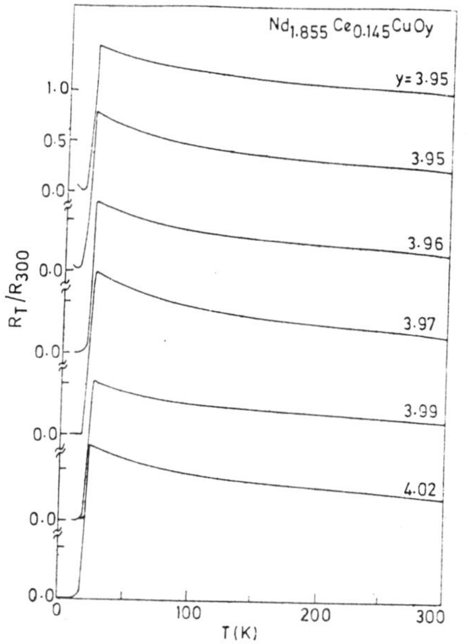

Fig.1. Normalized resistance, R_T/R_{300}, vs T plots of $Nd_{1.855}Ce_{0.145}CuO_y$ for different values of y.

~13K is observed for y=4.02. It is surprising to note that when y=4.01, R=0 is not observed while the other characteristics practically remain the same. These results therefore indicate that $T_C(R=0)$ is strongly dependent on oxygen content, y, and the double-transition feature consistently exists for all the possible values of y in the samples. Thus, the double-transition feature in the resistivity curves is a characteristic feature of these samples. These observations support the fact that the double-transition feature sets in at x=0.155 in the NCCO system. It is probably due to weak coupling among the grains and or the intrinsic property.

From the above resistivity data, it is apparent that it would be of interest to see the dependence of $T_C(R=0)$ on the oxygen content, y. So, we plot $T_C(R=0)$ versus y for NCCO, x=0.145 and 0.155 (Figure 3). It is apparent from the figure (curve a) that for x=0.145, $T_C(R=0)$ increases, attains a maximum (~17.5K) and decreases with the increase in y. In the case of x=0.155 (curve b), only two values of $T_C(R=0)$ are observed and these are also smaller than those of x=0.145 samples (curve a). For comparison, similar T_C vs y plot for x=0.150 samples [3] is shown in the inset of Figure 3. It is apparent that the samples NCCO, x=0.145 and 0.150, possess similar feature in $T_C(R=0)$ vs y plot while the maximum $T_C(R=0)$ for both the series of samples is ~17K. These results confirm that the compositions in the NCCO system for the highest T_C with narrowest ΔT lie in $0.145 \leq x \leq 0.150$ [3].

Consider the influence of oxygen content parameter, y, on the electrical transport properties of these samples. For any value of x in the NCCO system, the as-prepared samples show semiconducting-like behavior with no sign of superconductivity [6,7]. Further, these observations indicate that the conduction electrons in the as-prepared samples are somewhat localized. The conductivity, however, increases substantially and exhibits transition onset for some values of x in the system when these samples are quenched [1,3,6]. It is evident from the above results that conductivity increases more dramatically, and superconductivity sets in, improves and degrades when the samples are annealed in argon with progressively increasing duration (Fig.3). This is, in fact, due to

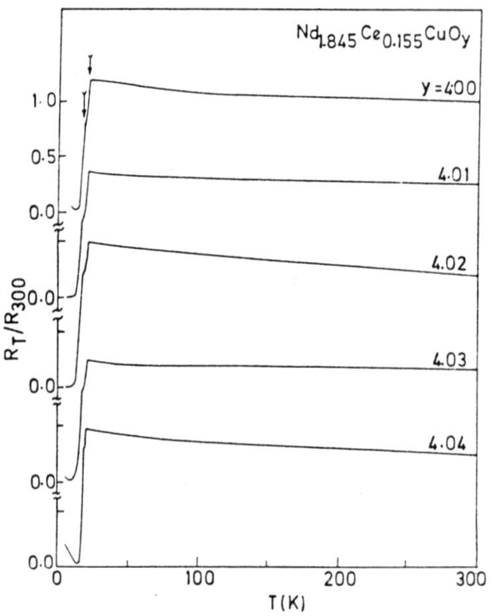

Fig.2. Normalized resistance, R_T/R_{300}, vs T plots of $Nd_{1.845}Ce_{0.155}CuO_y$ for different values of y. For both the 15th and 20th argon annealing, y=4.02. Arrows indicate the two transition onsets.

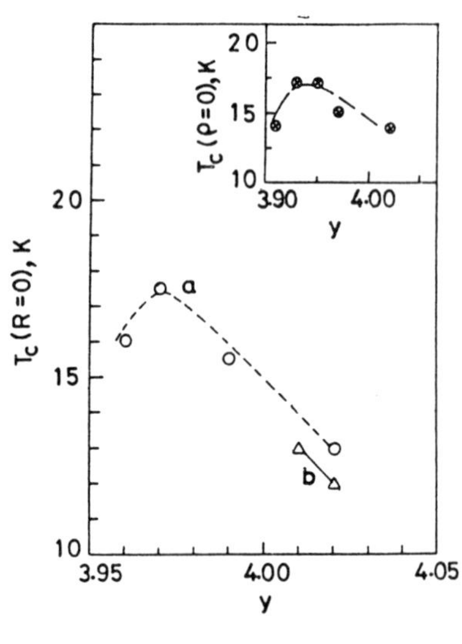

Fig.3. T_c vs y plots for $Nd_{2-x}Ce_xCuO_y$ (a) x=0.145 and (b) x=0.155 samples. The inset is for the x=0.150 sample adopted from ref. 1. The dashes are merely guide to eye.

partial removal of lattice oxygen (y) and thereby giving rise to donor level. Therefore, the role of oxygen in NCCO system for exhibiting SC behavior is crucial.

It may be emphasized on the basis of present study that the doping of mixed valence Ce alone does not give rise to superconductivity comparing well the available literature [2]. Further, as the annealing time exceeds beyond certain optimum value, weak impurity lines in the XRD spectra appear. Oxygen removal is not uniform [2]. If one considers these facts together for the case of a reduced sample of NCCO [1], bulk of the pellet can be considered as constituted of grains of this nominal composition with negligible intergrain impurities. Non-uniform distribution of oxygen vacancy in a grain [2], with the nominal composition, would give rise to variation of T_C which is expected to decrease towards the grain's interior and perhaps becoming zero at the core [2]. It may be stressed here that such a variation in oxygen (Δy) is negligible within the detection limit of XRD. Moreover, this (Δy) is not likely to give rise to problems regarding coupling among grains. However, the trace impurities detected for the samples subjected to prolonged annealing are expected to worsen the coupling among grains. Nevertheless, such worsening of coupling does not arise in exhibiting double-transition in the resistivity curves for the samples $x \leq 0.15$, in general, while it does so for the samples $x \geq 0.155$. This suggests that the granular nature of these two sets of compositions ($x \leq 0.15$ and $x \geq 0.155$) may be different in addition to the difficulty in easy removal of oxygen for $x \geq 0.155$ [1].

In conclusion, the T_C and its dependence on y in the samples NCCO, x=0.145 and 0.150 are similar. Both these compositions show no double-transition, indicating good coupling among the grains. The $Nd_{1.845}Ce_{0.155}CuO_y$ samples exhibit lower values of $T_C(R=0)$ than those of x=0.145 and 0.150, and R=0 is exhibited in a very narrow range of y. The persistence of double-transition feature in x=0.155 samples for any duration of argon annealing suggests that this feature prevails in samples with $0.155 \leq x \leq 0.175$ strengthening the finding in ref.1. The present study has therefore fully confirmed the fact that the optimum SC composition of the NCCO system lies in the very narrow range $0.145 \leq x \leq 0.150$.

Three of the authors, GSO, BDP and OP, gratefully acknowledge the partial support to this work by Department of Science and Technology, New Delhi.

REFERENCES

1. Okram G. Singh, B.D. Padalia, Om Prakash,K. Suba, A.V. Narlikar and L.C. Gupta, Physica C **219**, 156 (1994).
2. Okram G. Singh, C.S. Harendranath, Om Prakash and B.D. Padalia, Solid State Commun. (in press).
3. Okram G. Singh, Om Prakash, B.D. Padalia, D. Chandrasekharam, A.S. Tamhane and L.C. Gupta, Supercond. Sci. Technol. **5**, 561 (1992).
4. M. Brinkmann, T. Rex, H. Bach and K. Westerholt, Phys. Rev. Lett. **74**, 4927 (1995).
5. J.S. Kim and D.R. Gaskell, Physica C **224**, 256 (1993) and references therein.
6. Okram G. Singh, Ph.D. Thesis, Indian Instiute of Technology, Bombay (1994).
7. S. Uji and H. Aoki, Physica C **199**, 231 (1992).

OBSERVATION OF PARAMAGNETIC MEISSNER EFFECT IN GdBCO POWDER - A MICROWAVE ABSORPTION STUDY

P. V. Patanjali[1], V. Seshu Bai[1], R. M. Kadam[2] and M. D. Sastry[2]
[1]*School of Physics, University of Hyderabad, Hyderabad - 500 046, India.*
[2]*Radio Chemistry Division, BARC, Trombay, Mumbai - 400085, India.*

We have observed anomalous microwave absorption (MWA) representative of paramagnetic Meissner effect in GdBCO powder. The results are discussed in terms of π junction theory. The phase change in the MWA gives evidence for the change in behavior of π junction from para to dia state.

I. INTRODUCTION

The essential property of a superconductor is to become diamagnetic and expel magnetic field when cooled to below its transition temperature. However, some HTSC materials, when cooled in a small magnetic field acquire a net paramagnatic moment[1-9]. This paramagnetic moment increases as the sample is cooled to about 10K below T_c and then saturates. In some samples the onset of paramagnetism occurs a few degrees below T_c even when the T_c is changed by 20K by altering the oxygen concentration[3]. The paramagnetic moment seems to be superimposed on a conventional Meissner diamagnetism and is found to be frozen in if the external field is removed at low temperature. The possibility of this unusual paramagnetic moment originating due to paramagnetic impurities is ruled out on the grounds that no significant paramagnetism is observed in the normal state even at 1 Tesla[2] and the onset of paramagnetism always occurs a few degrees below T_c. The effect is well established and is known as paramagnetic Meissner effect (PME).

The signature of the PME is observed in the second harmonic susceptibility[10] and in microwave absorption[2,3,11] as anomalous responses. In the case of microwave absorption the samples which show PME give rise to maximum absorption at zero field and a reduction as the field is increased. In general, the microwave absorption in HTSCs is minimum in zero field and increases with the application of field. This increase is attributed to dissipation in the Josephson junctions or due to flux flow. Of the possible mechanisms suggested to explain the PME the idea that HTSCs for some reason exhibit an array of loops containing π junctions seems to be more appropriate.

Bulaevskii *et.al*[12] have shown that if the Josephson junction has a magnetic impurity then the current tunneling through the junction becomes negative. In such a case the phase difference across the junction would not be zero, as usual, but π. The phase difference depends on the inductance of the current loop and on J_c. A spontaneous flux exists if $\frac{2\pi L|J_c|}{\phi_0} > 1$. The state of lowest energy of such a loop in zero applied field is doubly degenerate with a trapped flux equal to approximately $\pm\phi_0/2$. Such a state with spontaneous orbital moments is called[13,14] a new condensed matter state with a critical field H_{c0}. It is also shown that for a π junction the magnetic response below T_c should be diamagnetic and paramagnetic currents appear when the J_c reaches a certain value. Thus a cross over from paramagnetic to diamagnetic behaviour of π juction is expected as the temperature is increased.

An alternative mechanism for generating an intrinsic π-phase shift in a Josephson junction is based on the idea of existence of an unconventional order parameter[15,16]. In this report we present the microwave absorption results of GdBCO superconductor ana provide experimental evidence for the change in behavior of π junctions from para to dia state as a function of temperature. We also discuss various possible origins of π junctions.

II. EXPERIMENTAL

$GdBa_2Cu_3O_7$ (GdBCO) sintered superconductor is made through solid state route. Part of the sample was used as a pellet and the ramaining part is crushed into powder with particle size $< 30\mu m$. The powder particles are well separated by suspending them in GE varnish paste. Both the samples when characterized by XRD showed single GdBCO orthorhombic phase without any impurity phases. Pellet sample showed zero resistance transition onset at 92K. The nonresonant microwave absorption (MWA) spectra were recorded using a Bruker ESR spectrometer which has a facility to vary field from -5 mT to 1.2 T. Field modulated microwave absorption (FMMWA) signals were recorded in the derivative form using phase sensitive detection. For all the experiments carried out using phase sensitive detection, the internal phase of the lock-in is set to zero. The temperature of the sample is varied using a closed cycle refrigerator.

III. RESULTS

Generally, the derivative signal (dP/dH) of the FMMWA in HTSCs has the form shown in Fig. 1a whose phase is opposite to the Gd^{3+} EPR signal. Figure 2 shows FMMWA signals recorded at 100 mOe modulation field as the dc field is scanned between $\pm$ 5 Oe on zero filed cooled (ZFC) GdBCO powder at various temperatures below T_c. The phase of the derivative signal at low temperatures is same as that of the Gd^{3+} EPR signal observed at 3.5 KOe (Fig. 1b). Hence, for the sake of convenience we call the signals which have the same phase as that of Gd^{3+} EPR signal in-phase signals and those with opposite phase out-of-phase signals. The in-phase signal (Fig. 1c) represents the PME response with maximum absorption in zero field followed by minima at $\pm$ 0.6 Oe. The flat base line above this field represents a sort of linear increase in the direct absorption (P(H)). As the temperature is increased the intensity of the in-phase signal reduces and at about 56K the signature of the out-of-phase signal, which is representative of absortion due to dissipation in the usual Josephson junctions, starts appearing. With further increase in the temperature the out-of-phase signal initially grows in intensity, and eventually dies of as T_c is approached. In the intermediate temperatures there is coexistence of both the in-phase and out-of-phase signals. When 1 Oe modulation field is set while recording FMMWA the anomalous in-phase signal is not observed, however, a weak out-of-phase signal whose intensity increases with increase in temperature is observed. As in the earlier case this signal also disappears as T_c is approached. For want of space we have not shown the figure. The anomalous in-phase signal is not seen in GdBCO pellet. We have confirmed this observation at various modulation fields and temperatures.

Figure 3 shows the in-phase signal recorded at 15K on GdBCO powder at various flux trapped states. To record the signal first a certain dc field (given in brackets in Fig.3) was applied at 15K on the ZFC sample and withdrawn. FMMWA response was then recorded between $\pm$ 5 Oe. From the figure one can see that there is an overall reduction in the intensity of the signal and disappearance at a flux trapped state corresponding to 125 Oe. In Fig. 4 changes in the direct MWA through transition for both GdBCO powder and pellet are shown. The broad transition width in the case of powder is due to loss of oxygen. The three different slopes in the transition region are reminiscent of three different slopes (a plateau in high temperatures, a fall and another plateau in the low temperature region) in the oxygen content dependent T_c variation curve of a typical 123 system[18,19].

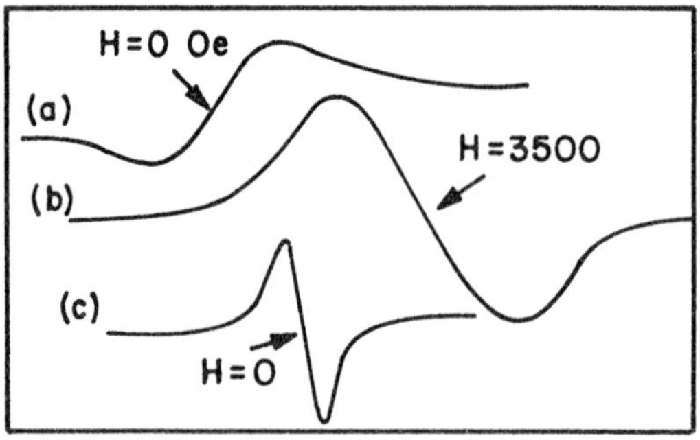

Fig. 1 Typical out-of-phase (a), Gd^{3+} ESR (b) and in-phase (c) signals. The pictures are not to the scale.

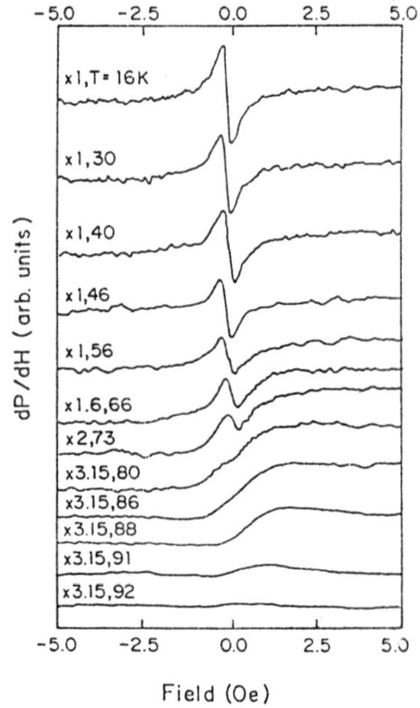

Fig. 2 FMMWA signals of ZFC GdBCO powder recorded when the dc field is scanned from -5 Oe to +5 Oe, at various temperatures.

We have also carried out field induced direct MWA at various temperatures in both pellet and powder. In the case of pellet we have observed a monotonic increase in absorption at low fields (0-30 mT), a saturation in the intermediate fields (30-100 mT) and a linear increase at high fields(0.1-1 T). These three segments are well explained[20] by invoking resistively shunted Josephson junction (RSJ) and flux flow models. In the case of powder no appreciable change in absorption is observed upto about 30 mT. But, at high fields an intense Gd^{3+} EPR response is observed. It is worth noting that the change in absorption in the low field region in the pellet is almost 1000 times that of the powder. Such a great reduction is possible due to the elimination of grain boundary weaklinks on powdering.

IV. DISCUSSION

As is already mentioned in the introduction, in the magnetization studies the PME is observed in field cooled samples alone. The reason being in the ZFC sample the spontaneous orbital moments are oriented randomly giving rise to zero net magnetization. In the field cooled samples these moments are polarized to give paramagnetism. However, in the microwave absorption the sample need not be field cooled to observe the PME, because irrespective of the orientation of the orbital currents it is the total dissipation in π junctions which contributes to the observed in-phase signal. In other words high frequency absorption techniques, namely, field modulated rf, microwave absorption, are best suited to study the virgin π junction responce. In fact, any application of field reduces the PME. We have shown this effect in Fig. 3 by exposing the sample to different magnetic fields before recording FMMWA. The overall reduction in the intensity of the absorption with higher flux trapping can be understood in terms of gradual destruction of orbital moments. And the disappearance of FMMWA at 125 Oe field suggests that the trapped field inside the sample is more than H_{c0}, the crossover field from para to dia state.

The absence of in-phase signal in the pellet could be for two reasons. **Firstly, the π junction response is overshadowed by a very strong dissipation in the normal Josephson junctions.** Secondly, the current in a junction network is decided by the weakest possible junction which has the lowest J_c. Whereas, the condition $\frac{2\pi L|J_c|}{\phi_0} > 1$ for a π junction demands that the product $L \mid J_c \mid$ be high enough, which is improbable in a pellet. This point is further validated by the fact that PME is seen in melt textured, single crystal and powder samples only[1-11]. J_c in all these three varieties of samples is much higher than in sintered samples.

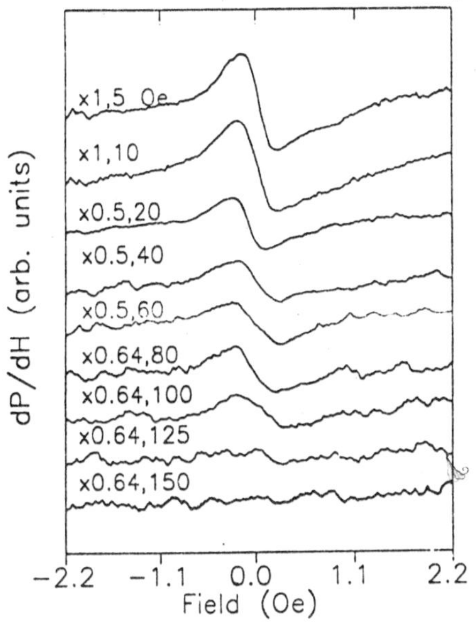

Fig. 3 FMMWA in-phase signal of ZFC GdBCO powder recorded at 15K when the field is scanned between -5 Oe and +5 Oe (the response is shown between ± 2.2 Oe).

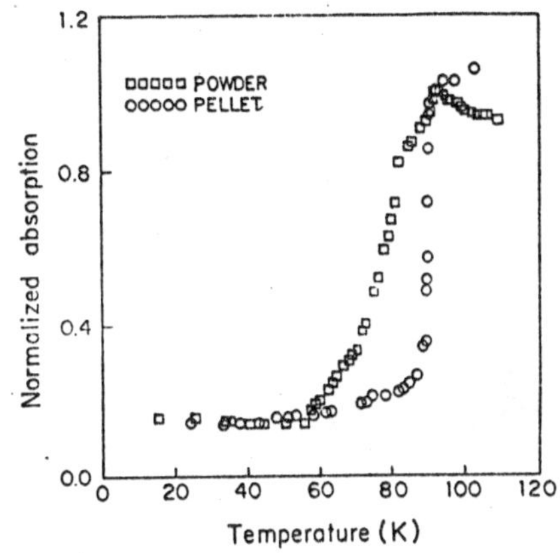

Fig. 4 Change in the direct MWA as a function of temperature of GdBCO pellet (a) and powder (b). Powder sample shows three different slopes in the transition region.

Kusmartsev [13,14] has shown that for a junction with coupling energy E_j, the temperature at which the orbital moments are destroyed is $T_0 \propto \frac{E_j}{n^2}$, n is the number of π junctions. (Incidentally, to create a spontaneous orbital moment it is necessary to have a loop with odd number of junction. On the ring with n=1 the orbital moment is not created at all as the quantum fluctuations destroy it. Hence, the minimum number of π contacts is three.) Since we do not have any a priori knowledge of E_j and n, it is only appropriate to say that depending on the distribution of E_j and n, T_0 has a distribution. Once the orbital moments are destroyed the π junction behaves as an ordinary Josephson junction. Keeping this in view and the fact that the out-of-phase signal develops at the expense of in-phase signal (Fig. 2) it is reasonable to say that the phase reversal observed could be due to change in behavior of π junction from para to dia state.

Qualitatively, the assumption of the existence of the spontaneous orbital currents and fluxes is enough to explain the experimantal observation. With regard to the microscopic origin of π junction in GdBCO powder, one possibility is Gd^{3+} ion acting as magnetic impurity thereby forming a π junction, if there is tunneling. Another possibility is Cu^{2+} itself acting as impurity. In GdBCO powder the broad transition (Fig. 4) suggests oxygen deficiency and the oxygen content varies from O_7 to $O_{6.6}$. The oxygen sites associated with the deficiency are the chain oxygen atoms in the charge reservoir layer[19]. The dominant effect of this variation in oxygen content is to change the oxydation state of chain copper atom resulting in a finite concentration of impurity Cu^{2+} ions with local moments (unpaired spins). In powder particles with randomly oriented grains the supercurrent that flows through Cu-O planes may be compelled, because of alignment of two adjacent grains, to tunnel across via the chains containing the local moments Gd^{3+} and/or Cu^{2+}, forming a π junction.

Yet another possibility for π junction formation is the grain boundary contact of unconventional pairing, $d_{x^2-y^2}^{15-17}$. In this case the internal phases of the order parameter which depend on the relative orientation of the two grains in contact determine the sign of J_c, so that for some particular orientations $J_c < 0$, therefore a π junction. Though our experimental results give evidence to the existence of π junctions in powder sample, it is not clear whether they originate due to paramagnetic impurities or $d_{x^2-y^2}$ symmetry expected in HTSCs. Estimation of the strength of paramagnetism or in-phase signal for both impurity moment and $d_{x^2-y^2}$ symmetry considerations may be considered important to resolve the ambiguity and for further understanding of this aspect.

REFERENCES

1. P.Svedlindh, K.Niskanen, P.Norling, P.Nordblad, L.Lundgren B.Lonnberg and T.Lundstrom, Physica C **162-164**,1365 (1989).
2. W. Braunisch, N.Knauf, V.Kataev, S.Neuhausen, A.Grutz, A.Kock B.Roden, D.Khomskii and D.Wohlleben, Phy. Rev. Lett. **68**, 1908 (1992).
3. W. Braunisch, N.Knauf, G.Bauer, A.Kock, A.Becker, B.Freitag, A.Grutz, V.Kataev, S.Neuhausen, B.Roden, D.Khomskii and D.Wohlleben, Phy. Rev. **B 48**, 4030 (1993).
4. B.Schliepe, M.Stindtmann, I.Nikolic and K.Baberschke, Phy. Rev. **B 47**, 8331 (1993).
4. S.Thoming, J.Kotzler, M.W.Pieper and A.Spirgatis, Verh. Dtsch. Phys. Ges. VI **27**, 912 (1992).
5. N.Knauf, W.Braunisch, G.Bauer, A.Kock, A.Becker, B.Freitag, V.Kataev, B.Roden and D.Khomskii, Physica **B 194-196**, 2229 (1994).
6. W.H.Lee, Y.T.Huang, S.W.Lu, K.Chen and P.T.Wu, Solid State Commun. **74**, 97 (1990).
7. S. Riedling, G.Brauchle, R.Lucht, K.Rohberg, H.V.Loneysen, H.Claus, A.Erb and G.Muller-Vogt, Phys. Rev. **B 49**, 13283 (1994).
8. M.D.Lan, J.Z.Liu and R.N.Shelton, Phys. Rev. **B 43** 12989 (1991).
9. Yusheng He, C.M.Muirhead and W.F.Vinen, Phys. Rev. **B 53**,12441 (1996).
10. Ch.Heinzel, Th. Theilig and P.Ziemann, Phys. Rev. **B 48**,3445 (1993).
11. S.V.Bhat, A.Rastogi, N.Kumar, R.Nagarajan and C.N.R.Rao, Physics **C 219**, 87 (1994).
12. L.N. Bulaevskii, V.V.Kuzii and A.A.Sobyanin, JETP Lett.**25**, 290(1977).
13. F.V.Kusmartsev, J. Supercond. **5**, 463 (1992).
14. F.V.Kusmartsev, Phys. Rev. Lett. **69**, 2268 (1992).
15. V.B.Geshkenbein et al. JETP Lett. **43**, 395 (1986).
16. M.Sigrist and T.M.Rice, J. Phys. Soc. Jpn.**61**, 4283(1992).
17. M.Sigrist and T.M.Rice, Rev. of Modern Phy. **67**, 503 (1995).
18. R.J.Cava, A.W.Hewat, E.A.Hewat, B.Batlogg, M.Marezio, K.M.Rabe J.J.Krajewski, W.F.Peck Jr, L.W.Rupp Jr, Physica **C 165**, 419 (1990).
19. J.D.Jorgensen, Physics Today **34**, June (1991).
20. P.V.Patanjali, V.Seshu Bai, R.M.Kadam and M.D.Sastry, unpublished.

THE VARIABLE VALENCE STATE OF MERCURY AND COPPER IN SUPERCONDUCTING $HgBa_2CuO_{4+\delta}$ AND $Hg_{0.7}Cr_{0.3}Sr_2CuO_{4+\delta}$

P. Selvam[1], J.V. Yakhmi[2], P. Veluchamy[3], H. Minoura[3], J.B. Mandal[4], B. Bandyopadhyay[4], and B. Ghosh[4]

[1]*Department of Chemistry, Indian Institute of Technology, Mumbai 400 076, India*
[2]*Chemistry Division, Bhabha Atomic Research Centre, Mumbai 400 085, India*
[3]*Department of Applied Chemistry, Gifu University Gifu 501 11, Japan*
[4]*Saha Institute of Nuclear Physics, Calcutta 700 064, India*

X-ray photoelectron spectroscopic studies of $HgBa_2CuO_{4+\delta}$ and $Hg_{0.7}Cr_{0.3}Sr_2CuO_{4+\delta}$ indicate the presence of unusual oxidation state of both mercury and copper ions. Hg(4f) core level spectra shows two peaks, one corresponding to divalent mercury and the other to a peculiar trivalent ion. Cu(2p) core level spectra suggest a significant reduction in the divalent copper and a noticeable increase in monovalent copper. Further, the absence of characteristic shake-up satellites in this region confirms this observation. The results are discussed in terms of charge transfer between Hg-O and $Cu\text{-}O_2$ layers by internal redox mechanism, via apical oxygen.

I. INTRODUCTION

The discovery of Hg/Ba-based cuprates[1] has led to extensive experimental and theoretical work. Successive reports[2-4] on these materials, $HgBa_2Ca_{1-n}Cu_nO_y$ (n = 1-5), with T_c ranging between 94-134 K and over 160 K under normal and high pressures respectively, intensified the high T_c research significantly. Nevertheless, the chemical instability and complexity in preparation of these phases pose serious problems. No superconducting phase has yet been isolated in the analogous Hg/Sr-based system, $HgSr_2Ca_{1-n}Cu_nO_y$ (n = 1-4). However, partial substitution of cations for mercury were found to be effective in stabilizing these phases with T_c 27-94 K.[5-7] On the one hand, $HgBa_2CuO_{4+\delta}$ and $Hg_{0.7}Cr_{0.3}Sr_2CuO_{4+\delta}$ attract much interest due to their substantially high transition temperature compared among single $Cu\text{-}O_2$ layer superconductors. On the other, the Hg/Sr-based compounds are of additional interest because of their improved chemical stability[8] and better flux-pinning properties[9] than the barium analog.

It is widely believed that excess oxygen affects the T_c of Hg-based conductors. However, the optimum doping level seems to be higher than most other high- T_c cuprates. The oxygen content in Hg-O plane and the charge carrier density have close relation with the valence of neighbouring Hg-ions. Therefore, it is important to have a knowledge of the oxidation state of mercury as well as the oxygen content in elucidating the density of charge carriers and the elecronic structure of the system. In addition, a couple of recent X-ray photoelectron spectroscopy (XPS) reports[10,11] on $HgBa_2CuO_{4+\delta}$ present conflicing results. Hence, we undertook the present XPS study on both $HgBa_2CuO_{4+\delta}$ and $Hg_{0.7}Cr_{0.3}Sr_2CuO_{4+\delta}$ systems.

II. EXPERIMENTAL

Samples with nominal compositions of $HgBa_2CuO_{4+\delta}$ were prepared by controlled vapour-solid technique. Ba_2CuO_y precursor was first prepared by firing a mixture of high purity BaO and CuO at 900°C for 20 h. Stoichiometeric amounts of HgO and Ba_2CuO_y were then mixed together, pressed into (composite) pellets, sealed in an evacuated quartz tube along with a precursor pellet. This was then heated at the rate of 160°C /h to 800°C and kept there for 5 h followed by a slow cooling to room temperature. The ratio of the precursor to the composite pellets was optimized to 0.3 which gave the best quality samples. Both the pellets show typical $HgBa_2CuO_{4+\delta}$ structure. The composite pellets were used for further studies. The solid solution phase with nominal compositions of $Hg_{0.7}Cr_{0.3}Sr_2CuO_{4+\delta}$ were prepared by a three-step solid-state reaction method[12]. Sr_2CuO_y was first prepared from high purity $Sr(NO_3)_2$ and $Cu(NO_3)_2.3H_2O$. In the second step, $Cr_{0.3}Sr_2CuO_y$ precursor was made by mixing appropriate amounts of Cr_2O_3 with Sr_2CuO_y and fired at 900°C for 24 h. Finally, HgO was added to the precursor, pressed into pellets and sealed in a quartz tube. To get the final desired product, it was sintered at 880°C for 8 h followed by quenching to room temperature. Well characterized samples of $HgBa_2CuO_{4+\delta}$ and $Hg_{0.7}Cr_{0.3}Sr_2CuO_{4+\delta}$ were used for the XPS measurements using Shimadzu 850 spectrometer with a base pressure of about 5×10^{-8} Pa. In order to assign the chemical states to mercury and copper, by comparison with reference compounds in which valence states are well established, spectra of high purity HgO, CuO, Cu_2O and $YBa_2Cu_3O_{7-\delta}$ were recorded using a VG scientific ESCALAB Mark II spectrometer with a base pressure of less than 10^{-9} mbar. The spectrometers were calibrated with Au (4f) at 83.8 eV and Ag (3d) at 368.5 eV. Mg K_α radiation was used as the excitation source. The surface was cleaned by sputtering and the binding energy values are given relative to C (1s) core level at 284.5 eV.

III. RESULTS AND DISCUSSION

X-ray powder diffraction patterns of both $HgBa_2CuO_{4+\delta}$ (a = 3.872(3) Å; c = 9.486(4) Å) and $Hg_{0.7}Cr_{0.3}Sr_2CuO_{4+\delta}$ (a = 3.845(2) Å; c = 8.715(6) Å) were indexed on the basis of tetragonal unit cell. T_c was measured by four-probe resistivity and dc-magnetic susceptibility and the values respectively were 93 and 53 K. Fig.1 shows the Hg(4f) core level XP spectra of mercury compounds. The broadening of the spectra (curves (b)-(d)) indicate the presence of at least two different chemical states of mercury. Deconvolution of the spectra confirms this observation, viz., one corresponding to divalent mercury (100.5 eV) and the other to a peculiar trivalent ion (101.7 eV). However, as can be seen from Fig.1d, the higher valent mercury was considerably decreased in the case of $Hg_{0.7}Cr_{0.3}Sr_2CuO_{4+\delta}$ possibly due to a partial substitution by chromium.

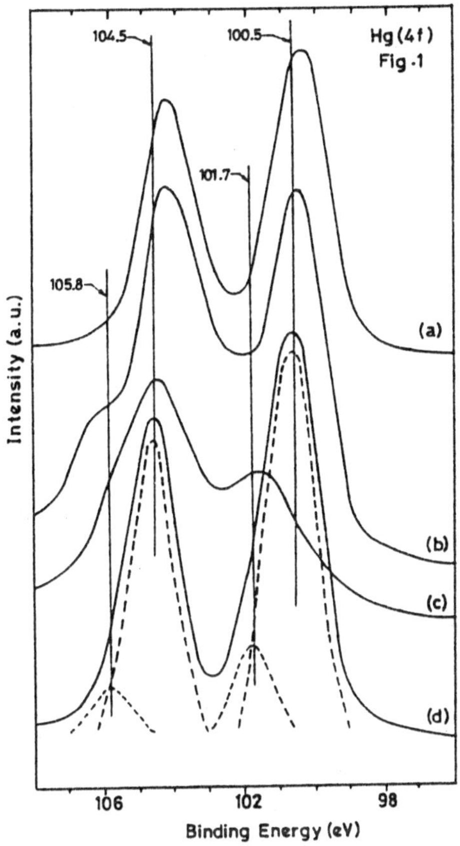

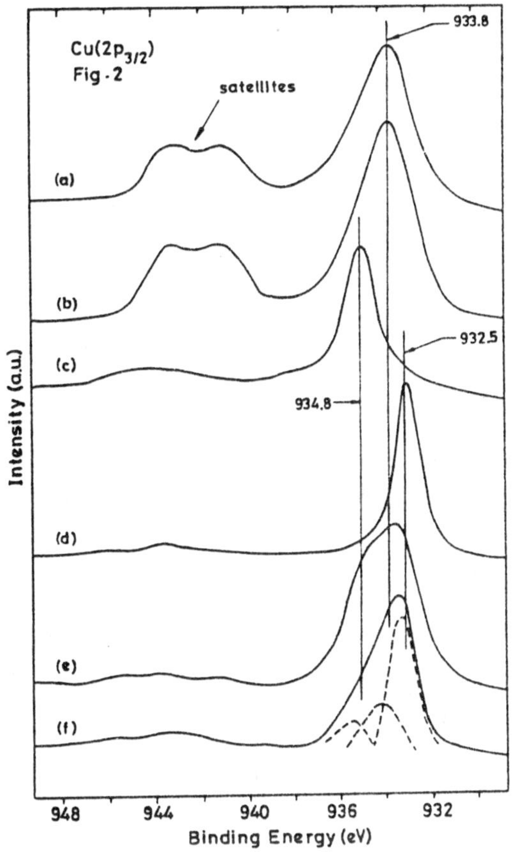

Fig.1. XPS of: (a) HgO; (b) $HgBa_2CuO_{4+\delta}$ [10]; (c) $HgBa_2CuO_{4+\delta}$; (d) $Hg_{0.7}Cr_{0.3}Sr_2CuO_{4+\delta}$.

Fig.2. XPS of: (a) $YBa_2Cu_3O_{7-\delta}$; (b) CuO; (c) $NaCuO_2$ [13]; (d) Cu_2O ; (e) $HgBa_2CuO_{4+\delta}$; (f) $Hg_{0.7}Cr_{0.3}Sr_2CuO_{4+\delta}$.

Furthermore, it is remarkable to note that there is much variation in the copper oxidation state in the Hg-based cuprates (Fig.2), in contrast with the other copper-based superconductors. A significant reduction in the divalent copper was observed while the respective monovalent ion was noticeably increased. The formation of the latter is, however, favored by the existence of the trivalent mercury ion. The absence of characteristic shake-up satellites in the $Cu(2p_{3/2})$ region demonstrate their very different behaviors. Upon deconvolution of the spectra, three peaks were obtained, the one at a low binding energy of 932.7 eV assigned to monovalent copper, the other at 934.0 eV to divalent copper and a third one at 935.0 eV can be ascribed to a possible trivalent nature of copper.

On the basis of simple ionic consideration, one can assume mercury, barium and oxygen to have a formal oxidation state of 2+, 2+ and 2- respectively, and expect an average oxidation state of copper be more than 2+ for hole-doped cuprate superconductors. Contray to this general expectations, we have observed a mixed valence for mercury (2+ and 3+) in $HgBa_2CuO_{4+\delta}$ and $Hg_{0.7}Cr_{0.3}Sr_2CuO_{4+\delta}$. As expected, copper also shows a multivalent character but, to our suprise, the oxidation states are quite different from that reported generally for high T_c cuprates. The short Hg-O bond length (1.794 Å)[12] compared to that in the compound HgO (2.05 Å) supports the interpretation and also explain the higher valent state. Such an exceptional trend in the valence states of the metal ions is likely to play an important role in the superconducting properties of these compounds, and we plan to take up further studies on (Hg,V)-series in this regard.

ACKNOWLEDGMENT

This work is supported by the Department of Atomic Energy (BRNS), Basic Sciences - Physical and Chemical Sciences, under Contract No.37/8/96-R&D-II (P.S. and J.V.Y.).

REFERENCES

1. S.N. Putilin, E.V. Atipov, O. Chmaissem and M. Marezio, Nature **362**, 226 (1993).
2. A. Schilling, M. Cantoni, J.D. Guo and H.R. Ott, Nature **363**, 56 (1993).
3. C.W. Chu. L. Gao, F. Chen, Z.J. Huang, R.L. Meng and Y.Y. Xue, Nature **363**, 323 (1993).
4. M. Nunez-Regueiro, J.-L. Tholence, E.V. Antipov, J.-J. Capponi and M. Marezio, Science **262**, 97 (1993).
5. J. Shimoyama, S. Hahakura, K. Kitazawa, K. Yamafuji and K. Kishio, Physica C **224**, 1 (1994), and references cited therein.
6. C. Michel, M. Hervieu, A. Maignan, D. Pelloquin, V. Badri and B. Raveau, Physica C, **241**, 1 (1995).
7. O. Chmaissem and Z.Z. Sheng, Physica C, **242**, 23 (1995).
8. J. Shimoyama, S. Hahakura, R. Kobayashi, K. Kitazawa, K. Yamafuji and K. Kishio, Physica C **235-240**, 2795 (1994).
9. K.E. Gray, Appl. Sucpercond. **2**, 295 (1994).
10. C.S. Gopinath and S. Subramanian, Physica C **232**, 222 (1994).
11. H.-D. Kim, S.-J. Oh, I.-S. Yang and N.H. Hur, Physica C **253**, 351 (1996).
12. J.B. Mandal, B. Bandyopadhyay, B. Ghosh, H. Rajagopalan, A. Sequeira and J.V. Yakhmi, J. Superconduc. **9**, 261 (1996).
13. P. Steiner, V. Kinsinger, I. Siegwart, S. Hufner, C. Politis, R. Hoppe and H.P. Muller, Z. Phys. B. **67**, 497 (1987).

PARACONDUCTIVITY OF Bi-Pb(2223) CERAMIC SUPERCONDUCTING SYSTEM

P. Subhas, G. Narsinga Rao and D. Suresh Babu

Department of Physics, M.S.R. L,.University College of Science
Osmania University, Hyderabad 500 007, INDIA

We report paraconductivity in the mean-field-like-region on polycrystalline $Bi_{1.7-x}Pb_{0.3}M_xSr_2Ca_2Cu_3O_y$ for x = 0 and 0.1 and M = V/Nb/Ta (nominal composition) superconducting system. The results have been analysed using Aslamzov-Larkrn (AL), Lawrance-Doniach (LD) and Viera and Vidal (VV) models, and intragrain and intergarain parameters have been estimated. The present results suggest that the addition of V-B group metal oxides in $Bi_{1.7}Pb_{0.3}Sr_2Ca_2Cu_3O_y$ system changes the intergrain properties of the system without changing their intragrain properties. It also suggests that the addition of V_2O_5 in $Bi_{1.7}Pb_{0.3}Sr_2Ca_2Cu_3O_y$ improves the coupling between the grains.

I. INTRODUCTION

The discovery of high T_c superconductors in metallic oxides has generated a great interest about the nature of the superconductivity in these materials[1]. Since the crystal structure of these metallic oxide superconductors is strongly anisotropic, the transport properties are expected to be dependent on the crystal orientation. Their anisotropic nature has been observed in resistivity[2], critical current density[3] and upper critical field measurements[4]. The fluctuations above the superconducting transition are expected to reflect the degree of anisotropy. The observation of large excess conductivity above the superconducting transition temperature due to thermodynamic fluctuations in the Yttrium based superconductors was first reported by Freitas et al[5]. The fluctuations induced excess conductivity studies have also been performed on Bi-based superconductors[6,7]. Apart from the high transition temperature, oxide superconductors differ from classical superconductors in a number of aspects. Their physical properties are closely linked with the carrier concentration. They are highly anisotropic and more specifically, the superconducting CuO_2 layers are separated by insulating blocks resulting in small coherence lengths, smaller than the unit cell dimension in the c direction. These properties can be modified by varying the carrier concentration by means of chemical doping.

In this paper, we report observations of paraconductivity in polycrystalline $Bi_{1.7-x}Pb_{0.3}M_xSr_2Ca_2Cu_3O_y$ for x = 0 and 0.1, and M = V/Nb/Ta superconducting system. The crossover temperatures from 2D to 3D fluctuations, T_0, were estimated from LD model. In addition, the intergrain and intragrain parameters were also estimated from the LD and VV models.

II. EXPERIMENTAL

The nominal compositions of the samples $Bi_{1.7-x}Pb_{0.3}M_xSr_2Ca_2Cu_3O_y$ for x = 0 and 0.1, M = V/Nb/Ta were prepared by the well known solid state reaction method. These samples were characterised using XRD, SEM, EDAX, ac-susceptibility, dc magnetisation, microwave absorption and dc electrical resistivity measurements[8]. XRD measurements showed that $Bi_{1.7}Pb_{0.3}Sr_2Ca_2Cu_3O_y$ (SI) sample had two phases i.e., 2212 and 2223. Whereas the other samples i.e., $Bi_{1.6}Pb_{0.3}V_{0.1}Sr_2Ca_2Cu_3O_y$ (S2), $Bi_{1.6}Pb_{0.3}Nb_{0.1}Sr_2Ca_2Cu_3O_y$ (S3) and $Bi_{1.6}Pb_{0.3}Ta_{0.1}Sr_2Ca_2Cu_3O_y$ (S4), had only 2223 phase along with $CaMO_y$ type impurity. Resistivity and ac susceptibility measurements showed a single superconducting transition in all the four samples, even though XRD studies of SI sample indicated presence of two phases. The diamagnetic transition temperature T_c is found to be 108 K for all these samples and T_c (R=0) values are found to be 108 K, 105 K, 97 K and 103 K for SI, S2, S3 and S4 samples, respectively. The temperature dependent ac susceptibility measurements on these samples showed that the coupling between the grains of S2 sample is stronger than the other samples. It may be pointed out that the EPR signal was not observed in these samples suggesting that the impurity phase is not paramagnetic. SEM measurements showed that the average grain size in all the samples were approximately 10 μm. The EDAX measurements of these samples showed that the dopants have not entered the 2223 phase and it was also observed in all these samples that there was a deficiency of Sr and Ca in the 2223 phase.

DC electrical resistivity measurements were carried out on a home built setup in the temperature range of 77 K to 300 K with a Vander pauw's four probe technique. The measuring instruments used were the Keithely Nanovoltmeter (model 181), Keithely constant current source (model 224) and Lakeshore temperature controller (model 805) along with a Si diode sensor, DT470, for measuring the temperature. All these instruments have been interfaced to PC-AT through IEEE-488 General Purpose Interface Card (GPIB).

III. RESULTS AND DISCUSSIONS

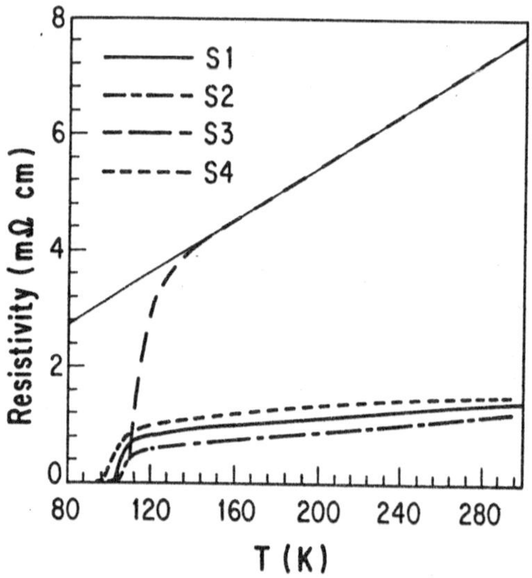

Fig. 1. Temperature dependence of dc electrical resistivity for sample S1, S2, S3 and S4. The fitted background resistivity is shown as solid line for S1

Figure 1 shows the electrical resistivity of the samples S1, S2, S3 and S4 respectively. The linear fits (a linear fit is shown in Fig. 1 for sample S1) to the experimental data is the background resistivity (ρ_B), which was estimated by fitting the measured resistivity $\rho(T)$ to the equation

$$\rho_B(T) = a + bT \tag{1}$$

in the temperature range $2T_{CMF}$ to 300 K, where T_{CMF} is the temperature at which the temperature derivative of the electrical resistivity exhibits a peak (as shown in Fig. 2 for the samples S1 and S2). The ρ(300 K) for the samples S1, S2, S3 and S4 are determined and given in Table 1. The ρ(300 K) of vanadium doped sample is found to be less compared to that of the other samples. The corresponding values of T_{CMF}, a and b of all these samples are also presented in Table 1. The excess conductivity was calculated by subtracting the extrapolated normal conductivity $[1/\rho_B(T)]$ from the measured conductivity $[1/\rho(T)]$ in Eq. 2[5],

Sample	Resistivity			Intergrain parameters			
	ρ(300) mΩcm	a 10^{-3} Ωcm	b 10^{-6} mΩcm/K	T_{CMF} K	ρ_c mΩcm	p	C_s
S1	7.64	0.94	0.22	112.5	0.94	160	0.33
S2	1.18	0.25	3.13	106.1	0.25	140.4	0.62
S3	1.5	0.9	2.2	100.5	0.9	165	0.42
S4	1.4	0.57	2.68	105.3	0.57	148.88	0.57

Table 1. List of parameters obtained by fitting the dc electrical resistivity data to the linear equation $\rho_B(T) = a + b\,T$, in addition to the intergrain parameters.

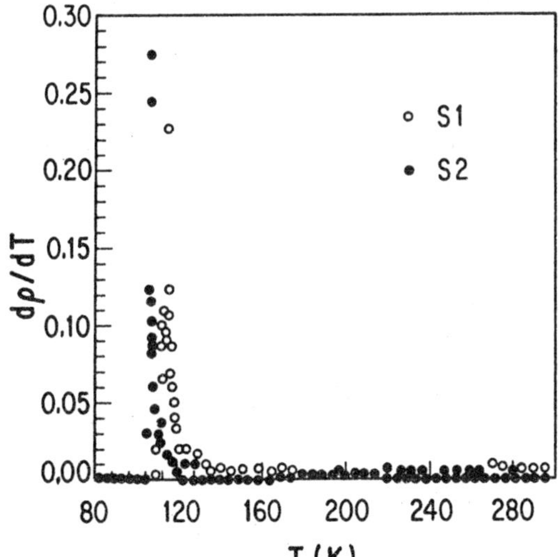

Fig.2. The temperature derivative of dc electrical resistivity of the samples S1 and S2.

$$\Delta\sigma = 1/\rho(T) - 1/\rho_B(T) \tag{2}$$

The experimental excess conductivity data calculated using the above equation was analysed in terms of LD theory in the NTR region. However, in Bi-based superconductors the cross-over between the 2D to 3D is expected to be quite close to T_c because the interlayer coupling is weak. The Bi-based superconductors could be described as a set of discrete superconducting layers whose order parameters are coupled by Josephson interactions[9]. The interaction between the layers is weak when $\xi(o) << d$, but increases as T decreases towards T_{CMF}. According to the temperature dependence $\xi_c(T) = \xi_c(o)\,\varepsilon^{-1/2}$, $\xi_c(T)$ may become equal to or larger than d for small values of ε, and a cross over from 2D to 3D behaviour is expected.

The experimental excess conductivity data was fitted to the expression $\Delta\sigma(T)/\sigma_0 = A/\varepsilon\,[1+B/\varepsilon]^{-1/2}$ where $A = e^2/16\,h\,d$ and $B = [2\xi_c(o)/d]^2$ are free parameters of fit, for the samples S1, S2, S3 and S4, respectively. The corresponding range of fits i.e., ε^1 (lower range) and ε^h (upper range), for LD fit, for all these samples is shown in Table 2. By substituting the values of $[2\xi_c(o) / d]^2$ and using following Eq.

$$T_0 = T_{cmf} \exp (2\xi_c(o)/d)^2 \tag{3}$$

the crossover temperature, T_0, from the 2D to 3D fluctuations were estimated for all these samples. The values of $\xi_c(o)$ were determined from the values of $[2\xi_c(o) / d]^2$ by substituting d as the c axis length of the unit cell. Using these values of $\xi_c(o)$ and $J = \xi_c(o)/d)^2$, the interlayer coupling constant, J, were also estimated. The estimated values of A, B, T_0, $\xi_c(o)$ and J for all these samples are given in Table 2.

Sample	Intragrain parameters					MFR		
	C2 mΩcm/K	A (Ωcm)-1	B 10-3	ξc (o) °A	J 10-3	T_0 K	In(ε^{λ})	In(λ^{h})
S1	0.32	40.5	6.09	1.45	1.53	113.2	-4.63	-3.4
S2	0.33	40.5	6.51	1.5	1.57	106.9	-4.75	-3.4
S3	0.33	40.5	6.25	1.47	1.56	101.1	-3.65	-3.1
S4	0.33	40.5	6.51	1.5	1.56	106	-4.2	-3.4

Table 2. List of intragrain and MFR region parameters obtained by fitting the experimental excess conductivity to the LD expression.

The experimental excess conductivity data are fitted to AL-2D ($\Delta\sigma(T)/\sigma_0= [e^2 / 16\ h\ d]\varepsilon^{-\lambda}$, with $\lambda = 1$) and AL-3D ($\Delta\sigma(T)/\sigma_0 = [e^2 / 32\ h\xi(o)]\varepsilon^{-\lambda}$ with $\lambda = 1/2$), with C and λ as free parameters of fit for samples S1, S2, S3 and S4. The values of λ for AL-2D are 1.2, 0.85, 1.18 and 0.82 for the samples S1, S2, S3 and S4, respectively. Similarly the values of λ for AL-3D are 0.52, 0.52, 0.42 and 0.43 for the samples S1, S2, S3 and S4, respectively. These results allow us to confirm the 2D nature of the SCOPF in the MFR region and 3D nature in the critical region for all these samples.

Oh et al[10] used a scaling factor C_s for the coefficient of reduced temperature, ε, in the expressions $C = C_s [e^2 / 16\ h\ d]$ for AL-2D and $C = C_s [e^2 / 32\ h\ \xi(o)]$ for AL-3D. The parameter C_s is to account for the non-uniform current path because of microcracks, poor grain boundaries etc. If $C_s = 1$, the magnitude of experimental $\Delta\sigma$ will be the same as the theoretically predicted value for a particular value of d. From the fitted values of C and taking d as the c-axis unit length (37.16 Å) as suggested by Oh et. al.[10], the C_s values were estimated and are given in Table 1. The C_s values of these samples suggest that these sintered samples exhibit non-uniform current path which may be due to poor grain boundaries in these samples. The C_s value of sample S2 is found to be large compared to that of the other samples, suggesting that the coupling between the grains of this sample is stronger compared to that of the other samples, supporting our earlier results[8]. The order parameter fluctuations are changing from 2-dimensional to 3-dimensional, i.e, from the mean field region to critical region as the temperature decreases towards T_c . The crossover from 2D to 3D fluctuations was found to be very close to T_c. The crossover temperatures for the samples S1, S2, S3 and S4 are found to be 113.6 K, 107.1 K, 101.5 K and 106.5 K, respectively.

IV. CONCLUSIONS

The paraconductivity studies in the MFR region in the zero applied magnetic field were performed on the samples S1, S2, S3 and S4. These results were analysed using the theories of LD and VV. The parameters were identified in terms of intergrain and intragrain. It was seen for all these samples that the intragranular parameters did not change appreciably confirming that the V/Nb/Ta has not entered the lattice but present at the grain boundaries as $CaMO_Y$ type impurity. All these results suggest that the addition of V-B group metal oxides to (Bi,Pb)-2223 superconducting system changes the intergrain properties appreciably. In addition, from the AL fitting, we confirm the 2D nature of the SCOPF in the WR region and 3D nature in the critical region.

ACKNOWLEDGMENTS

Authors would like to thank the Department of Science and Technology, Government of India, New Delhi for the financial support. One of the authors (PS) would like to thank the CSIR, Government of India, New Delhi for awarding a fellowship (SRF).

REFERENCES

1. J.G.Bednorz and K.A.Muller, Z Phys. **B64,** 189 (1986).
2. S.W. Tozer, A.W. Kleinsasser, T. Penney, D. Kaiser and F. Holtzberg, Phys. Rev. Lett. **59,** 1768 (1987).
3. T.R. Dinger, T.K. Worthington, W.J. Gallagher and R.L. Sandtrom, Phys.Rev. Lett. **58,** 2687 (1987).
4. Y. Hidaka, Y. Enomoto, M.Suzuki, M. Oda, A. Katsui and T. Murakami, Jpn. J. Appl. Phys. **26,** L726 (1987).
5. P. Freitas, C.C. Tsuel and T.S. Plaskett, Phys. Rev. **B36,** 833 (1987).
6. F. Vidal, J.A. Veira, J. Maza, J.J. Ponte, F. Garcia-Alvarado, J. Moran LAmador, C. Cascales, A. Castro, M.T. Casais and I. Rasines, Physica **C156,** 807 (1988).
7. S.H. Han and O. Rapp, Solid State Commun. **94,** 661 (1995).
8. P. Subhash, G. Narsinga Rao and D. Suresh Babu, Mat. Sci. Engg. B (unpublished).
9. R. Kleiner, F. Steinmeyer, G. Kunkel and P. Muller, Phys. Rev. Lett. **68,** 2394 (1992).
10. B. Oh, K. Char, A.D. Kent, M. Naito, M.R. Beasley, T.H. Geballe, R.H. Hammond, A. Kapitulnik and J.M. Graybeal, Phys. Rev. **B37,** 7861 (1988).

SURFACE MORPHOLOGY STUDIES OF Zr ADDED $YBa_2Cu_3O_{7-\delta}$ SINGLE CRYSTALS

A.Thamizhavel, D.Prabhakaran, R.Jayavel and C.Subramanian
Crystal Growth Centre, Anna University, Madras 600 025, India

Single crystals of Zr added $YBa_2Cu_3O_{7-\delta}$ were grown using excess BaO-CuO as flux for Zr concentration 0.1, 0.2 and 0.3 wt%. The growth mechanism was found to vary for pure and Zr added single crystals. The optical micrograph of the grown crystals reveals that the spiral growth mechanism is observed in the case of pure YBCO single crystals. A possible explanation for the shape and formation of the spirals has been given by the Periodic Bond Chain (PBC) theory. The Zr addition greatly affects the spiral growth patterns and the growth is facilitated by layer growth mechanism and hence, slightly thicker crystals have been obtained. The susceptibility measurement showed a sharp transition at 71 K for Zr 0.1wt% added single crystals, without any further oxygen treatment.

I. INTRODUCTION

Single crystals are essential to investigate the nature of high T_c superconductivity and to measure the various physical properties[1]. A great deal of research has been carried out into crystal growth of high temperature superconductors by the flux method because of their incongruent melting nature[2]. Recently, Erb et al.[3] have shown that high quality crystals have been obtained with the use of $BaZrO_3$ crucible because of its low reactivity with the melt during the growth process. In the present study, we report on the single crystal growth of pure and Zr added $YBa_2Cu_3O_{7-\delta}$ (YBCO) for 0.1, 0.2 and 0.3 wt% single crystals. From the optical microscopic pictures it has been observed that the growth mechanism changes for the pure and Zr added crystals. In the case of pure YBCO single crystals, the spiral growth pattern is predominant, whereas it is not so in the case of Zr added YBCO single crystals. The growth takes place by layer growth mechanism and hence, slightly thicker crystals have been obtained. No impurity peaks due to $BaZrO_3$ were observed from the powder X-ray diffraction (XRD) of the Zr added single crystals. The susceptibility measurement of the as grown crystals showed a sharp transition at 71K for the Zr 0.1 wt% added single crystals, without any further oxygenation.

II. EXPERIMENTAL

The composition of the starting materials used in the present study is the same as that we used for the growth of large size twin free single crystals[2]. Three batches were prepared with the addition of ZrO_2 (0.1, 0.2 and 0.3 wt%). The oxides and carbonates were calcined at 750°C for two to three times. The calcined powders were then heated to 1050°C and kept at this temperature for 24 hours in order to attain proper homogenization. When the temperature is lowered to the peritectic one, the liquid has a composition in equilibrium with the Y_2BaCuO_5 and the $YBa_2Cu_3O_{7-\delta}$, with no driving force for $YBa_2Cu_3O_{7-\delta}$ formation. On further cooling the 211 particles began to disappear from the liquid and the crystallization of 123 is generated. As the 211 particles get dissolved in the melt, the solution becomes richer in Ba and Cu thereby becoming undercooled with respect to $YBa_2Cu_3O_{7-\delta}$ and hence, grows the YBCO crystal. This has been observed by hot stage optical microscopy by Nishimura et al[4]. The local heterogeneities in the melt, like impurities and inclusions, act as nucleation centers apart from the rough inner wall of the crucible and thus resulting in the growth of large number of tiny crystals with square platelet morphology along the a-b plane. As the supersaturation of the 123 phase seemed to be quite high because of the dissolution of the 211 particles, the crystals were found to have many dislocations and defects. Hence low supersaturation is required for the growth of good quality single crystals.

III. RESULTS AND DISCUSSION

The micrograph in Fig. 1 shows the spiral growth pattern observed in pure YBCO single crystal. The origin of the spiral growth could be due to the dislocations crossing the surface with Burgers vectors making different angles with the dislocations. In the case of pure YBCO single crystals, the screw dislocation growth mechanism was dominant. Different shapes of growth spirals were observed on the majority of the crystals.

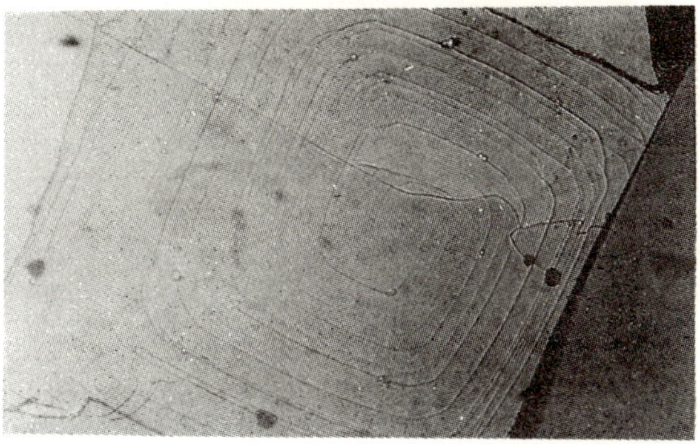

Fig.1. Spiral Growth pattern observed on the a-b plane of pure YBCO single crystal.

The morphology of the growth spiral is principally decided by the lateral roughness of the spiral step, that is, the kink density[5]. Usually the kink density can be described by Jackson's α factor[6] defined as $\alpha=\varepsilon L/kT_o$ where ε is related to the crystallographic anisotropy, L is the latent heat of melting, k is the Boltzmann constant and T_o is the equilibrium temperature. Here α is the measure of roughness of the interface. When the interface is rough, the steps of the spirals on the (001) surface can grow independently of the crystallographic directions and their shape is near circular. These crystallographic directions are decided by the periodic bond chains (PBC) directions[7]. Here we briefly recall the Periodic Bond Chain (PBC) theory. The PBC are uninterrupted sequences of periodically repeated strong bonds in crystallographic directions and that three categories of faces are distinguished: flat (F) containing two or more non-parallel PBCs in a layer of thickness d_{hkl}, stepped (S) containing only one PBC and kinked (K) containing one PBC. The kinked faces do not require nucleation and are normally not present in final morphology; the stepped faces grow by a one-dimensional nucleation, they are corrugated in a direction perpendicular to the PBC direction and sometimes appear so on a microscopic scale. The flat faces are slow growth faces, which grow by a two dimensional nucleation or by a spiral growth mechanism, and are usually the only ones to determine the final habit of the crystal. When the interface becomes rougher and rougher these steps are forced to follow the crystallographic directions, the next strongest <110> PBC direction where the kink density is lower and strongest <010>/<100> PBC directions where the kink density is even lower[8]. So on decreasing the roughness, there is a transition from circular to octagonal to square in the morphology of the spirals.

The bunching of steps and some macrokinks are observed on the singular face of the crystal of Zr added YBCO single crystal as shown in Fig 2. A possible explanation for the bunching of steps and the macrokinks is given by Sangwal et al[9]. The bunching of steps is due to the perturbation in the motion of steps. Due to the anisotropy in the surface energy, the surface energy per unit length of the shock waves of a crystal increases with their thickness and thereby inducing an instability in the bunches of shock waves. When the thickness of shock wave exceeds a certain value, it tends to attain a stable configuration. These stable configurations are either macrosteps or disintegrated segments of the wave. The macrosteps composed of rises formed by faces, while disintegrated steps appear in the form of macrokinks as shown.

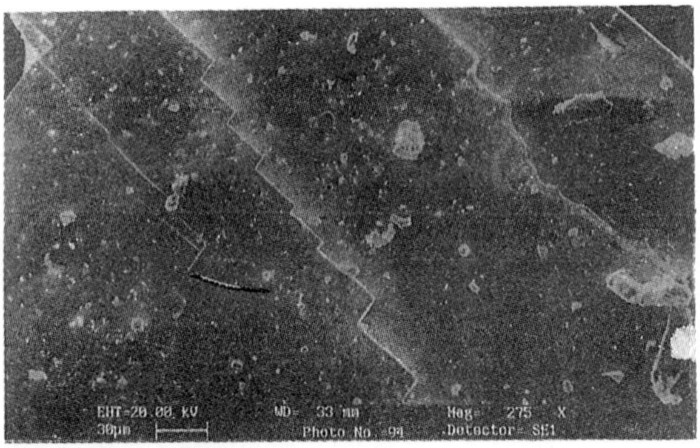

Fig.2. SEM picture showing macrosteps observed on Zr 0.1 wt% added YBCO crystal.

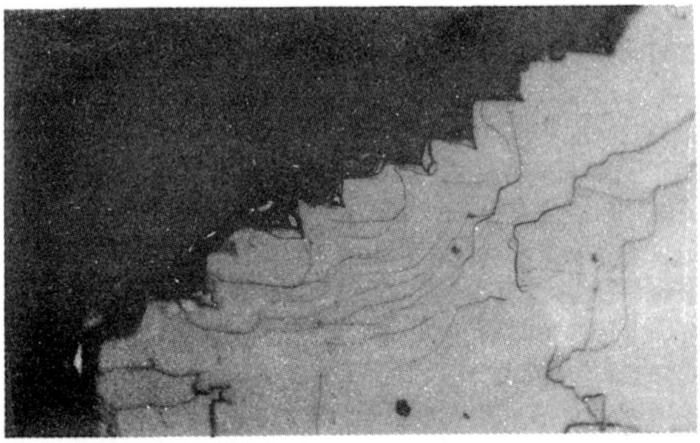

Fig.3. The external form of the crystal showing the K face for Zr 0.3wt% added YBCO.

The external form of the crystal is determined by various growth parameters like supersaturation, temperature, impurities etc. At high supersaturation, the additive Zr is adsorbed in the kink sites which in turn decreases the growth rate for all values of the impurity concentration and thereby retaining the K face in the final growth form as shown in Fig 3. The susceptibility measurement of the grown crystal for Zr 0.1 wt%, showed a sharp superconducting transition at 71 K without any further oxygen treatment. The unannealed pure YBCO did not show any superconducting transition down to 14 K. This implies that a small amount of Zr addition tremendously increases the ability of oxygen absorption.

IV. CONCLUSION

In conclusion, pure and Zr added YBCO single crystals have been grown from BaO-CuO flux using alumina crucibles. The spiral growth mechanism observed in the pure YBCO single crystals has been explained by means of the Periodic Bond Chain theory. The final growth form of the crystal changed in the case of Zr added single crystals and several macrokinks are observed because of the addition of Zr in the kink sites; the growth rate is decreased, and hence, the K face is retained in the final growth form.

ACKNOWLEDGMENT

The authors A.T and D.P gratefully acknowledge the financial assistance provided by CSIR, Govt. of India in the form of a fellowship.

REFERENCES

1. U.Welp, M.Grimsditch, H.You, W.K.Kwok, M.M.Fang, G.W.Crabtree and J.Z.Liu, Physica C **161,** 1 (1989).
2. R. Jayavel, A.Thamizhavel, P.Murugakoothan, C.Subramanian and P.Ramasamy J. Cryst. Growth **137,** 429 (1993).
3. Erb, E Walker and R.Flukiger,Physica C **245**, 245 (1995).
4. Y. Nishimura,Y.Yasuhara, S.Miyashita and H.Komatsu, J. Cryst. Growth **158,** 255 (1996).
5. I. Sunagawa and P .Bennema, *Preparation and Properties of Solid State Materials,* Vol.7, Ed. W.R. Wilcox , p. 129 (1982).
6. K.A. Jackson, in *Liquid Metals and Solidifications,* American Society for metals, Cleveland,OH, p. 147 (1958).
7. P.Hartman and W.G.Perdok Acta Crystallogr. **8,** 49 (1955).
8. B.N.Sun, R.Boutellier and H.Schmid, Physica C **157,** 18 (1989).
9. Sangwal , Prog. Cryst. Growth and Characterization **19,** 189 (1989).

STABILISATION OF OXYGEN IN SUBSTITUTED $YBa_2Cu_3O_{7-\delta}$

A.M. Umarji, N.S. Kini, P. Somasundaram, V.V.S.S.Saisundar
Materials Research Centre., Indian Institute of Science., Bangalore 560 012

The compositions in the system $YBa_{2-x}Ln_xCu_3O_{7-\delta}$ (Ln=La, Nd, and Gd and x = 0.0 - 1.0) prepared by the solid state method have been investigated for formation and improved superconducting properties. The importance of oxygen stoichiometry on the superconducting properties of 123 phase is widely recognised and an effort has been made to chemically fix the oxygen in the lattice by partially replacing Ba^{2+} with La^{3+}. The changes in the physical properties of 123 phase due to the stabilisation of oxygen is demonstrated in $YBa_{2-x}La_xCu_3O_{7-\delta}$ system. Results of acoustic emission measurements, dilatometry TGA and bulk modules measurement carried out by us are also summarised.

I. INTRODUCTION

It is now realised that the practical application of $YBa_2Cu_3O_{7-\delta}$ (123 phase) family of superconductors is hinged upon achieving high J_c values in bulk. Oxygen stoichiometry plays an important role in controlling the phase transition temperatures, getting good superconducting and mechanical properties. The oxygen tends to diffuse out of the lattice at higher temperatures during formation, processing and use, which is detrimental to the behaviour of the material and the device. The engineering applications of HTSC are restricted due to low strength and brittleness.[1] The 123 phase undergoes extensive microcracking and loses its strength during slow cooling when it undergoes structural phase transitions.[2] This lowers the transport J_c. The nonsuperconducting $YBa_2Cu_3O_{6.5}$ (i.e. $\delta > 0.5$) on the other hand, which does not undergo this transition, records comparatively higher mechanical properties.[3] Though high J_c values are reported for thin films, bulk materials always had low values until the melt texturing processes was developed.[4] Realisation of clean grain boundaries and incorporation of flux pinning centres has resulted in J_c values of the order of 10^6 amp/cm^2 even in bulk material.[5]

The substitution of La^{3+} at Ba^{2+} site in $YBa_2Cu_3O_{7-\delta}$ has been reported by us and others to be advantageous in many ways.[6] The oxygen content is stabilised, higher sintering temperature can be achieved and better mechanical properties are observed.[7,8] The solubility of Ln^{3+} at the Ba site has been shown to be determined by the size consideration in the study of $Ln_{3-x}Ba_{3+x}Cu_6O_{14}$ system.[9] This phase, refereed to as 336 phase, is closely related to 123 phase where, the ordered arrangement between Y and Ba sites is removed. However, these materials are semiconducting in nature. Demonstration of high J_c values in the melt textured Nd123 material,[10] where the flux pinning centres are Nd336 type of phase, has renewed our interest in these materials. In the present study attempt has been made to correlate various thermal properties of the substituted $YBa_2Cu_3O_{7-\delta}$ along with brief review of our earlier work. Preliminary results of acoustic emission experiments in $YBa_{2-x}La_xCu_3O_{7-\delta}$ system are presented.

II. EXPERIMENTAL

The compositions in the system $YBa_{2-x}Ln_xCu_3O_{7-\delta}$ (x = 0.0 - 1.0 and Ln = La, Nd and Gd) were prepared by the ceramic method, the details of which have been published elsewhere[7]. The samples were characterised by x-ray diffraction and superconducting phase transition temperature. TG/DTA experiments were carried out in the range 30 °C - 1200°C using a simultaneous thermal analyser (STA - 1500). The thermal expansion data was recorded in an indigenously built LVDT based dilatometer (30 °C - 850°C). Acoustic emission due to microcracking was recorded while cooling from 900°C. An alumina rod was used as wave guide to transmit the acoustic signals from hot zone to the piezoelectric sensor (175 kHz central frequency). The cumulative counts were recorded by acoustic emission spectrometer (Physical Acoustics model AE201), integrated over 25°C and plotted as a function of temperature.

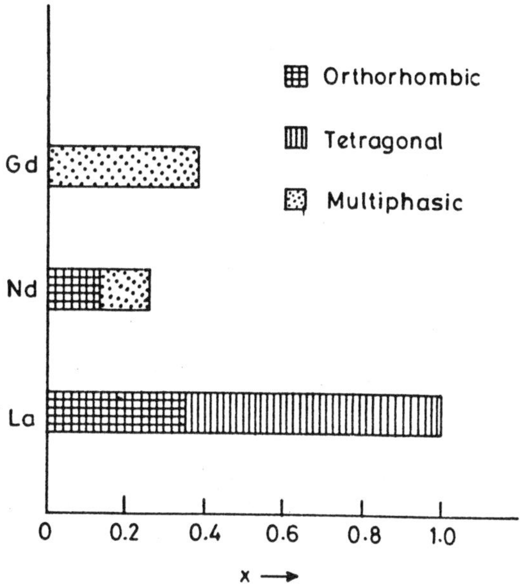

Fig.1. Range of monophasic compositions in $YBa_{2-x}Ln_xCu_3O_{7-\delta}$ system.

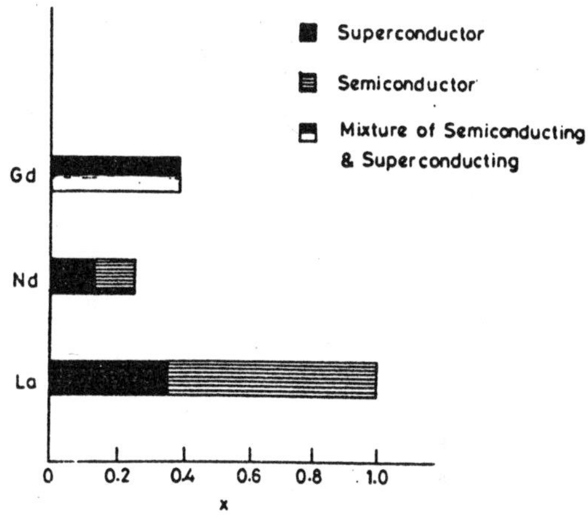

Fig. 2. Variation of electrical properties as a function of composition in $YBa_{2-x}Ln_xCu_3O_{7-\delta}$

III. RESULTS AND DISCUSSIONS

***Solubility of Ln at Ba site in $YBa_{2-x}Ln_xCu_3O_{7-\delta}$*:** The range of 123 phase formation as function of composition is shown in Fig. 1. The limit of monophasic range is seemingly limited by the size considerations. This is consistent with the observation for $Ln_{3-x}Ba_{3+x}Cu_6O_{14}$ system.[9] For Ln = La this limit is 50%; for Nd it is 15% and for Gd it is 0%. (the effective ionic radii for twelvefold co-ordination are Ba^{2+}=1.61A, La^{3+}=1.34A, Nd $^{3+}$=1.37A, Gd $^{3+}$=1.22A). For Ln = La, the lattice parameters show smooth variation as a function of x.[7] The structure changes from orthorhombic to tetragonal at x = 0.35. The cell volume is constant up to x = 0.35 and decreases beyond, indicating the δ to be positive for x > 0.35. For Ln = Gd, due to insolubility of Gd at Ba site, the phase formed from the nominal composition is likely to be

$$\underset{\text{(nominal)}}{YBa_{2-x}Gd_xCu_3O_{7-\delta}} \dashrightarrow \underset{\text{(actual)}}{Y_{1-x}Gd_xBa_2Cu_3O_{7-\delta}} + \underset{\text{(green phase)}}{Y_2BaCuO_5} + \ldots$$

All the x-ray patterns in this series were multiphasic, with the amount of green phase increasing with the value of x. The Nd system showed intermediate behaviour.

Electrical properties of $YBa_{2-x}Ln_xCu_3O_{7-\delta}$ (Ln = La, Nd, Gd) system: The nature of electrical properties are summarised in Fig.2. Monophasic compositions show superconductivity in the orthorhombic region. For Ln = La, the T_c gradually decreases upto x = 0.325 and semiconductor behaviour was observed for compositions with x > 0.35.[7] For small concentrations of La (0.02 < x < 0.1) the superconducting transition temperatures are very sharp. Among the Nd containing compounds only the x = 0.125 composition was superconducting, while the x = 0.25 composition was semiconducting. For Ln = Gd, all compositions show superconducting transition due to the presence of solid solution $Y_{1-x}Gd_xBa_2Cu_3O_{7-\delta}$.

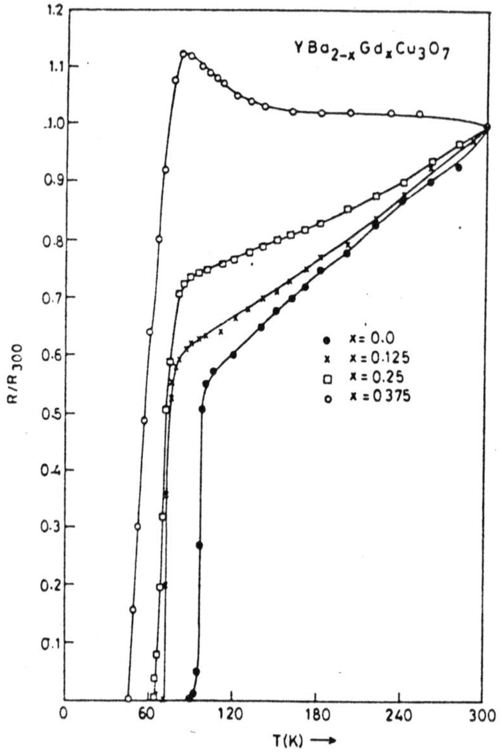

Fig.3. Variation of electrical resistivity as a function of temperature for $YBa_{2-x}Gd_xCu_3O_{7-\delta}$ system.

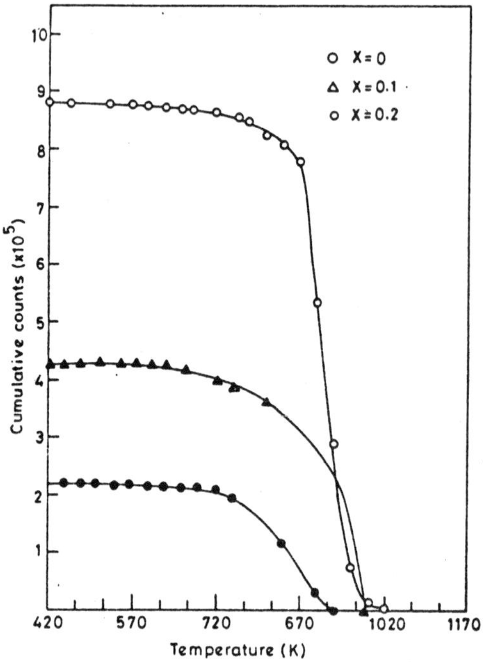

Fig.4. Cumulative acoustic counts recorded during cooling form 900°C in $YBa_{2-x}La_xCu_3O_{7-\delta}$

Increasing amounts of Y_2BaCuO_5 is reflected in the normal state resistivity shown in Fig. 3. This contrasting behaviour between La and Gd system gives an indirect evidence to show that La actually substitutes Ba and does not go to the Y site. The x-ray studies also support this conclusion. A detailed structural analysis is needed to confirm these observations.

Properties of $YBa_{2-x}La_xCu_3O_{7-\delta}$: These compositions were found to withstand higher sintering temperatures. Sintering temperature for undoped 123 is limited to 930°C due to decomposition to 211 phase and $BaCuO_2$ phase (which melts). The relative stability of 123 phase compared to 211 phase decreases as the size of the Ln atom decreases. In fact Lu123 is not known to exist and Yb123 is difficult to prepare in single phase form.[11] The reduction in the size of the Ba site (by substituting La) adds towards the stability of the 123 phase. This prevents the decomposition to 211 phase+ $BaCuO_2$ and hence, allows sintering at higher temperatures. Additionally the presence of 3+ ion at Ba site stabilises the O1 oxygen in the structure. This delays the loss of oxygen from the lattice and adds to the stability of oxygen. The high temperature x-ray studies indicate that the orthorhombic to tetragonal phase transition is pushed to higher temperatures with the increase in x.[12] The value of δ is close to 0 for $x<0.35$ irrespective of oxygen treatment. It was observed that oxygen treatment made little difference on the superconducting transition temperature. The T_c of compositions with x = 0.02 to 0.05 are reported to be a couple of degrees higher than the parent Y123.

Acoustic Emission in $YBa_{2-x}La_xCu_3O_{7-\delta}$: Microdeformations caused by the structural phase transitions during cooling of a sintered pellet causes microcracking which results in the release of acoustic signals. Results of acoustic emission experiments on $YBa_{2-x}La_xCu_3O_{7-\delta}$ are shown in Fig. 4. The cumulative counts recorded during cooling form 900°C indicate progressively lower acoustic activity between 650 °C to 350 °C as x value increases. The result is indicative of changes in the nature of oxygen in the lattice and associated phase transitions. Our earlier observation that the load bearing capacity increases with La substitution supports this result.[8] Bulk modules derived from unloading portion of the curve shows an increase from 70 GPa. (x = 0) to 110 GPa (x = 0.2). This increase may be attributed to the solid solution effect or change in the microstructure. It is expected that the La substituted material will have better superconducting properties like higher J_c.

TG/DTA and Dilatometric Studies in $YBa_{2-x}La_xCu_3O_{7-\delta}$ System: The preliminary thermal analysis results are consistent with the above observations. DTA experiments carried out up to 1200°C to study the decomposition behaviour clearly indicate a shift in the endothermic peak at ≈960°C to higher temperatures upon La substitution. The thermogravimetric curves indicate delayed oxygen loss and higher weight recovery upon cooling in air. Detailed experiments are being carried out in various atmospheres and will be published elsewhere. The thermal strains for $YBa_{2-x}La_xCu_3O_{7-\delta}$ system obtained during heating clearly show the effect of La substitution. The loss of oxygen results in the production of Cu^+ species (which is bigger in size) causing extra strain between 400 °C to 550 °C. The α value of pure 123 compound is about 15 x 10^{-6}/ °C up to 340 °C, above which a peak value of 30 x 10^{-6}/°C at 592 °C is observed. This coincides with the temperature of loss of oxygen and transformation to tetragonal phase. Upon La substitution, this peak value comes down to 23 x 10^{-6}/°C indicating substantially lower oxygen loss. Interesting step like behaviour is seen between 150 °C to 250 °C which may be related to structural changes involving oxygen rearrangement.

CONCLUSIONS

The Ln solubility at Ba site in $YBa_{2-x}Ln_xCu_3O_{7-\delta}$ is determined by size considerations. The decomposition of 123 phase to 211 phase is delayed by La substitution leading to the possible use of higher sintering temperatures. The nature of oxygen in La substituted material is altered and the materail appears to be more suitable for practical applications. A marginal decrease in T_c is compensated by the improvement of physical properties, like higher mechanical strength and reduced microcracking. TG/DTA and dilatometric experiments support these conclusions and indicate changes in the phase transition temperatures.

REFERENCES

1. I.K. Parker and H.G. Kim, Solid State Commun., **75**, 907 (1990).
2. T.J. Richardson and L.C. De Jonghe, J. Mater. Res., **5**, 2066 (1990).
3. J.H. Yin, X. Wang, Y. Sun, Fanchun and J. Du, Mat. Sci. Engg., **B7**, 227 (1990).
4. M. Murakami, Mod. Phys. Lett., **B4**,163 (1990); M. Murakami, M. Morita, K. Doi, K. Miyamoto and H. Hanada, Jpn. J. Appl. Phys. **28**, 1332 (1989).
5. A.S. Parikh, B. Mayer and K. Salama, Supercond. Sci and Tech. **7**, 455 (1994); K. Salama, V. Selvamanickm, L. Goo and K. Sun, Appl. Phys. Lett. **54**, 2352 (1989).
6. A.M. Umarji, Ind. J. Tech. **28**, 320 (1990).
7. P. Somasundaram, K.S. Nanjundaswamy, A.M. Umarji and C.N.R. Rao, Mat. Res. Bull. **23**, 1139 (1988).
8. S.A. Syed Asif, P. Somasundaram, A.M. Umarji and S.K. Biswas, Mat. Res. Bull., **28**, 73, (1993).
9. L. Er-Rakho, C. Michel, J. Provost and B. Raveau, J. Solid State Chem., **37**, 151 (1981).
10. T. Murama, S. Goshima, N. Sakai, S.I. Yoo and M. Murakami, this conference proceedings.
11. P. Somasundaram, R.A.Mohanram, A.M. Uamrji and C.N.R. Rao, Mat. Res. Bull. **25**, 331 (1990).
12. A.M. Uamrji, P. Somasundaram and S. Senbhagaraman, Mod. Phys. Lett., **B3**, 735 (1989).

ON THE SYNTHESIS AND OCCURRENCE OF CURIOUS LONG PERIOD POLYTYPE LIKE STRUCTURE OF Hg(Tl,Pb)-Ba-Ca-Cu-O HIGH TEMPERATURE SUPERCONDUCTOR

G.D.Verma, A.K.Srivastava and O.N.Srivastava
Department of Physics
Banaras Hindu University, Varanasi-221005, INDIA

In the present investigation, we have synthesised Hg(Tl,Pb)-Ba-Ca-Cu-O high T_c system with nominal compositions $Hg_{0.50}Tl_{0.50-x}Pb_xBa_2Ca_2Cu_3O_{8+\delta}$ (where x = 0.05, 0.10, 0.20). The as-synthesised materials showed transition temperature in the range of 110 to 120K. TEM explorations of these (Tl,Pb) substituted Hg-bearing HTSC materials revealed the occurrence of curious long period polytype (LPP) like structures. The possible influence of the occurrence of LPP on the superconducting parameters T_c and J_c has been outlined.

I. INTRODUCTION

The discovery of a transition temperature of ~134K in $HgBa_2Ca_2Cu_3O_{8+\delta}$, the new mercury cuprate high temperature superconductors (HTSC),[1-3] which broke the barrier held by $Tl_2Ba_2Ca_2Cu_3O_{10}$ with Tc ~125K[4] for more than five years during 1988 to 1993, led to a spurt of studies on the Hg-Ba-Ca-Cu-O and related HTSCs. In these compounds the great flexibility of the rock salt (Hg-O) layers permits the introduction of foreign elements e.g. suitable (size and valencies) cations to replace part of mercury in the Hg-O layers. In the present investigations, we have synthesised Hg(Tl,Pb)-Ba-Ca-Cu-O HTSC system with different nominal compositions of (Hg,Tl,Pb) and explored their structural features by employing X-ray diffraction and transmission electron microscopic (TEM) technique.

II. EXPERIMENTAL

For synthesis of HTSC phase of Hg(Tl,Pb)-Ba-Ca-Cu-O system, first high purity powders of HgO, BaO, CaO, CuO, Tl_2O_3 and PbO were mixed to form a mixture with nominal compositions $Hg_{0.50}Tl_{0.50-x}Pb_xBa_2Ca_2Cu_3O_{8+\delta}$ (where x = 0.05, 0.10, 0.20; the samples with these compositions have been designated as sample A, B and C, respectively) and ground thoroughly in a closed glove box having P_2O_5, NaOH and Ar/N_2 atmosphere. After grinding, the resulting mixture was immediately pelletised and the pellets were enclosed inside a silver box. The silver box containing the pellets was sealed inside a silica tube evacuated to 10^{-2} torr. The sealed tube was enclosed inside the steel container to avoid the possible explosion due to high vapour pressure of Hg at high temperature (~700°C). The steel container was kept inside the furnace for thermal treatment. The temperature of the furnace was raised at the rate of 200°C/hr upto 500°C and thereafter at the rate of 60°C/hr upto 720°C and held at this temperature for 5-6 hrs. Finally it was cooled at the rate of 60°C/hr upto room temperature.

III. RESULTS AND DISCUSSION

The variation of resistance with temperature of the as-synthesised Hg(Tl,Pb)-Ba-Ca-Cu-O HTSC samples was measured by the standard four probe method employing the Vander Pauw technique. The $T_c(R=0)$ for samples A, B, and C were found to be ~120K, ~113K and ~110K, respectively.

Curious structural features observed in the Hg-bearing pure and doped (e.g. Hg(Tl,Pb) HTSC cuprate) system observed recently relate to (a) growth of rather large number of CuO_2

layers bearing unit cells[5], (b) occurrence of long period polytype (LPP) structure which often are formed through intergrowth of smaller basic structural units e.g. 1212 and 1223 or 1234 and 1245[6,7]. In the present investigation, studies involving substitution of Tl and Pb at Hg sites have been carried out. This has led to the occurrence of the long period polytypes (LPP) resulting from intergrowth of smaller basic structures and the structures employing long range ordering of (Tl,Pb) atoms. These are described and discussed in the following.

Occurrence of Long Period Polytypic Structures in the Hg(Tl,Pb)-Ba-Ca-Cu-O, HTSC Materials

(a) $Hg_{0.50}Tl_{0.45}Pb_{0.05}Ba_2Ca_2Cu_3O_{8+\delta}$ (Sample A)

Investigation of the HTSC material with Tl/Pb doping in Hg-Ba-Ca-Cu-O leading to $Hg_{0.50}Tl_{0.45}Pb_{0.05}Ba_2Ca_2Cu_3O_{8+\delta}$ (T_c ~120K) revealed the existence of interesting structural transitions resulting in long periodicity structures. Figure 1 represents a typical electron diffraction pattern and corresponding direct lattice resolution micrograph. The dominant c spacing discernible from the diffraction pattern was found to be 18.98Å. However, the presence of subsidiary spots and streaking along c, suggested the presence of other long period polytype structures. Analysis of high resolution electron micrographs (HREM) revealed that LPP structural units (slabs) with c-periodicities of ~31.58Å, ~63.16Å, ~107Å and ~139Å were the most commonly encountered polytypic periodicities. The occurrence of some other polytypic slabs were also noticed but frequency of their occurrence was rather small. A noteworthy feature regarding the observed polytypic structure is that their distribution did not exhibit long range ordering, i.e., these slabs were not found to repeat to discernible distances. Instead they were found to be present in localised regions. Explorations of the origin of these LPP structures revealed that their genesis becomes readily intelligible in terms of the intergrowth of small period basic structures.

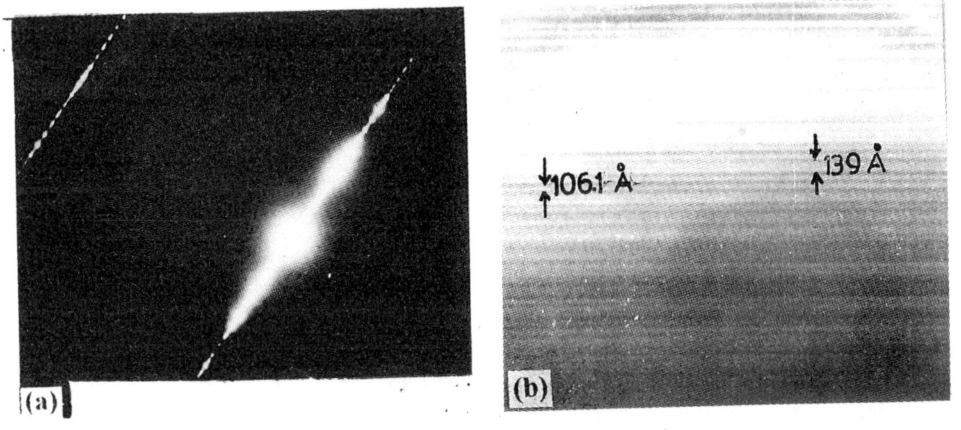

Fig.1. (a) Electron diffraction pattern, and (b) HREM of sample A.

(b) $Hg_{0.50}Tl_{0.40}Pb_{0.10}Ba_2Ca_2Cu_3O_{8+\delta}$ (Sample B)

Figure 2 (a) and (b) brings out representative examples of SAD pattern and HERM of the HTSC phase corresponding to sample B. The direct lattice resolution micrograph (Fig.2b) reveals that unlike the previous (Pb:0.05) sample, the (Pb:0.10) sample shows several c periodicities embodying individual structural slabs, polytypic slabs repeating over large extents, and hence, having ordered nature. These ordered LPP structural units with c ~246Å are easily discernible in Fig.2(b). An interesting aspect regarding the ordered polytypic slabs is that each individual slab contains three subsidiary units with perodicity c ~82.12Å. The fine structure of the individual polytypic slab in terms of subsidiary slabs are clearly visible in the main slab and can become intelligible in terms of intergrowth of (1212) + 2(1223) + 2(1234) = 12.60 + 2(15.79) + 2(18.97) = 82.12Å ~82Å. However, the formation of LPP having c-periodicity of 3(82) = 246Å cannot be understood in terms of any type of intergrowth.

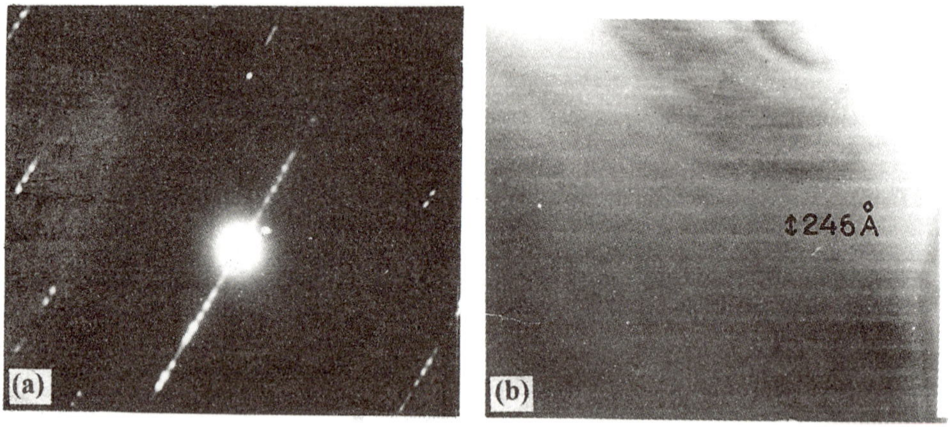

Fig.2. Sample B: (a) electron diffraction pattern; and (b) HREM.

(c) $Hg_{0.50}Tl_{0.30}Pb_{0.20}Ba_2Ca_2Cu_3O_{8+\delta}$ (Sample C)

Figure 3 exhibits HERM of representative long period polytypic structures obtained through structural transition in sample C. A close examination of the micrograph shown in Fig.3 brings out that all the fringes have the periodicity of ~18.93Å which corresponds to the 1234 structure. However, an additional feature easily discernible in Fig.3 is the ordering of 4(1234) + 4(1234) + 2(1234) (Å) leading to a ~189Å wide polytypic slab. This LPP with 189Å periodicity then repeats upto several units covering a span of more than ~1μm. It is interesting to point out that the creation of a LPP with ~189Å periodicity cannot be produced through any intergrowth of small period basic structures. However, the formation of LPP having c-periodicity ~246Å (sample B) and ~189Å (sample C) can be explained based on ordering of Tl and Pb atoms (cations) over distances corresponding to the respective c-periodicities.

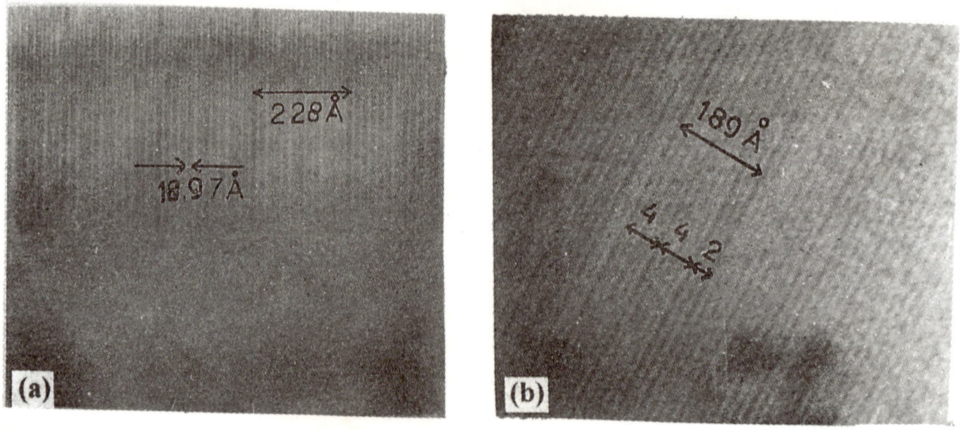

Fig.3. HREM of (a) sample B, and (b) sample C.

In regards to the influence of the occurrence of the LPP on the superconducting transition temperature T_c is concerned, it may be mentioned that except for hole concentration optimization effects originating from Tl, Pb substitution on Hg sites, the intergrowth of short period basic structures or the ordering of Tl, Pb atoms in the Hg(Tl,Pb)-O layers are not expected to influence T_c. This is in keeping with the present observed results where T_c is found to lie between ~110K and ~120K for the LPP structures and hence, does not vary significantly. However, the structural transitions resulting from Tl,Pb substitutions would produce interfaces and stacking faults and these can influence the critical current density. Further studies relating to the influence of the said defects associated with the presence of LPP, on J_c are required to elucidate this aspect.

CONCLUSIONS

In summary it can, therefore, be said that the present investigations aimed at studying the structural transitions orginating from Tl, Pb substitutions on the Hg-O layer in regard to the creation of new structural variants, has led to the observation of curious long period polytypes. It has been found that in addition to the known process of formation of long period polytypes in terms of intergrowth of short period basic structures a new mechanism of creation of LPP through long range ordering of Tl, Pb atoms on the Hg(Tl,Pb)-O layers is also possible. This mechanism noticed and discussed for the first time in the present investigation results in the occurrence of LPP structures with c-periodicities which are integral multiples of the c-periodicities of the small/intermediate period basic structures or intergrowth induced moderate period polytype structures. Thus LPP with c ~ 246 Å (3 times the intermediate period polytype periodicity of ~ 82 Å); and c ~ 228 Å (12 times the c-period (18.97Å)) have been found for the HTSC material with the composition $Hg_{0.50}Tl_{0.40}Pb_{0.10}Ba_2Ca_2Cu_3O_{8+\delta}$ (sample B). Similarly for the composition $Hg_{0.50}Tl_{0.30}Pb_{0.20}Ba_2Ca_2Cu_3O_{8+\delta}$ (sample C), the observed LPP corresponds to c = 189Å which is integral multiple of 18.93Å, the c-periodicity of the 1234 structure. This LPP gets formed through ordering of Tl, Pb with a 4, 4, 2 sequence of 1234 basic structure. The possible influence of the occurrence of LPP on the superconducting parameters T_c and J_c has been outlined.

ACKNOWLEDGMENTS

The authors are grateful to Professors A.R.Verma, C.N.R.Rao, R.P.Rastogi, H.P.Gautam, A.V.Narlikar, T.V.Ramakrishnan and R.K.Singh for encouragement. Two of the authors (AKS, GDV) are grateful to C.S.I.R. (New Delhi) for providing financial support.

REFERENCES

1. A. Schilling, M. Contoni, J.D. Guo and H.R. Ott, Nature **363**, 56 (1993).
2. S.N. Putilin, E.V. Antipov and M. Marezio, Physica C **212**, 266(1993).
3. Z.J. Huang, R.L. Meng, X.D. Qiu, Y.Y. Sun, J. Kulik, Y.Y. Xue and C.W. Chu, Physica C **217**, 11 (1993).
4. S.S.P. Parkin, V.Y. Lee, A.I. Nazzal, R. Savoy, R. Beyers and S.J. La Placa, Phys. Rev. Lett. **61**, 750 (1988).
5. J. Karpinski, H. Schwer, I. Mangelschots, K. Conder, A. Morawski, T. Lada and A. Paszewin, Physica C **234**, 10 (1994).
6. H.M. Shao, L.J. Shen, X.Y. Hua, P.F. Yuan and X.X. Yao Physica C **232**, 5 (1994).
7. C.C. Lam, et al., Physica C **247**, 43 (1995).

LONGITUDINAL CURRENTS AND 'C' AXIS RESISTIVITY IN SUPERCONDUCTORS

A.M.Campbell

IRC in Superconductivity, Madingley Road, Cambridge, UK

A survey is given of experiments on low T_c superconductors. Although it is generally agreed that the ease of flux cutting determines critical currents, vortex instabilities lead to complex structures for which there is no satisfactory theory. In high T_c superconductors there has been little evidence of the effects of flux cutting. This is probably due both to the presence of weak links in granular materials and the layered structure of single crystals. In both YBCO and BSCCO the material behaves above the irreversibility line as a standard anisotropic resistive material. However the fact that the c-axis and ab activation energies are very similar, and also much greater than kT gives important clues as to the transport mechanism.

1. INTRODUCTION

The electrodynamics of flux lines in a perpendicular field are well understood. The Lorentz force tends to push them in the direction of lowest field, but they do not move until the Lorentz force exceeds the pinning force. However, as soon as there is a component of magnetic field parallel to the current we do not have much idea how to calculate the field distributions or critical current, even without the complication of pinning centres. The results depend on the ability of flux vortices to cut each other, but a satisfactory quantitative theory is not yet available. Experiments on low T_c superconductors usually used wires in a parallel field. This type of experiment has been carried out on high T_c superconductors with rather different results. However the most interesting physics is probably to be found in the c-axis resistivity, or critical current, measurements in thin films and single crystals in a perpendicular field. This is a very different geometry from that investigated in low T_c wires, and yet the same problems arise if we view the system in terms of flux lines. If we use Josephson junction theory the presence of perpendicular flux lines is frequently ignored, but we know from low T_c experiments that this is not justified. Many scientists have contributed to the subject of longitudinal currents over the last thirty five years but it remains a major challenge. The subject is not entirely academic. Force free winding of magnets may be inefficient in terms of materials but it removes an upper limit on magnetic fields imposed by the strength of materials. There is an added bonus if superconductors are used in that the critical current density is higher if the field is parallel to the conductor. A number of experiments on high T_c materials benefit from comparison with early work on low T_c materials. So in this paper I will bring together a summary of some of the low T_c work in historical order, and show to what extent it is relevant to high T_c measurements.

Perhaps the first relevant experiment was carried out by Pippard[1] around 1960. He put a magnet at the end of a Type II cylinder in an axial field and rotated it about the axis in an attempt to wind the vortices into helices. Since pinning effects proved dominant the results were not interesting and therefore not published. A few years later Josephson[2] pointed out the paradox of a voltage in longitudinal geometry with cylindrical symmetry, which is that we need a continuous flow of flux to the axis, and to avoid the corresponding increase in flux density the vortices must cut each other.

A number of experiments and theoretical developments followed,[3,4]. A more complete list of references will be found in ref. 5. If flux is cutting we are not bound by the Josephson equation **E=vxB**. In this case the velocity of the flux lines at a point is not uniquely defined so the equation is meaningless. The connecting theme of this paper is the influence of flux cutting on a number of experiments and the extent to which low T_c experimental results can be applied to high T_c materials.

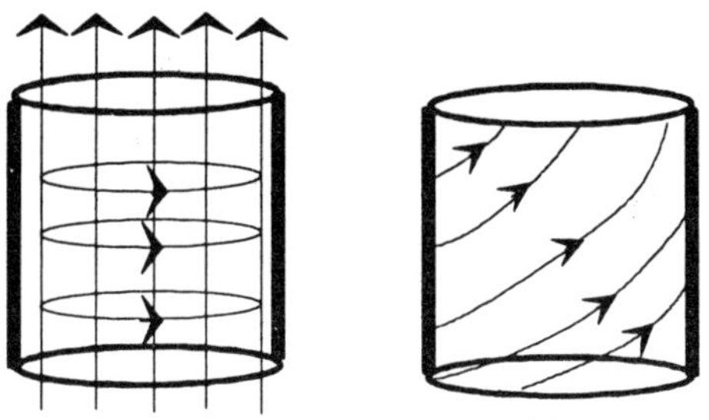

(a) (b)

Fig.1. (a) is a superposition of an axial and an azimuthal field. The lowest energy of the flux lines is a helical field as in (b).

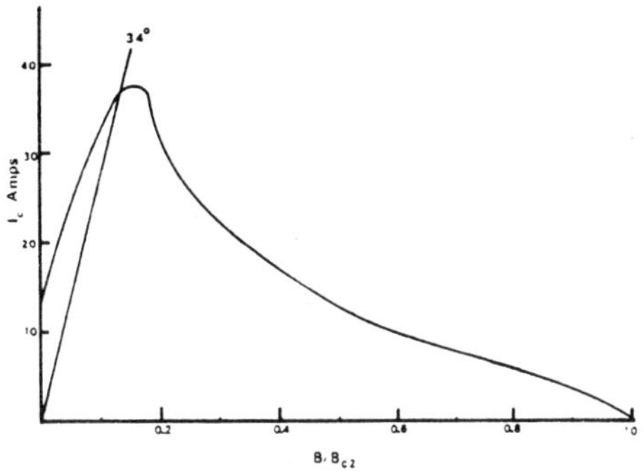

Fig.2. The longitudinal critical current density in PbIn, from ref. 10.

II. FLUX DISTRIBUTIONS IN WIRES

If a type II superconducting cylinder is in zero field an axial transport current will generate flux rings. If the critical current is exceeded these rings will collapse to the axis and vanish continuously, creating an axial voltage. If we add an axial field and consider a linear system we can imagine the straight flux lines being independent of the flux rings due to the current, so that there is no interaction between them, and the V-I characteristics are unaltered (Fig.1a). It is not easy to see how this works on a microscopic scale since the flux lines must be helices, as in Fig.1b., but this must be the general picture just below B_{c2}, since there is no discontinuous change on crossing to just above B_{c2}. The normal state can be regarded as one that requires no force or energy for flux cutting. As we go below B_{c2} the discrete nature of the vortices becomes manifest so that it becomes more and more difficult for the two systems to cross each other and we see the effects associated with force free current flows.

The flux picture can also be applied to Josephson Junctions, where a voltage is associated with the collapse of Josephson vortices or phase slip. There is nothing special about Josephson coupling which distinguishes Josephson junctions from pinned flux[6]. Josephson effects will appear in any superconductor small compared with the penetration depth whether it is Josephson vortices or Abrikosov vortices which are moving.

A current parallel to a field is force free and satisfies the equation $\mathbf{B}=k\ \mathrm{curl}(\mathbf{B})$ where k is any function of position. For planar geometry this reduces to $|\mathbf{B}|$=constant. The angle of the field can change in any way we like. It is not necessary to make any effort to set up a force free configuration. If we start with an axial field, and apply an axial current, helical vortices will try to enter from the surface. If they cannot cut through the axial flux, a force free configuration is automatically set up in the surface layer.

Experimentally the critical current curves look quite different from those in a transverse field. In particular, after an initial small drop, there is a rise in critical current density with applied field. The critical current goes through a maximum at about one third B_{c2} before going to zero at B_{c2} (Fig.2). A clear indication that this peak effect is due to the difficulty of flux cutting came in an experiment by Berlanger and Le Blanc[7] who showed that trapping axial flux lines in a cylinder led to an increase in the critical current of a wire. It was also found that the helical flux compressed the axial flux leading to a paramagnetic moment[8].

A number of experiments showed that although the geometry appears simple the situation is in fact very complex. Irie et. al.[9] put voltage contacts round a wire and concluded that there was a helical pattern of flux flow.

However Cave et al.[10] used a slider to measure the voltage along a rod and found a very complicated pattern, including sections of reversed electric field (Fig.3). We know from plasma theory that a purely longitudinal current is unstable with respect to the formation of helices and a model based on this has been used to predict a critical current[11]. If a straight flux structure is slightly curved the magnetic pressure inside the curve increases and tends to increase the curvature. This is resisted by the line tension if the ends are pinned, but we might expect the critical current to occur when the axial lines have moved right across the sample so that cutting can occur in free space, as shown in Fig. 4b.

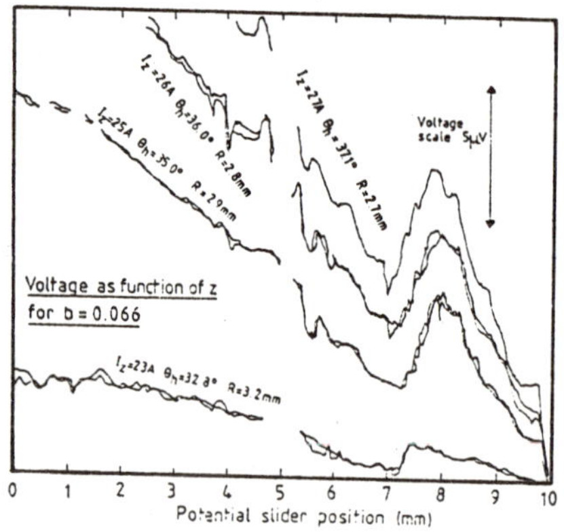

Fig.3. Complex voltage patterns along a rod.

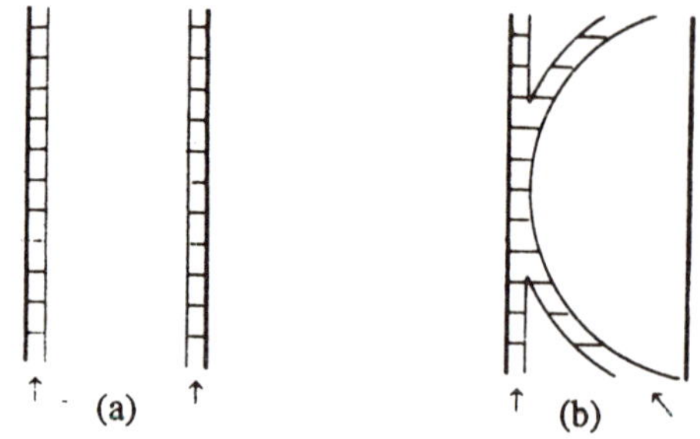

Fig.4. An axial current makes flux unstable with respect to bending.

This predicts I_c when the angle of the field at the surface is 34 degrees, and this line is drawn in Fig.2. There is only agreement over a small field range. However this is the only model which predicts fields which are not uniform along the length of the sample and is a useful qualitative picture. It illustrates the difficulty in interpreting experiments, such as those of Le Blanc, designed to show flux cutting directly[12].

Le Blanc trapped azimuthal flux in a cylinder by applying a current through a wire up the axis, and then applied an axial external field. The appearance of flux inside the cylinder showed that the axial flux must have crossed the azimuthal flux to get there. However this did not necessarily take place inside the superconductor. Flux could be pushed out of the cylinder wall at some point and the cutting take place in free space.

Another experiment of Cave et al.[13] measured the change in axial flux and azimuthal flux when an ac current was applied. If we imagine vortices generated as helices on the surface, and moving in to the centre of the sample, the ratio of the two fluxes is just the tangent of the field angle at the surface. Experimentally it was found that the ratio was lower than this by a factor of 1000. The best model which explained most of Cave's results was to regard the axial field lines as extra pinning centres, (see Fig.1a) while the azimuthal flux generated by the current moved through them building up the standard critical state. However this does not explain what is going on a microscopic scale, and is not consistent with the longitudinal inhomogeneity.

Although old results, these demonstrate two terms introduced in the theory of high T_c superconductors, to be discussed later. The first is entanglement of flux lines. The extremely complex electric fields in cylinders must be caused by complex helices of entangled vortices. The second is nonlocality[14]. Although the relation between flux lines and the atomic lattice is local on a sub micron scale, and the geometry is cylindricaly symmetric, the macroscopic electric field is inhomogeneous on a millimetre scale. However the concepts were introduced to explain data above the irreversibility line, where the entanglement is thermally induced, and the resistivities linear.

III. OTHER GEOMETRIES

Although most experiments were done in cylindrical symmetry many other geometries involve a longitudinal component of the current. One of these is the penetration of a rotating field into the surface of a disc. These results have been interpreted with a phenomenological model which supposes that there is a critical angle for flux cutting as well as a critical pinning forces[15].

An even simpler example is a fully reversible cylinder at an angle to an applied field. It is commonly thought that a reversible Type II material behaves much like a conventional conductor with $\mathbf{E}=\rho\mathbf{J}$ where ρ is the flux flow resistivity. However a standard application of Faraday's law shows that the induced electric field is not then perpendicular to the applied field, so that the current has a component along the field. We can have $\mathbf{E}=\rho\mathbf{J}$ or $\mathbf{E}=\mathbf{v}\mathrm{x}\mathbf{B}$ but not both.

IV. THEORY

In trying to construct a theory the basic problem is how to get the azimuthal flux into the sample. One route is by adjacent rows of vortices coming together, twisting, and separating, thus effectively passing a ring between them[16].A calculation of the flux cutting force has been made by Clem et al.[17]. Another possibility is to twist the vortex lattice, or nucleate a helical instability[18,19]. The problem with this is that it passes the flux rings along the length of the wire so that the critical current would be very dependent on length, and it does not seem probable in long wires of low T_c materials. However the thin films and crystals used in experiments on high T_c materials could favour this process. Another route applicable to layered structures is the passage of flux rings between the layers where the coupling is weak, and this is considered in more detail below.

V. HIGH T_c MATERIALS

With this background in low T_c materials let us see what differences high T_c materials have brought. There are two main new features. One is the presence of weak links or Josephson Junctions, the other the irreversibility line.

Longitudinal Currents in Josephson Junctions

Let us look first at the effect of the junctions. Although Josephson junctions are normally treated using tunneling theory, this gets pretty complicated with arrays of junctions and in the presence of vortices. However, a Josephson vortex is not very different from an Abrikosov vortex, and many of the properties of junctions are more easily understood by starting from our knowledge of Type II superconductors[6]. In this picture the critical current of a junction is determined by the surface barrier to the nucleation of a vortex ring at the edge of a junction. Once nucleated this collapses to the centre like the flux rings in a wire, phase slip in junction terms. Any irregularity in the junction can lead to the pinning of Josephson vortices just as for Abrikosov vortices in bulk superconductors, so that the graph of J_c against B no longer has the simple diffraction pattern of a single slit.

In low T_c materials the distinction between flux pinning and junctions is usually clear, but in high T_c materials it is much less obvious. Grain boundaries are large compared with the Josephson penetration depth, and the planes in BSCCO are far more closely coupled than any array of junctions.

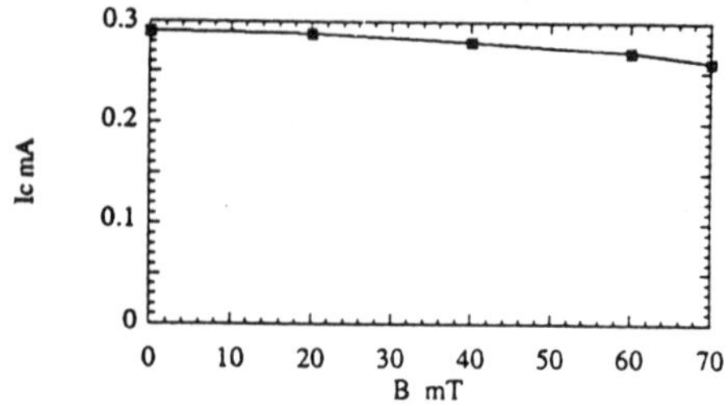

Fig.5. The reduction in the I_c of a Josephson junction with applied parallel field.

The Josephson equations are unaltered by the presence of a field parallel to the current, but the flux line picture suggests that there might be an interaction. Kumar[20] has looked for this by applying a field parallel to an electron beam cut junction in YBCO. He was able to measure diffraction patterns largely unaltered up to applied fields of 70 millitesla, at which field the I_c was reduced by about 10%, (Fig.5). The results show no increase in J_c with applied field indicating that flux cutting is extremely easy in a junction. This is not too surprising. The Josephson vortex has a much larger length scale than the Abrikosov vortices it is trying to cross, so that the interaction force will be low. (It is forces rather than energies which are important here). The implication is that where currents are determined by weak links we do not expect to see flux cutting effects.

However the field certainly affects the resistivity as was shown by Kim et. al.[21]. They monitored the resistance of a Nb junction in a perpendicular field as it went through the transition temperature. They found that the activation energy of the resistivity was that of a junction of size equal to the vortex spacing. One interpretation of this result is that the junction current, which is parallel to the field, generates the azimuthal flux by the helical instability of each vortex, as suggested by Clem for low T_c materials.

Granular Materials

Many experiments have shown that there is little Lorentz force dependence of J_c in granular superconductors. This can be attributed either to the fact the weak links will not show one, or that the current path meanders through the sample so that there are always sections perpendicular to the current. Probably both are true.

Le Blanc et. al have looked for flux cutting in cylinders by applying a field in one direction, and then a perpendicular field[22]. Instead of compressing the field as would occur if flux cutting were difficult, as happened with low T_c materials, the perpendicular field wipes out the initial magnetic moment. This may be understood if J_c is independent of Lorentz force. If the magnitude of J_c is constant the direction must rotate to screen the new applied field.

Single crystals and Melt Processed YBCO

More variable are experiments on single crystals. Park et al. found that c-axis flux increased a-b hysteresis at low temperatures,[23]. However, Hasanain et al. found that at 77K c-axis remanent flux increases the ease of penetration of a-b flux[24]. Clear evidence of force free configurations has been found by André et.al. using optical methods[25].

However, most of the work has been done using transport currents and an array of voltage contacts. These experiments can be divided into those above the irreversibility line and those below. Most work has been conducted above the irreversibility line with fields parallel to the c direction, where the system is linear and low current densities can be used. This regime does not occur to any significant extent in low T_c materials and a wide range of explanations has been offered. The irreversibility line is associated with the reduction in a-b plane resistivity[26]. Also the most important factor in the position of the irreversibility line is the coupling between planes[27]. This means that c-axis conductivity is central to the problem and therefore involves the little understood force free configuration. As always the problem can be put in terms of either currents or flux lines, so that Josephson coupling between flux pancakes in adjacent layers will be closely associated with c-axis conductivity. A purely magnetic interaction on the other hand will not be associated with c-axis supercurrents and must be regarded as a type of Giaever transformer. It might seem that if decoupling is the criterion pinning centres are irrelevant. However, it is well established that columnar defects will increase the irreversibility field. The mechanism may be due to the correlation of pancakes in adjacent layers by pinning helping the Josephson coupling between planes. The subject is still wide open.

The regime above the irreversibility line means that flux lines can wander about and become entangled in an equilibrium phase due to thermal agitation,[14], as opposed to low T_c materials where the entangled helices were produced by a transport current. However there is evidence that the irreversibility line is associated with decoupling between planes, in which case flux cutting becomes extremely easy, and any entanglement would be rapidly removed. The equilibrium structure will depend on the competition between these two processes. It is therefore of considerable importance to determine whether flux cutting is easy in high T_c superconductors. It should be noted that as for pinning effects there will be two regimes. Below what might be called the flux cutting irreversibility line it requires a force to make vortices cut. This means there will be a critical current and the V-I characteristics will be non-linear. Above this irreversibility line, which is where the cutting energy becomes comparable with kT, the flux cutting will occur by thermal activation and the characteristics will be linear. The material will then behave as a standard anisotropic resistive material.

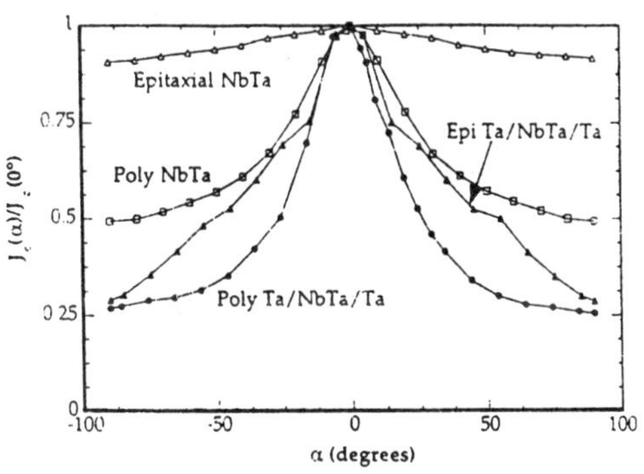

Fig.6. Lorentz force dependence of J_c in thin films.

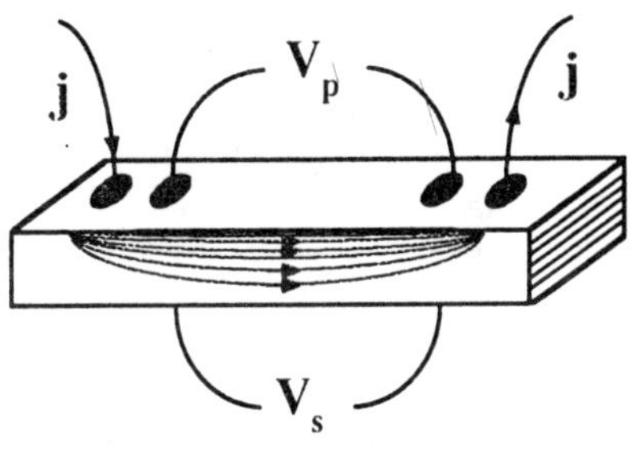

Fig.7. The contact configuration.

There is a considerable body of evidence for a pancake model which simplifies the problem. The principle is that at a certain field and temperature the planes of the crystal decouple, possibly with a phase transition. This would allow flux rings to penetrate between the layers so that force free effects would disappear at the decoupling temperature, at the same time as pinning effects. If this is true we might expect to see the effects of force free configurations and flux cutting, below the irreversibility line, but not above. Since flux pancakes will have lower activation energies than flux lines, the pinning irreversibility line will coincide with the flux cutting irreversibility line. To establish that force free effects can be used to probe the dimensionality of a system Kumar[28] did some experiments on the Lorentz force dependence of J_c in Nb Ta thin films (Fig. 6). Three of these were dominated by bulk pinning. These were a polycrystalline film, a polycrystalline film with the surface barrier removed by grading the concentration, and an epitaxial film with the surface removed. All these showed a marked Lorentz force dependence. The only film to show almost no Lorentz force dependence of J_c was an epitaxial film with intact surfaces, i.e. with two dimensional pinning on the top and bottom surfaces (Fig.6). We conclude that lack of Lorentz force effects is the mark of a two dimensional system.

Transformer Configuration

The largest number of experiments are measurements of the V-I characteristics of a crystal measured by contacts on various faces of the crystal. The most striking results appear in the so-called 'transformer' configuration in which the current is passed into the top surface and the voltages on top and bottom surfaces compared[30]. The configuration is shown in Fig.7. The top voltage is called the primary, Vp and the bottom voltage the secondary, Vs. At first sight this might not seem to have anything to do with force free configurations, but since the current is injected in the top plane it must have a component in the c direction if it is to spread through the crystal. A typical result is shown in Fig.8, which is for a BSCCO crystal in the linear regime. The voltage on the top contacts goes smoothly to zero but that on the bottom contacts reappears after disappearing into the noise and joins up with the voltage on the upper contacts. It has been claimed that the coincidence of the two voltages implies that the flux lines are moving as complete lines, so that the top vortices are pulling the bottom ends behind them with their line tension. However, this conclusion is very dependent on the resolution of the voltmeters.

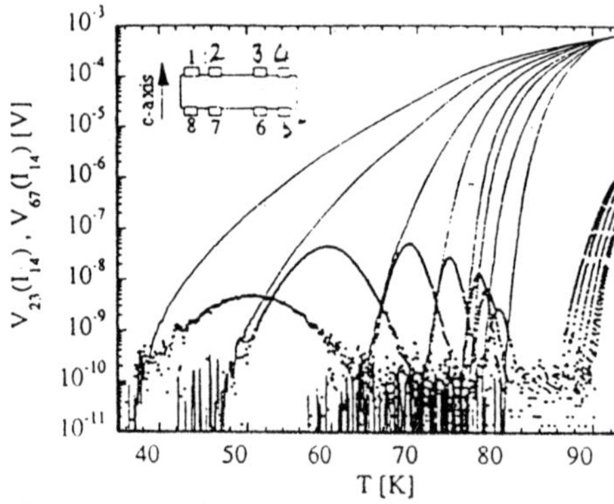

Fig.8. Vp and Vs in a BSCCO crystal in several fields.

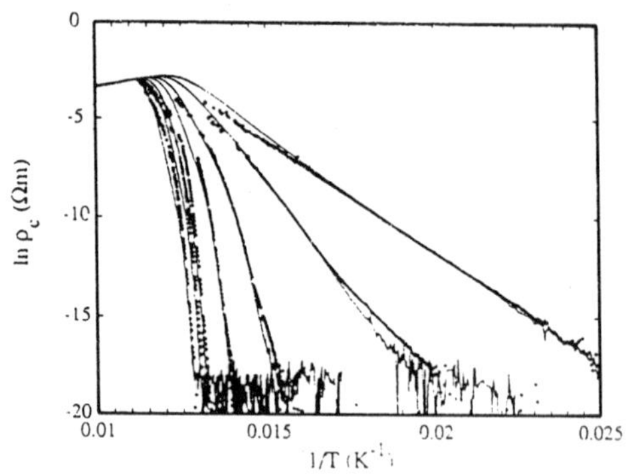

Fig.9. The c-axis resistivity deduced from the data of Fig.8 compared with the values measured.

The effect has been termed non-local resistivity and is certainly a possible explanation. However, the coincidence might be only due to the fact that the Lorentz force is the same in the two planes so that the pancake velocity is the same. This can happen without any connection along a single line. It only requires that the c-axis resistivity be low so that the current spreads uniformly through the crystal.

The connection between this picture of vortex movement and the force free configurations above is illustrated in Fig.10. This shows how a phase slippage of the top pancakes with respect to the bottom ones is equivalent to passing a perpendicular vortex through the array. If the planes are decoupled this will be easy.

No conclusions about non-locality can be drawn from measurements of two voltages only. Two resistivities can always be extracted from the two voltages. The essential test is to extract these resistivities from the voltages in the transformer configuration, and then compare them with the resistivities extracted by passing current directly down the c-axis. If the resistances are different we infer non local behaviour.

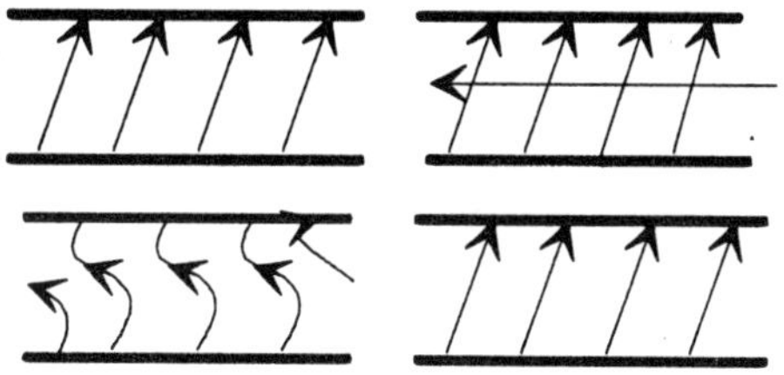

Fig.10. A flux ring crossing an array allows the top vortices to slip one spacing.

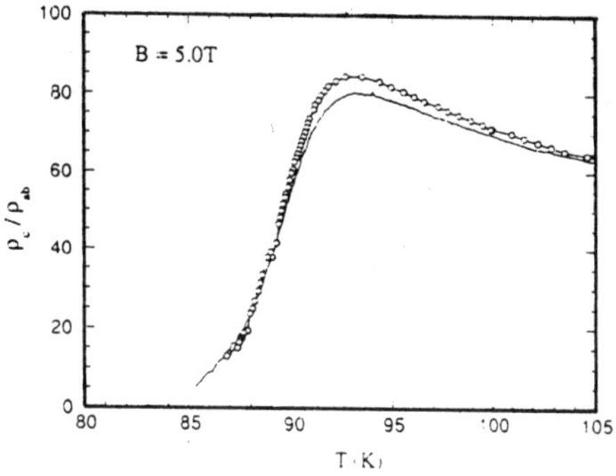

Fig.11. ρ_{ab}/ρ_c *in YBCO.*

It is therefore sensible to ask whether the results obtained so far can be explained by the standard theory of anisotropic conductors. Although a well known subject the implications of the equations in a material as anisotropic as BSCCO were not initially appreciated by the superconducting community. The central result is that if current is injected into the top surface of a rectangle it decays downward exponentially with a characteristic distance $d(\rho_{ab}/\rho_c)^{1/2}\pi$ where d is the distance between current contacts (Fig.7). The ratio of the bottom voltage to the top voltage depends exponentially on this distance which can be only a few atomic spacings. (Fig.8) Now the resistivities are thermally activated so that they depend exponentially on temperature. The result of these double exponentials is that huge changes in the ratio of the primary and secondary voltages can be explained by quite simple and continuous changes in the resistivities due to the different activation energies.

Figure 9 shows the results of extracting the c-axis resistivity from Fig.8. The continuous lines are the values directly measured. It can be seen that there is close agreement showing that the local equations work[29]. Three other features appear. Firstly the behaviour is Arrhenius type over the whole temperature range. There is no evidence of different regimes. Secondly the activation energy is much larger than kT. If this were a flux line activation energy we would expect strong flux pinning with slow creep so clearly the activation energy is not that of flux lines but must reflect the flux flow resistivity. Thirdly the a-b activation energies and the c activation energy are found to be within 10% of each other, in spite of the very large anisotropy. This suggests the same mechanism for both but can have two interpretations. Either cutting of the connection between planes releases the pancakes to move in the a-b plane or alternatively the movement of vortices in the a-b plane determines how easily electrons can tunnel between layers.

It is known that YBCO has much stronger interlayer coupling than BSCCO so we are more likely to see flux cutting effects and we may expect different results in YBCO.

Safar et. al. have claimed evidence for non locality by plotting the ratio of ρ_{ab} to ρ_c obtained for the two current directions[31]. This ratio diverged close to the irreversibility line, but the voltages were also very close to the noise level. Eltsev et. al. found no such divergence[32] and our results agree with theirs. Figure 11 shows the ratios and it can be seen that they agree closely showing local behaviour in YBCO.

VI. CONCLUSIONS

1. The problem of force free configurations remains a challenging and largely unsolved problem with important practical applications.
2. In Josephson Junctions flux cutting is easy.
3. The absence of flux cutting effects is an indication of two dimensional behaviour.

4. Above the irreversibility line both YBCO and BSCCO behave as conventional anisotropic resistive materials.
5. Most results are consistent with decoupling of the planes at the irreversibility line.

ACKNOWLEDGMENTS

I would like to thank Dr. R. Doyle and W-S Seow for many useful conversations, and permission to reproduce Fig 11.

REFERENCES

1. A.B.Pippard, private communication.
2. B.D.Josephson, Phys.Rev A **152**, 211 (1966).
3. D.G. Walmsley, J.Phys. F **2**, 510 (1972).
4. A.M.Cambell and J.E.Evetts, Adv. Phys. **21**, (1972).
5. J.R.Clem, *Concise Ecyclopaedia of Magnetic and Superconducting Materials*, ed. J.E.Evetts, Pergamon, 1992, p612.
6. A.M.Campbell, Proc. **LT 18**, Kyoto (1987).
7. B.C Belanger and M.A.R.Le Blanc, Appl. Phys. Lett, **10**, 298 (1967).
8. M.A.R Le Blanc, Phys.Rev. **143**, 220 (1966).
9. F. Irie, T. Izaki and K. Yamafuji, IEEE Trans. Magn. **11**, 332 (1975).
10. J.R.Cave and J.E Evetts, Phil Mag. B **37**, 111 (1978).
11. A.M.Campbell, Helv. Phys. Act. **53** (1980).
12. G.Fillion, R. Gauthier, and M.A.R. Le Blanc, Phys. Rev. Lett., **43**, 86 (1979).
13. J.R.Cave, J.E.Evetts and A.M.Campbell, J.Phys. **39**, 614 (1978).
14. M.C.Marchetti and D.R.Nelson, Physics C. **174**, 40 (1991)
15. J.R.Clem and A.Perez-Gonzalez, Phys. Rev. B **33**, 1601 (1986).
16. M.G.Blamire and J.E.Evetts, Phys. Rev. B **33**, 5131 (1986).
17. J.R.Clem and S.Yeh, J.Low Temp. Phys. **39**, 173 (1980).
18. E.H.Brandt, J. Low Temp Phys. **44**, 33, (1981).
19. J.R. Clem, Phys. Rev. Lett, **38**, 1425 (1977).
20. D. Kumar, Ph.D. thesis, Cambridge 1994.
21. D.H.Kim, K.E.Gray, and J.H.Kang, Phys.Rev.B, **45**, 7563 (1962).
22. M.A.R. Le Blanc, S.Celebi, S.X.Wang, and V.Plechacek, Phys. Rev. Lett. **71**, 3367 (1993).
23. S.J.Park, J.S.Kouvel, H.B.Radousky, and J.Z.Lin, Phys.Rev.B, **48**, 13998 (1993).
24. S.K.Hasanain, S.Manzoor, and A. Amirabadizadeh, Supercond. Sci. Tech. **8**, 519 (1995).
25. M.-O. Andre, G.D'Anna, M.V.Indenbohm, and W.Benoit, Proc. 7th Critical Currents Workshop, ed. H Weber, World Scientific, 276 (1994).
26. F. Supple, A.M.Campbell, and J.R.Cooper, Physica C **242**, 233 (1995).
27. D.N.Zheng, A.M.Campbell, R.S.Liu, and P.P.Edwards, Cryogenics **46**, 46 (1993).
28. D. Kumar, M.G.Blamire and A.M.Campbell and J.E.Evetts, J.Appl. Phys. **76**, 2361 (1994).
29. R.A.Doyle, W.S.Seow, Y.Yan, A.M.Campbell, T.Mochiku, K.Kadowaki and G.Wirth, Phys. Rev. Lett. **77**, 1155 (1996).
30. R. Busch, G.Ries, H. Werthner, G.Kreiselmeyer, and G. Sceimann Ishenko, Phys. Rev. Lett. **69**, 522 (1992).
31. H.Safar, P.L.Gammel, D.A.Huse, S.N.Majumdar, L.F.Schneemeyer, D.J.Bishop, D.Lopez, G.Nieva and F. de la Cruz, Phys Rev. Lett. **72**, 1272 (1994).
32. Yu Eltsev and Ö Rapp, Phys. Rev. Lett. **75**, (comments), 2446 (1995); Yu Eltsev, and Ö Rapp, Phys. Rev. B **51**, 9419 (1995).

THE MINIMUM-FLUX-CHANGE CRITERION FOR THE ELECTROMAGNETIC RESPONSE OF HARD SUPERCONDUCTORS

P. Chaddah[1] and K.V. Bhagwat[2]

[1]*Low Temperature Physics Group*
Centre for Advanced Technology, Indore 452 013, India
[2]*Technical Physics and Prototype Engineering Division*
Bhabha Atomic Research Centre, Mumbai 400 085, India

The critical state model (CSM) predicts the field and current distributions inside a hard superconductor when it is exposed to external magnetic fields or transport current. The CSM can be solved exactly when sample shapes have zero demagnetisation factor, and we establish that the same current/field distributions are obtained if one applies a minimum-flux-change criterion. We discuss the case of sample shapes having non-zero demagnetisation factor, and present our recent results for the response of thin strips to changing external magnetic fields.

I. INTRODUCTION

The critical state model (CSM) was first proposed by Bean [1] to relate the transport critical current density to hysteresis in the isothermal M-H curves. The model in its common usage assumes that the sample is at zero temperature (so that thermally activated creep can be ignored) and that its $\kappa = H_{c2}/H_{c1}$ is very large (so that equilibrium magnetisation can be ignored). Under these assumptions a hard type II superconductor can be approximated by a non-ohmic conductor whose resistivity ρ is zero below $J= J_c$, and is infinite above J_c. In the last one year or so Brandt [2] has considered hard type II superconductors as a non-ohmic conductor with resistivity $\rho = \rho_0(J/J_c)^n$, and has made numerical calculations for n varying between 19 and 99, the limit $n \rightarrow \infty$ gives the Bean model. For a long cylinder of infinite length along the applied field direction, the demagnetisation factor is N=0. An application of Lenz's law along with Bean's step-function $\rho(J)$ approximation provides the steady state screening current distribution that will be set up in response to any change in the external field. Any change in the local B(r) sets up screening currents, as in any conductor, in accordance with Lenz's Law. While the screening currents start with an infinite J, the very large ρ for $J > J_c$ rapidly reduces the current density to J_c. Since ρ is now zero, these screening currents become persistent at $J = J_c$. This is in accordance with Bean's statement of the CSM, which is enunciated as "Any change in $\vec{B}(\vec{r})$ however small, induces J_c to flow locally at that r. The direction of J_c is perpendicular to the field axis and its direction depends on the sense of the last change in B(r), in accordance with Lenz's law". We shall come back to this view of a hard superconductor as a non-ohmic conductor in the next section. We just state here that analytic solutions of Bean's CSM have been possible for samples with demagnetisation factor N=0, i.e. for cylinders of constant cross-section having infinite length along the applied field direction. For sample shapes with non-zero N, analytic exact solutions have not been possible.

Chaddah [3] has stated that the response of a hard superconductor to external electromagnetic fields is to set up a distribution of persistent currents (with $J(r) \leq (J_c)$) such that the change of flux penetrating the sample is minimised. This produces current distributions indentical to Bean's CSM for all N=0 geometries. This is illustrated in Fig. 1 for the simple case of an N=0 slab with field independent J_c.The situations illustrated are (a) a virgin zero-field-cooled (ZFC) slab is subjected to an external field B_a; (b)a ZFC slab during the field-decreasing hysteresis loop; (c) a ZFC slab carrying a transport current $I < I_c$; (d) a ZFC slab in the remanent state after application of a large magnetic field, is then made to carry a transport current $I < I_c$; and (e) a ZFC slab in the remanent state after carrying a current $I > I_c$, is then subjected to an external magnetic field B_a. The Fig. 1 assumes that the external field B_a is along z-axis, the currents are along y and the finite dimension of the slab is along x. In this N=0 geometry, the slope of the field profile is directly proportional to the screening current flowing at that point. The field profiles B(x) shown by the solid line correspond to minimum flux change (and this flux change is proportional to the shaded area), and are identical to Bean's CSM solutions. The profiles shown by dashed lines correspond to other possible persistent current distributions, and clearly correspond to a larger change in the flux penetrating the sample.

The illustration in Fig. 1 exemplifies the rigorous assertion that for all N=0 samples-shapes, the minimum flux change criterion and Bean's CSM have the same solution [3]. We shall now come to non-zero N shapes where exact solutions of Bean's CSM are not yet available.

Fig. 1

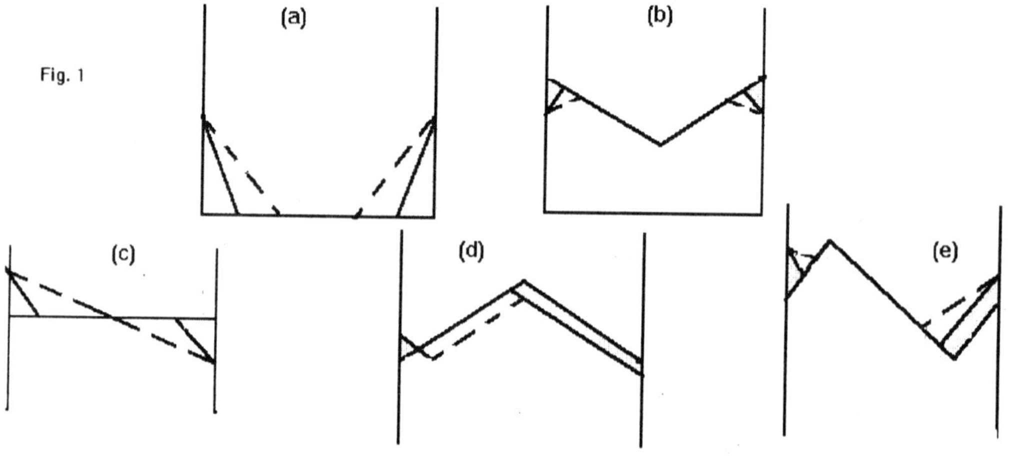

II. SPHERICAL SAMPLES

In this section we shall consider the case of a non-zero N sample and motivate the need for an alternate to Bean's CSM. We shall argue that the use of Lenz's law on non-zero N samples requires successive iterations on the dynamics of setting up the critical state. The minimum-flux-change criterion, on the other hand, compares the possible steady-state persistent current distributions, and chooses the critical state from amongst them by a variational criterion.

In figure 2 we illustrate the dynamics of setting up the critical state for an N=0 slab and an $N = \frac{1}{3}$ sphere. We shall be viewing the samples as non-ohmic conductors and, following Brandt [2], shall replace Bean's step-function resistivity by the power-law form $\rho(J) = (J/J_C)^n$. In this form the screening current density decreases from infinity to J_C in a short time (say ϵ), and this allows us to have a physical picture of how the critical state is set up.

If a small positive δB is applied at the surface ($X = X_1$) of the slab at $t = 0$, then since $J(X_1) = -\infty\hat{y}$ at t=0 we have $\delta B(X_2) = 0$ at a neighbouring internal surface. At $t = \epsilon$, the surface J has decayed to $-J_C\hat{y}$, $\delta B(X_2, t = \epsilon) = +ve$ and $J(X_2, t = \epsilon) = -\infty\hat{y}$. The currents set up at X_2 do not create a field at X_1 because current shells with N=0 do not create a field external to themselves. The setting up of shielding currents moves progressively, and monotonically, inwards until X_0 where $\delta B(X_0)$ remains zero.

Let us now consider a sphere where the surface r=R sees the field δB at t=0. Because N is non-zero, we treat separately (r,0) on the equator and (r, θ_0) close to the pole. At t=0, we have $J(R, 0) = -\infty\hat{e}_\phi$ and $J(R, \theta_0) = -\infty\hat{e}_\phi$, and $\delta B(r_1, 0)$ and $\delta B(r_1, \theta_0)$ are both zero. At $t = \epsilon$, J(R,0) and J(R,θ_0) have decayed to $-J_C\hat{e}_\phi$, and the external field penetrates upto r_1 so that $\delta B(r_1, 0, t = \epsilon)$ is +ve and $\delta B(r_1, \theta_0, t = \epsilon)$ is +ve. At $t = \epsilon$, both $J(r_1, 0)$ and $J(r_1, \theta_0)$ are $-\infty\hat{e}_\phi$. In contrast to the slab geometry, the currents at r_1 do create a field at R, and the sign of this new field change, caused at $t = \epsilon$, is positive at (R, 0) but is negative at (R, θ_0). Thus at $t = \epsilon$, screening currents set up at (R,0) are $-\infty\hat{e}_\phi$, but those set up at (R, θ_0) are $+\infty\hat{e}_\phi$. The sign of J near the poles has thus switched, and the dynamics of setting up of the critical state are complicated and need to be iterated to consistency. We mention here that some regions of the sample may carry $+J_C\hat{e}_\phi$ when the final critical state is set up. Just as in our macroscopic picture of the N=0 sample unidirectional J(X) are obtained by averaging the circulating currents flowing around unevenly spaced vortices, averaging over larger length scales will sum over the $+J_C\hat{e}_\phi$ and $-J_C\hat{e}_\phi$ regions to give a unidirection $-J\hat{e}_\phi$, but with $J < J_C$.

The minimum-flux-change criterion does not look at the dynamics, and will just compare between persistent current distributions satisfying $|J(r)| \leq J_C$. Our inability to iterate the dynamics to consistency and obtain the critical state current distribution, motivates us to use the minimum-flux-change criterion for sample shapes with non-zero N. Since this is a variational criterion, the solution obtained may be an approximation if we have not exhausted all possible persistent current distributions.

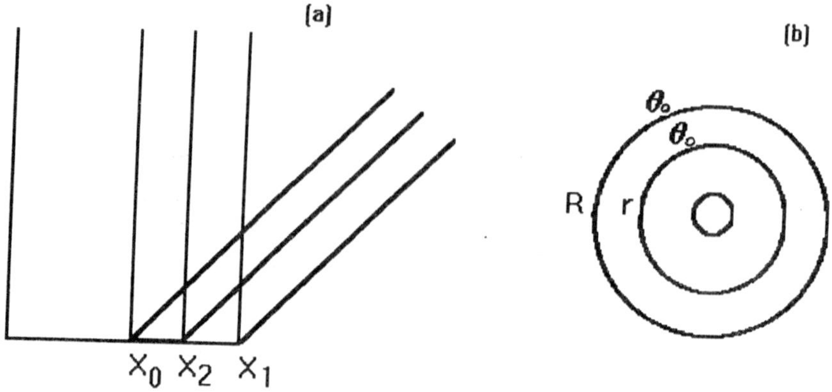

Fig. 2

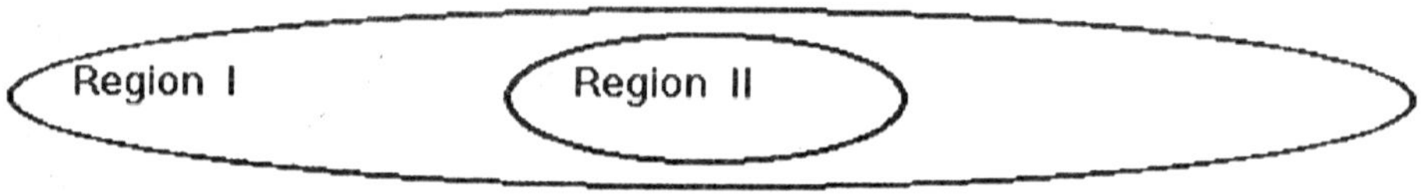

Fig. 3

III. ELLIPTICAL CYLINDERS & THIN STRIPS

The magnetic response of thin films of high-T_C superconductors (HTSC) is of great experimental interest, but these samples have a large demagnetisation factor ($N \rightarrow 1$) when the magnetic field is perpendicular, or the electrical field is parallel, to the plane of the film. Long thin strips can be approximated by elliptic cylinders while square films can be approximated by oblate spheroids. Bhagwat and Chaddah [4] have considered both these shapes, but for conciseness we will consider here only thin strips.

Let us consider first an elliptic cylinder with axes 2c and 2a, and apply a magnetic field $B_a\hat{z}$ parallel to c. Flux will enter the sample upto a surface called the flux-front, shielding currents will flow in the region (say region I) contained within the sample surface and flux-front, and in the region (say region II) contained within the flux-front we will have B=0 and J=0. The problem is to obtain J(r,θ) in region I, and to obtain the shape of the flux-front $r = f(\xi_0, \theta)$, for various values of the parameter ξ_0 (which is directly related to B_a).

Since we must have B=0 in region II, we will assume that the current distribution J(ξ ,θ) on each shell $r = f(\xi, \theta)$ is such that it creates a constant field in region II. If we choose the shells to be elliptic cylinders with axis a(ξ) and c(ξ), then we know that a constant field will be created in the interior by a surface current distribution

$$I(\theta) = \frac{I(\xi)\cos\theta}{\sqrt{1+\left(\frac{a^2(\xi)}{c^2(\xi)}-1\right)\sin^2\theta}} \tag{1}$$

Here I(ξ) is the surface current at $\theta = 0$ and can be looked upon as a weight function W(ξ). These elliptic shells now have to be given a (θ-dependent) thickness such that the space in region I is filled up, and the surface currents will be appropriately converted into volume current densities. If we rewrite $\frac{c(\xi)}{a(\xi)} = \frac{1}{\sqrt{1+p(\xi)}}$ and use the convention that $B_a = 0$ corresponds to $\xi = 1$, then the sample surface is the flux-front for $\xi = 1$ and we have the boundary condition $p(1) = \frac{a^2}{c^2} - 1$. The volume current density in region I is then given by

$$J(\xi,\theta) = W(\xi)\frac{\sqrt{1+p(\xi)\cos\theta}}{1+\left[p(\xi)-\frac{\xi}{2}\frac{\partial p}{\partial\xi}\sin^2\theta\right]}\cos\theta \tag{2}$$

We have a freedom in choosing how the ellipticity of the flux-front changes as B_a rises, and we consequently have a freedom in choosing p(ξ), subject to $p(1) = \frac{a^2}{c^2} - 1$. The shape of the flux-front is that it is an ellipse of axes aξ and $a\xi/(\sqrt{1+p(\xi)})$, so that it can be written as

$$r = f(\xi,\theta) = \frac{a\xi}{\sqrt{1+p(\xi)\sin^2\theta}} \tag{3}$$

Given any choice of p(ξ), and a choice of W(ξ), it is now possible to calculate [4] $B_a(\xi_0)$, the virgin magnetisation $M_v(\xi)$, and the total flux penetrating the equator (or a-axes) of the cylinder. We shall now choose p(ξ) so as to minimise the flux penetrating the sample, and this yields [4] $p(\xi) = \frac{a^2}{c^2}\xi - 1$. Equations (1) tells us that the largest value of $J(\xi,\theta)$ is W(ξ) at $\theta = 0$, and we put $W(\xi) = J_C$ independent of ξ . It has been shown[4] that $p(\xi) = \frac{a^2}{c^2}\xi - 1$ makes the decay of $J(\xi,\theta)$ with rising θ as small as possible. Thus, the $J(\xi,\theta)$ are as close to J_C as possible, and $|M_V|$ has been maximised.

Though in the above description we have assumed elliptic current shells, we have actually minimised flux penetration amongst all J(r,θ) that will yield a constant Amperian magnetisation in region II. For minimisation of flux-change in this space of J(r,θ), Bhagwat and Chaddah [4] have provided virgin and hysteresis magnetisation curves for arbitrary c and a.

To approximate thin strips, we have to consider c << a. Mikheenko and Kuzovlev [5] initiated the idea that for very small c, we should not put constraints on J(r,θ), but on the line current density $I(x) = \int J(x,y)dy$, and we should impose $I(x) = \int J_C dy$, rather than requiring $J(x,y) \leq J_C$. For the transverse elliptic cylinder, we are requiring that we choose W(ξ) such that, in region I,

$$\int J(x,y)dy = 2cJ_C\sqrt{1-\frac{x^2}{a^2}} \tag{4}$$

and we obtain, after some algebra [6],

$$W(\xi) = \frac{2J_C}{\pi\sqrt{\xi}}E(\sqrt{1-\xi^2}) \tag{5}$$

With this weight function the J(x,y) will, at some points, exceed J_C. For $c/a = 10^{-3}$, the saturation magnetisation obtained is within 5% of that obtained when $W(\xi) = J_C$. Also, the line current density, and the equatorial fields, are almost indistinguishable on a linear plot [6]. The original solution [4] with $W(\xi) = J_C$, which satisfies $|J(\xi)| \leq J_C$, thus provides results which are very close to those obtained by using the Mikheenko-Kuzovlev prescription of imposing $\int Jdy = J_C \int dy$.

We now come to the more interesting case of a rectangular strip. Here the weight has to be chosen to satisfy, for x in region I, $\int J(x,y)dy = 2cJ_C$.

This yields, after some algebra [6],

$$W(\xi) = \frac{2J_C}{\pi\sqrt{\xi}\sqrt{1-\xi^2}} \tag{6}$$

and we have analytic solutions of J(ξ,θ), $\xi_0(B_a)$, M_V and hysteresis loops, for thin rectangular strips of arbitrary c/a. The detailed results are given in ref [6]. The line current density is given by

$$I(x) = \begin{cases} 2cJ_C & \text{for x in region I} \\ \frac{2cJ_C}{\pi}\tan^{-1}[\frac{x}{a}\sqrt{\frac{a^2-X_0^2}{X_0-x^2}}] & \text{for } x < X_0\text{, i.e. in region II} \end{cases}$$

where we follow the conversion $x = a\xi$. This result is in agreement with that being quoted in literature [5,7,8] as valid for strips with c$\rightarrow$ 0. In this limit of zero thickness but with cJ_C fixed at a finite value, we have

$$B_a(\xi_0) = \frac{2\mu_0 cJ_C}{\pi}\cosh^{-1}\left(\frac{1}{\xi_0}\right) \tag{7}$$

and

$$B_{equator}(\xi,\xi_0) = \frac{2\mu_0 J_C a}{\pi}\ln\left[\frac{\xi\sqrt{1-\xi_0^2}+\sqrt{\xi^2-\xi_0^2}}{\xi_0\sqrt{1-\xi^2}}\right] \tag{8}$$

Equations (4) and (5) give $B_a(\xi=0)=\infty$, and $B_{equator}(\xi=1)=\infty$ for all ξ_0. $B_a(0)$ is infinite because as c$\rightarrow$ 0 , $J_C \rightarrow \infty$. Thus $B_a(0)$ cannot be infinite for finite (but small) c, and our exact results [6] give

$$B_a(0) = \frac{2\mu_0 J_C a}{\pi}\int_0^1 \frac{d\xi}{(1+\frac{a}{c}\sqrt{\xi})\sqrt{\xi}\sqrt{1-\xi^2}} \tag{9}$$

which is finite. Also $B_{equator}(\xi=1)$ is infinite for c$\rightarrow$0 because the demagnetisation factor is then exactly one. This, again, will not be true once c is finite (though small) and that is what we find[6]. To sum up, *the results we have obtained for thin strips using the minimum flux change criterion agree with literature [5, 7, 8] when c$\rightarrow$ 0, and also provide analytic solutions when c is small but non-zero.*

IV. FIELD-DEPENDENT J_C

The results in Section 3 persist with Bean's original assumption that J_C is field-independent. It has been accepted for some time now that J_C(B) must decrease with increasing B, and the commonly used functional forms are the Kim-Anderson model [9] and the exponential model [10]. The former was used extensively for conventional superconductors, where many other functional forms have also been used [11]. The exponential form was briefly explored [12] for conventional superconductors, but has been widely exploited to fit the behaviour of high-T_C superconductors. Ikuta et al [13] have shown that only the exponential model can fit the magnetostriction data in these materials, and Kobayashi et al [14] have shown that M-H loops of underdoped $(La_{1-x}Sr_x)_2CuO_4$ fit the exponential model in preference to the Irie-Yamafuji models. Bhagwat and Chaddah [15] have calculated the M-H loops for thin strips and oblate spheroids using the exponential model. We briefly outline their results.

In our formulation in Section 3, the J(r) is largest at $\theta = 0$, and we also have analytic expressions for the equatorial field $B_{eq}(\xi, \xi_0)$ in terms of W(ξ). We now choose W(ξ) to depend on ξ_0, and put

$$W(\xi, \xi_0) = J_C(0) \exp\left[-B_{eq}(\xi, \xi_0)/B_0\right] \tag{10}$$

We then have an integral equation for $B_{eq}(\xi, \xi_0)$ which we solve by a numerical iteration scheme. Once W(ξ, ξ_0) is fixed, we obtain $B_a(\xi, \xi_0)$ and $M_V(\xi_0)$ as in Section 3. The hysteresis curves so obtained show a sharp peak in the hysteresis curve close to the H=0 line. This peak, as we have argued earlier for a spheroid [16], moves towards the H=0 line as the aspect ratio increases.

V. CONCLUSIONS

We have presented here a minimum-flux-change criterion to determine the electromagnetic response of thin strips and films. We have justified this criterion on the grounds that it gives the same results as Bean's CSM for all N=0 shapes, and have motivated its need for non-zero N shapes. *We have obtained analytic solutions for rectangular strips that agree with literature in the 2-dimensional $c \to 0$ limit, and provide analytic solutions for small but non-zero c.* We have argued that the $c = 0$ solutions cannot be per se used for small but non-zero c because they have the unphysical features of (i) infinite field at the sample surface (ξ= x/a = 1), and (ii) infinite applied field for the flux-front to collapse to a point ($\xi_0 = 0$). We have finally presented results for J_C(B) varying exponentially with B.

REFERENCES

1. C.P. Bean, Phys. Rev. Lett. **8**, 250 (1962); Rev. Mod. Phys. **36**, 31 (1964).
2. E.H. Brandt, Phys. Rev. **B54**, 4246 (1996), and references therein.
3. P. Chaddah, Pramana - J. Phys. **36**, 353 (1991); in *Studies of High Temperature Superconductors* ed. A.V. Narlikar, Vol. **14** (Nova Science, 1994) p. 245 and references therein. Also, K.V. Bhagwat, S.V. Nair and P. Chaddah, Physica **C227**, 176(1994).
4. K.V. Bhagwat and P. Chaddah, Pramana - J. Phys. **33**, 521 (1989); Physica **C166**, 1 (1990); Physica **C190**, 444 (1992).
5. P.N. Mikheenko and Yu.E. Kuzovlev, Physica **C204**, 229 (1993).
6. K.V. Bhagwat and P. Chaddah, Physica **C254**, 143 (1995).
7. E. Zeldov et al. Phys. Rev. **B49**, 9802 (1994).
8. E.H. Brandt and M.V. Indenbom, Phys. Rev. **B48**, 12893 (1993).
9. Y.B. Kim et al, Phys. Rev. **129**, 528 (1963).
10. G. Ravikumar and P. Chaddah, Phys. Rev. **B39**, 4704 (1989).
11. See A.M. Campbell and J.E. Evetts, Adv. in Phys. **21**, 199 (1972).
12. See D.X. Chen and R.B. Goldfarb, J. Appl. Phys. **66**, 2510 (1089).
13. H. Ikuta et al., Phys. Rev. Lett. **70**, 2166 (1993); Physica **C235**, 237 (1994); J. Appl. Phys. **76**, 4776 (1994).
14. T. Kobayashi et. al., Physica **C254**, 213 (1995).
15. K.V. Bhagwat and P. Chaddah (to be published).
16. K.V. Bhagwat and P. Chaddah, Physica **C224**, 155 (1994).

TRANSPORT BEHAVIOR IN SUPERCONDUCTORS AT EXTREME DISSIPATION LEVELS

M. N. Kunchur[1,2], D. K. Christen[3] and B. I. Ivlev[4]

[1] *WL/POOX-2, Bldg. 450, Area B, Wright Patterson Air Force Base, OH 45433*
[2] *Dept. of Physics and Astronomy, University of South Carolina, Columbia, SC 29208*
[3] *Solid State Division, Oak Ridge National Lab, P. O. Box 2008, Oak Ridge, TN 37831*
[4] *Landau Institute for Theoretical Physics, Kosygin St. 2, 117940 Moscow, Russia.*

A number of fundamental physical phenomena unfold in the mixed state of superconductors, when subjected to enormous current and power-dissipation levels. A sufficiently large current can destroy the superconducting state itself—the so-called pair-breaking effect. At intermediate current densities, below the onset of pair-breaking, one expects to see the free viscous flow of flux vortices. In the present work a pulsed-current technique was used to explore this dissipative regime of high-T_c superconductors. verifying both free flux flow and the pair-breaking effect, as predicted by traditional theories. This paper concentrates on the dissipation and Hall behavior in the free flux flow state.

In this work, we have employed short-duration ($\sim 1\mu$s) current pulses to explore superconducting transport behavior at current levels above the onset of dissipation J_c and beyond the depairing limit J_d[1]. It was shown[2] that by applying a sufficiently large Lorentz driving force on flux vortices, and driving their motion into the state of free flux flow, simple Bardeen-Stephen[3] behavior could be verified for the first time in any superconductor: the transport characteristic becomes Ohmic (total resistivity $\rho = E/J$ is current independent) and the dissipation obeys:

$$\rho_{FFF}/\rho_n \approx B/H_{c2}, \tag{1}$$

where ρ_n and H_{c2} are the normal-state resistivity and the upper-critical field respectively *for that temperature*. Fig. 1 shows longitudinal resisitivity in free flux flow for two YBCO samples. Recently, similar saturation of $\rho(J)$ leading to an Ohmic plateau was also observed in the low-T$_c$ low-pinning materials 2H-NbSe$_2$ by Higgins et al.[4] and in $Mo_{77}Ge_{23}$ by Hellerqvist et al.[5].

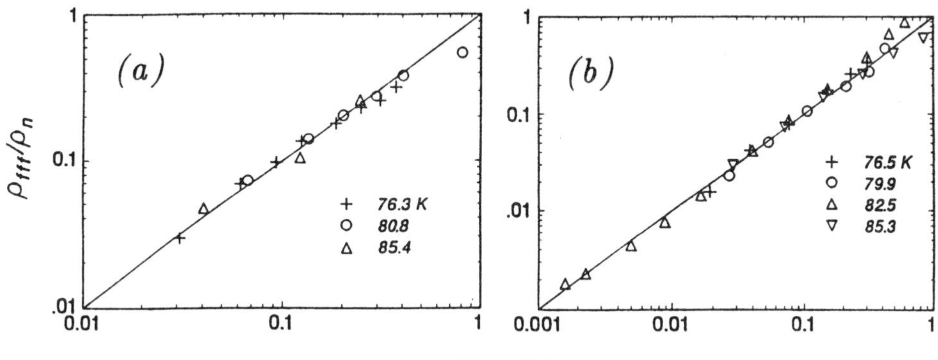

Figure 1: Normalized free-flux-flow resistivity versus H for two YBCO specimens.

At even higher current densities we were able to observe current-induced pair-breaking[6], where T_c was found to shift as:

$$\Delta T_c(H,J)/T_c(H,0) = [J/J_0]^{\frac{2}{3}}.$$

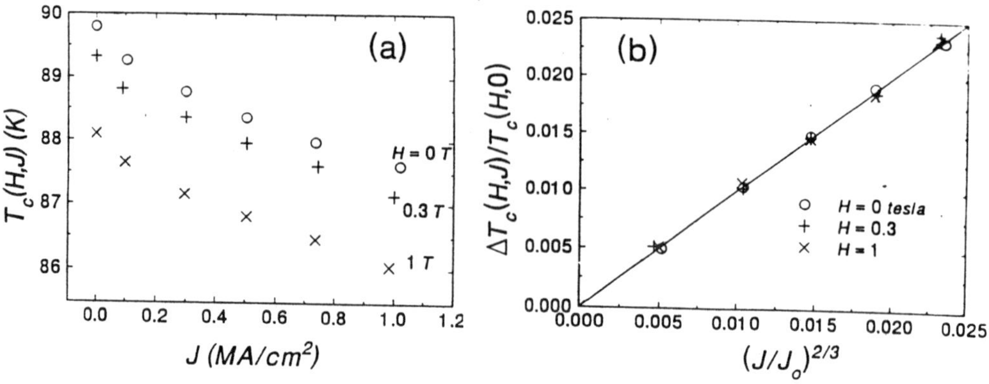

Figure 2: (a) Mid-transition T_c's at different H and J values for a YBCO film. (b) Normalized shifts in T_c as a function of $(J/J_0)^{2/3}$. $J_0 \approx 4 \times 10^8$ A/cm^2 is a field-independent constant.

This is shown in Fig. 2.

The mixed-state Hall effect has been a topic of great current interest and one surrounded by controversy. Central to the controversy is the observed sign reversal of the tangent of the Hall angle $\tan\alpha = \rho_{xy}/\rho_{xx}$, where ρ_{xx} and ρ_{xy} are the longitudinal and transverse (Hall) resistivities respectively[1]. One of the main questions that has plagued the interpretation of the Hall effect is whether the sign reversal is caused by pinning[7]. Our previous work on dissipation and free flux flow had shown the effectiveness of high currents in suppressing the effect of pinning. We were able to apply this same principle to investigate the role of pinning in the Hall effect. Fig. 3 shows the Hall angle versus temperature at different fixed magnetic fields[8]. The upper panel (a) was measured at low continuous currents (5 kA/cm^2), the lower panel at much higher pulsed currents ($\sim$ 1 MA/cm^2). As is evident, there is a dramatic enhancement of the sign reversal showing that pinning hinders the sign reversal rather than causes it. The data in Fig. 3 was one of the earliest and most decisive proofs that the sign reversal is not caused by pinning. Recently, Kang et al.[9] were able to show qualitatively similar behavior by varying the pinning by introducing artificial defects into their samples. An advantage of overcoming pinning using a high current as opposed to varying the defect concentration is that the process is reversible and continuously variable. Also there is little change in other parameters, such as the carrier concentration, which can however be dependent on defect concentration.

In addition to the sign reversal, there is another feature that has been observed by us and other groups[10, 11, 12, 8]. This is the decomposition of the Hall angle, α, into a field-proportional component α_n, and a field-independent component α_M. The field-proportional component arises from the Hall effect of the normal core of vortices, whereas the field-independent component α_M, can arise from the hydrodynamic (Magnus) force on the "body" of the vortex. The latter can be treated through the time-dependent-Ginzburg-Landau (TDGL) theory of the dynamics of the superconducting order parameter (as shown in more detail below). α_M can be estimated by subtracting from the total Hall angle, the normal-core component $\alpha_n(T)$ found by fitting to its $\sim H/T^2$ behavior above T_c and extrapolating this to temperatures below T_c. Fig. 4 shows α_M for our data measured on a YBCO film for 3 different fields[8]. Notice that as the temperature is lowered below T_c, as long as pinning is not influential, α_M seems to saturate to a value that does not depend on the field or current (As long as these are large enough to ensure free flux flow.)

[1] Since the Hall angle is typically small, we will not always distinguish between the angle and its tangent in the remainder of this discussion.

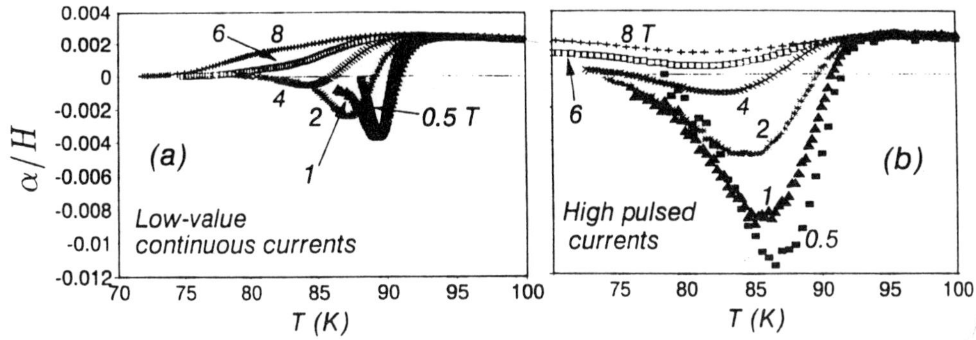

Figure 3: (a) Temperature dependence of α/H for indicated fixed fields measured at a continuous current of J=5 kA/cm^2 in a YBCO film. (b) Similar data measured with pulsed current at $J = 0.7MA/cm^2$ (H = 0.5, 1, 2, and 4 T), $J = 1.1MA/cm^2$ ($H = 6T$) and $J = 1.5MA/cm^2$ ($H = 8T$).

Moreover the saturation or maximum negative value (about 0.017 in our Fig. 4) is about the same order as that reported by other groups in various materials: 0.005–0.007 for $Tl_2Ba_2CaCu_2O_8$ (references [10] and [11]) and 0.009–0.017 for YBCO (references [12] and [8]).

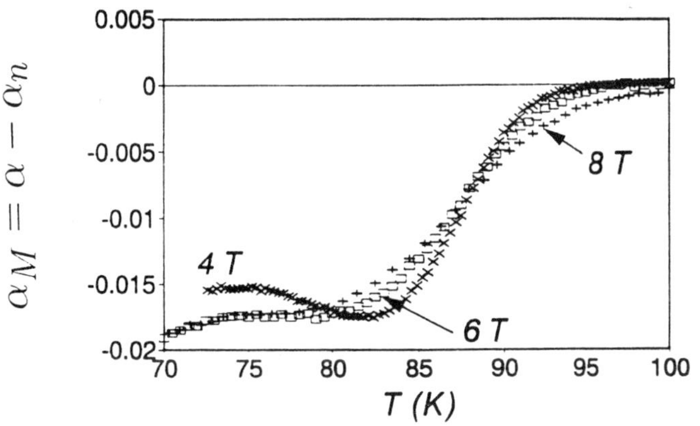

Figure 4: Magnus-force component of the high-field Hall angle $\alpha_M = \alpha - \alpha_n$, for the data of Fig. 3. Here, $\alpha_n = 21.9 \times H/T^2$ is the observed T- and H-dependent normal-state contribution.

We now show that the decomposition $\tan\alpha \approx \tan\alpha_n + \tan\alpha_M$ follows from the dynamics of the superconducting order parameter as calculated in the framework of time-dependent-Ginzburg-Landau (TDGL) theory[13]. Moreover the magnitude of $\tan\alpha_M$ obtained in this way (0.014) is in rough accord with the experimental observations ($\sim$ 0.005–0.02).

The vortex viscosity η, that governs the longitudinal conductivity consists of two parts: the Bardeen-Stephen component, which arises from dissipation of normal carriers inside the vortex core, and the Tinkham component, which arises from the dynamics of the order parameter. The net longitudinal conductivity in free flux flow is [13]

$$\sigma \approx \sigma_n \left(\frac{\alpha u}{2}\right)\left(\frac{H_{c2}}{B}\right), \tag{3}$$

where σ_n is the normal-state conductivity. Similarly, the transverse conductivity can be written as

$$\sigma^H \approx C\sigma_n^H \left(\frac{\alpha u}{2}\right)\left(\frac{H_{c2}}{B}\right) + sign(e)\sigma_n \left(\frac{\zeta\beta u}{2}\right)\left(\frac{H_{c2}}{B}\right), \tag{4}$$

where C is a constant of order unity. The first term arises from the Hall effect of normal carriers in the core, and the second term is the contribution from order-parameter dynamics. There is no contribution from quasiparticles embedded in the superfluid outside the core at zero frequency. $\alpha \sim 0.0.5$, $\beta \sim 0.27$, and $\zeta \approx T_c/E_F \sim 90/4000$ are constants of the TDGL theory[13] (E_F was taken from Wolf and Kresin [14].) The constant α should be distinguished from the Hall angle α.

From Eqs. 3 and 4 one then gets

$$\tan\alpha = \frac{\sigma^H}{\sigma} = C\frac{\sigma_n^H}{\sigma_n} + sign(e)\frac{\zeta\beta}{\alpha} \sim \alpha_n + 0.014. \tag{5}$$

This expression shows a Hall angle that decomposes into a field-proportional "normal-core" component α_n and a field-independent term $\alpha_M \sim 0.014$, which is comparable to the experimental observations (~ 0.005–0.02).

We'd like to point out that in a recent Letter[15] Khomskii and Freimuth present an interesting mechanism based on charging of the vortex core that explains the sign-reversing component of the mixed-state Hall effect and also predicts the qualitative decomposition $\tan\alpha \sim \tan\alpha_n + \tan\alpha_M$. Their predicted estimate of α_M (which they call α_q), about 10^{-4}, is however lower than the experimental results by about a factor of 100.

To summarize, we have reviewed our past work that explored the transport behavior of superconductors at current densities well above the critical value, where flux motion converts to free flow and eventually leads into the current-induced pair-breaking regime. In addition we present results on the mixed-state Hall effect, where high J was once again used to decisively rule out pinning as the cause of the sign reversal. In the free-flux-flow limit, we find that the Hall angle shows a decomposition $\tan\alpha \sim \tan\alpha_n + \tan\alpha_M$, also seen in other superconductors. An explanation, based on TDGL theory, describes the qualitative features of the data and also shows rough quantitative agreement.

Some of this research was sponsored by the Division of Materials Sciences, U. S. Department of Energy under Contract No. DE-AC05-84OR21400, with Martin-Marietta Energy Systems, Inc. Partial support for the program was administered by the Oak Ridge Institute for Science and Education.

REFERENCES

1. M. N. Kunchur, Mod. Phys. Lett. B **9**, 399 (1995).
2. M. N. Kunchur, D. K. Christen, and J. M. Phillips, Phys. Rev. Lett. **70**, 998 (1993).
3. J. Bardeen and M. J. Stephen, Phys. Rev. **140**, A1197 (1965).
4. M.J Higgins, S. Bhattacharya, and T. V. Ramakrishnan, Bull. Am. Phys. Soc. **39**, 730 (1994).
5. Hellerqvist et al., Phys. Rev. Lett. **76**, 4022 (1996).
6. M. N. Kunchur et al., Phys. Rev. Lett. **72**, 752 (1994).
7. Z.D. Wang and T.S. Ting, Phys. Rev. B **46**, 284 (1992); Phys. Rev. Lett. **67**, 3618 (1991).
8. M. N. Kunchur, D. K. Christen, C. E. Klabunde and J. M. Phillips, Phys. Rev. Lett. **72**, 2259 (1994).
9. Kang et al., Phys. Rev. Lett. **76**, 2993 (1995).
10. S. J. Hagen, C. J. Lobb, R. L. Greene and M. Eddy, Phys. Rev. B **43**, 6246 (1991).
11. A. V. Samoilov, Z. G. Ivanov, and L.-G. Johansson, Phys. Rev. B **49**, 3667 (1994).
12. A. V. Samoilov, Phys. Rev. B **49**, 1246 (1994).
13. N. B. Kopnin, B. I. Ivlev and V. A. Kalatsky, J. Low Temp. Phys. **90**, 1 (1993).
14. S. A. Wolf and V. Z. Kresin, IEEE Trans. Mag. **27**, 852 (1991).
15. D. I. Khomskii and A. Freimuth, Phys. Rev. Lett. **75**, 1384 (1995).

HIGH CURRENT POLYCRYSTAL YBCO SUPERCONDUCTOR

K. Salama, S. Sathyamurthy and A.S. Parikh
Texas Center for Superconductivity
University of Houston, TX 77204

Using the liquid phase removal method, melt textured polycrystalline YBCO bars with 'clean' grain boundaries free of any secondary phases are processed. These bars, with dimensions ranging from 50 to 100 mm in length and 30 mm^2 cross section, are found to exhibit a low resistance transition from the zero resistance superconducting state. Typical voltage drops registered with these bars range between 20 to 50 μ V for a current of 1000 A at 77 K. The same effect persisted even in the presence of an external magnetic field of up to 1.5T perpendicular to the length of the specimen. These samples contain a large number of grain boundaries with high misorientation angles. This gives promise to fabrication of polycrystalline YBCO bars capable of carrying high current.

I. INTRODUCTION

A major impediment to the practical application of YBCO superconductors in bulk polycrystalline form is the poor current transport properties of these materials. Good transport properties measured on single crystals and textured materials suggest that improving the characteristics of the grain boundaries may hold the key to improving the current transport characteristics of bulk polycrystalline materials[1]. The main reasons for the "weak link" nature of the grain boundaries are (i) solidification of non superconducting phases along the grain boundaries, (ii) microcracks which interfere with the supercurrent paths, and (iii) atomic disorder at the grain boundaries which leads to poor coupling of the copper oxide planes across the boundary. The first two reasons are mainly caused by poor processing conditions and have been shown to be rectified by using the liquid phase removal method of processing[2,3]. The main reason for atomic disorder at the grain boundaries is the misorientation between the adjoining grains. It has been demonstrated that in epitaxial thin films of YBCO, the J_c is quite sensitive to the misorientation between the adjoining grains[4]. These results suggest that the J_c decreases drastically if the misorientation angle is greater than 10°. The results obtained by Babcock et al[5]., however, show that some large angle boundaries in YBCO bicrystals are capable of carrying high current. These results were explained in terms of the presence of an inherent periodicity at the grain boundaries at certain misorientation leading to low grain boundary energy configurations. Later, Zhu et al.[6], in a study involving over 200 boundaries reported the existence of preferred misorientation with low energy boundaries in melt textured YBCO. These results show that it is indeed possible for bulk polycrystalline materials to have good transport properties if they have clean low energy boundaries. This paper deals with the processing of long polycrystalline YBCO bars using the liquid phase removal method, and their I-V characteristics.

EXPERIMENTAL

Two Types of powders, $YBa_2Cu_3O_{7-x}$ (123) and Y_2BaCuO_5 (211), are used in this study. The $YBa_2Cu_3O_{7-x}$ (123) and the Y_2BaCuO_5 (211) powders were processed by calcination of Y_2O_3, $BaCO_3$, and CuO at 880 °C, 900°C and 920°C for 24 hours at each temperature, with intermittent grinding. The 123 powder was then uniaxially compacted into bars which were placed on a high density alumina plate and heated to the sintering temperature of 930°C at a rate of 30°C/hr. The compacted bar was held at the sintering temperature for 24 hours before cooling down to room temperature at a rate of 30°C/hr.

The 211 powder was then compacted and cut into small blocks. These blocks were used as absorbants for the liquid removal during melt texturing. The sintered 123 bar was supported over the 211 blocks placed on a high density alumina plate and loaded into a box furnace for melt texturing as shown in fig.1. During melt texturing, the samples were heated to 1100°C and held at this temperature for some time for melting to set in. The 123 decomposes into 211 and a liquid phase. A part of the liquid phase formed gets absorbed into the 211 blocks supporting the 123. The amount of liquid removed from the melt is controlled by the weight of the 211 used with respect to the weight of the 123 bar. The amount of 211 absorbant required to remove the optimum fraction of liquid phase from the sample was studied by Willén[7] and was found to be 90% of the weight of the 123 bar. From 1100°C, the partially melted mass was rapidly cooled to 1023°C. From this temperature, the melt is slowly cooled at a rate of 0.5°C/hr to 1.5°C/hr to 930°C, to allow for the recombination of the 211 and the liquid. The system is then cooled down to room temperature at a rate of 30°C/hr. The melt textured samples were then prepared for oxygen annealing.

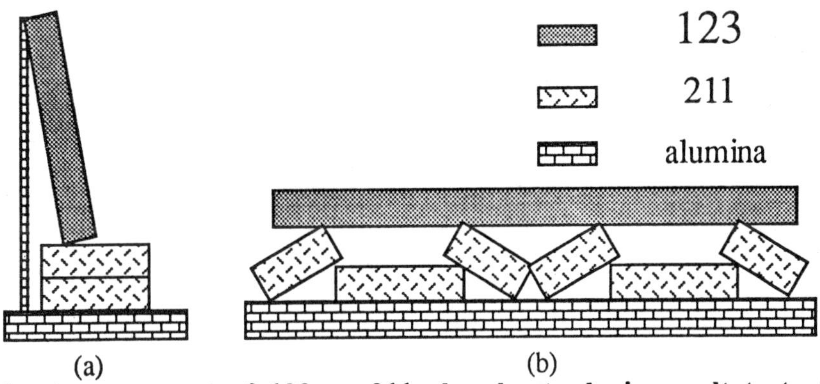

Fig.1. Arrangement of 123 on 211 absorbants during melt texturing. (a) Vertical arrangement, (b) Horizontal arrangement.

The critical current characterization of these samples was performed using the four point probe method. Four silver leads were pasted on to the sample surface using colloidal silver paste. The samples were then heated in flowing oxygen at a rate of 60°C/hr to 500°C, held at this temperature for 144 hours, and then cooled down to 400°C at a rate of 10°C/hr. After a dwell time of 24 hours at 400°C, the samples were cooled down to room temperature at a rate of 60°C/hr. These samples contain a large number of grain boundaries with high misorientation angles.

The I-V characteristics of the annealed samples were determined using a pulsed DC supply with a capacity of 1200A. The current and voltage leads in the annealed samples were connected to the Pulsed DC supply through a special sample holder. The sample along with the holder was placed in a dewar and cooled to 77 K using liquid nitrogen. Current pulses of 10 μs duration ranging from 1 A to 1000 A were passed through the sample, and the corresponding voltage drop were measured to get the I-V characteristics of the sample.

RESULTS AND DISCUSSION

Samples of two sizes, namely 55 mm and 100 mm, were processed using the liquid phase removal method. For the short samples (55 mm), 40 grams of 123 were used for the compaction and sintering, and 36 grams of 211 were used for the absorbant. The samples were supported vertically on the 211 absorbants. In the melt texturing cycle, a hold time (t_h) of 0.0 hours and a cooling rate (R) of 0.5°C/hr were used. All the samples lost 40 % of their weight in the form of liquid. The I-V characteristics obtained for three samples (S.1, S.2, and S.3) are shown in fig.2 along with the calculated I-V curve for pure copper at the same temperature. All the samples show a low resistance transition, with a resistance less than 1 μΩ, from a zero resistance superconducting state to the normal state where the resistance is typically greater than 1 mΩ. This effect can be clearly seen in the pulsed I-V curves at 77 K. Sample S.1 showed a voltage drop of only 25 μV at 700 A, and sample S.2 showed a voltage drop of only 17 μV at 800 A. The same effect persisted in the presence of an external magnetic field of upto 1.5 T applied perpendicular to the length of the specimen. The effect of external magnetic field on the I-V characteristics of sample S.1 is shown in fig.3. The results were obtained at fields of 0.5, 1.0, and 1.5T. A DC power supply of maximum 250A was used in these measurements.

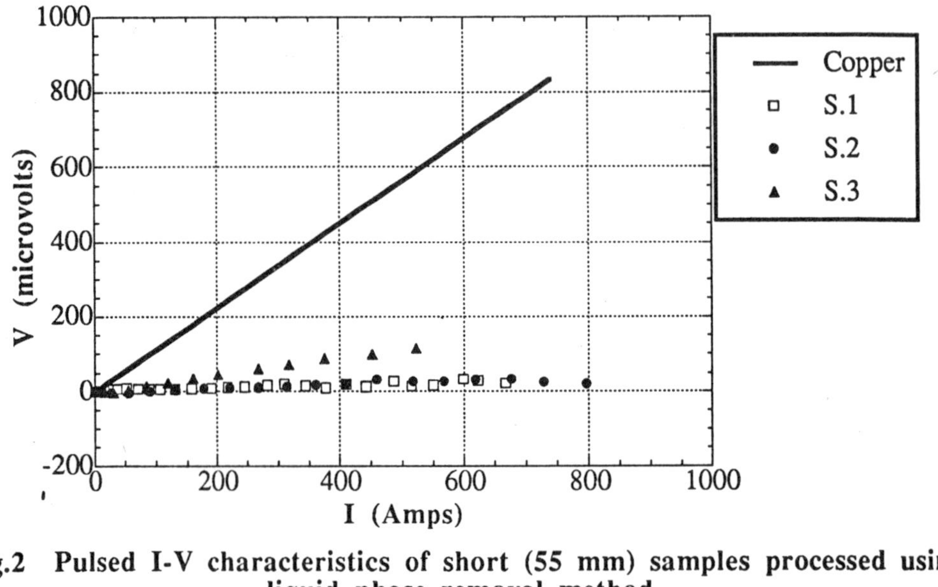

Fig.2 Pulsed I-V characteristics of short (55 mm) samples processed using liquid phase removal method.

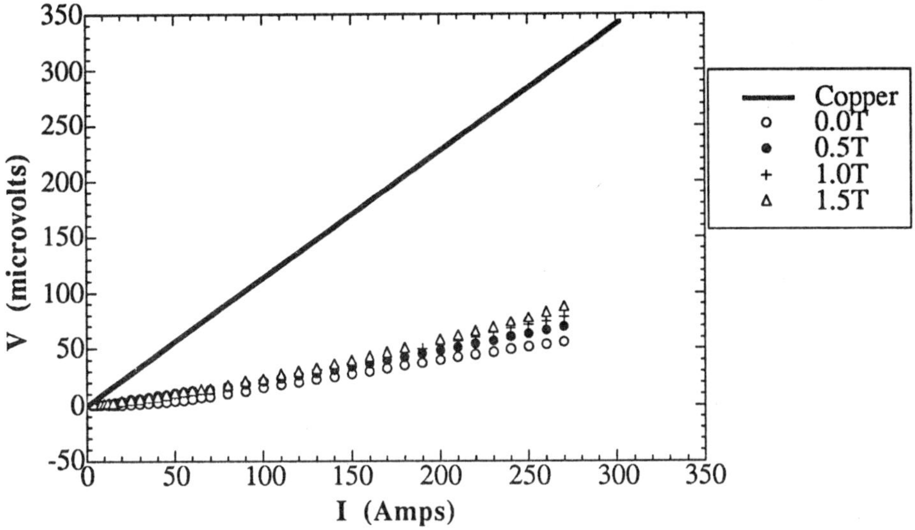

Fig.3. Effect of external magnetic field on the I-V characteristics of S.1.

For the long (100 mm) samples, 50 grams of 123 was used for compaction and sintering, and 45 grams of 211 was used for the absorbant. The 123 was placed horizontally on the 211 absorbants. In the melt texturing cycle, two hold times (t_h = 0.1 and 0.3 hours), and two cooling rates (R = 1.0 and 1.5 °C/hr) were used. All the samples lost about 40 % of their weight in the form of liquid. The I-V characteristics obtained for six samples (H.1 to H.6), shown in fig.4, also exhibited a low resistance transition similar to that observed using the short samples. The sample H.1 showed a voltage drop of 40 μV at 850 A, the sample H.2 showed a drop of 35 μV at 850 A, sample H.5 showed a drop of 45 μV at 800 A, and sample H.6 showed a voltage drop of 100 μV at 800 A.

Table.1. Hold time and cooling rate used for samples H.1 to H.6

Sample	Hold time (t_h hr)	Cooling rate (R °C/hr)
H.1	0.3	1.0
H.2	0.3	1.0
H,3	0.3	1.5
H.4	0.1	1.5
H.5	0.1	1.5
H.6	0.1	1.5

In conventional superconductors, there is a sharp transition from the superconducting state to the normal state. This is mainly due to the very high pinning forces in these materials. The resistance of the material changes from zero in the superconducting state to a few mΩ in the normal state. In the case of samples processed by the liquid phase removal method, the transition from the superconducting to the normal state is not sharp. The resistance in the transition state is less than 1 μΩ as seen in the I-V characteristics of these samples in figures 2, 3, and 4. This behavior can be explained by considering a dissipative model developed by M.N.Kunchur[8].

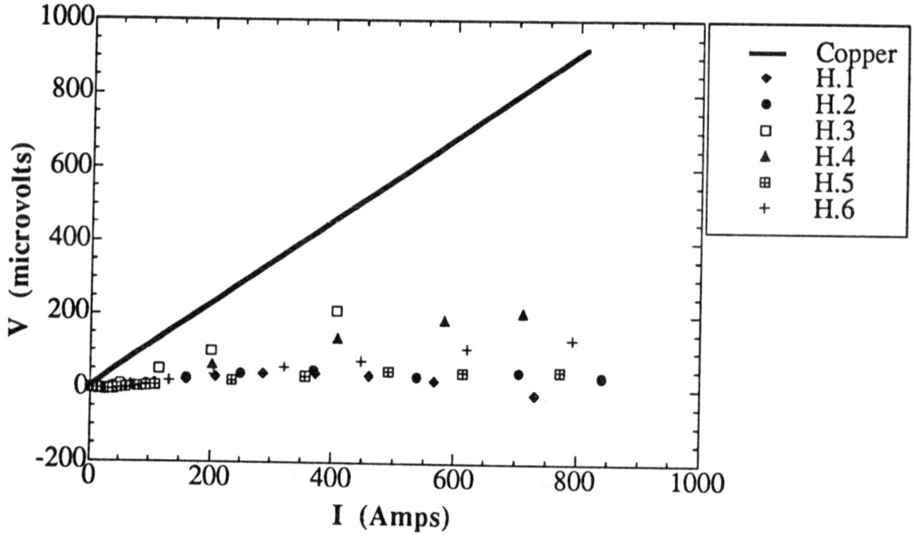

Fig.4. Pulsed I-V characteristics of long (100 mm) samples processed using liquid phase removal method

As the transport current through a superconductor is increased above its critical value, I_c, dissipation sets in. In this state, the superconductor has some finite resistivity. However, the superconducting state still persists up to a point where the kinetic energy associated with the current becomes greater than the condensation energy of the electron-phonon pairs causing pair breaking. The current density corresponding to this state is J_d. Thus, the transport behavior of the superconductor will show three broad regimes. The first is the superconducting state with zero resistance at current densities below J_c, followed by the dissipative regime between J_c and J_d where the superconductor has a nonzero resistance, and then the normal state above J_d[9].

The grain boundary misorientation was characterized using optical micrography. Specimens, approximately 30 mm long with 1 mm × 1 mm cross section, were cut from the bulk sample. This was done in such a way that when reassembled they will cover the whole length of the bulk specimen. The number of boundaries differed from one sample to another. All the four faces along the length of the specimen were polished and observed under cross-polarized light. The trace of each boundary on two adjacent faces of the specimen was photographed at a suitable magnification to help observe clearly the platelet cracks on either side of the boundary. The angles made by the platelet cracks at the two adjacent faces with the common edge was measured from the photographs, and the kink angle was calculated using these values[10]. Each of the angles were measured a number of times and an average was obtained to calculate the actual kink angle. The standard deviations calculated from these measurements are typically less than 3°. The error will be large if the four angles used in the calculation of the kink angle are close to 0° (or 180°). Since these instances were infrequent, the error in these angle measurements was taken to be about 3°.

The small specimens were cut from the bulk sample in a way that when reassembled they will cover the whole length of the bulk specimen. The number of boundaries differed from sample to sample. Two small specimens, 2.7 cm and 2.8 cm in length, were cut out from the bulk sample S1, 5.5 cm in length. The 2.8 cm long specimen, S1.1, has seven boundaries cutting across the cross section as shown in Figure 5(a). Similarly, the 2.7 cm long specimen, S1.2, has eleven boundaries across the cross section with large kink angles as shown in Figure 5(b). Table 2 and Table 3 list the misorientation angles of these boundaries in specimens S1.1 and S1.2 respectively.

All the samples on which the grain boundary analysis was performed, demonstrated good I-V characteristics and almost all the boundaries measured have large kink angles (> 15°). This is contrary to the widespread belief that high angle boundaries have poor current carrying capacity[4]. One explanation for this can be that high angle boundaries are not inherently weak links. This suggestion is consistent with the results obtained by Babcock et al[5] and Zhu et al[6]. The reason for the strong linkage across some high angle boundaries can be understood using the coincidence site lattice (CSL) theory. According to this theory, at certain misorientations there is a high degree of periodicity at the grain boundary which is characterized by the value Σ and as the periodicity increases (Σ decreases), the grain boundary energy decreases. Grain boundaries of high periodicity or low energy are expected to have strong linkages.

The CSL theory can be directly applied only to cubic systems, but can be extended to other crystal systems using some additional constraints. In their study involving the misorientations of over 200 pairs of adjacent grains separated by grain boundaries, Zhu et al.[6] have used a constrained coincident cite lattice (CCSL) theory to textured YBCO. Their results show a good agreement between the observed misorientations and the misorientations predicted by CCSL theory. Thus it can be inferred that textured materials with clean boundaries have a clear tendency to form low energy grain boundaries which can carry high currents.

CONCLUSION

The grain boundaries in high current YBCO samples, processed by melt texturing with partial liquid phase removal, have been characterized. The I-V characteristics demonstrate the ability of these samples to carry very large currents without developing a high voltage drop (low resistance transition). These samples contain a large number of grain boundaries. The misorientation of the a-b planes across the boundary (kink angle) is measured using the platelet cracks observed under an optical microscope and used as a parameter to characterize the boundary. A predominantly large number (>70) of high angle grain boundaries with kink angle values greater than 15° existed in these samples. This, along with the high current capacity (> 1,000 Amps) of these samples suggests strong grain boundary coupling in these bulk materials.

Table 2. Grain boundary kink angles in S1.1

S1.1.x	F(1)	T(1)	F(2)	T(2)	φ
1	127	82	126.5	47.5	30
2	109.5	65	155.5	34.5	41
3	171	21.5	6.5	22	140 (=40)
4	9	31.5	21	26	24
5	24	23.5	174	11	103 (=77)
6	178	16.5	177.5	3	38
7	174	175.5	14	71	128 (=52)

Table 3. Grain boundary kink angles in S1.2

S1.2.x	F(1)	T(1)	F(2)	T(2)	φ
1	114	36	80	109.5	77
2	79	111.5	48.5	118	28
3	48	114.5	85.5	29.5	85
4	81.5	22.5	84.5	105	82
5	86	112.5	118	131.5	31
6	120	126.5	35.5	38.5	105 (=75)
7	140.5	128	110.5	99	34
8	113	101.5	87.5	71	39
9	158.5	60	150	46	16
10	155	40.5	97	48.5	50
11	94	56	114	22	34

Bulk Sample : S1 Length : 5.5 cm I-V : 20 μV at 800 A

Sample : S1.1 Length : 2.8 cm Cross section : 1-2 mm^2

(a)

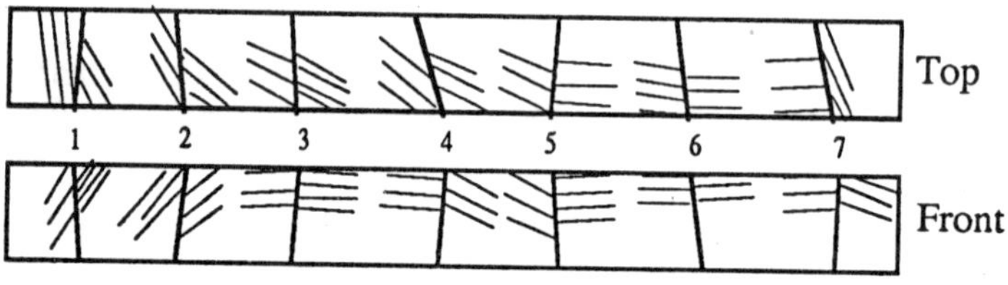

Sample : S1.2 Length : 2.7 cm Cross section : 1-2 mm^2

(b)

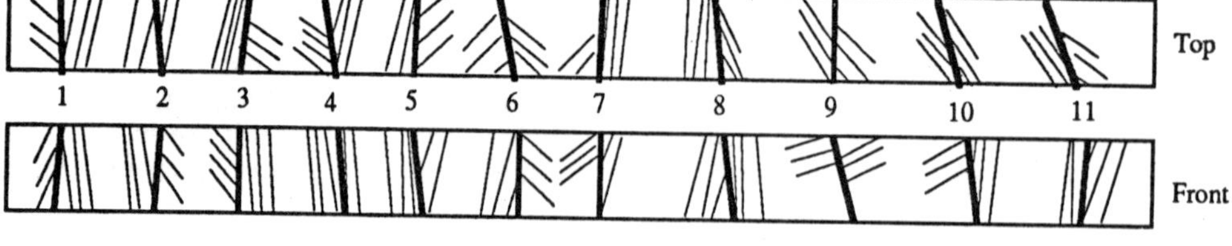

Figure 5 Schematic illustration of the microcrack pattern (a) S1.1, and (b) S1.2.

ACKNOWLEDGMENT

This work was supported by the Texas Center for Superconductivity at the University of Houston and the State of Texas.

REFERENCES

1. K. Salama, V. Selvamanickam and D.F. Lee, in *Processing and Properties of High T_c superconductors*, p. 155 ed. S. Jin, World Scientific (1993).
2. A.S. Parikh, B. Meyer and K. Salama, Superconductor Science and Technology **7**, 455 (1994).
3. D.W.A. Willen and K. Salama, Physica **C201**, 331 (1992).
4. D. Dimos, P. Choudhary and J. Mannhart, Phys. Rev. **B41**, 4038 (1990).
5. S.E. Babcock, X.Y. Cai, D.L. Kaiser and D.C. Larbalestier, Nature **347**, 167 (1990).
6. Y. Zhu, H. Zhang, H. Wang and M. Suenaga, J. Mater, Res. **6**, 2507 (1991).
7. D.W.A. Willen, M.S. Thesis, University of Houston (1991).
8. M.N. Kunchur, Mod. Phys. Lett. **B9**, 399 (1995).
9. S. Sathyamurthy, A.S. Parikh and K. Salama, Physica C (in press).
10. S. Sathyamurthy, A.S. Parikh and K. Salama, Mod. Phys. Lett. B (in press).

IRREVERSIBILITY CHARACTERISTICS OF Tl-MONOLAYER `1212' CUPRATE SUPERCONDUCTORS

I.K. Gopalakrishnan and J.V. Yakhmi
Chemistry Division, Bhabha Atomic Research Centre,
Mumbai 400 085, India.

Magnetic irreversibility data are reported for the Tl-monolayer superconductors $(Tl_{0.5}Pb_{0.5})CaSr_{1.8}La_{0.8}Cu_2O_7$ and $TlBaSrCaCu_2O_7$ from the DC magnetization measurements at several applied magnetic fields. The irreversibility line (IL) conforms to the power law $H=A(1-T/T_c)^n$ with a value of the exponent n = 2.5. The IL is located along the other Tl monolayer cuprates and above Tl bilayer phases.

I. INTRODUCTION

A unique feature in the H-T phase diagram of high-T_c oxides is the "irreversibility line" (IL) which separates a magnetically irreversible, zero-resistance state situated below it from a reversible state with dissipative electrical transport properties arising from a very weak nature of vortex pinning[1,2]. The concept of IL was introduced after the discovery of high temperature superconductors even though it holds for conventional low-T_c superconductors as well[3,4] for which the IL is either indistinguishable from H_{c2} line or lies extremely close to it. For high-T_c oxides, however, a combination of not completely unrelated factors viz., higher transition temperatures, shorter coherence length, large magnetic penetration length and quasi-two-dimensionality conspire to enhance the effects of thermal fluctuations which push the reversible regime down to rather low temperatures. A simple power law of the type

$$H_{irr} = A(1-T/T_c)^n \qquad (1)$$

approximately describes the field-temperature dependence boundary in high-T_c materials[1,2]. Even though the underlying mechanism responsible for the irreversibility is still ambiguous[5-9], the position of IL is of great interest for applications because it determines the pinning efficiency responsible for the limitation of critical current. The cuprate family of Tl-1212 superconductors are very interesting because of their low anisotropy which can be attributed to the presence of a single insulating Tl-O layer. Their higher T_c in comparison with $YBa_2Cu_3O_{7-\delta}$ suggests that these superconductors may be an excellent choice for applications.In this paper we have investigated the irreversibility characteristics of Tl-monolayer superconductors $(Tl_{0.5}Pb_{0.5})CaSr_{1.8}La_{0.8}Cu_2O_7$ and $TlCaSrBaCu_2O_7$.

II. EXPERIMENTAL

The polycrystalline samples used in the present study were prepared by a matrix reaction method[10]. DC magnetization measurements were carried out by employing E.G. & G.P.A.R. Model 4500 Vibrating Sample Magnetometer (VSM). The IL was determined from zero field cooled (ZFC) and field cooled (FC) magnetization measurements at several applied magnetic fields, the irreversibility temperature being the temperature where the ZFC and FC magnetization meet. The ZFC and FC curves were recorded for fields ranging from 10 Oe to 7000 Oe.

III. RESULTS AND DISCUSSION

The superconducting transition temperatures (T_c) as determined by AC and DC magnetic susceptibility and electrical resistivity measurements are 103 and 104 K, respectively. The X-ray diffraction analyses show that the samples are single phase with tetragonal(space group P4/mmm) lattice parameters a=3.789(1) Å and c=12.070(5) Å, and a= 3.827(1) Å and c=12.123(3) Å, respectively. The enhancement of T_c in $(Tl_{0.5}Pb_{0.5})$ $CaSr_{1.8}La_{0.8}Cu_2O_7$ as compared to $(Tl_{0.5}Pb_{0.5})CaSr_2Cu_2O_7$ indicates that the substitution of La^{3+} for Sr^{2+} optimizes the hole content in the former. The irreversibility line for these two samples was determined by DC magnetization measurements, at several applied fields upto 1 Tesla. Fig. 1 shows the irreversibility lines in the H-T plane and Fig.2 shows the log-log plot of applied magnetic field H versus $1\text{-}T/T_c$ for two samples.

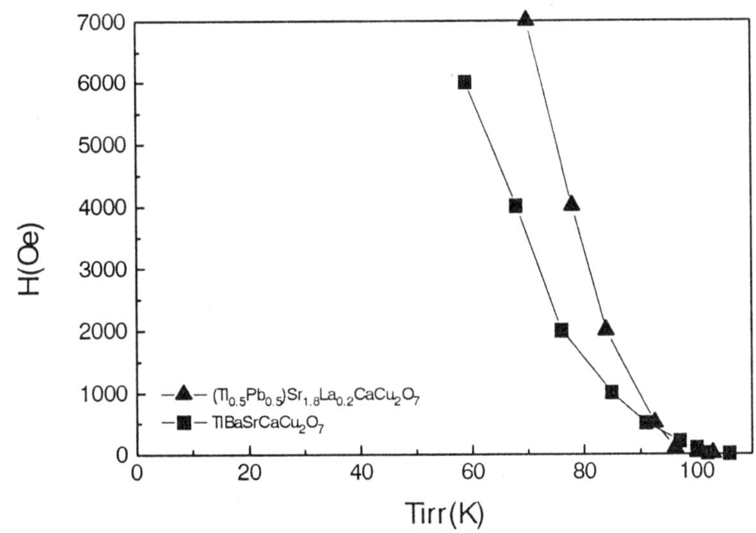

Fig.1. Irreversibility lines in the H-T plane.

The symbols represent the measured data points while the continuous lines is linear fit to these data points with slope equals to 2.5. IL of both compounds was found to obey the power law $H_{irr}= A(1-T/T_c)^n$ with a single value of the exponent n=2.5 and was located in the same region of the H-T phase diagram where other Tl-monolayer superconductors are. The analysis of the data with that reported in the literature for other cuprate superconductors indicates that there exists a correlation between the structural anisotropy and magnetic irreversibility.

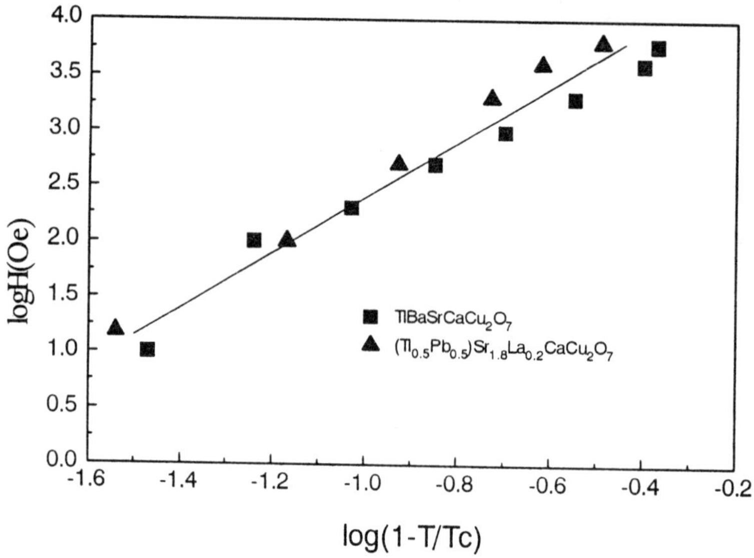

Fig.2. Log-Log plot of applied field vs. reduced temperture.

In literature, the exponent n has been shown to have values ranging from 1.5 to 5.5 viz. n=1.5 for Y-123, 2.5 for single layer cuprates of Hg and Tl and 5.5 for double layer cuprates of Tl and Bi[4,13]. The value of n is known to increase with increase in the anisotropy of the structure and there have been some attempts to correlate the anisotropy with the spacing between CuO_5 pyramid sheets sandwiching a unit of charge reservoir block[11,14]. The Tl-2223 and Bi-2223 phases have double Tl-O or Bi-O layers which are non superconducting. On the other hand, Tl monolayer, Hg monolayer and Y-123 systems have single TlO, HgO or CuO layer, respectively. The distance d_i between the adjacent superconducting Cu-O planes across the charge reservoir blocks in Y-123, Tl1223, Tl2223 and Bi-2212 compounds are 8.3, 8.7, 11.4 and 12.1A, respectively. Nabatame et al[14] found an exponential dependence of d_i on the irreversibility field H_{irr} for different high-T_c systems. Isawa et al[13] have extended this to Hg-based cuprate superconductors like Hg-1223. H_{irr} has thus been shown to decrease in the order Y-123 > Tl-1223 > Hg-1223 > Tl-2223 > Bi-2212. These authors suggest that this exponential dependence of H_{irr} on d_i is consistent with the concept of Josephson coupling between the Cu-O layers. The conducting CuO double chains have been suggested to short circuit the Josephson interlayer coupling in the case of 124 structure[1], because of which 10.2A may not be the true value of d_i for Dy(Tb)-124. Thus our data is by and large consistent with the current understanding that a correlation exists between the crystalline anisotropy and the strength of pinning and support the conjecture that the thickness and nature of the charge reservoir layer sandwiching the CuO_5 pyramids is the key factor that affects the irreversibility line and hence the flux pinning properties[1].

REFERENCES

1. K.A. Muller, M. Takasighe and J.K. Bednorz, Phys. Rev. Lett. **58**, 1143 (1987).
2. Y. Yeshurun and A.P. Malozemoff, Phys. Rev. Lett. **60**, 2202 (1988).
3. J.R. Fraser, T.R. Finlayson and T.F. Smith, Physica C **159**, 70 (1989).
4. C. Rossel, E. Sandvold, M. Sergent, R. Chevrel, Physica C **165**, 233 (1990).
5. P.L.Gammel, L.F.Schneemeyer, J.V. Waszczak and D.J. Bishop, Phys. Rev. Lett. **60**, 1666 (1988).
6. P.W. Anderson and Y.B. Kim, Rev. Mod. Phys. **36**, 39 (1964).
7. A.I. Larkin, Sov. Phys. JETP **31**, 784 (1970).
8. M.P.A. Fisher,Phys. Rev. Lett. **62**, 1415 (1989).
9. E.H. Brandt, Physica C, **195**, 1 (1992).
10. I.K. Gopalakrishnan, J.V. Yakhmi and R.M. Iyer, Physica C **172**, 450 (1991).
11. D.H. Kim, K.E. Gray, R.T. Kampwirth, J.C. Smith, D.S. Richeson, T.J. Marks, J.H. Kang, J. Talvacchio and M. Eddy, Physica C **177**, 431 (1991).
12. Y.C. Kim, J.R. Thompson, D.K. Christen, Y.R. Sun, M. Paranthaman and E.D. Specht, Phys. Rev.B **52**, 4438 (1995).
13. K. Isawa, T. Higuchi, T. Machi, A.Tokiwa-Yamamoto, S. Adachi, M. Murakami, and H. Yamauchi, Appl. Phys. Lett. **64**, 1301 (1994).
14. T. Nabatame, J. Sato, Y. Saito, K. Aihara, T. Kamo and S. Matsuda, Physica C **193**, 390 (1992).

CORRELATION OF MICROSTRUCTURE TO INTERGRANULAR MAGNETIC RESPONSE OF HTSCS FROM FLUX PROFILES

N. Harish Kumar, E. Isaac Samuel, A. Radhika Devi and V. Seshu Bai
School of Physics, University of Hyderabad, Hyderabad - 500 046

The flux profiles are derived from the low field M-H loops recorded at 77K and 33 Hertz using lock-in wideband detection method in sintered and press-sintered BSCCO 110 K phase samples and in melt grown YBCO samples with and without Y_2BaCuO_5 and Ag additions. The changes in the flux profile are correlated to microstructural alterations introduced by various processing methods.

I. INTRODUCTION

The low field magnetic response of high T_c ceramic superconductors is decided by the nature of intergranular region linking the superconducting grains. These regions have a low critical current density than the superconducting grains and hence become operative for flux entry at low applied magnetic fields. The coupling strength of these intergranular links and their J_c dependence on field gets altered due to different processing methods used to synthesize them which alter their microstructure. Hence a study of low field magnetic flux penetration in conjunction with the microstructure of the material will help in microstructural optimization to get the desired physical properties.

II. EXPERIMENTAL

The Magnetic flux penetration into a superconductor in the mixed state can be studied using the flux profile technique developed by *Campbell*[1]. We have slightly modified this method to get the flux profiles. First the magnetic hysteresis loops are recorded using the lock-in flat-band detection technique[2-4]. The magnetization (M_{peak}) corresponding to the peak field(H_m) values are plotted against the field amplitude. The derivative of this curve is then plotted against the field amplitude to get the flux profiles.

The normalized penetration depth is given by

$$\frac{p}{R} = 1 - \sqrt{-\frac{dM_{peak}}{dH_m}}$$

Where p and R are the penetration depth in the mixed state and radius of the cylindrical superconductor respectively.

We have recorded the field variation of isothermal ac M-H loops at 77 K and 33 Hz on sintered (A) and press sintered (B) $Bi_{1.2}Pb_{0.3}Sr_{1.5}Ca_2Cu_3O_y$(BSCCO) 110 K phase samples[5] and melt grown $YBa_2Cu_3O_7$(YBCO) samples[6] with different amounts of Y_2BaCuO_5(211) and Ag inclusions : 0 mole% 211 + 0 wt% Ag (C), 20 mole % 211 + 10 wt% Ag (D), 28 mole % 211 + 5 wt% Ag (E), and 28 mole% 211 + 2 wt% Ag (F), and then constructed the flux profiles from them.

III. RESULTS and DISCUSSION

Figs. 1a-2b show the microstructures of the samples A,B,C and E. While the sintered BSCCO sample shows a random distribution of grains with 50 % porosity, the microstructure of the press sintered sample shows a reduction in porosity to 15 % and in grain size to 1 μ m from 10 μ m. On the otherhand the melt grown YBCO samples show an entirely different microstructure. They exhibit a randomly oriented domain structure with aligned platelets inside the domains separated by microcracks. Besides these there are also voids and unreacted liquid phases found in the microstructure of the meltgrown YBCO samples without any additions. The sample with 211 and Ag addition shows a better microstructure with less crackwidth, clean domain boundaries (no segregation of Ag at domain boundaries) and no voids or liquid phases.

Fig.1. The microstructure of the fractured surface of (a) Sintered and (b) Press sintered BSCCO-110 K phase showing reduction in porosity and grain size in the later.

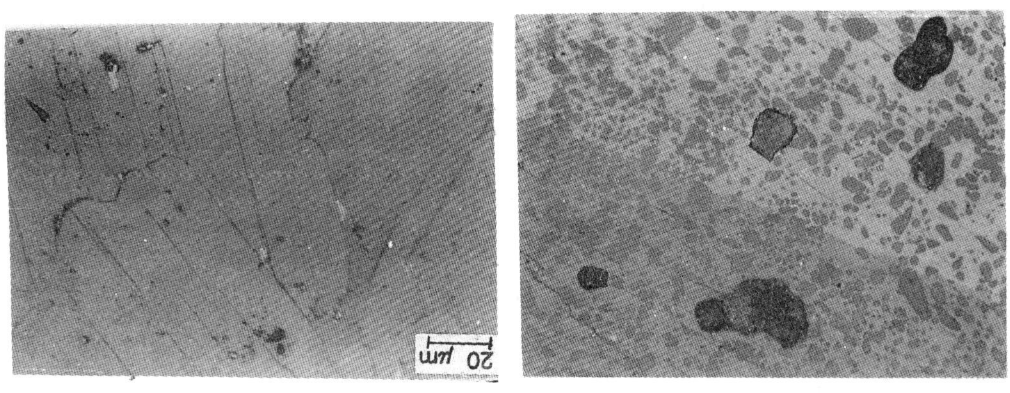

Fig.2 The optical micrograph of the polished surface of a meltgrown YBCO sample (a) without additions showing the domain structure with aligned platelets inside the domain separated by microcracks and (b) with 28 mole% 211 and 5 wt% Ag inclusions showing reduction in crack-width and clean domain boundaries with a uniform distribution of 211 particles.

Figures 3a & 3b show the flux profiles derived from the loops. The sintered sample exhibits a non linear increase in penetration above 1 Oe with a sharp nose like structure around 8 Oe followed by a region of constant penetration in the field range 15 to 55 Oe. The press sintered sample also exhibits a similar behaviour but with a less pronounced nose around 10 Oe and the region of constant penetration extends upto 100 Oe beyond which it increases once again. While the nose occurs around the full penetration field H_{pi} of the matrix, the field upto which the constant region of penetration extends before the second rise in penetration starts, gives the apparent lower critical field of the grains H_{c1g}. From the flux profiles one can see that not only the intergranular Jc but also the H_{c1g} has increased on press sintering. While the increase in J_{ci} may be due to the strengthening of the coupling between the grains, the rise in H_{c1g} may have its origin in surface barrier effects[4]. The non linear flux profiles suggest a strong field dependence of J_c.

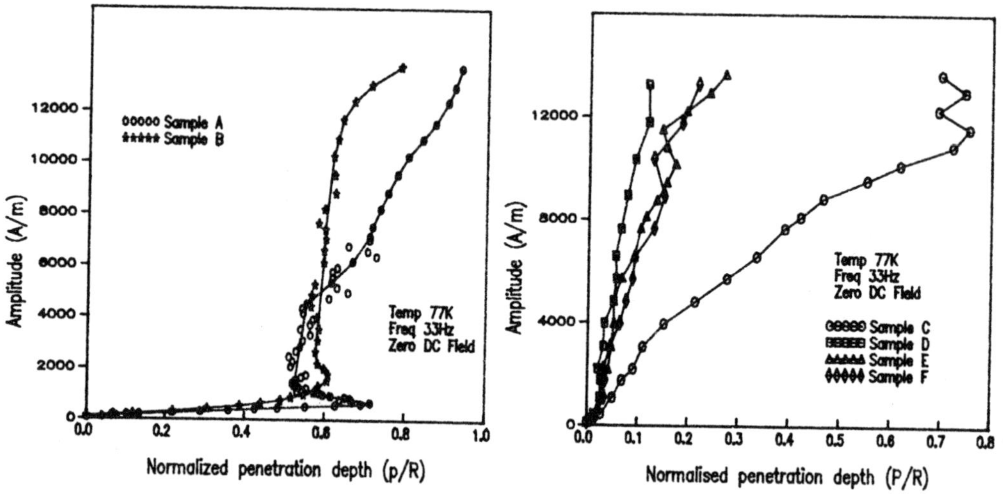

Fig.3. Flux profiles in (a) sintered and press sintered BSCCO - 110 K phase and in (b) meltgrown YBCO samples with and without additions. The solid lines are a guide to the eye.

On the otherhand the flux profile of the melt grown sample(C) without 211 and Ag additions is linear with distinct slope changes at certain fields. The linearity in the flux profiles suggests lack of field dependence of J_c and the distinct slope changes indicate the presence of regions with distinctly different J_cs and having a narrow distribution. The flux penetration is nearly 70 % at 150 Oe which is comparable to that of a sintered sample. In contrast the flux profiles of melt grown samples with 211 and Ag show < 10 % penetration even at 150 Oe. While the first rise in p/R occurs around 25 Oe in sample C, it occurs only above 60-80 Oe in the case of the other samples with pinning centers indicating an enhancement of H_{c1} on melt processing and addition of 211 and Ag. The flux penetrates the region of lowest J_c followed by next higher one and so on. The different J_c regions could be the residual weaklinks (impurities, unreacted liquid phases, voids etc), the domain boundaries and the platelet boundaries. While the residual weaklinks constitute the region having the lowest J_c which becomes operative for the flux entry process first, the domain boundaries could be the region that allows flux entry next. The platelet boundaries allow flux entry only at higher fields and do not contribute to the low field magnetic response. The slope of the flux profile is a measure of the J_c associated with the region of flux entry.

IV. CONCLUSIONS

1. The Press sintering process improves the coupling between the grains and increases the intergranular J_c and intragranular lower critical field (H_{c1g}). Non-linearity of the profiles at low fields indicate J_c to be field dependent and it is seen that the field dependence gets weaker on press sintering.

2. Though the melt processing technique results in a large reduction in the density of weaklinks, the current density depends on the microstructural features. The J_c values are limited by the strength of the domain boundaries and platelet boundaries. Further the presence of residual weak links also reduces the J_c. The addition of 211 and Ag is seen to eliminate residual weeklinks and strengthen the domain boundaries resulting in higher H_{c1} and J_c.

3. The linear flux profiles exhibited by melt grown samples suggest lack of field dependance of J_c. The distinct slope changes in the flux profile of sample C are attributed to the presence of regions with distinctly different J_cs and having a narrow distribution which become operative for flux entry at different applied fields.

ACKNOWLEDGEMENT

The authors wish to acknowledge the Department of Science and Technology, Govt. of India for financial support. NHK acknowledges the Council of Scientific and Industrial Research and EIS and ARD acknowledge the University Grants Commission for fellowship.

REFERENCES

1. A. M. Campbell, J. Phys. **C 2**, 1492 (1969).
2. F. Gömöry, Rev. Sci. Instrum. **62**, 2019 (1991).
3. G. E. Gough *et al.*, Cryogenics **31**, 119 (1991).
4. N. Harish Kumar and V. Seshu Bai, Phys. Rev. **B 53**, 15281 (1996).
5. V. Seshu Bai *et al.*, J. Appl. Phys. **70**, 4378 (1991).
6. N. Harish Kumar, Ph.D. thesis, University of Hyderabad (1996).

DEPENDENCE OF CRITICAL CURRENT (I_c) ON WEAK LINKS IN METAL/METAL OXIDE ADDED Bi-2212

K.S. Ramesh, P. Sumana Prabhu and M.S. Ramachandra Rao*
*Materials Science Research Centre *and Department of Physics, Indian Institute of Technology, Madras-600 036, INDIA*

The presence of weak links in high T_c superconductors is a major drawback in achieving high critical current densities. The present study deals with the effect of additives (Ag, Ag_2O and $BaBiO_3$) on the weak link behaviour of Bi-2212 in bulk form. It is observed that the addition of Ag upto 6 mole% improves the proximity between the grains and thereby enhances the current carrying capability even at temperatures very close to T_c. De Gennes and Clarke models have been invoked to establish that insulating additives ($BaBiO_3$ and Ag_2O) show a superconductor-insulator-superconductor (S-I-S) behaviour and thereby enhance the weak link network whereas metallic Ag shows a superconductor-normal metal-superconductor (S-N-S) behaviour and is therefore a potential candidate for large scale applications. Further support to this data is obtained from ac susceptibility studies carried out at various fields (0.1-4 Oe).

I. INTRODUCTION

One of the main problems which affects the J_c of polycrystalline oxide superconductors is the presence of grain boundary weak links[1] that hinder the applications of high T_c materials (HTSC) in devices. "Weak-link-behaviour" is a term used to describe what happens to a supercurrent as it crosses a non-superconducting region such as a grain boundary. Classically, a supercurrent would not cross such a barrier. However, quantum mechanics permits the supercurrent to tunnel through the barrier. In doing so, the supercurrent experiences an exponential attenuation which is proportional to the barrier thickness and inversely proportional to the coherence length and the attenuation increases with increasing magnetic field. Since the coherence length (ξ) in the HTSC's is very low (typically 3-30Å), attenuation can be quite large for a barrier of a few atoms thick. Weak-link behaviour is responsible for the steep fall off in transport current particularly in the presence of an applied magnetic field, in bulk samples. Weak links arise due to various reasons such as compositional non-stoichiometry, intergrowth of related phases, impurities etc[2,3]. The elimination of weak links is the major goal of research concerned with making high quality HTSC bulk materials. One way to overcome this is to add metal/metal oxide additives that may improve the connectivity between the grains by proximity effect. The present work aims at observing the changes in weak link network of Bi-2212 superconductor by adding metallic silver powder and two insulating oxides viz., Ag_2O and $BaBiO_3$.

II. EXPERIMENTAL

About 30g of Bi-2212 having an off-stoichiometric composition, $Bi_{2.1}Sr_{1.93}Ca_{0.97}Cu_2O_8$ was prepared using solid state reaction method[4]. The additives Ag, Ag_2O and $BaBiO_3$ were added in steps of 2, 4 and 6 mole% to a known amount of Bi-2212. The mixture was heated at 875°C for 24h, pelletised and sintered at 870°C for 24h. and finally annealed at 805°C in a preheated furnace and quenched in air. The samples were characterised by powder X-ray diffraction (SEIFERT, Germany). Resistivity and I-V measurements were carried out by the four probe van der Pauw method in the temperature range 300-13K using a closed cycle He refrigerator (Leybold Heraues, Germany). ac susceptibility measurements were made on the samples at a field of 0.1Oe and frequency ~313 Hz in the temperature range 300-13K using the SUMITOMO Superconducting Materials Property Measurement equipment (Model SCR 204-T, JAPAN). The microstructure was determined using a scanning electron microscope (SEM; JEOL, Japan).

III. RESULTS AND DISCUSSION

XRD studies indicate that all the compositions upto 6 mole% of the additives are single phase. Resistivity (ρ-T) measurements on Bi-2212 show consistent $T_{c,zero}$ values in the range 74-76K. With increase in concentration of insulating additives, the superconducting transition temperature ($T_{c,zero}$) systematically decreases and the ΔT_c ($T_c(90\%)$-$T_c(10\%)$) values progressively increase. From the real part (χ'- T) of ac susceptibility measured at 0.1 Oe and 313 Hz, it is observed that the transition width (difference between the intra and intergranular T_c) of the Ag-doped samples is smaller than that of pure Bi-2212 signifying a better connectivity in the grain network (Fig.1). However, the compounds containing $BaBiO_3$ and Ag_2O (Fig.2), show a broadening of the transition width indicating that the added oxide materials segregate between the grains and bulk superconductivity occurs at a lower temperature.

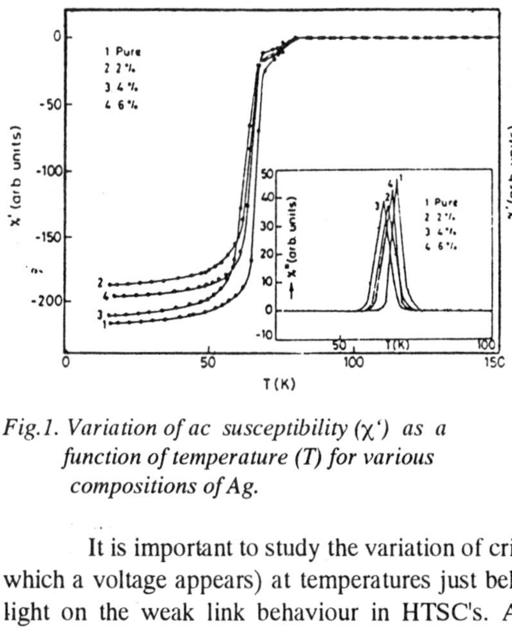

Fig.1. Variation of ac susceptibility (χ') as a function of temperature (T) for various compositions of Ag.

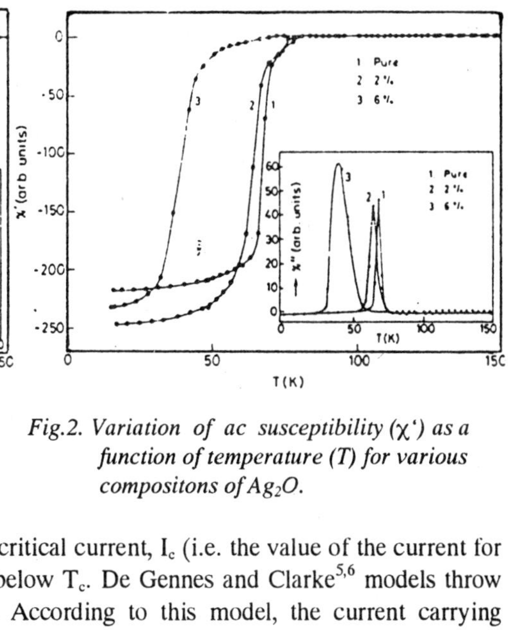

Fig.2. Variation of ac susceptibility (χ') as a function of temperature (T) for various compositons of Ag_2O.

It is important to study the variation of critical current, I_c (i.e. the value of the current for which a voltage appears) at temperatures just below T_c. De Gennes and Clarke[5,6] models throw light on the weak link behaviour in HTSC's. According to this model, the current carrying capability of the sample depends on the nature of the weak link. When a non-magnetic normal metal (N) is in good electrical contact with a superconductor (S), cooper pairs diffuse into the normal metal. If the electron-electron interaction in N is zero (V_N=0), and in the absence of magnetic impurities and applied magnetic fields, cooper pairs in N are only destroyed by the thermal energy (k_BT). In addition, the presence of the normal metal tends to lower the cooper pair density in the superconductor in the vicinity of the contact. This phenomenon is known as "proximity effect". When a normal metal is sandwiched between two superconductors provided the normal metal is not too thick, a supercurrent can pass between the two superconductors.

Under appropriate conditions the sandwiches are able to sustain a supercurrent, but a voltage appears across the junction once a certain critical current is exceeded. The sandwiches may behave, phenomenologically, as Josephson junctions in that the supercurrent is related to the difference in the phase of the order parameter ø across the junction by an equation of the form,

$$I = I_c \sin ø \tag{1}$$

For a superconductor-normal metal-superconductor (S-N-S) junction, it has been shown that

$$I_c \propto (T_c - T)^2 \tag{2}$$

whereas in the case of a superconductor-insulator-superconductor (S-I-S) junction,

$$I_c \propto (T_c - T) \tag{3}$$

In the present study, the critical current I_c was determined at various temperatures below T_c for all compositions of additives. Ag-added samples show an increase in I_c with increasing concentration at temperatures close to T_c and an S-N-S type of behaviour as seen from the $(I_c)^{0.5}$ vs $(T_c$-T) plot (Fig.3). Interestingly, pure Bi-2212 clearly shows a deviation from the S-N-S type of behaviour. With increase in Ag content, the slopes systematically increase indicating decrease in grain boundary width. Thus, the addition of metallic silver enhances the proximity effect between the grains and proves to be useful in overcoming the weak link problem.

However, both the Ag_2O and $BaBiO_3$ added samples showed a decrease in I_c with increasing concentration and do not obey the S-N-S type behaviour (Inset of Fig. 4). On the contrary, they fit very well into (S-I-S) type behaviour which is evident from linear plots of I_c vs $(T_c$-T) (Fig.4). The slopes systematically decrease with increasing concentration of the added oxides indicating an enhanced weak link network. Bi-2212 which shows a deviation from S-N-S type behaviour shows an S-I-S type of behaviour which clearly reflects the poor connectivity between the grains.

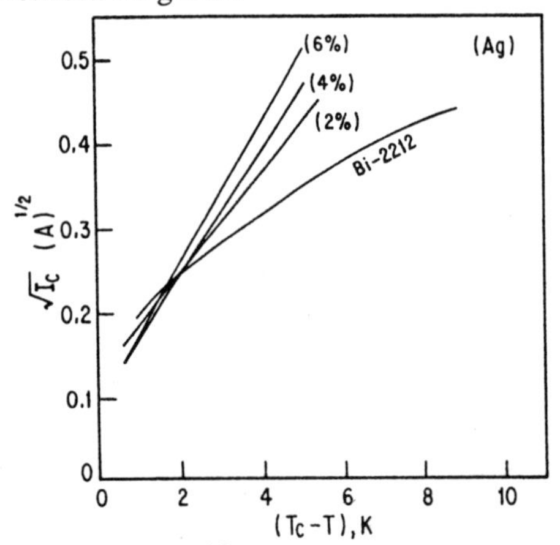

Fig.3. Plot of $(I_c)^{0.5}$ vs $(T_c$-T) for Ag-added Bi-2212. The straight line fit indicates S-N-S behaviour. Pure Bi-2212 shows clear deviation from this behaviour.

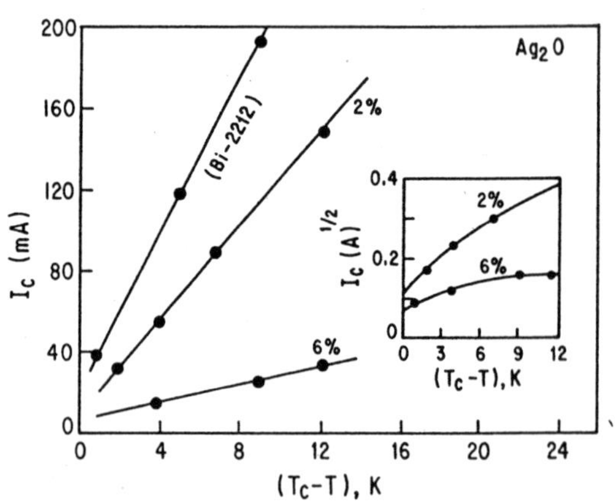

Fig.4. Plot of I_c vs $(T_c$-T) for Ag_2O added Bi-2212. The straight line fits indicate S-I-S behaviour. Inset shows plot of $(I_c)^{0.5}$ vs $(T_c$-T).

The additives preferentially concentrate in the grain boundary region. This has been shown from the ac susceptibility measurements. Ag, being a good conductor, improves the connectivity between the grains by the proximity effect whereas the insulating oxides Ag_2O and $BaBiO_3$ further enhance the weak link behaviour.

Scanning Electron Micrographs (SEM) taken on the additive samples indicate changes in the microstructure. The 6 mole% Ag-added compound shows a well connected grain network, whereas in the case of Ag_2O and $BaBiO_3$, the microstructure shows an enhanced weak link network. These results corroborate very well with the observations made in resistivity and ac susceptibility studies. These studies have laid foundation for our future work on the fabrication of high quality bulk samples of Ag added Bi-2212 current leads.

CONCLUSIONS

Studies on the effect of oxide additives on $Bi_{2.1}Sr_{1.93}Ca_{0.97}Cu_2O_8$ have indicated changes in the granular nature and hence microstructure of Bi-2212. Resistivity (ρ-T), I-V measurements and ac susceptibility (χ'-T) studies indicate that the addition of insulating oxide materials viz., $BaBiO_3$ and Ag_2O progressively increase the transition width (ΔT_c) with increasing concentration. Significant changes in superconducting properties are observed with field variation. SEM studies have indicated changes in the microstructure. Ag addition gives the best microstructure with lesser defects as compared to the insulating oxide added Bi-2212. The most significant result of the present study is that Ag addition improves the connectivity between the grains and thus is a potential candidate for overcoming the weak link problem.

ACKNOWLEDGMENTS

MSR would like to thank DST for the sanction of Young-Scientist Project. PSP wishes to thank CSIR for the award of a research fellowship.

REFERENCES

1. J.F. Kwak, E.L. Venturini, P.J. Nigrey and D.S. Ginley, Phys. Rev. B. **37**, 9749 (1988).
2. M. Kawasaki, P. Chaudhary and A. Gupta, Phys. Rev. Lett. **68**, 1065 (1992).
3. S.E. Russek, D.K. Lathrop, B.H. Moekly, R.A. Buhrman, D.H. Shin and J. Silcox, Appl. Phys. Lett. **57**, 1155 (1991).
4. A. Ono, Jpn. J. Appl. Phys. **28**, L1372 (1989).
5. P.G. De Gennes, Rev. Mod. Phys. **36**, 225 (1964).
6. J. Clarke, Proc. of Royal Soc. A **308**, 447 (1969).

ENHANCEMENT OF CRITICAL CURRENT DENSITY IN $Y_1Ba_2Cu_4O_8$ AND $Y_2Ba_4Cu_7O_{15-\delta}$ HTSC PHASES BY THERMAL DISSOCIATION

A.K. Srivastava[1], T.V.C. Rao[2], K.V. Gopal Krishnan[3] and O.N. Srivastava[1]

[1]*Department of Physics, B.H.U. Varanasi 221 005, India*

[2]*Solid State Physics Division, B.A.R.C., Mumbai 400 085, India*

[3]*Solid State Physics Division, T.I.F.R., Mumbai 400 005, India*

In the present investigation, the 124 and 247 YBaCuO phases were thermally dissociated for different time spans and at higher temperature in order to create flux pinning centres. The 124 phase dissociated for time spans of 10, 30 and 60 s at a temperature 915°C has been investigated. Samples dissociated for 30 s shows an enhancement of J_c approximately 11 folds and approximately 7 folds at 77K and 15K, respectively at H=0.9 T. The transformation covering a span of 30 s was found to result in the creation of 125 and interfaces corresponding to 124/125 and 124/123 besides the native 124 phases. In addition to these, stacking faults are also present. Similar studies have been carried out for 247 phase. It has been found that sample dissociated for 30 s shows highest J_c i.e., 1.02×10^5 A/cm^2 at 77K and 1.20×10^6 A/cm^2 at 15K at H=0.5T. For this sample, highest density of stacking faults and interfaces has been observed.

I. INTRODUCTION

One of the main problems with HTSC's inhibiting their ready application in devices is their low intragrain (J_c) and also low transport (J_{ct}) critical current densities. Low J_c can be overcome by developing appropriate flux pinning centres. The present study seeks to investigate simultaneously the details of phase dissociation of 124[1] and 247[2] phases by thermal decomposition (for creating flux pinning centres) through electron microscopic techniques and J_c enhancements through magnetization hysteresis measurements. In the present study, the thermal dissociation of 124 has been so planned (915°C for 10, 30 and 60 sec) that the transformed phase is invariably based on the 124 matrix. It has been found that J_c for the 30 sec decomposed 124 samples are ~2.15×10^5 A/cm^2 at 77K and ~3.9×10^6 A/cm^2 at 15K; at H=0.9T. Enhancement of J_c was correlated with microstructural characteristics. Similar studies were also carried out on 247 phase. It has been found that the J_c for the 30 sec dissociated samples are 1.02×10^5 A/cm^2 and 1.2×10^6 A/cm^2 at temperature 77K and 15K respectively (both at H=0.5 T) which is ~2.4 times and ~2.0 times higher than the as-synthesised samples. The details on the enhancement of J_c and its correlation with the microstructural characteristics will be outlined in the succeeding sections.

II. EXPERIMENTAL DETAILS

The synthesis of 124 and 247 was carried out using solid state reaction route. Its thermal dissociation was carried out at an elevated temperature in the phase stability region of 123. The dissociation process involves the heating of pellets to ~915°C at a rate of 10°C/min and holding the pellets for different time spans (i.e. 10, 30 and 60 sec. for 124 and 10, 30 and 50 sec. for 247). The pellets were quenched in air and finally annealed at 550°C in flowing oxygen for 12 hrs and then slowly cooled to room temperature. Magnetic measurements have been carried out in order to evaluate the J_c by employing DC SQUID magnetometer. Magnetization hysteresis measurements were made corresponding to ZFC conditions. The J_c was calculated by using simplified Bean's critical state model : (J_c= -30 M/d) where M is difference in magnetization in increasing and decreasing magnetic field, d is the average grain size and J_c is critical current density in Amp/cm^2. Electron microscopic investigations were carried out in order to bring out the possible influence of the microstructure on the enhancement of J_c.

III. RESULTS AND DISCUSSION

3.1 Electrical, Magnetic and Microstructural Studies on 124

Electrical characterisation of the 124 phase shows increase in T_c from 80K to 91K for dissociated samples. The J_c's of the as-synthesised and dissociated 124 are listed in Table 1.

Fig.1(a) depicts the HREM for the as-synthesized samples. A lattice fringe spacing of 13.6Å showing the 124 phase can easily be seen. For the as-synthesized 124 samples the presence of stacking faults could not be made out. This confirms the nearly defect free as-synthesized 124 phase.

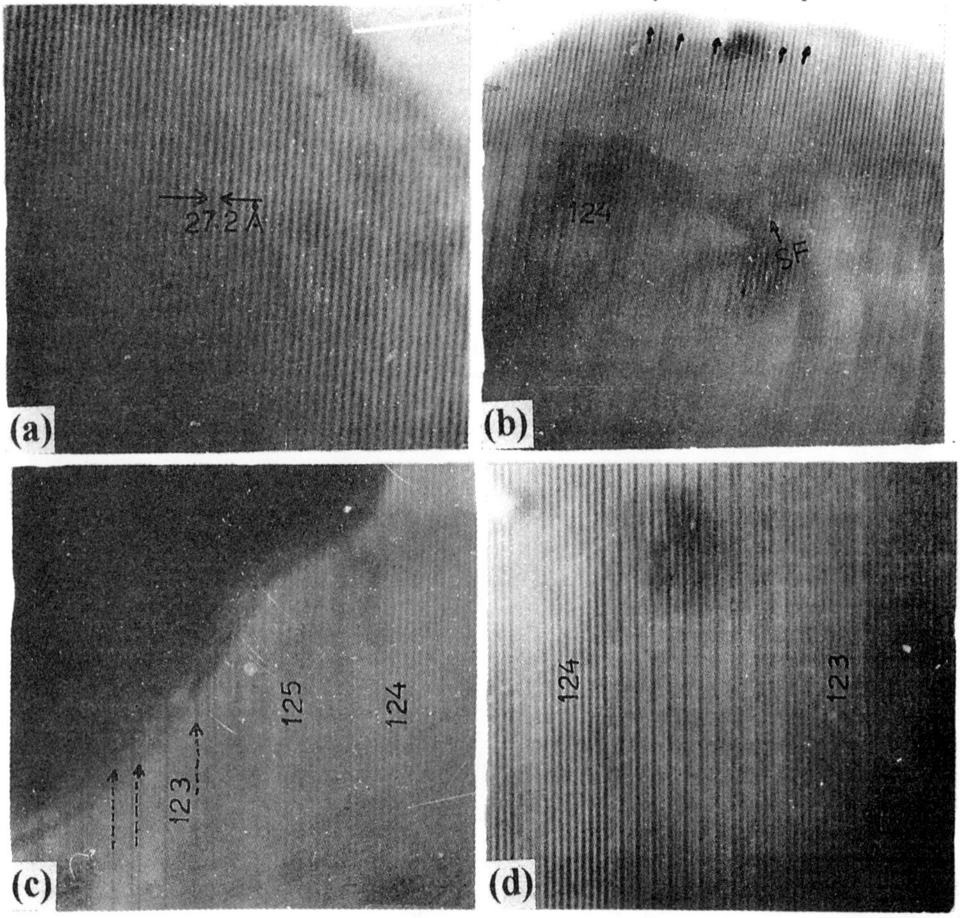

Fig.1(b) shows a representative HREM for the 124 dissociated for 10 sec. In addition to the 124 regions, several stacking fault configurations can easily be seen. These regions are marked by arrows with the label SF. The faulted regions extend from one to a few unit cells and hence span very narrow regions of the order of 15-45 Å. As revealed by the TEM micrograph shown in Fig.1(b) some regions corresponds to the local presence of 123 regions (c=11.67Å); these are shown by arrows. The presence of the 123 regions adjacent to 124 would produce 124/123 interfaces, which would be very narrow and presumably would act as pinning centres. It is thus clear that the dissociation of 124 for 10 sec produces stacking defects and 123 regions of very narrow extent (10-40Å), which are compatible with the coherent length of the order of 30 Å and narrow 124/123 interfaces, therefore, form flux pinning centres and hence results in the enhancement of J_c .

Fig.1(c) shows that even after 30 sec dissociation, the as-synthesized 124 phase is still the dominant phase even though the 123, 125 and other interfaces appear. Prominent 124 (c/2 = 13.6Å, c = 27.2Å) and 123 (c=11.67Å) regions have been outlined. In the dissociated samples besides the 123 phase the higher-CuO-content phase 125 (c=15.6Å) is also found to be present. Narrow regions of these phases, particularly of 123 (about four unit cells thick), were also discernible. In addition to these phases, interfaces bordering 124/125 and 124/123 regions (marked by dotted arrows) and also stacking faults within 124 regions were also found to be present. Since the 125 regions have extents of 60Å, stacking faults have extents ranging between 10Å and 20Å and interfaces would also be narrow regions, these would be compatible with the coherence length (~30Å along the ab plane), it is expected that these would act as effective flux pinning centres. Since the observed J_c ~2.15x10^5 A/cm^2 is one of the highest known J_c values for dissociated 124 it appears that for the 30 sec dissociated samples all the phases i.e. the 125, the interfaces and stacking faults, may act as flux pinning centres.

Fig.1(d) shows a typical HREM from a 60 sec dissociated sample. It was found that in these samples again 124 was the dominant phase; in addition, 123 in the form of a minority phase is also present. Stacking faults, even though sparse, are discernible. Based on the occurrence of the observed microstructural characteristics, it can be said that for the 60 sec dissociated sample the defect and interface densities, as compared to 30 sec sample, have decreased. This may be the cause of the decrease of J_c to comparatively lower values for the 60 sec dissociated sample.

3.2. Electrical, Magnetic and Microstructural Studies on 247

The electrical characterization of the as-synthesised samples shows the T_c (onset) ~95 K and T_c (R=0) ~91K. It has been observed that on dissociation the transition temperature does not change, it remains ~91 ± 1K. The J_c's of the as-synthesised and dissociated 247 are listed in Table1.

Fig.2(a) shows the HREM of the as-synthesised sample. A lattice fringe spacing of 25.3Å has been observed which corresponds to the c/2 periodicity of 247 phase. This observation identifies the phase to correspond to 247.

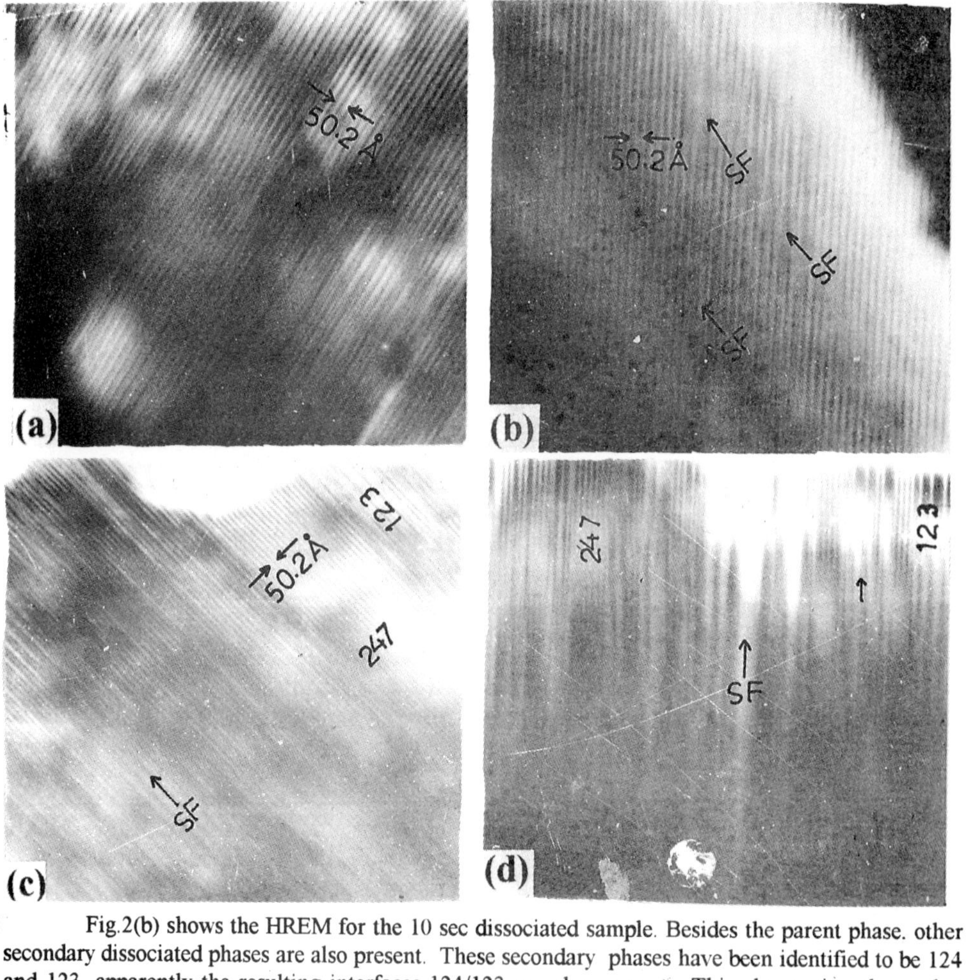

Fig.2(b) shows the HREM for the 10 sec dissociated sample. Besides the parent phase, other secondary dissociated phases are also present. These secondary phases have been identified to be 124 and 123, apparently the resulting interfaces 124/123 are also present. This observation shows that after 10 sec of dissociation at 915°C, the 247 phase remains dominant albeit other phases viz. 123 and 124 also start getting formed. The enhancement in J_c is thought to result due to the interfaces since these interfaces will be normal regions and the extents are comparable to the coherence length (30Å along a-b direction) and hence are expected to act as effective flux pinning centres.

Fig.2(c) shows the HREM of the 30 sec dissociated sample. It reveals the presence of 123 and 124 dispersed in the 247 matrix. The density of 123 and 124 phases in these samples is higher as compared to the 10 sec dissociated sample. This becomes evident through a comparison of Fig.2(b) and 2(c). These interfaces have extents of about 30 to 50 Å and have higher densities compared to the samples dissociated through time periods different than 30 sec and their extents being comparable to coherence length render them to be suitable flux pinning centres. Hence, it is expected that with the 247 samples dissociated for 30 sec would exhibit the highest J_c. The J_c for the 30 sec dissociated sample which is 1.02×10^5 A/cm^2 at 77K and 1.20×10^6 A/cm^2 at 15K (both at 0.5 T) is higher than any other 247 (as-synthesised or dissociated) sample investigated in the present study.

HREM observations as revealed in Fig.2(d) the 247 parent phase has nearly completely dissociated into subphases 123 and 124. The decrease in J_c may be attributed to the decrease in interfaces and stacking faults. Due to the extensive formation of 123 and 124 phases, the density of 247 parent phase is quite small. As a matter of fact the J_c here may be representative of the HTSC phase 123, enhanced through 124/123 interfaces.

TABLE - 1

Sample	J_c 77K (A/cm^2)	J_c 15K (A/cm^2)	Magnetic field (T)
As-synthesised 124	1.9×10^4	5.7×10^5	0.9
10 sec dissociated 124	9.6×10^4	1.93×10^6	0.9
30 sec dissociated 124	2.15×10^5	3.94×10^6	0.9
60 sec dissociated 124	1.25×10^5	2.75×10^6	0.9
As-synthensised 247	4.21×10^4	6.02×10^5	0.5
10 sec dissociated 247	1.76×10^4	7.22×10^5	0.5
30 sec dissociated 247	1.02×10^5	1.20×10^6	0.5
50 sec dissociated 247	5.04×10^4	7.01×10^5	0.5

IV. SUMMARY

In the present investigation, the 124 phase dissociated for time spans of 10, 30 and 60 sec at a temperature of 915°C. Samples dissociated for 30 sec shows a highest enhancement of J_c approximately 11-fold and approximately seven-fold at 77K and 15K respectively at H=0.9 T. The transformation covering a span of 30 sec was found to result in the creation of 125 and interfaces 124/123 besides the native 124 phases. In addition to these, stacking faults are also present. The dissociated 247 samples have been found to embody other related HTSC variants e.g. 124 and 123 as sub-phases, stacking faults and 247/124, 247/123 and 124/123 interfaces. The observed subphases, stacking faults and interfaces have been found to be correlated with the observed J_c's. Thus the highest J_c of 1.02×10^5 A/cm^2 (77K) and 1.20×10^6 A/cm^2 (15K) (both at H=0.5T) for 247 dissociated for 30 sec have been found to be explicable in terms of the presence of the highest density of interfaces and stacking faults for these samples.

ACKNOWLEDGMENTS

The authors are thankful to Dr. B.A. Dasannacharya and Dr. K.R. Rao of BARC (Mumbai) and Prof. R. Vijayaraghavan and K.V. Gopal Krishnan of TIFR (Mumbai) for extending the use of the SQUID magnetometer facility. One of the authors (AKS) is grateful to CSIR for financial support.

REFERENCES

1. P. Bordet, C. Chaillout, J. Chenarans, J.W. Hodean, M. Marezio, J. Karpinski and E. Kaldis, Nature **334**, 596 (1988).
2. H.W. Zanabergen, R. Gronsky, K. Wang and G. Thomas, Nature **105**, 207 (1988).

INTRAGRAIN CRITICAL CURRENT DENSITY & LOWER CRITICAL FIELD STUDIES ON Bi-2223 SUPERCONDUCTING SYSTEM

P. Subhash, G. Narsinga Rao and D. Suresh Babu

Materials Science Research Laboratory, Department of Physics
Osmania University, Hyderabad 500 007, India

A detailed magnetic hysteresis studies were carried out on samples with nominal compositions $Bi_{1.6}Pb_{0.3}V_{0.1}Sr_2Ca_2Cu_3O_y$ and $Bi_{1.6}Pb_{0.3}Ta_{0.1}Sr_2Ca_2Cu_3O_y$ in the range 5 K and 80 K. An empirical phase diagram was derived from the temperature and field dependent critical current density measurements. The region with non-zero critical current densities exhibit both single and collective pinning behaviour. The temperature dependent critical current density analysis suggests two types of pinning centres. Below 30 K, a large number of pinning centres with low activation barriers may be responsible for the enhancement of J_C and above 30 K deep-lying pinning centres with larger activation energies are active resulting in weak-temperature-dependent J_C. The temperature dependent lower critical field, H_{C1}, shows a rapid decay below 20 K, with increasing temperature. The power law dependence between H_{C1} and pinning force density, F_{pin}, describes an existence of surface pinning effects yielding the enhancement of H_{C1} at low temperature.

1. INTRODUCTION

The main disadvantage of high T_C superconductors in many fields of application is the drastic decrease of the critical current density with increasing temperature. Since the critical current is closely related to the properties of the vortex or critical state, it is necessary to study the vortex state and the related pinning behaviour in detail. The magnetic studies especially the temperature dependent magnetic hysteresis measurements are quite useful in order to get more information about the behaviour of the vortex lattice. Since one can estimate the dependence of critical current density J_C on field and temperature, which leads to determine the pinning force density F_{pin} and its dependence on temperature directly from the width of the hysteresis loop.

In continuation of our previous work[1,2] on the magnetic hysteresis measurements on Bi-high T_C (2223) phase superconducting system, in this paper, we report detailed magnetization measurements on $Bi_{1.6}Pb_{0.3}M_{0.1}Sr_2Ca_2Cu_3O_Y$ with M = V and Ta samples, as a function of temperature and applied field. Temperature and field dependent magnetization data for samples S1 and S2 are found to be similar. Hence we have presented a detailed analysis of the sample S1. We discuss the variation of critical current density J_C and the pinning force density F_{pin} in terms of different pinning mechanisms. In addition we have presented the temperature dependent lower critical field, H_{C1}. We also presented magnetic phase diagram of the vortex state.

2. EXPERIMENTAL

The samples with nominal compositions $Bi_{1.6}Pb_{0.3}V_{0.1}Sr_2Ca_2Cu_3O_Y$ (S1) and $Bi_{1.6}Pb_{0.3}Ta_{0.1}Sr_2Ca_2Cu_3O_Y$ (S2) were prepared by two step solid state reaction method . These samples were characterised using XRD, AC susceptibility and Electron Energy Dispersive X-ray Spectrometry measurements. It was found that samples S1 and S2 both had single superconducting 2223 phase along with $CaMO_Y$ type impurity [3]. From AC susceptibility measurements T_C of the samples S1 and S2 were found to be 108 K. The average grain size of the 2223 phase of samples S1 and S2 were estimated from the SEM micrographs as 10μm. EDAX measurements showed that the V/Ta has not entered the lattice but is present at the grain boundaries. The magnetic hysteresis measurements were performed on a commercial SQUID Magnetometer (Quantum Design) in fields upto 5 T at different fixed temperatures between 5 K and 80 K.

3. RESULTS AND DISCUSSION

Fig. 1. shows the magnetic hysteresis curves for the sample S1, similar curves was observed for sample S2. The intragrain critical current density J_C (A/cm^2) was estimated from the hysteresis loop using an expression

$$J_C = 30\ \Delta M / d \qquad (1)$$

derived from Bean's model[4], where d is the diameter of the grain (in cms) and ΔM is the difference between the magnetization (in emu/cm^3) in a field H obtained by increasing and decreasing the field. In our calculations we used d = 10μm. This choice of a different d would only change the absolute value of J_C, not altering the main results of this paper. We disregarded any correction due to the demagnetisation effects as the geometric effects are dominated by the grain shape, which is inhomogeneous but with a small aspect ratio. Despite the uncertainty in the knowledge of the actual values, it is of course useful to study the field and temperature dependence of J_C and related parameters like pinning force density F_{pin}.

Fig.2 shows the temperature dependent critical current density J_C obtained from the magnetization curves at H = 0, for the sample S1. Below 30 K, the critical current density J_C shows a sharp decay with the increase in temperature. Above 30 K much weaker temperature dependence is observed. Nevertheless, at 80 K, J_C still reaches about 10^5 A/ cm^2. The large critical current density at lower temperatures hint at a strongly - enhanced number of active pinning centres, in comparison with the temperature range above 30 K. The drastic decay of J_C upto 30 K might originate from rather low values of the activation energy for these large number of pinning centres. At higher temperatures, the large number of pinning centres which enhance J_C below 30 K are no more active. The lower and only weakly-temperature-dependent J_C value above 30 K emphasise that another type of pinning centre is active, involving deeper-lying pinning centres with larger activation energy. This transition indicate a change in the behaviour of the structure of the vortex lattice.

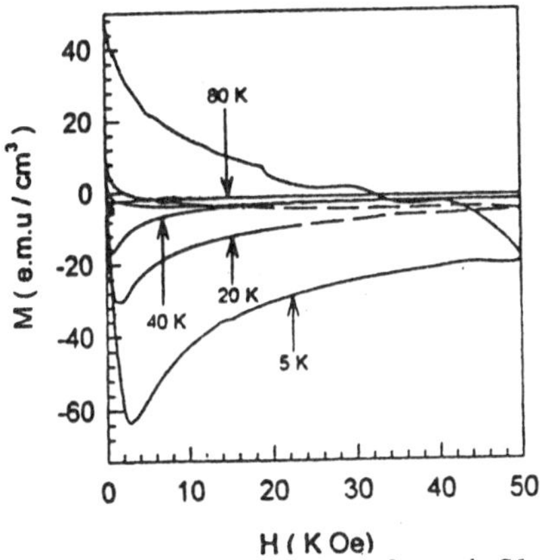

Fig1. A set of magnetization curves of sample S1 at different fixed temperatures between 5K and 80K.

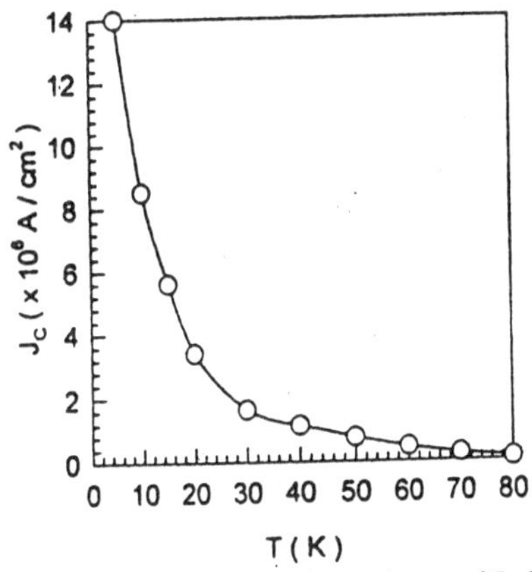

Fig.2. Temeprature dependence of J_c for sample S1 obtained from magnetization curves with H = 0 with solid line representing Eq. 4.

It was found, from $\ln(J_C)$ versus $\ln(1\text{-}T/T_C)$ plot, that the temperature dependent J_c, follows a power law for the temperatures higher than 30 K, . i.e.

$$J_C(T) = J_C(o)\,(1 - T/T_C)^n, \qquad T \geq 30\text{ K} \qquad (2)$$

The value of $J_C(o)$ and the exponent n are found to be 0.4 x 10^6 A/cm^2, 0.39 x 10^6 A/cm^2 and 2.65, 2.38 for the samples S1 and S2 respectively, and such a power law is often observed [5,6]. Assuming the enhancement of J_C due to independent pinning mechanism, which is mainly active in the low temperature region and is additive to the power law dependence of J_C, in the high temperature region, one can subtract Eqn. (2) from experimental data and obtain the enhanced part of the critical current density (δJ_C). Below 20 K, the temperature dependent δJ_C was found to follow the empirical relation

$$\delta J_C(T) = \delta J_C(o) \exp(-mT^2), \qquad T < 30\text{ K} \qquad (3)$$

The values of $\delta J_C(o)$ and m were obtained from the δJ_C Vs T^2 plot and found to be 1.34 x 10^6 A/cm^2, 1.02 x 10^6 A/cm^2 and 0.009, 0.008 for the samples S1 and S2 respectively. Therefore observed temperature dependence of J_C in the range between 5 K and 80 K and our assumption of two different pinning mechanisms can be described by an equation

$$J_C(T) = J_C(o)\,(1 - T/T_C)^n + \delta J_C(o) \exp(-mT^2), \qquad (4)$$

which was obtained by combining Eqn. (2) and (3). The solid line in fig.2 represents Eqn. (4) with the calculated parameters. similar results were observed by Narsinga Rao et al[1,2] and Job and Rosenberg[7] in Bi-high T_C ceramic samples.

3.1 FIELD DEPENDENT CRITICAL CURRENT DENSITY

More information about the pinning behaviour or mechanism can be obtained by analysing the field dependence of the critical current density. The pinning force density is given by

$$F_{pin} = J_C H \qquad (5)$$

This formula holds good in the case of stationary vortex lattice. Using Eqn.(5) one can estimate F_{pin} directly from the plots of J_C^{-1} Vs H, since F_{pin} is equal to the reciprocal slope of the linear dependence.

It was observed that the F_{pin} decreases drastically with temperature. The value of F_{pin} at T = 0 K, F_{pin} (o), was obtained by a linear extrapolation in a double logarithmic plot of F_{pin} Vs ($1 - T/T_C$) for the samples S1 and S2 and are found to be F_{pin} (o) = 4×10^{10} N/m³ for sample S1 and F_{pin} (o) = 3×10^{10} N/m³ for sample S2.

In the case of single vortex pinning where the vortex lines are pinned independently of each other, Larkin and Ovchinnikov[6] found the behaviour

$$F_{pin} = J_C H; \qquad \text{i.e. } J_C \propto H^{-1} \qquad (6)$$

Temperature and field region of the SVP behaviour in the vortex lattice for sample S1 is shown in fig.3. It was found that the SVP region for sample S1 is large compared to that of sample S2.

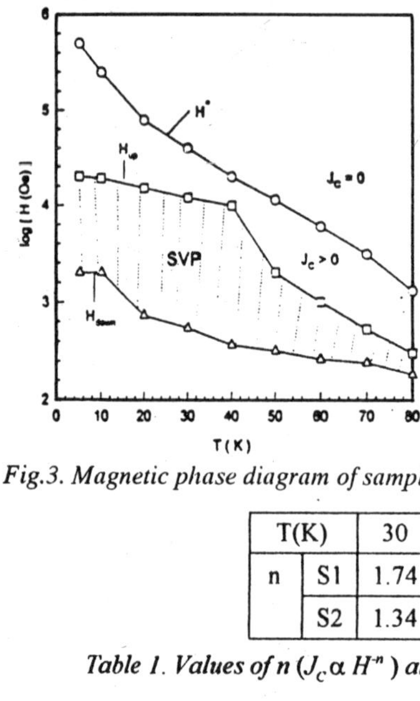

Fig.3. Magnetic phase diagram of sample S1.

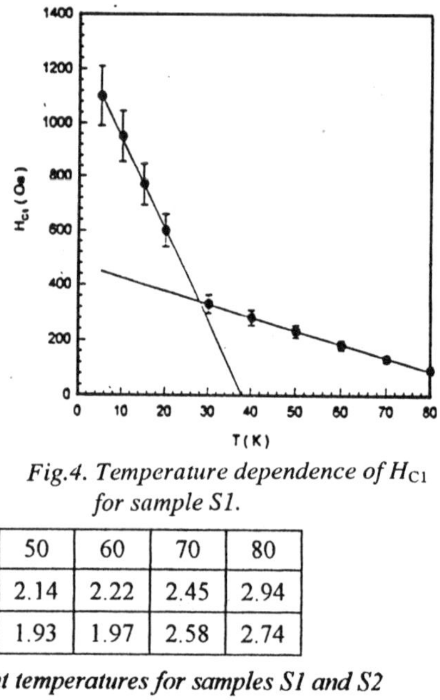

Fig.4. Temperature dependence of H_{C1} for sample S1.

T(K)		30	40	50	60	70	80
n	S1	1.74	2.08	2.14	2.22	2.45	2.94
	S2	1.34	1.61	1.93	1.97	2.58	2.74

Table 1. Values of n ($J_C \propto H^{-n}$) at different temperatures for samples S1 and S2

According to Kramer's theory[8] based on the shearing of two-dimensional vortex lattice, F_{pin} becomes zero at sufficiently large fields because of the reduction of the shear modulus of the vortex lattice. Kramer predicted the field dependence of F_{pin} as

$$F_{pin} \propto h^{1/2} (1 - h^2) \qquad (7)$$

where $h = H/H^*$ (H^* is the field at which $F_{pin} = 0$). This relation can be used to estimate H^*. By making use of $F_{pin} = J_C H$, Eqn.(6) can be written as

$$J_C^{1/2} H^{1/2} \propto (1 - h) \qquad (8)$$

showing that J_C becomes zero for $H = H^*$. H^* was estimated by linear extrapolation of $J_C^{1/2} H^{1/2}$ Vs. H plot (Eqn.(8)) and shown in fig.3 for sample S1.

The analysis of field dependent critical current density can be understood as follows. In low temperature region (≤ 20 K), $J_C \propto H^{-1}$ relation is satisfied and it is true for single vortex pinning as described earlier, in which F_{pin} (= J_C H) is independent of applied field. In the high temperatures (> 30 K), $J_C \propto H^{-n}$. The exponent n was obtained by the slope of the linear curves between ln(J_C) and ln(H) for sufficiently large fields at temperatures between 30 K and 80 K, for the samples S1 and S2 respectively and are given in table 1, it was found that the value of exponent n > 1, suggests the pinning behaviour in high temperatures is different from that in low temperatures and it may be collective pinning behaviour[6].

From the above results one can draw a magnetic phase diagram for the sample S1 is shown in fig.3, similar magnetic phase diagram was also determined for the sample S2. For sufficiently large fields (> $H^*(T)$ line), critical current density J_C vanishes. High value of $H^*(T)$ at low temperatures infers that in the samples S1 and S2, flux line lattice seems to be strongly pinned. Below the $H^*(T)$ line, critical current density has a non-zero value. In this region, different pinning behaviours can occur. We found single vortex pinning in the region between $H_{up}(T)$ and $H_{down}(T)$. In the region between $H^*(T)$ line and $H_{up}(T)$ line, the pinning behaviour maybe collective. At fields smaller than H_{up}, all the pinning interactions are broken giving rise to single vortex pinning, and at fields larger than H_{up}, a collective pinning behaviour is expected by assuming that the strong pinning centres do not break and the flux line lattice shears around those pinning centres. In the region below $H_{down}(T)$ phase line, the sample is not completely in critical state

3.2 TEMPERATURE DEPENDENT LOWER CRITICAL FIELD (H_{C1})

The lower critical field H_{C1} of the high temperature superconductors is an intrinsic material property which depends upon temperature and crystallographic axes. Several methods are proposed in the literature[9-11] for the determination of the lower critical field. For this work we choose the classical method which consists of determining $H_{C1}(T)$ from the deviation from linearity in M(H) curves.

Fig. 4 shows the temperature dependence of the lower critical field H_{C1} for the samples S1. The uncertainty of H_{C1} was found to be about 10 %. Below 20 K the lower critical field H_{C1} shows a sharp decay with increasing temperatures. Similar results were observed by Narsinga Rao et al[1,2] for W doped (Bi,Pb)-2223 and V doped Bi-2223 ceramic samples. Different explanations are available in literature for this anomalous behaviour of H_{C1}[12,13].

Recently McElfresh et al[14] have explained anomalous increase of H_{C1} at low temperatures by incorporating surface barriers in an extended Bean's model. Since, at high temperatures surface barrier can be overcome more easily by thermal activation, they claimed that the appearance of surface barrier effect at low temperatures is possible (rather than at high temperatures). Due to surface barriers flux lines are able to penetrate the crystal only when the applied field is greater than the intrinsic H_{C1}. The more perfect the surface the higher the applied field. In case of our sample surface effects cannot be excluded because well developed grains are observed. Therefore the anomalous increase of H_{C1} at low temperatures may be due to blocking effect of the surface barriers.

The power law dependence of H_{C1} Vs F_{pin} was observed for the samples S1 and S2 and it describes an existence of surface pinning effects yielding the enhancement of H_{C1} at low temperatures (i.e. $T \leq 30$ K), in agreement with the above mentioned theoretical models.

ACKNOWLEDGEMENTS

The authors would like to thank the Department of Science and Technology, Govt. of India, New Delhi for the financial support. One of the authors (PS) would like to thank CSIR, Govt. of India, New Delhi, for awarding Senior Research Fellowship (SRF)

REFERENCES

[1] G.Narsinga Rao, P.Molinie, M.Ganne and D.Suresh Babu, *Mod. Phys. Lett.* **B7**, 999 (1993).

[2] G.Narsinga Rao, P.Molinie, M.Ganne and D.Suresh Babu, *Mod. Phys. Lett.* **B9**, 1377 (1995).

[3] P.Subhash, G.Narsinga Rao and D.Suresh Babu, *Mat. Sci & Engg.* **B** (submitted).

[4] C.P.Bean, *Rev. Mod. Phys.* **36**, 31 (1964).

[5] A.P.Malozemoff, In *Physical Properties Of High Temperature Superconductors* Ed. By D.M.Ginzberg (World Scientific, Singapore, 1989) **Vol 1,** P 71

[6] A.I.Larkin and Y.N.Ovchinnikov, *J. Low Temp. Phys.* **34**, 409 (1979).

[7] R.Job And M.Rosenberg, *Supercond. Sci. & Tech.* **5**, 7 (1992).

[8] E.J.Kramer, *Appl. Phys.* **44,** 1360 (1973).

[9] Y.Yeshurun, A.P.Malozemoff, F.Holtzberg and T.R.Dinger, *Phys.Rev.* **B 38**, 11828 (1988).

[10] L.Krusin-Elbaum, A.P.Malozemoff, Y.Yeshurun and F.Holtzberg, *Phys. Rev.* **B 39**, 2936 (1989).

[11] A.Umezawa, G.W.Crabtree, L.Z.Liu, T.J.Moran, S.K.Malik, L.H.Nunez, W.L.Kowak and C.H.Sowers, *Phys. Rev.* **B38**, 2843 (1988).

[12] T.Koyama, N.Takezawa and M.Tachiki, *Physica* **C168**, 69 (1990).

[13] G.Deutscher and P.G.De Gannes, In *Superconductivity,* Edited By R.J.Parks (Marcel Dekker Inc, New York, 1969) **Vol 2,** P 1005

[14] M.W.Mcelfresh, Y.Yeshurun, A.P.Malozemoff and F.Holtzberg, *Physica* **C168**, 308 (1990).

IRREVERSIBILITY LINE STUDIES ON Bi-2223 SUPERCONDUCTING SYSTEM

P. Subhash, G. Narsinga Rao and D. Suresh Babu
Materials Science Research Laboratory, Department of Physics
Osmania University, Hyderabad 500 007, India

We report the Irreversibility Line (IL) studies using AC susceptibility and Zero-field-cooled (ZFC) and Field-cooled (FC) DC magnetisation measurements on the samples $Bi_{1.6}Pb_{0.3}V_{0.1}Sr_2Ca_2Cu_3O_Y$ (S1) and $Bi_{1.6}Pb_{0.3}Ta_{0.1}Sr_2Ca_2Cu_3O_Y$ (S2) (nominal composition), a single phase 2223 superconducting system. The ZFC and FC measurements, in low field region the IL behaves like a flux pinning model whereas in the high field region an exponential dependence of the IL gives a better fit to the experimental data. From this exponential dependence the characteristic parameters like Coherence length, ξ_0, the Penetration depth, $\lambda_0(0)$, the Fermi velocity in normal region, V_{FN}, the Fermi velocity in superconducting region, V_{FS} were estimated. The IL was also identified through AC susceptibility measurements and it was found to follow a power law. The AC IL was found to be far below the DC IL and it is suggested that the AC IL is due to the intergrain J_C and the DC IL is due to the intragrain J_C.

1. INTRODUCTION

The high T_C superconductors show a wide range in their H-T phase diagram. Below some H(T) line, called the irreversibility line (I L), irreversibility behaviour is observed. This was first noticed in the La(Ba)-cuprates by Müller et al[1], in DC magnetization measurements. A true thermodynamic phase transition i.e., flux line melting was proposed[2]. A cross over from flux creep to thermally assisted flux flow (TAFF) can be considered as well[3].

The widely used method is based on the detection of energy losses in the sample through AC susceptibility measurements. The onset in the loss peak or higher harmonics is taken as the criterion for the irreversibility[4-6], although some authors use the peak in the imaginary susceptibility, χ'', as the criteria. According to Müller et al[4], χ'' peak occurs at a temperature T_P where the hysteresis loop area is maximum. The definition of taking T_P as the T_{IRR} places irreversibility well within the region of maximum hysteretic loss. Clearly in the reversible region, between T_{IRR} and T_C, there must be zero loss since the pinning force density is effectively zero. Hence it is appropriate to define T_{IRR} at the onset of the hysteretic loss[4,7]. Yeshurun and Malozemoff[3] have shown that similar line (IL) exists in the YBaCuO system. They have explained this behaviour in terms of thermally activated flux creep. De Rango et al[8] reported their magnetic study of ceramic $Bi_{1.7}Pb_{0.3}Sr_2Ca_2Cu_3O_Y$ sample, that in low temperatures (high fields) the irreversibility line follows an exponential decrease with the increase in temperature. They explained this behaviour based on proximity effect[9]. Weak superconductivity might be induced into the normal regions of BiO and SrO blocks by the proximity effect and this weak superconductivity could be destroyed in a relatively small field leading to a superlattice of superconducting blocks composed of CuO layers[7].

In this paper we report the Irreversibility Line studies of $Bi_{1.6}Pb_{0.3}V_{0.1}Sr_2Ca_2Cu_3O_Y$ (S1) and $Bi_{1.6}Pb_{0.3}Ta_{0.1}Sr_2Ca_2Cu_3O_Y$ (S2), a single phase (Bi,Pb)-2223 superconductors, using AC susceptibility and ZFC, FC magnetisation measurements.

2. EXPERIMENTAL

The samples with nominal compositions $Bi_{1.6}Pb_{0.3}V_{0.1}Sr_2Ca_2Cu_3O_Y$ (S1) and $Bi_{1.6}Pb_{0.3}Ta_{0.1}Sr_2Ca_2Cu_3O_Y$ (S2) were prepared by two step solid state reaction method . These samples were characterised using XRD, AC susceptibility and Electron Energy Dispersive X-ray Spectrometry measurements. It was found that samples S1 and S2 both had single superconducting 2223 phase along with $CaMO_Y$ type impurity[10]. From AC susceptibility measurements T_C of the samples S1 and S2 were found to be 108 K. The average grain size of the 2223 phase of samples S1 and S2 were estimated from the SEM micrographs as 10μm. EDAX measurements showed that the V/Ta has not entered the lattice but is present at the grain boundaries. The AC susceptibility measurements were carried out on the Lakeshore AC Susceptometer 7000 on a cylindrical slab specimen (radius = 1mm, length = 5mm), in the AC magnetic fields ranging from 0.01 Oe to 10 Oe, at a frequency of 666.7 Hz (the applied field is along the length of the specimen). Absolute values of χ' and χ'' were determined based on an assumption that at temperatures substantially below T_C, where $d\chi'/dT = 0$, is $\chi' = -1$. The Field-cooled and zero-field-cooled measurements were performed on a commercial SQUID Magnetometer (Quantum Design) in the fields between 5 Oe to 10 kOe.

3. RESULTS AND DISCUSSION

Fig. 1 show the typical temperature dependent imaginary (χ'') part of the AC susceptibility at different applied AC magnetic fields (H_{AC}), for the sample S1.The diamagnetic onsets for the sample S1 in the real part (χ') of the AC susceptibility was found to be 108 K and is independent of the applied magnetic field. As the amplitude of the applied magnetic field increases, the position of χ'' peak shifts to lower temperatures, the height and the width of the χ''-peak increases. Similar variation of AC susceptibility data was observed for sample S2. The shift of the χ''-peak due to change in applied magnetic field can be explained by critical state model. In the critical state model, the AC losses are solely hysteretic due to pinning of the flux line in type II superconductors. The maximum loss always takes place when the magnetic flux penetrates to the centre of the sample. As the driving field amplitude increases, larger screening currents are required to shield the applied magnetic field, as a result the position of χ''-peak moves to lower temperatures. But the critical state model is unable to explain the increase of the height and width of the χ''-peak. According to Critical state model the height of the χ''- peak is independent of the applied field and having a maximum value of $\simeq 0.24$. As it can be seen from Fig. 1, the height of the χ''-peak increased with decrease in the amplitude of the applied AC field and approaches the Critical state model value of χ''-peak (i.e. $\simeq 0.24$). This can be explained by the Campbell model of reversible fluxoid motion[11].

Fig.2 shows the field dependence of T_P(χ''-peak) for the two specimens S1 and S2 respectively, determined at frequency of 666.7 Hz. The obtained behaviour has been interpreted to yield information on the strength of flux pinning in the intergranular regions[12,13]. For fields below 0.1 Oe, T_P is almost independent of field and approaches the transition temperature T_C. For fields above 1 Oe there is a considerable shift of T_P to low temperatures. This indicates that for typical polycrystalline superconductors the pinning of magnetic flux in the intergranular regions is very weak. The maxima in the χ'' curves originates from the hysteretic loss as the AC field moves in and out of the superconductor. According to Bean's model this loss becomes maximum when the superconducting volume is just penetrated by the applied magnetic field, i.e., when the shielding current is equal to the critical current. The data for T_P Versus field can thus be used to determine the intergrain critical current density.

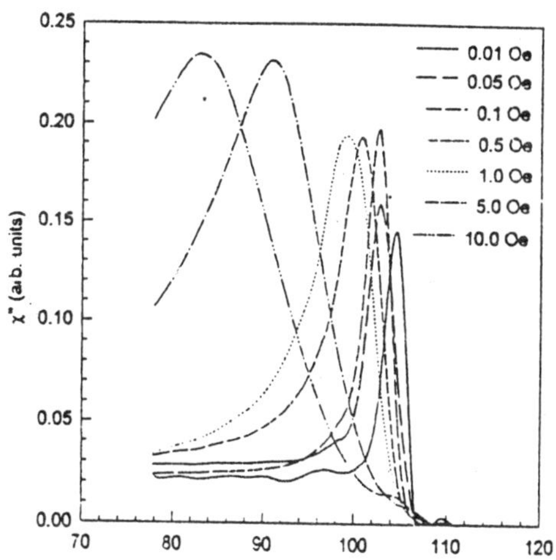

Fig.1. Temperature dependent imaginary part (χ'') of the AC susceptibility at a frequency of 666.7 Hz for sample S1.

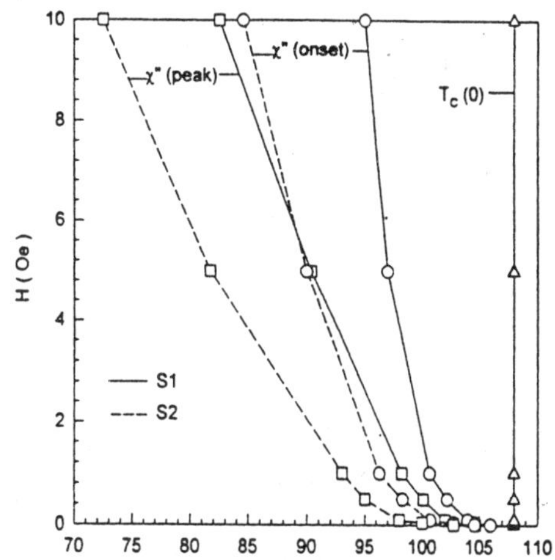

Fig.2. Irreversibility lines for samples S1 and S2 determined from χ''-onset, and the lines determined using the χ'' peak temperature.

As described earlier, we have determined the irreversibility line by considering χ''-onset temperature as the irreversibility temperature[4]. Fig. 2 shows the irreversibility line for the two samples S1 and S2. Also shown in the Figure are the line determined using the χ''-peak temperature, T_P. In our measurements the χ''-onset can be easily determined since intragrain losses are absent in low fields. The irreversibility line is steeper compared to the line based on χ''-peak and is moved to higher temperatures so that the reversible region becomes quite narrow. In these low fields the irreversibility line follows a power law ($H^*=H_o(1-T/T_c)^n$) and the values of n and H_o were found to be 3.78 & 3.7 and 2.8 x 10^4 Oe & 3.5 x 10^3 Oe respectively for the samples S1 and S2. It was also found that the line representing the χ''-peak also follows the power law with the values of n being similar to, and the prefactor H_o differing from, that of the irreversibility line.

Fig.3 show some of the temperature dependence of field cooled (FC) and zero field cooled (ZFC) magnetization curves for the samples S1, in temperature range 5 K to 120 K, similar curves were also obtained for sample S2. The irreversibility line was determined from ZFC (Shielding) and FC (Meissner) susceptibility measurements at different applied magnetic fields. The samples were cooled in an applied field through T_C down to 5 K for recording Meissner signal where as the shielding measurement were carried out by first cooling the samples in zero field down to 5 K and recorded the data while warming up in the applied field. At temperature T^* where both ZFC and FC curves merge and the corresponding value of the magnetic field (H^*) yields a point on the irreversibility line. Fig.4 show the irreversibility line for the samples S1 and it can be fitted to an exponential form for fields above 100 Oe. i.e. $H^* = H_O \exp(-T^*/T_O)$ the parameters H_O and T_O were found and are presented in table 1. It is in good agreement with the observation of de Rango et al[8], in the high field region. The break down field (the field at which the induced superconductivity in the normal regions of the unit cell breaks down) varies exponentially with temperature following the relation[8] i. e. $H_b = H_O \exp[-2\pi d_n k_B T/\hbar V_{FN}]$ where d_n is the thickness of the normal region (i.e. almost equal to the distance between the CuO layers in the unit cell, 12.16 Å) and V_{FN} is the Fermi velocity in the normal region. By identifying H^* with H_b^*, and comparing eqn.(4) with eqn.(3), one can determine Fermi velocity in the normal regions. By taking d ≈ 12.16 Å, the Fermi velocity in the normal region, V_{FN}, for samples S1 and S2 were found and are given in table 1. In the fields below 100 Oe the data could be fitted to the simple power law ($H^*=H_o(1-T/T)^n$). The parameters H_O and n found from this fit is given in table1.

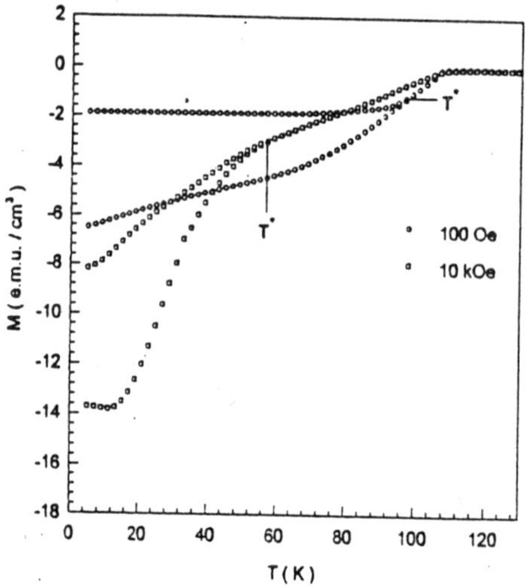

Fig.3. Zero-field-cooled (ZFC) and Field-cooled (FC) magnetization curves of sample S1, for fields 100 Oe and 10 kOe.

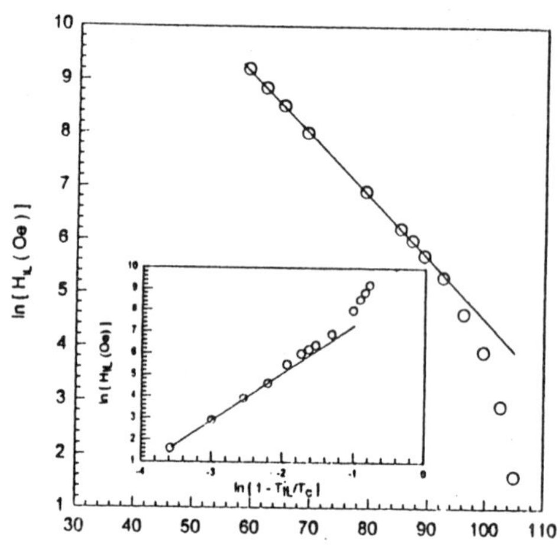

Fig.4. Irreversibility line of sample S1 showing the exponential fit for fields >100 Oe; inset shows the power law fit for fields ≤ 100 Oe.

sample	≥ 100 Oe H_O (10^4 Oe)	≥ 100 Oe T_O	V_{FN} (10^6 cm/s)	≤ 100 Oe H_O (10^4 Oe)	≤ 100 Oe n
S1	63.4	9	1.02	1.35	2.2
S2	35	10	1	0.85	2.86

Table. 1. Parameters obtained from the exponential & power law fits for samples S1 & S2.

The characteristic parameters like Coherence length at zero Kelvin (ξ_o), the Penetration depth at zero Kelvin ($\lambda_o(0)$), the Fermi velocity in normal region, V_{FN}, the Fermi velocity in superconducting region, V_{FS} were estimated for sampls S1 and S2 by using the procedure described in our previous work[15] and found to be ξ_o=16.18Å, 5.7Å, λ(o) =1,820 Å, 1,800 Å and V_{FS} =1.27 x 10^7 Oe, 0.44x 10^7 Oe for the samples S1 and S2 respectively.

A detailed magnetic hysteresis studies were also carried out on these samples in the temperature range between 5 K and 80 K and presented elsewhere[15]. From the field and temperature dependent critical current densities J_C, an empirical magnetic phase diagram was derived giving evidence for a depinning line, i.e., $H^*(T)$ line, where J_C becomes zero. This depinning line is found to be analogous to the IL determined from the FC & ZFC magnetisation measurements. This result favours the interpretation of the IL as a depinning line rather than a flux lattice melting or vortex glass-vortex liquid transition[16]. For our ceramic samples, in the low field range, the IL determined by the AC measurements does not coincide at all with the IL derived from the FC & ZFC magnetisation data. Comparing the AC IL with the DC IL (for example at 5 Oe field, in the low field range, for sample S1, $T_{irr\ ac}$ = 0.89T_C and $T_{irr\ dc}$ = 0.97T_C and for sample S2, $T_{irr\ ac}$ = 0.83T_C and $T_{irr\ dc}$ = 0.91T_C), it is found that the AC IL is below the DC IL. It can be suggested that the AC IL could be due to the intergrain J_C and the DC IL is due to intragrain J_C.

ACKNOWLEDGEMENTS

The authors would like to thank the department of Science and Technology, Govt. of India, New Delhi for the financial support. One of the authors (PS) would like to thank Council for Scientific and Industrial Research (CSIR), Govt. of India, New Delhi, for awarding Senior Research Fellowship (SRF)

REFERENCES

[1] K.A.Müller, M.Takashige and J.G.Bednorz, *Phys. Rev. Lett.*, **58**,1143 (1987).

[2] M.P.Fisher, *Phys. Rev. Lett.*, **62**, 1415 (1989).

[3] Y.Yeshurun and A.P.Malozemoff, *Phys. Rev. Lett.*, **60**, 2202 (1988).

[4] K.H.Müller, *Physica* **C 185-189**,1609 (1991).

[5] E.R.Yacobi, A.Shaulov, Y.Yeshurun, M.Konczykowski & F.Rullier-Albenque, *Physica* **C199**,15 (1992).

[6] J.R.Clem, In *Magnetic Susceptibility of Superconductors and other Spin Systems* (Edited by R.A.Hein, T.L.Francavilla and D.H.Liebenberg), p.177 Plenum, New York (1992).

[7] N.Savvides, A.Katsaros, C.Andrikidis and K.H.Muller, *Physica* **C197**, 267 (1992).

[8] P. de Rango, B. Giordanengo, R. Tournier, A. Sulpice, J. Chaussi, G.Deutsher, J. L. Genicon, P. Lejay, R. Retoux and B. Raveau, *J.Phys.* (Paris) **50**, 2857 (1989).

[9] G. Deutscher and P. G. de Gannes, In *Superconductivity* Vol.**2** (Edited by R.J. Parks), p.1005 Marcel Dekker, New York (1969).

[10] P.Subhash, G.Narsinga Rao and D.Suresh Babu, *Mat. Sci. & Engg.* **B** (submitted)

[11] A.M.Campbell, *J. Phys.* **C4**, 3186 (1971).

[12] P.Subhash, G.Narsinga Rao and D.Suresh Babu, In *Superconductivity : Theoretical and Experimental effects* (Edited by K.N.Srivastava), p.260 Novo Science Pub., New York (1995).

[13] R.Navaro, F.Lera, A.Badia, C.Rillo and J.Bartolome, *Physica* **C 183**, 73 (1991).

[14] R. Job and M. Rosenberg *Physica* **C172**, 391 (1991).

[15] P. Subhash, G.Narsinga Rao, P.Molinie, M.Ganne and D.Suresh Babu, *Pramana J. Phys* (submitted)

[16] Y.Xu and M.Suenaga, *Phys. Rev.* **B43**, 5516 (1991).

INTRINSIC JOSEPHSON EFFECT IN HIGH TEMPERATURE SUPERCONDUCTORS

Reinhold Kleiner and Paul Müller

Physikalisches Institut III, Universität Erlangen - Nürnberg, Erwin Rommel Str. 1
D-91058 Erlangen, Germany

Many high temperature superconductors form natural superconducting multilayers where adjacent superconducting layers are weakly coupled by the Josephson effect. As a result, single crystals act intrinsically as vertical stacks of hundreds of Josephson junctions. In this paper we outline the basic physics of intrinsic Josephson junctions and give some examples of the present status of research regarding theoretical understanding of the intrinsic Josephson effect and fabrication of high quality junctions.

INTRODUCTION

The essential universal property of the many-particle wave function of the superconducting state is the phase coherence over macroscopic distances. This phenomenon was discovered about 60 years ago by W. Meißner and R. Ochsenfeld. Manifestations are the Ochsenfeld-Meißner effect[1], the quantization of flux[2] and the Josephson effect[3]. The coherence length of a classical superconductor like Pb extends over many lattice constants. Peculiarities of the crystallographic structure or lattice defects influence the superconducting state only marginally via the structure of the Fermi surface. The superconducting order parameter is essentially homogeneous and only weakly anisotropic. In contrast, high temperature superconductors exhibit large anisotropy and extremely short coherence lengths[4]. They form natural multilayers where CuO_2 single-, double- or triple-layers alternate with layers like BiO and SrO in the case of $Bi_2Sr_2CaCu_2O_{8+\delta}$ and TlO and SrO in the case of $Tl_2Ba_2Ca_2Cu_3O_{10+\delta}$. The large anisotropy and the small coherence lengths suggest that this layered crystal structure is mapped onto a periodic modulation of the superconducting order parameter. At least for the most anisotropic high temperature superconductors this idea leads to the concept of superconducting sheets (the CuO_2 layers) that are weakly coupled by the Josephson effect; every pair of adjacent copper oxide double or triple layers, together with the intervening nonsuperconducting layers forms an intrinsic Josephson junction with a thickness of about 15Å. A single crystal of, say, 1μm thickness acts as a natural vertical stack of 700 Josephson junctions. The Josephson effects that can be expected if the above picture holds have clearly been observed in c-axis transport measurements[5,6]. At present, intrinsic Josephson junctions are successfully fabricated and investigated by a large number of groups. In this paper we will discuss some of the activities in the field. Of course, by no means a complete review of all activities is intended.

In the next section we motivate the intrinsic Josephson effect by a Ginzburg-Landau description of layered superconducting structures. We then discuss the basic physics of intrinsic Josephson junctions using a system of coupled sine-Gordon equations. The fourth part will address the present junction technology. Finally, in section 5 we will discuss subgap features that were found in the current-voltage characteristics of intrinsic junctions.

CURRENT-PHASE RELATION IN LAYERED SUPERCONDUCTORS

In this chapter we show that the intrinsic Josephson effect can be obtained from a Ginzburg-Landau description of layered superconductors[7]. For parameters that are reasonable for the most anisotropic systems like $Bi_2Sr_2CaCu_2O_{8+\delta}$ and $Tl_2Ba_2Ca_2Cu_3O_{10+\delta}$ we will find a sinussoidal current-phase relation for currents flowing perpendicular to the layers. The model used below is more general than the Lawrence-Doniach formalism[8] since it does not a priori include Josephson coupling. A Ginzburg-Landau energy, suitable for layered superconductors has been discussed earlier by Doll[9] and by Takezawa et al.[10]. It has been used to discuss the upper and lower critical fields in these structures. The basic idea is to make the Ginzburg-Landau parameters α, β and the effective mass m periodic functions of the z-coordinate. Variation of the free energy leads to the Ginzburg-Landau equations

$$g(z)\,(\nabla - i\frac{2\pi s}{\Phi_0}\mathbf{A})^2\Psi + \frac{\partial g(z)}{\partial z}\,(\frac{\partial}{\partial z} - i\frac{2\pi s}{\Phi_0}A_z)\,\Psi = a(z)\,\Psi + b\,|\Psi|^2\Psi$$

$$\mathbf{j} = \frac{2e\hbar}{m_0 s}\,g(z)\{\frac{1}{2i}\,(\Psi^*\nabla\Psi - \Psi\nabla\Psi^* - \frac{2\pi s}{\Phi_0}\,\mathbf{A}\,|\Psi|^2)\}$$

In the above equations we have used some normalized units, $g(z)=m_0/m(z)$, $a=2m_0s^2\alpha/\hbar^2$ and $b=2m_0s^2\beta/\hbar^2$, where m_0 is the the value of the effective mass taken at the center of the superconducting layers. Φ_0 is the flux quantum and s is the layer period. The Nabla operator is normalized to s. Note that the second term on the right hand side of the first equation is a consequence of the z dependence of g. It vanishes in the standard Ginzburg-Landau equations. The gradient term g(z) is a function which reaches its maximum value of 1 in the middle of the superconducting electrodes. It is natural to use a large effective mass, 10^3m_0 or larger for the non-superconducting layers. g will thus be very small in these regions. We use the form $g(z)=g_{min}+(g_{min}-1)p(z)$, where p(z) is a periodic test function varying between 0 and -1. This function used is shown in Fig. 1. The second z-dependent function is a(z). It has to be negative in the superconducting layers and positive or slightly negative in the non- or only weakly superconducting layers. Using the same periodicity as for g(z) we have $a(z)=a_0-(a_0-a_1)p(z)$, where a_1 is the (negative) value of a(z) in the superconducting layers and a_0 is the value of a(z) in the interlayers. The repulsive term, b(z), can actually be taken constant without changing the qualitative behaviour of the systems. Reasonable values for the parameters a_0, a_1 and g_{min} can for example be found by calculating upper critical fields H_{c2} parallel and perpendicular to the layers. In order to calculate the current-phase relation we first write Ψ as $\Psi=\Psi_0\exp(i\varphi)$ with a real amplitude Ψ_0. This yields for the z-component of the current density

$$j_z = \frac{2e\hbar}{m_0 s}\,g(z)\,\Psi_0^2(d\varphi/dz - \frac{2\pi s}{\Phi_0}\,A_z).$$

Note that the first Josephson equation reads $j_z=j_c\sin(\varphi_1-\varphi_0-\frac{2\pi s}{\Phi_0}\int_0^1 A_z\,dz) =: j_c\sin\gamma$. Here, the integral has to be taken between the centers of the electrodes 0 and 1 of the junction. Introducing the continuous quantity γ(z) in analogy to the gauge invariant phase γ of the first Josephson equation,

$$\gamma(z) = \varphi(z) - \varphi(0) - \frac{2\pi s}{\Phi_0}\int_0^z A_z\,dz,$$

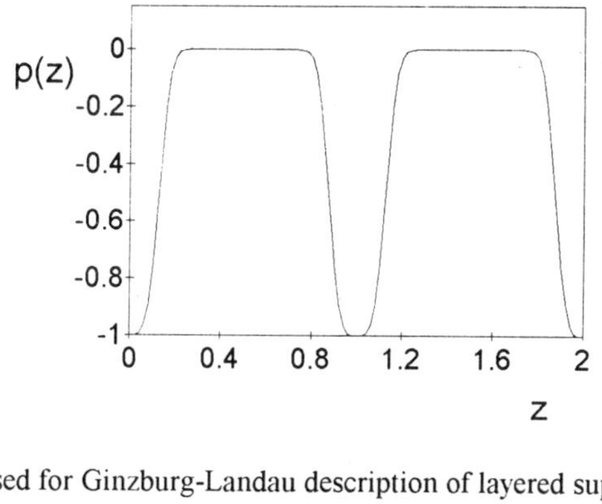

Fig. 1: Test function p(z) used for Ginzburg-Landau description of layered superconductors.

we can calculate $j_z(\gamma)$ with the boundary conditions $\gamma(z=1) = \gamma_0$ and $\gamma(z=0) = 0$. The function $j_z(\gamma_0)$ is the current-phase relation we are looking for. For a Josephson junction it has to have the form $j_z(\gamma_0) = j_c \sin\gamma_0$. In Fig. 2 we show results for a system with $a_1=-1$, $a_0=0.2$ and $g_{min}=10^{-3}$. These parameter values yielded values of H_{c2} that are reasonable for $Bi_2Sr_2CaCu_2O_{8+\delta}$ and $Tl_2Ba_2Ca_2Cu_3O_{10+\delta}$. Fig. 2(a) shows $\Psi_0(z)$ for $\gamma_0/2\pi=0$, ¼ and ½. $\Psi(z)$ is normalized to the value Ψ_∞ of a homogeneous, non-modulated system. Already for $\gamma_0/2\pi=0$, i. e. for zero transport current, Ψ_0 is strongly modulated. It is maximum at z=-1, 0 and 1, i. e. in the centers of the superconducting layers. In the non-superconducting layers Ψ_0 is strongly suppressed. For $\gamma_0/2\pi$ = ¼ and ½, i. e. for nonzero transport current Ψ_0 is further suppressed in the nonsuperconducting layers but is unchanged in the superconducting layers. Fig. 2(b) shows the gauge invariant phase $\gamma(z)$ for the above values of γ_0. It exhibits a steep rise at z=-0.5 and 0.5. This functional dependence of Ψ_0 and γ is very close to what one would expect for stacked Josephson junctions. Here, Ψ_0 should be constant and independent of transport current in the superconducting electrodes and drop to very small values in the barrier. The gauge-invariant phase $\gamma(z)$ should also be a constant in the superconducting layers and it should exhibit a jump in the barrier. Fig. 2(c) shows the calculated current density $j_z(\gamma_0)$, normalized to the critical current density j_{ch} parallel to the layers. It is of the form $j_z(\gamma_0) = j_c \sin\gamma_0$, as required for a system that exhibits the Josephson efffect.

THE INTRINSIC JOSEPHSON EFFECT

Having motivated the Josephson nature of the interlayer supercurrents we now turn to a much more simplified model of intrinsic Josephson junctions. Fig. 3 illustrates the basic concept of intrinsic Josephson junctions in $Bi_2Sr_2CaCu_2O_8$. Each CuO_2 double layer forms a superconducting slab of 3 Å thickness. The intervening BiO- and SrO- layers are considered as non-superconducting 12 Å thick barrier layers. An intrinsic junction with a total thickness of 15 Å is thus formed by two adjacent CuO_2 double layers and the intervening layers. The supercurrent density perpendicular to the layers across the n-th junction is described by the Josephson relations $j_{s,n}=j_c \sin\gamma_n$ and $\hbar\, d\gamma_n/dt = 2eU_n$. Here, γ_n denotes the difference of the phases of the order parameters in the electrodes of the n-th junction and U_n is the voltage drop across the junction. The intrinsic Josephson effect has been directly confirmed in c-axis transport measurements[5,6]. Measurements included I-V characteristics, the magnetic field and temperature dependence of the critical current, Shapiro steps occurring in external microwave fields and microwave radiation according to the Josephson frequency-voltage relation f=2eU/h. As an example,

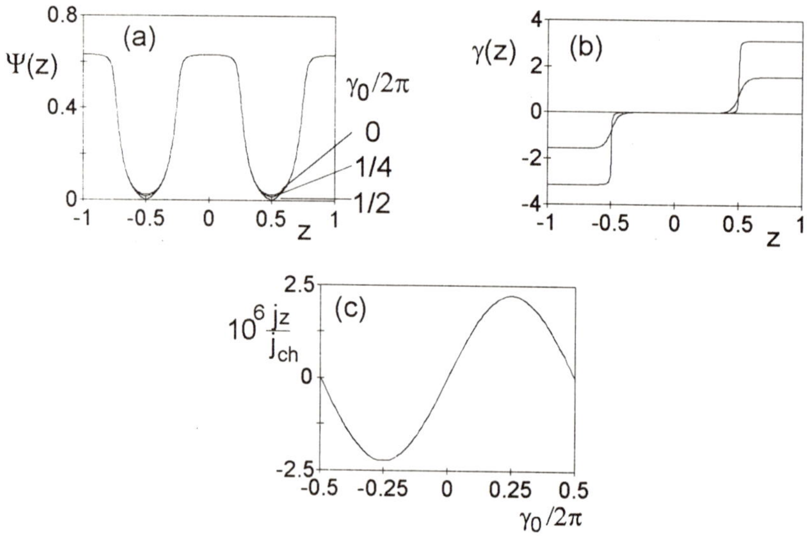

Fig. 2: Numerically calculated spatial dependence of order parameter $\Psi_0(z)$ (a), gauge invariant phase $\gamma(z)$ (b) and resulting current-phase relation $j_z(z)$ (c). Model parameters are $a_1=-1$, $a_0=0.2$ and $g_{min}=10^{-3}$.

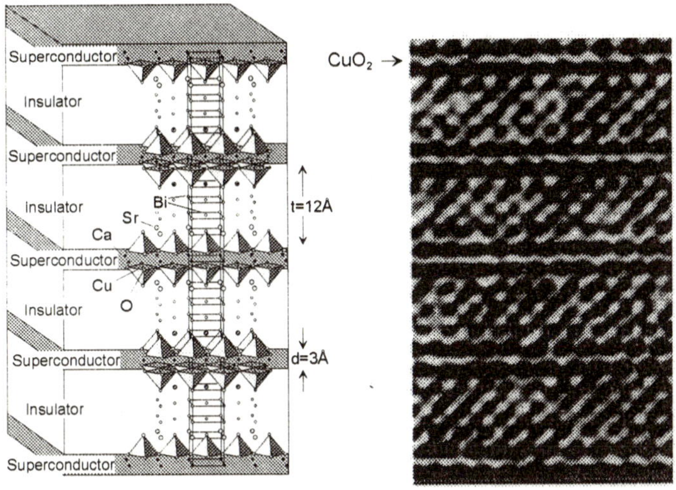

Fig. 3: Superposition of the crystal structure of $Bi_2Sr_2CaCu_2O_8$ (BSCCO) and the model of slabs (left) together with TEM image of BSCCO (right).

the shape of the I-V characteristics shall be described here. Fig. 4 shows a typical curve. For the measurement, a small single crystal of $30x30\mu m^2$ in size in ab-direction was mounted between two contact rods. The crystal thickness was $3\mu m$ corresponding to a stack of 2000 intrinsic junctions. At small values of bias current the whole sample is superconducting. For bias currents above 12mA the voltage across the crystal jumps to increasingly larger values (horizontal arrows in Fig. 4). After each jump the voltage across the crystal can be continuously decreased by lowering the bias current (double arrows in the figure). With this procedure hundreds of branches can be traced out up to voltages of several volt. This structure of the I-V characteristic is typical for intrinsic Josephson junctions: Each junction has an hysteretic I-V characteristic consisting of a cuperconducting and a resistive branch similar to standard SIS tunnel junction. Below the critical current each junction can be either superconducting or resistive. The total I-V characteristic of a stack of N junctions thus exhibits N branches in the resistive state, with adjacent branches differing by one resistive junction. In Fig. 4, the voltage jumps V_c at the critical current are about 15 mV; for the best junctions they are up to 25 mV. For a standard, say Nb, SIS tunnel junction, V_c would equal twice the energy gap Δ. For BSCCO, Δ values of about 20 mV are reported so that $V_c \approx \Delta$. This fact has been noted and discussed by Yurgens et. al.[11]. Intrinsic Josephson effects have also been observed in $Tl_2Ba_2Ca_2Cu_3O_{10}$ (TBCCO)[5,6], where the superconducting sheets consist of CuO_2 triple layers, in the n-doped single layer compound $(PrCe)_2CuO_4$ [12] and, very recently, also in $YBa_2Cu_3O_{7-\delta}$ (YBCO) [13]. In terms of applications the latter material might turn out to be very important since thin film technology is most advanced here. In addition to the above high T_c compounds also some organic superconductors like κ-(BEDT-TTF)$_2$Cu(NCS)$_2$ show the effect.

So far, we have described intrinsic Josephson junctions essentially as independent series arrays of junction. There are, however, peculiar features associated with the fact that these junctions form a closely packed stack with mutual interactions between the junctions. Interactions for example arise from currents flowing along the superconducting layers. Since the thickness of the superconducting layer is only about 3Å, much less than the London penetration depth $\lambda_{ab} \approx 1700$ Å, this leads to a strong (inductive) coupling between adjacent junctions sharing one superconducting layer[14-17]. As a consequence, a Josephson vortex located in one junction will induce screening currents in essentially all other junctions. In time dependent situations charges in a given electrode or flux changes in a given junction will lead to electrical fields also in other junctions. Another type of coupling arises from dissipation due to relaxation of the superconducting order parameter and from dissipation due to in-plane quasiparticle currents. Suitable equations describing the dynamics of stacked Josephson junctions

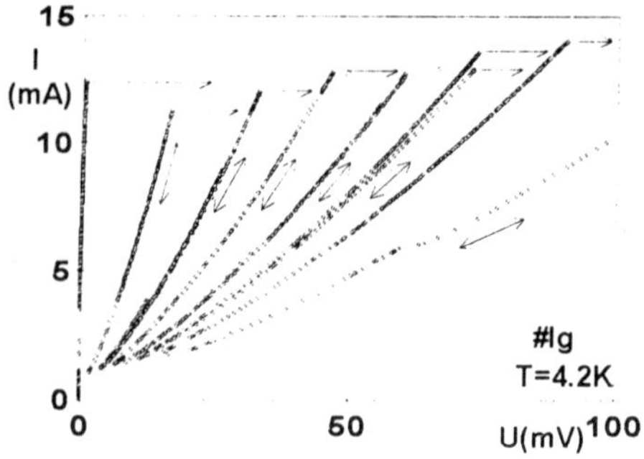

Fig. 4: I-V characteristic of a BSCCO single crystal at 4.2K showing multiple branches in the resistive state.

can be found from the Lawrence-Doniach Theory[8]. The resulting equations generalize the sine-Gordon equation for single Josephson junctions to multilayer systems[14-17]. The equations contain two important length scales. the c-axis magnetic penetration depth $\lambda_c \approx 100\mu m$, and the Josephson penetration depth $\lambda_J \approx 1\mu m$. Josephson vortices can form if the stack dimensions in ab-directions are well above λ_J. This is fullfiled in almost all experiments. There are predictions for two different kinds of collective resonant modes, the collective plasma mode involving small oscillations of the phases γ_n around some equilibrium value[15] and a set of modes involving collective movement of Josephson vortices[16,17]. The latter occurs in nonzero magnetic fields at nonzero bias currents. Josephson-vortices entering a stack of width b move coherently through the stack. The motion is triggered by collective stack resonances forming a standing wave along x and z direction. The frequency spectrum is $\omega(k_x,k_z) = \omega_{pl}\ (\lambda_k/b)\pi k_x\{(\lambda_k/\lambda_c)^2+2[1-\cos(\pi k_z/(N+1))]\}^{-1/2}$, with $\lambda_k=[\Phi_0 d_{eff}/(2\pi\mu_0\lambda_{ab}^2\ j_c)]^{-1/2}$ and $\lambda_c=[\Phi_0/(2\pi\mu_0 t_{eff}\ j_c)]^{-1/2}$. ω_{pl} is the Josephson plasma frequency and t_{eff} and d_{eff} are effective lengths determined by λ_{ab} and the thicknesses of the

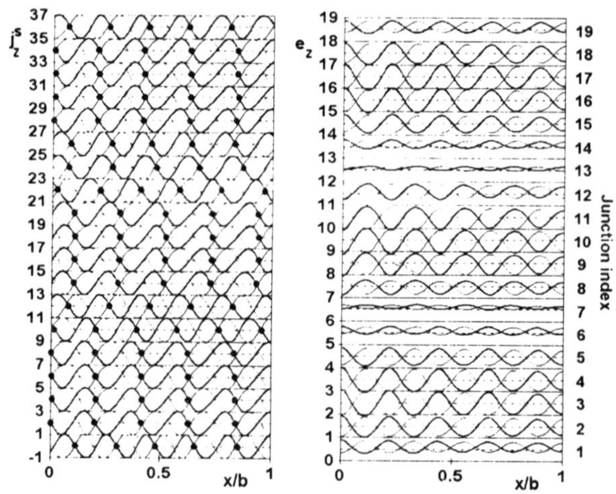

Fig. 5: Dynamic simulation of a stack of 19 Josephson junction in an external magnetic field of 5 flux quanta per junction and a bias current density of $0.6j_c$. Field orientation is parallel to the layers. Left plot shows distribution of supercurrent density along c-direction in each junction. Curves are vertically offset by the junction index in order to map the stack geometry. Thin and thick curves correspond to current distributions at two different values of time t_1 and t_2. For t_1, vortex centers are marked by black dots showing that vortices move in three aligned domains. Right plot shows the electrical field distribution across each junction. The figure shows that a standing wave pattern has developed with nodes both along x and z-direction.

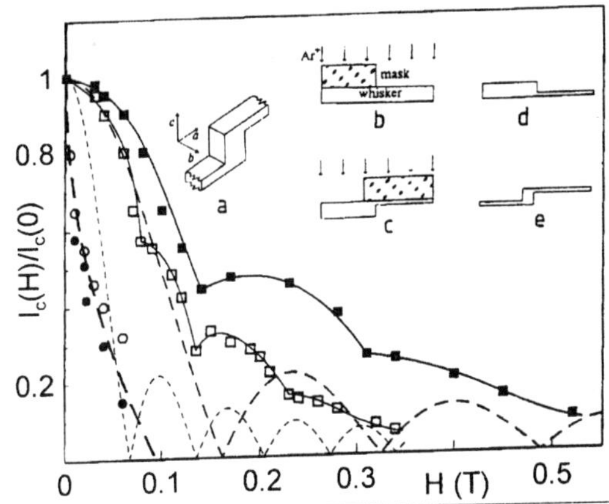

Fig. 6: Critical current vs. temperature of three stacks of intrinsic Josephson junctions fabricated from BSCCO single crystal whiskers. Data points are for stack sizes of 8μm (solid squares), 20μm (open squares), 40μm (solid circles) and 200μm (open circles). Solid lines are guides to the eye, dashed lines comparisons with theory. For small junction sizes, a clear Fraunhofer pattern is visible. Inset shows steps of ion etching procedure.

superconducting and nonsuperconducting layers; k_x and k_z denote the number of halfwaves in x and z-direction. An example with k_x=9 and k_z=3 is shown in Fig. 5. So far, this mode has not been confirmed in high T_c superconductors. However, very recently, this mode has been discovered in stacks of Nb Josephson tunnel junctions[18].

The collective Josephson plasma resonance has been predicted by Bulaevskii et. al.[15]. It can be measured by microwave absorption. Until now it has not been seen in zero magnetic field since (in BSCCO) its frequency is around f=250 GHz. However, the frequency of the plasma resonance can be decreased by a magnetic field oriented in c-direction. In this way, the plasma resonance was observed by Tsui et. al.[19,20] and by Matsuda et. al. [21].

JUNCTION TECHNOLOGY

The first experiments on intrinsic Josephson junctions have been performed by mounting small single crystals between two contact rods[6]. Crystals contained typically 2000 junctions. Clearly, in view of applications like local oscillators or mixers, a more advanced, if possible planar junction technology is required. Latyshev et. al.[22] pattern stacks of typically 100 intrinsic junctions from BSCCO single crystal whiskers. Stacks are patterned by photolithography and ion milling. Lateral sizes are as small as 5 μm. A typical measurement of critical current vs. magnetic field, together with the geometry of the whisker,

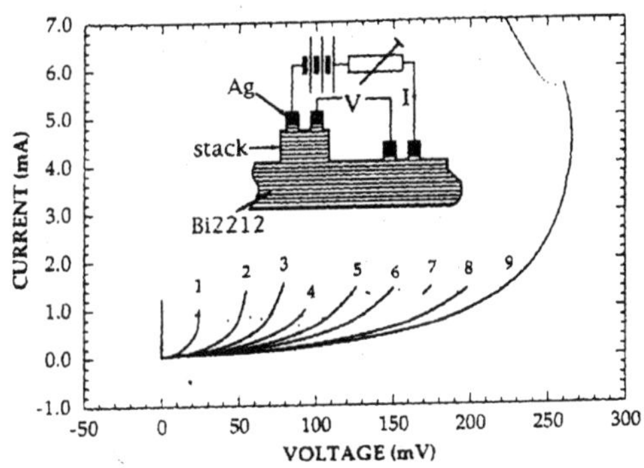

Fig. 7: I-V characteristic of mesa structure patterned on top of BSCCO single crystal. The stack consists of 9 intrinsic Josephson junctions.

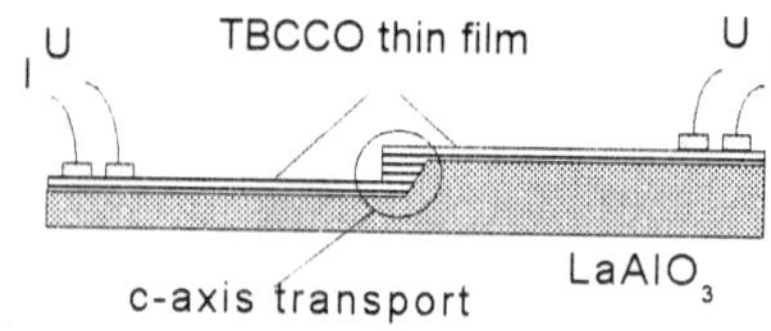

Fig. 8: Schematic view of TBCCO step stack junctions.

is shown in Fig. 6. A major step towards planar fabrication has been achieved by Regi et. al. who fabricated small stacks (mesa structures) on top of BSCCO single crystals[23]. Their diameter is about 50μm and their thickness is several 100 nm corresponding to about 100 active intrinsic junctions. A similar technology is used by Schmidl et. al. to prepare mesas on thin films of TBCCO and BSCCO[24] with junction sizes down to $3x3\mu m^2$ and mesa thicknesses down to 50nm. Rapp et. al.[13] use this method for patterning mesas on YBCO thin films. Here, junction sizes have to be as small as 1μm. In order to further decrease the number of junctions, Yurgens et. al.[11] monitor I-V characteristics *in situ* at ≈80K during ion milling. This allows to fabricate even single intrinsic junctions. An I-V characteristic for a stack of 9 junctions, together with a schematic view of the mesa, is shown in Fig. 7. A different approach is described in ref. 25. The authors pattern a bridge of a TBCCO thin film grown over an artificial substrate step. By this method, a stack of typically 50 intrinsic junctions forms at the step ("step stack"), as shown in Fig. 8.

SUBGAP STRUCTURES

Tunnel junctions are a widely used tool to investigate the quasiparticle density of states of superconductors. However, junctions grown on the surface of a high T_c superconductor can only probe the outermost few Angstroms and are sensitive to surface degradation. In contrast, intrinsic Josephson junctions are perfectly epitaxial layer by and can provide local information about quasiparticle conductivity between pairs of CuO_2 bi- or trilayers located in the bulk of the sample. Therefore, these junctions might be a unique tool to study the intrinsic properties of high T_c superconductors. Recently, subgap structures have been found in BSCCO and TBCCO[26,27]. Structures appear as additional sub-branches at bias currents well below the critical current and at voltages well below the gap (Fig.9). The voltage position of these structures is slightly different in BSCCO and in TBCCO, but, for a given material, does neither depend on junction geometry nor on temperature up to $0.5T_c$ nor on magnetic fields up to 3 Tesla. So far the physical origin of these structures is not clarified. However, the fact that the structures are observed on every branch of the I-V characteristics strongly indicates that they are of intrinsic origin and thus important for the microscopics of high T_c superconductors.

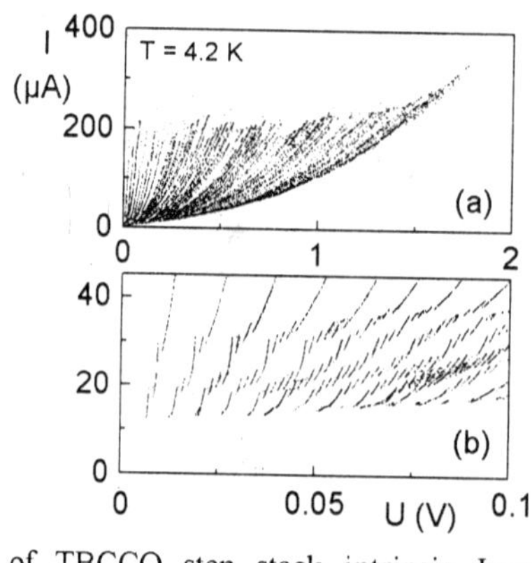

Fig. 9 (a) I-V characteristic of TBCCO step stack intrinsic Josephson junctions. (b) same I-V characteristic on expanded scales showing subgap structures at low bias currents.

SUMMARY
During the last years there have been major improvements both in fabrication of intrinsic Josephson junctions and theoretical understanding of the intrinsic Josephson effect. It has been shown that the effect can be found in a variety of high T_c superconductors as well as in other layered compounds like organic superconducors. The physics of stacked, interacting Josephson junctions contains new effects like collective modes associated with plasma oscillations or vortex movement or the appearance of subgap structures. Intrinsic Josephson junctions can be fabricated in stacks containing between one and thousands of junctions and are thus promising candidates for a variety of high frequncy applications.

ACKNOWLEDGMENTS
The authors like to thank A. Ustinov, G. Hechtfischer, G. Kunkel, K. Schlenga and W. Walkenhorst for valuable contributions. Partial financial support by the Bayerische Forschungsstiftung via the FORSUPRA consortium, and by the Bundesministerium für Forschung und Technologie is gratefully acknowledged.

REFERENCES

1. W. Meißner, R. Ochsenfeld, Naturwissenschaften **21**, 787 (1933).
2. R. Doll, M. Näbauer, Phys. Rev. Lett. **7**, 51 (1961).
3. B. D. Josephson, Phys. Lett. **1**, 251 (1962).
4. M. J. Naughton, R.. C. Yu, P. K. Davies, J. E. Fisher, R. V. Chamberlin, Z. Z. Wang, T. W. Jing, N. P. Ong, P. M. Chaikin, Phys. Rev. B **38**, 9280 (1988).
5. R. Kleiner, F. Steinmeyer, G. Kunkel, and P. Müller, Phys. Rev. Lett. **68**, 2394 (1992).
6. R. Kleiner, P. Müller, Phys. Rev. B **49**, 1327 (1994).
7. R. Kleiner, unpublished (1994).
8. W. E. Lawrence, S. Doniach, in *Proceedings of the 12th International Conference on Low Temperature Physics*, E. Kanda (ed.), Academic Press, Kyoto, 361 (1971).
9. R. Doll, unpublished (1985).
10. N. Takezawa, T. Koyama, M. Tachiki, Physica C **207**, 231 (1993).
11. A. Yurgens, D. Winkler, N.V. Zavaritsky, T. Claeson, Phys. Rev. B **53**, R8887 (1996).
12. K. Schlenga et. al., Physica C **235**, 3273 (1994).
13. M. Rapp, A. Murk, R. Semerad, W. Prusseit, Phys. Rev. Lett. **77**, 928 (1996).
14. S. Sakai, P. Bodin, N. F. Pedersen, J. Appl. Phys. **73**, 2411 (1993).
15. L. N. Bulaevskii, D. Dominguez, M. P. Maley, A. R. Bishop, B. I. Ivley, to appear in Phys. Rev. B.
16. R. Kleiner, Phys. Rev. B. **50**, 6919 (1994).
17. S. Sakai, A. V. Ustinov, H. Kohlstedt, A. Petraglia, N. F. Pedersen, Phys. Rev. B **50**, 12905 (1994).
18. N. Thyssen, A. V. Ustinov, H. Kohlstedt, preprint.
19. Ophelia K. C. Tsui, N. P. Ong, Y. Matsuda, Y. F. Yan, J. B. Peterson, Phys. Rev. Lett. **73**, 724 (1994).
20. Ophelia K. C. Tsui, N. P. Ong, J. B. Peterson, Phys. Rev. Lett. **76**, 819 (1995)
21. Y. Matsuda, M. B. Gaifullin, K. Kumagai, K. Kadowaki, T. Mochiku, Phys. Rev. Lett. **75**, 4512 (1995).
22. Yu. I. Latyshev, J. E. Nevelskaya, P. Monceau, Phys. Rev. Lett. **77**, 932 (1996).
23. F. X. Regi, J. Schneck, J. F. Palmier, H. Savary, J. Appl. Phys. **76**, 4426 (1994)
24. F. Schmidl, A. Pfuch, H. Schneidewind, E. Heinz, L. Dörrer, A. Matthes, P. Seidl, U. Hübner, M. Veith, E. Steinbeiß, Supercond. Sci. Technol. **8**, 740 (1995).
25. K. Schlenga, G. Hechtfischer, W. Walkenhorst, P. Müller, F. X. Regi, H. Savary, J. Schneck, M. Veith, W. Brodkorb, E. Steinbeiß, IEEE. Trans. Appl. Supercond. **5**, 3272 (1995).
26. K. Schlenga, G. Hechtfischer, R. Kleiner, W. Walkenhorst, P. Müller, H. L. Johnnson, M. Veith, W. Brodkorb, E. Steinbeiss, Phys. Rev. Lett. **76**, 4943 (1996).
27. K. Schlenga, G. Hechtfischer, R. Kleiner, W. Walkenhorst, P. Müller, H. L. Johnnson, *Proceedings of the LT 21*, Praque, Czechoslovak Journal of Physics, 1277 (1996).

SUPERCONDUCTING FILTERS USING SLOW WAVE TRANSMISSION LINES

M. J. Lancaster, F. Huang, A. Porch, J.-S. Hong, D. Hung
Birmingham University, School of Electronic and Electrical Engineering, Edgbaston, Birmingham, UK. B15 2TT

Superconductors can help substantially in the development of high performance miniature microwave filters. This paper describes experimental work on using slow wave transmission lines to reduce the size of filters. Examples are given of a ladder line microstrip filter with a comparison of the miniaturised structure with a conventional microstrip filter of the same specification. In addition a new type of slow wave line is discussed. Experimental results are given of this fractal slow wave line when it is incorporated into a miniature resonator. The velocity reduction and the quality factor of the resonator are discussed.

I. INTRODUCTION

In the 10 years since the discovery of high temperature superconductors (HTS) there has been an enormous effort in the development of passive microwave devices for many application areas. There are a number of circuits now going into systems which result in substantially improved performance. Examples include mobile radio base station front end filters and aircraft electronic systems. During this development of HTS passive microwave circuits there has been a large number of innovative new ideas many of which are still under development. In fact there is still enormous scope for the invention of new, improved performance microwave circuits based on the unique properties of superconductors.

Miniaturisation of microwave filters is one area where superconductors can help. As a filter is reduced in size the current density increases. This leads to increased dissipation resulting in a higher insertion loss and a lowering of the Q of each resonant section within the filter. By using superconductors the dissipation can be substantially reduced as compared with the losses obtained when using normal conductors. Although eventually, even with the superconductors, as the filter reduces in size dissipation will increase and the filter performance will deteriorate to unacceptable levels. There is thus a trade off between size and performance no matter what material is used. It should also be noted that as the filter size reduces the power handing capability also gets worse due to the increased current density. Small HTS filters may not, therefore be applicable to high power applications.

It may not necessarily be a simple matter to reduce the size of a filter. If the substrate material can not be changed then the only way to reduce filter sizes is to change the shape of the device. There are five ways to reduce the size of a filter as listed in Table 1.

1. Use of High dielectric constant substrates leads to reduction of circuit velocity.
2. Thinning of substrate leads to narrower lines in microstrip. This gives lower coupling and tighter line spacing
3. Use of kinetic inductance allows an increase in transmission line inductance without the usual associated increase in capacitance. Thus giving a lower velocity.
4. Use of lumped elements, which are required to be much less than a wavelength in length.
5. Use of slow wave transmission lines to give a reduced velocity. These are made of discreet inductors and capacitors value.

Table 1. Methods of miniaturising HTS filters

Methods (1)-(4) will not be discussed further here but further information can be obtained in ref 1 and 2. The rest of this article is devoted to the last entry in Table 1.

II. SLOW WAVE TRANSMISSION LINES

The phase velocity of an electromagnetic wave travelling along a conventional transmission line such as a coaxial cable or microstrip line is just given by

$$v_p = \frac{1}{\sqrt{LC}}$$

where L and C are the inductance and capacitance per unit length respectively. By increasing either L or C it is therefore appears possible to reduce the phase velocity along the transmission line. However, for conventional transmission lines such a coaxial cable or microstrip, increasing the capacitance per unit length by a change in the geometry of the line also reduces the inductance per unit length proportionally. The same applies to an increase in the inductance per unit length. However, the capacitance (or inductance) can be increased by the addition of discreet capacitors (or inductors) along the length of the transmission line. An example of this for a microstrip type transmission line is shown in Fig.1.

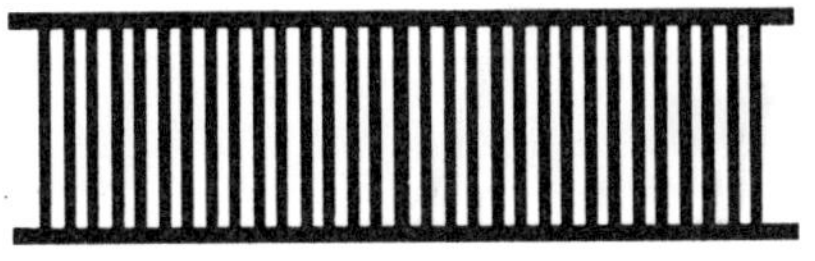

Fig.1. The ladder line slow wave transmission line.

This structure consists of two parallel lines with connections of conductor between. This is situated directly above a ground plane as in conventional microstrip. The presence of the ladder network causes additional capacitance without an associated increase in inductance, and therefore reduces the phase velocity of this type of transmission line over a conventional microstrip line of the same size. These lines have been studied as resonators[3,4] in order to ascertain the reduction in phase velocity. For example a YBCO ladder line resonator measuring 4mm by 1mm with 195 fingers resonates at 10.4GHz with a Q of 1200. Ladder line transmission lines have been demonstrated in a filter geometry as shown in Fig.2. The diagrams of the filters are all to the same scale so the size reduction can be seen. This size reduction is obtained with very little difference in the performance of the filters as demonstrated in the figure.

Although there is a reduction in the size of the filter using this ladder line, it is difficult to provide very large reductions in size by this method. However, coplanar line offers a potentially larger size reduction simply because it is physically easier to increase the capacitance from the signal line to the ground plane. There are a number of ways to do this with coplanar circuits. One method is to use an interdigital finger structure, however, we present here a new coplanar slow wave structure using a fractal geometry. This transmission line is shown in Fig. 3. Capacitive loading is formed by the interleaved nature of the transmission line. The smaller the line width available gives a smaller the gap between the conducting lines and hence a large capacitance thereby reducing the transmission line phase velocity. This reduction can be achieved to an arbitrary extent, limited only by the smallest line width which can be patterned.

In order to demonstrate the principle and compare practical transmission line performance with simulations, a superconducting fractal slow wave line has been constructed and tested. The transmission line is made in the form of a resonator with open circuits at both ends.

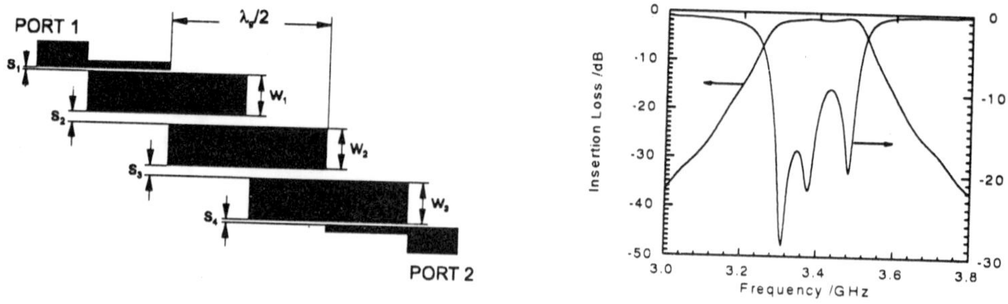

(a)

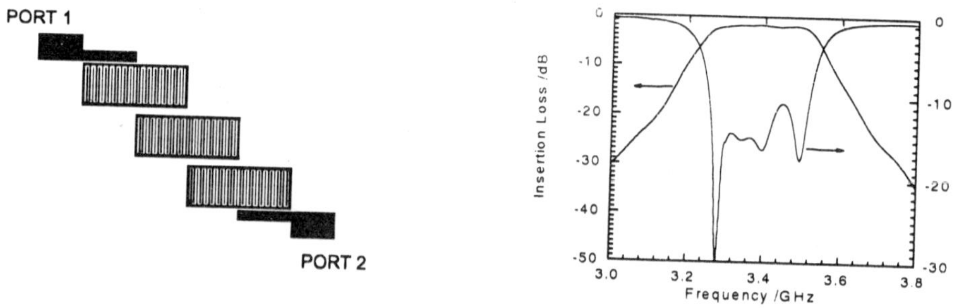

(b)

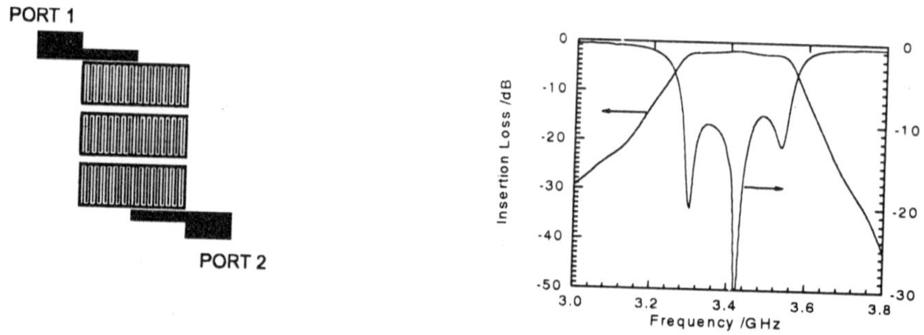

(c)

Fig.2. Comparison of copper filters constructed out of conventional microstrip and ladder line transmission lines. The performance of each filter is on the right of the filter design. (a) Conventional microstrip (b) ladder line filter (c) ladder line with resonators in-line.

Energy is coupled into the device by optimal position of the pins of a standard K-connector. The central fractal structure is 7.96mm long by 1.08mm wide with 20μm line widths and gaps. The whole YBCO device fits on a 1cm square MgO substrate. Fig.4 gives the frequency response of the resonator showing the first six modes of resonance. There are two main limitations on the maximum frequency of operation of this (and any other) slow wave lines. Namely a half wave resonance which occurs normal to the length of the resonator and also when the wavelength approached the periodicity of the line. In this latter case it essentially behaves as a low pass filter. Dispersion occurs throughout the whole frequency range because of this effect, but becomes more predominant at the higher frequencies.

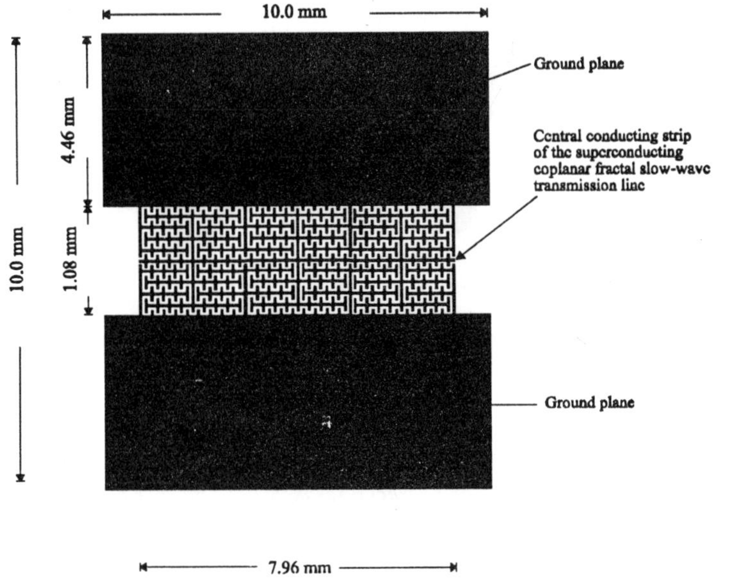

Fig. 3. The fractal slow wave transmission line.

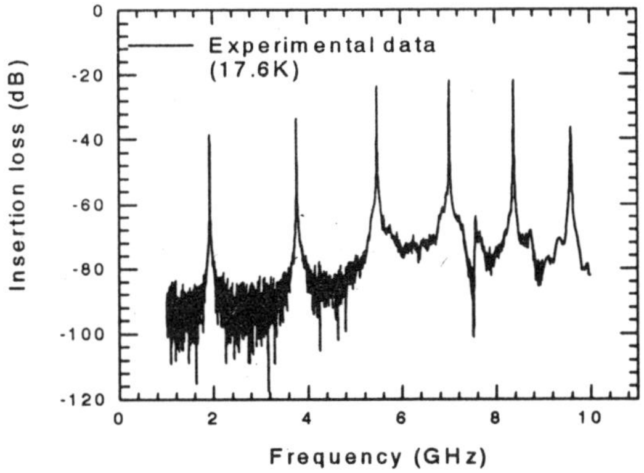

Fig.4. Frequency response of the fractal slow wave transmission line resonator.

Fig. 5 shows a similar frequency response only this time obtained at a number of different temperatures. It can be seen that the shift in frequency due to kinetic inductance effects can be quite substantial. This of course is general for all miniature filters, as they shrink in size kinetic inductance effects become more exaggerated. It should be noted that despite the complex nature of the fractal structure, excellent symmetrical high Q resonances can be obtained all with a large dynamic range.

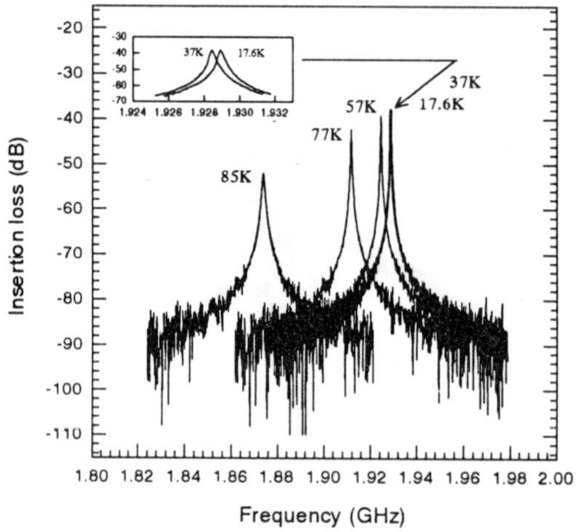

Fig. 5. The frequency response of the fractal resonator at a number of different temperatures.

The resonant frequencies of the device can be translated into a velocity of the electromagnetic wave along the transmission line. Assuming each resonance is a half wavelength or multiples of half wavelength for the harmonics other than the fundamental; then Fig.6 can be produced. This figure gives the velocity reduction over free space as a function of frequency at a number of different temperatures. It can be seen that the line is dispersive as mentioned above.

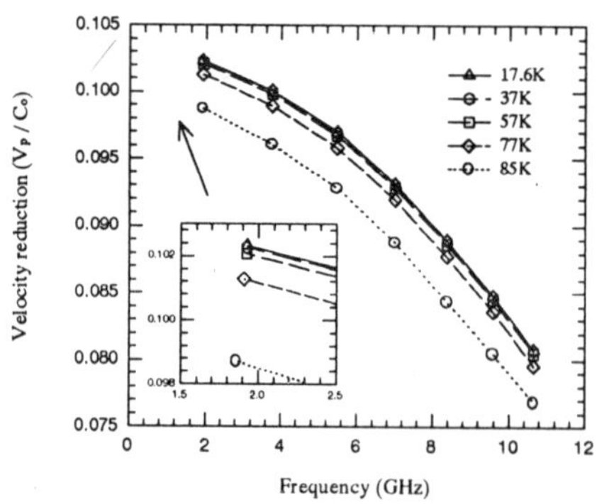

Fig.6. Velocity reduction of the free space velocity.

Finally the quality factor of the fundamental mode of the resonator is given as a function of temperature in Fig. 7. This approaches about 8000 as the temperature is reduced and represents a good high value for application purposes.

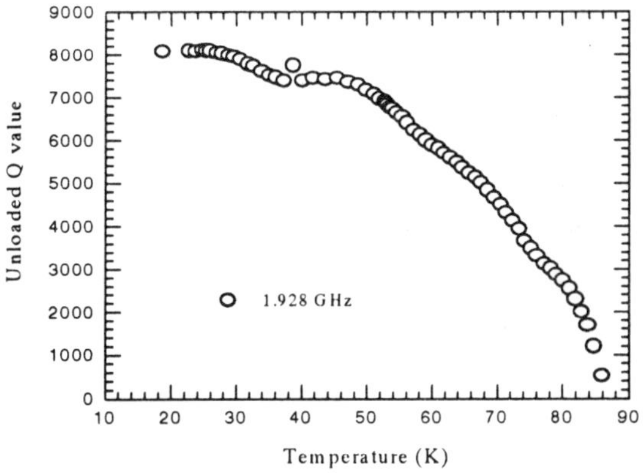

Fig. 7. The quality factor for the fundamental mode.

CONCLUSIONS

A new type of slow wave transmission line has been demonstrated with good slow wave properties. This coplanar line can now be included into miniature filters. Such filters have applications in a number of areas where space is limited and high performance is required. Because of the reduced size of the filter the cooling requirements are also reduced to a certain extent. Although these filters can not be used for high power applications the substantial space savings warrant further investigation. The velocity reductions quoted here are by no means minimised and with smaller line width and/or different geometry, further substantial reductions can be made.

ACKNOWLEDGMENTS

We would like to thank Dr. F Wellhofer and Dr. P Woodall for the production of the thin film material. Also, we wish to acknowledge the support and help from members of the superconductivity group in the school of electronic and electrical engineering as well as our fellow workers in the rest of the university of Birmingham. This work was supported by the UK, EPSRC.

REFERENCES

1. M. J. Lancaster, *Passive Microwave Device Applications of High T_c Superconductors* Cambridge University Press. Cambridge UK (1996).
2. M.J. Lancaster, F. Huang, A. Porch, B. Avenhaus, J. S. Hong, D. Hung, IEEE Trans. Microwave Theory Tech. **MTT44** (1996).
3. J. S. Hong and M. J. Lancaster, Microwave Opt. Technol. Lett. **V9,** 207 (1995).
4. J. S. Hong and M. J. Lancaster, IEEE Microwave Guided Wave Lett. **V5,** 87 (1995).

OPTICAL EFFECTS IN SUPERCONDUCTING HETEROSTRUCTURES

S.H. Pawar, L.D. Jadhav, T.S. Desai and S.B. Kulkarni
Energy Studies Laboratory, Shivaji University, Kolhapur 416004, India.

A new method for synthesis of high T_c superconducting films and wires via electrochemical route has been developed by the author and considerable amount of work has been carried out during last eight years. The present paper summarizes the synthesis of superconductor/metal/semiconductor heterostructure via electrochemical route. The quaternary alloyed films of Bi-Sr-Ca-CuO are deposited onto silver substrate using electrodeposition technique. These films are electrochemically oxidized in 1N KOH solution. On these superconducting films, CdSe semiconducting films are deposited using the same technique. Thus, Ag/Bi-Sr-Ca-CuO/CdSe or CdSe/Ag/Bi-Sr-Ca-CuO heterostructures are fabricated. Photoinduced characteristics of these heterostructures are studied by illuminating either CdSe or Bi-Sr-Ca-CuO at different intensities at room and at low temperatures. I-V characteristics in dark and under illumination are measured at low as well as at room temperatures. The contact potential is also sensitive to light and is a function of temperature. Optical detection of T_c has also been carried out. The details about the properties of heterostructure will be presented.

I. INTRODUCTION

Since the discovery of high T_c superconductors in 1986 by Bendorz and Muller[1], a number of materials with higher T_c have been synthesized with variety of techniques. The transport properties of these materials have been studied to a great extent. However, very little attention has been paid to their optical properties. The optical properties studied mainly include absorption of radiation, emission of radiation (luminescence), photoinduced superconductivity, photo-inducting and Raman scattering; Pawar and Todakar[2] have reviewed the luminescence of high T_c superconductors. The data on other optical properties of high T_c superconductors is limited. In recent years, however, there is considerable interest in photoinduced effects in high T_c superconductors due to photo-doping and possibility of increasing T_c[3]. A number of experimental studies have been performed on the changes that high T_c oxide materials undergo during illumination with white light or laser radiation. Studies have shown that there is increase in T_c[4] with illumination.

A lot of attention has been paid to fabricate SNS and Josephson junctions and study them theoretically[5,6]. Recently, superconducting heterostructures[7] using $YBa_2Cu_3O_{7-\delta}$ and $KTaO_3$ have been fabricated for microwave applications using laser ablation method. But, optical effects in superconducting heterostructures have been studied very little compared to optical effects in high T_c superconductors.

The present paper summerises the synthesis of metal/superconductor/semiconductor heterostructure via room temperature electrochemical synthesis route. The Ag/Bi-Sr-Ca-CuO/CdSe and Ag/Bi-Sr-Ca-CuO heterostructures are fabricated. Photoinduced characteristics of these heterostructures are studied by illuminating at different intensities and at different temperatures, when the Bi-Sr-Ca-CuO is in normal as well as in superconducting state.

II EXPERIMENTAL

The quaternary alloyed films of Bi-Sr-Ca-CuO were deposited onto silver substrate using electrodeposition technique[8]. These films were then electrochemically oxidized at room

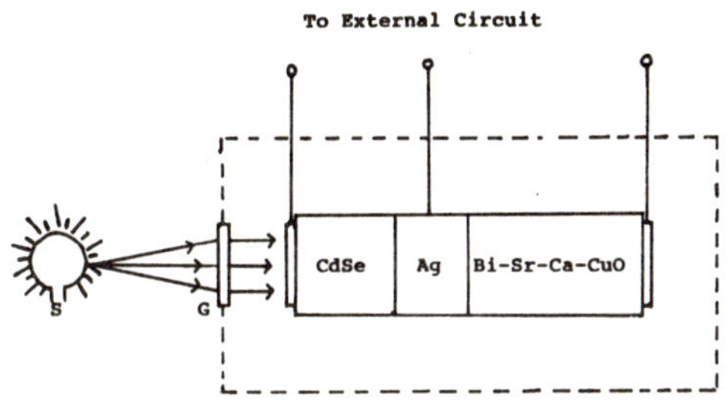

Fig.1. Schematic of experimantal setup; S: source, G: optical window

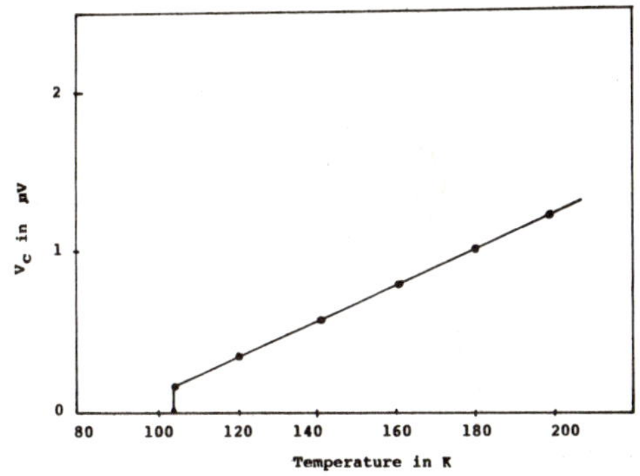

Fig.2. Plot of V_c vs temperature for Ag/ Bi-Sr-Ca-CuO interface.

temperature[9] in 1N KOH solution. This geometry results in Ag/Bi-Sr-Ca-Cu-O heterostructure. To fabricate Ag/Bi-Sr-Ca-Cu-O/CdSe heterostructure, Bi-Sr-Ca-CuO layer onto silver substrate was used as substrate for the deposition of CdSe film using the same technique and this geometry results in the Ag/Bi-Sr-Ca-CuO/CdSe heterostructure. To characterize these heterostructures ring type contacts with silver paint were made on the front surface. The samples were mounted in APD cryogenic system having an optical window. A flourescent lamp was used to irradiate the sample and the junctions were studied at different intensities and at different temperatures. Eight and half digit nanovoltmeter was used to measure the voltage across the junction. A schematic diagram of experimental set up is shown in Fig. 1.

III RESULTS AND DISCUSSION

Ag/Bi-Sr-Ca-CuO interface

The contact potential, a potential at zero current across the junction formed between superconducting Bi-Sr-Ca-CuO film and Ag, is measured with precision nanovoltmeter. The change in contact potential, measured in dark and light conditions is a function of temperature. The change in contact potential, ΔV_c, decreases with decrease in temperature and at critical temperature of Bi-Sr-Ca-CuO film the change in ΔV_c is found to be zero. This effect is shown in Fig. 2. The equation for photovoltaic signal, is

$$\Delta V_c = \frac{nkt}{e} \ln\left(1 + \frac{I_{sc}}{I_o}\right) \tag{1}$$

which may be modified for metal-superconductor junction as,

$$\Delta V_c = \frac{nk(T - T_c)}{e} \ln\left(1 + \frac{I_{sc}}{I_o}\right) \tag{2}$$

Thus, at $T = T_c$, $\Delta V_c = 0$.

Bi-Sr-Ca-CuO film goes into superconducting state below T_c and it does not respond to photosignal. This leads to optical detection of critical temperature, 104K, which is in agreement

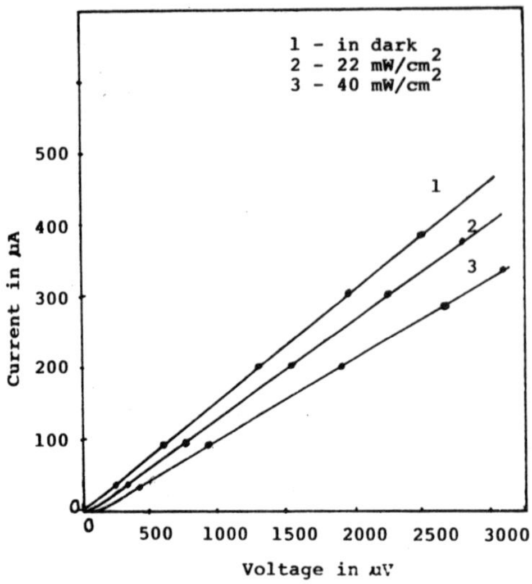

Fig.3. I-V characteristics for Ag/ Bi(2223) contact in dark and with light at 300 K.

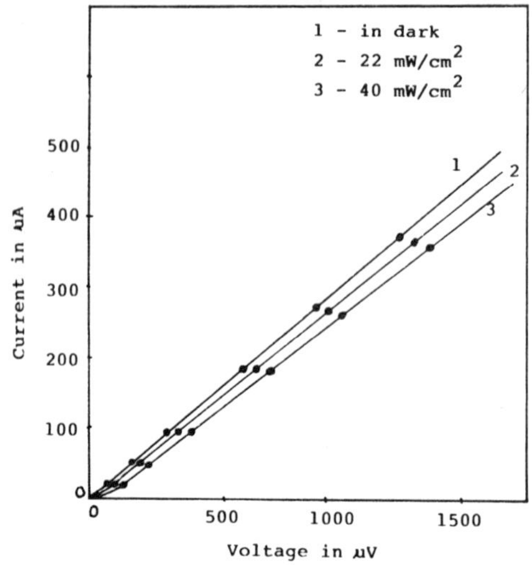

Fig.4. I-V characteristics for Ag/ Bi(2223) contact in dark and with light at 175 K.

with that found by four probe resistivity method (103K). I - V characteristics are measured in dark and light conditions at two different intensities, $I_1 = 40$ mw cm^{-2} and $I_2 = 22$ mw cm^{-2}. As seen from Fig. 3, under illumination the curve is shifted towards higher voltage and shift is higher for higher intensity; I - V characteristics at room temperature are similar to that of a junction diode in dark and under illumination.

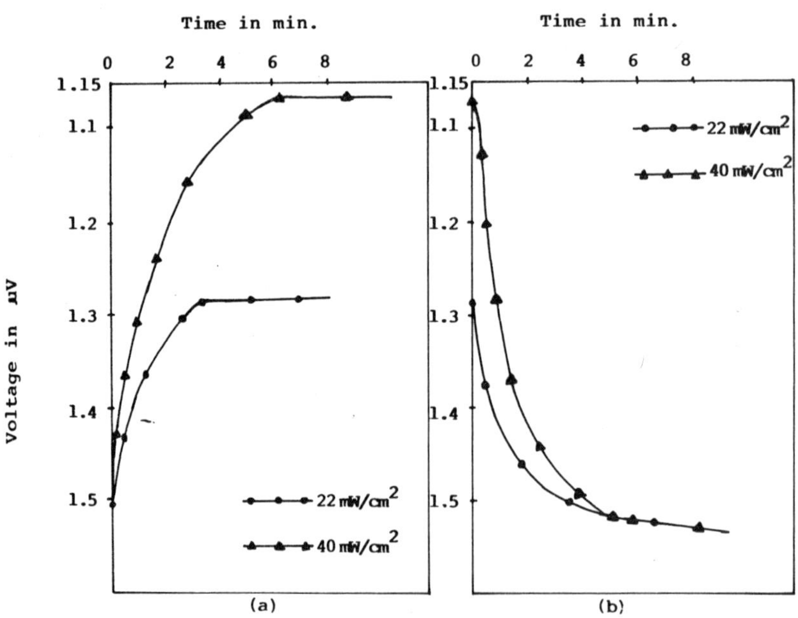

Fig.5. Effect of light on contact potential of Ag/Bi(2223)/CdSe junction at 300K; (a) charging, and (b) discharging.

The energy gap in the range of 1.5 - 2 eV for undoped insulating samples, called charge transfer (CT) gap, is recognized as resulting from transition from Cu 3d sites[10,13]. When the sample is illuminated, the charge carriers are excited and probed in the CuO_2 planes[14]. The charge carriers generated by photoexcitation are driven towards the junction by applied electric field and there is increase in voltage. At 175K as shown in Fig. 4, there is no more change in characteristics since, Bi-Sr-Ca-CuO sample gives a very little response to photosignal. Below T_c there is no change in characteristics, since it does not respond to photosignal.

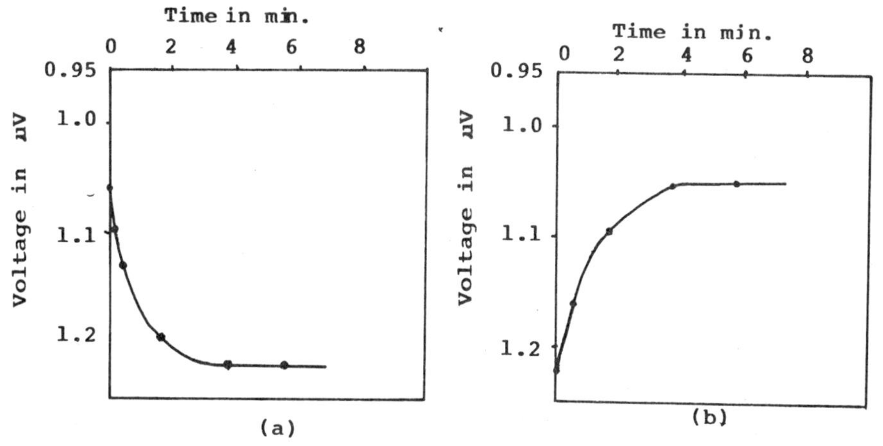

Fig.6. Effect of light on contact potential of Ag//Bi(2223)/CdSe junction at 100K; (a) charging, and (b) discharging.

Photoinduced Characteristics of Ag/Bi-Sr-Ca-CuO/CdSe

Silver paste window contact is given to CdSe film and contacts are taken from CdSe and Ag. For these heterostructures, central superconducting Br-Sr-Ca-CuO film provides path for charge carriers since it is very thin (1.5 µm). Contact Potential recorded on precision nanovoltmeter with Ag connected to positive terminal is due to the difference in Fermi levels of Ag and CdSe. Fermi level of Ag is at higher position with respect to CdSe Fermi level (work function of Ag = 4.31 eV and of CdSe = 5.1 eV). At room temperature, the difference in Fermi levels decreases when CdSe is exposed to light and there is negative decrease in contact potential V_c as shown in Fig. 5(a). Fig. 5(b) gives discharging nature. Effect of light on contact potential is less at 175K. While at 100K contact potential increases in negative direction as shown in Fig. 6. This happens as the Fermi level of Ag is lowered with temperature while that of CdSe is raised. I - V characteristics in dark and light conditions at room temperature are shown in Fig. 7. Under illumination the curve is lowered, which may be due to generation current opposite to total current. At 175K, the curve is lowered with less amount of shift. This happens as CdSe gives poor response to photosignal. At 100K, however, effect of light is reversed as shown in Fig. 8, which may be due to shifting of Fermi levels of Ag and CdSe with temperature. Also at 100K, Bi-Sr-Ca-CuO is in superconducting state and provides resistanceless path to charge carriers thus increasing the current.

IV CONCLUSION

In conclusion, we have successfully fabricated heterostructures with superconductor Bi-Sr-Ca-CuO and semi conductor CdSe. The junctions formed were studied under illumination with fluorescent lamp at room as well as low temperatures.

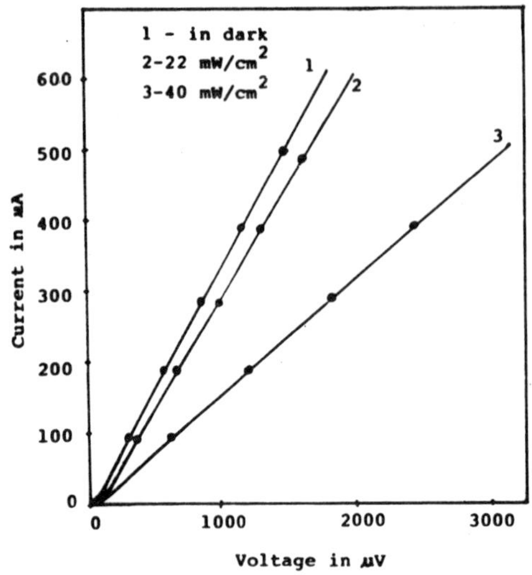

Fig.7. I-V characteristics in dark and in light for Ag/Bi(2223)/CdSe at 300K.

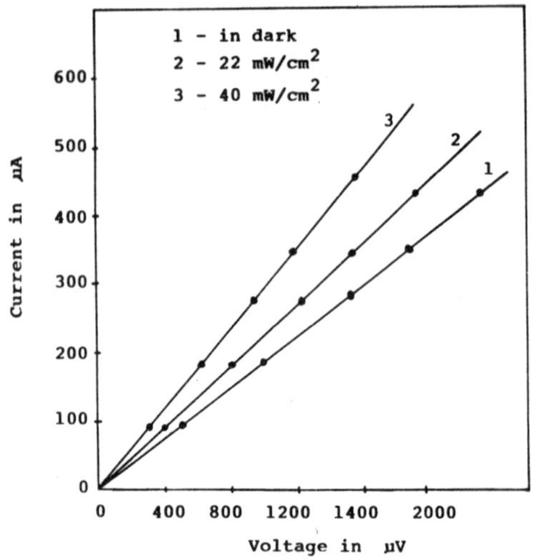

Fig.8. I-V characteristics in dark and in light for Ag/Bi(2223)/CdSe at 100K.

ACKNOWLEDGMENT

The authors wish to thank the Department of Science and Technology, New Delhi for financial support and Dr.A.V.Narlikar, National Physical Laboratory, New Delhi for constant encouragement.

REFERENCES

1. J.G.Bednorz and K.A.Muller, Z.Phys. **B 64**, 189 (1996).
2. S.H.Pawar and B.M.Todkar, Ind.J.Pure and Appl. Phys. **30**, 562 (1992).
3. K.P.Shina, Physica **C 212**, 128 (1993).
4. G.Nieva, E.Osquiguli, J.Guimpel, M.Maenhoudt, B.Wuyts, Y.Bruynsererade, M.B.Maple and Ivan K. Schuller, Appl. Phys. Lett. **60**, 2159 (1992).
5. R.L.Kautz, J.Appl. Phys. **76**, 5538 (1994).
6. S.P.Benz, Appl. Phys. Lett, **67**, 2714 (1995).
7. Yu.A.Boikov, Z.G.Ivanov, A.L.Vasiliev, I.Pronin, E.Olsson and T.Claeson, Appl. Phys. Lett. **67**, 2708 (1995).
8. S.H.Pawar and H.A.Mujawar, Mat. Chem. Phys. **28**, 259 (1991).
9. S.H.Pawar, S.B.Kulkarni and M.J.Ubale, Phil. Mag. Lett. **70**, 327 (1994).
10. T. Timusk and D.B. Tanner in *Physical Properties of High Temperature Superconductors I*, ed. D.M. Ginsberg (World Scientific, Singapore, p. 339, 1989).
11. D.B. Tanner and J. Timusk in *Physical Properties of High Temperature Superconductors II*, ed. D.M. Ginsberg (World Scientific, Singapore, p. 363, 1989).
12. C.M. Varma, S. Schmitt-Rink and E. Abrahams, Solid state coummun. **62**, 681 (1987).
13. J. Orenstein, G.A. Thomas, A.J. Mills, S.L. Cooper, D.H. Rapkine, T. Timsuk, L.F. Schneemeyer and J.V. Waszczak, Phys. Rev. **B 42**, 6342 (1990).
14. G. Yu, C.H. Lee, A.J. Heeger and S.W. Cheong, Physica **C 203**, 419 (1992).

MICROSTRUCTURE AND MICROWAVE PROPERTIES OF LASER ABLATED Ag- AND Au-DOPED $YBa_2Cu_3O_{7-\delta}$ THIN FILMS

R. Pinto

Tata Institute of Fundamental Research, Homi Bhabha Road, Mumbai 400 005, India

Microwave transmission properties such as surface resistance, R_s and magnetic penetration depth, λ of high temperature superconductor (HTS) thin films have been found to critically depend upon the microstructure of films. In the absence of a true epitaxy, in-plane alignment determines the transport properties of HTS films. Extensive work carried out on laser ablated Ag-doped $YBa_2Cu_3O_{7-\delta}$ films has shown that grain enlargement and alignment caused due to the enhanced oxygenation and surfactant action of AgO during insitu growth, significantly improves the critical current density, J_c (highest on $SrTiO_3$: 1.4 x 10^7 Acm^{-2} at 77K) and reduces R_s (lowest on $LaAlO_3$: 210μΩ at 77K, 10GHz). Recent results on Au-doping of $YBa_2Cu_3O_{7-\delta}$ films have shown much improved film stability. Another interesting observation is that the film properties begin to deteriorate above a certain thickness. This has been studied using atomic force microscopy (AFM). AFM has shown increased granularity with film thickness > 3000Å (with certain growth parameters) which has been found to increase both R_s and λ. The exact value of thickness at which film properties degrade depends upon the growth rate and other growth parameters; grain enlargement and alignment throughout the film is the key to the realization of improved microwave properties.

I. INTRODUCTION

High temperature superconductors (HTS) present a number of challenging material technology problems. The main reason for this is their granularity which makes their physical properties more extrinsic in nature.[1,2] Most HTSs, with the exception of high quality single crystals and epitaxial thin films, are granular with varying types of grain boundaries. In most thin films, the material in the grain boundary regions is weakly superconducting or normal metal type which results in superconductor-normal metal-superconductor (S-N-S) type of Josephson junctions.[3-5] These weaklink junctions are highly sensitive to magnetic fields, H due to the fact that magnetic fields of even a few gauss are generally strong enough to introduce a flux quantum into the weaklink tunnel barrier and to shift the phase difference along the barrier. This reduces critical current density, J_c substantially and the J_c (H) characteristics are degraded significantly.[6]

The electromagnetic response of HTSs has also been found to critically depend upon their intrinsic as well as extrinsic properties. Therefore, microwave studies of HTSs have become an important area of research since microwaves can probe both the intrinsic and extrinsic properties of these short coherence length superconductors.[7,9] From technological point of view, microwave surface resistance, R_s, has become an important characterization parameter. Even for fundamental studies, R_s is important since it depends on a number of physical parameters such as penetration depth (λ), coherence length (ξ), energy gap (Δ) and mean free path of cooper pairs.

While the difficulty in growing reasonably large single crystals has hampered both the understanding and applications of HTSs, high quality thin films have made significant progress.[10,12] The main advantage with HTS thin films is that they can be realized in near epitaxial form on single crystal lattice-matched substrates such as $SrTiO_3$ and $LaAlO_3$ using techniques such as off-axis sputtering and pulsed laser ablated deposition (PLAD). However, HTS films in general are not truely epitaxial; their microstructure varies with grain size and orientation, and obviously with grain boundaries which critically depend upon the complexity of the growth process and the quality of the substrate used. Even the highly textured films with c-axis orientation may have considerable in-plane misalignment, which depends not only upon the type of insitu growth techniques used but also on their variations. Evidence has shown that while J_c itself may be affected to a lesser extent due to in-plane misalignment (probably due to a large density of pinning centres in thin films), R_s is affected significantly.

This paper reviews some of the work carried out by us on microstructural improvement of $YBa_2Cu_3O_{7-\delta}$ (YBCO) thin films insitu grown by PLAD and its effect on microwave properties of YBCO thin films. The microstructural improvement was realized by the addition of optimum amount of Ag or Au in the target used for PLAD. Extensive experimentation has shown a significant improvement in J_c of Ag-YBCO films with values in the range 6-8 x 10^6 Acm^{-2} on <100> $LaAlO_3$ and 0.8-1.4 x 10^7 Acm^{-2} on <100> $SrTiO_3$ both at 77K and in self-field conditions.[12] Equally important is the observation of substantial reduction in R_s of Ag-YBCO films with values as low as 210μΩ at 77K, 10 GHz which is the lowest reported R_s value in YBCO thin films.[13] Although Au-YBCO films have shown less improvement, they have been found to be much more stable than the undoped or Ag-YBCO films.[14] However, reduction in λ (T) with Ag-addition has been found to be marginal.[15]

II. YBCO THIN FILM GROWTH

The two crucial aspects of high quality YBCO thin film growth are the substrates and the insitu growth technique used for the realization of films. The substrate qualifications are well discussed in the literature.[9] Some of the most important parameters of YBCO compatible substrates are lattice match, thermal expansion match and chemical compatibility with YBCO. Further, for microwave device applications, dielectric constant and loss tangent are also important. <100> $SrTiO_3$ and <100> $LaAlO_3$ have been extensively used. Most of the YBCO film growth in the author's laboratory was carried out on $LaAlO_3$ since it is microwave compatible with low loss tangent and a near perfect lattice match (lattice mismatch is ~ 2%) with YBCO.

The realization of state-of-the-art quality HTS films by PLAD involves the optimization of various PLAD parameters.[9,16] However, HTS thin film growth optimization during PLAD has relied mostly on empirical methods. Nevertheless, many experiments have been carried out to understand the nature of laser generated plume using spectroscopic techniques such as emission, absorption and laser-induced fluorescence. Mass spectroscopy has also been used to study the plume dynamics which depends upon many factors such as target density, operating pressure, laser beam fluence and spot size, laser wavelength with pulse width and repetition rate and target substrate distance. For insitu growth, substrate temperature and oxygen pressure during growth are important parameters.

The PLAD system used in the author's laboratory for the insitu growth of undoped, Ag-doped and Au-doped YBCO thin films has a stainless steel chamber pumped by a turbomolecular pump having a quartz window for laser beam entry and a target assembly. The targets (undoped and both Ag and Au-doped) were subjected to oscillatory movement during ablation using a stepper motor drive. The laser used is a 248 nm, 1200mJ KrF excimer laser from Lambda Physik with the beam focused to an optimum spot size of 3 mm x 0.8 mm using a 200 mm focal length quartz lens. The pulse repetition rate is 10 Hz with laser fluence in the range 2.5-3.5 Jcm^{-2}. Typical target-substrate distance is 5 cm with oxygen pressure of 200 m Torr during growth. The substrate heater which is a critical part of the PLAD system is a resistance leater capable of operation at temperatures as high as 850°C in oxygen ambient. Most of the film growth was carried out at 720-750 °C substrate temperature. Details of YBCO film growth by PLAD and optimization are reported elsewhere.[12,16]

III. FILM MICROSTRUCTURE

In granular superconducting films such as YBCO films, the microstructure plays an important role. The size of the grains and the grain boundaries between them critically control the film transport properties. Hence, grain enlargement and alignment are the two important issues in the improvement of HTS thin film microstructure. The various growth parameters discussed earlier such as laser spot size, target-substrate distance, substrate temperature, oxygen pressure and film growth rate determine film microstructure critically. The substrate lattice match and surface conditions (smoothness) also critically determine the films microstructure.

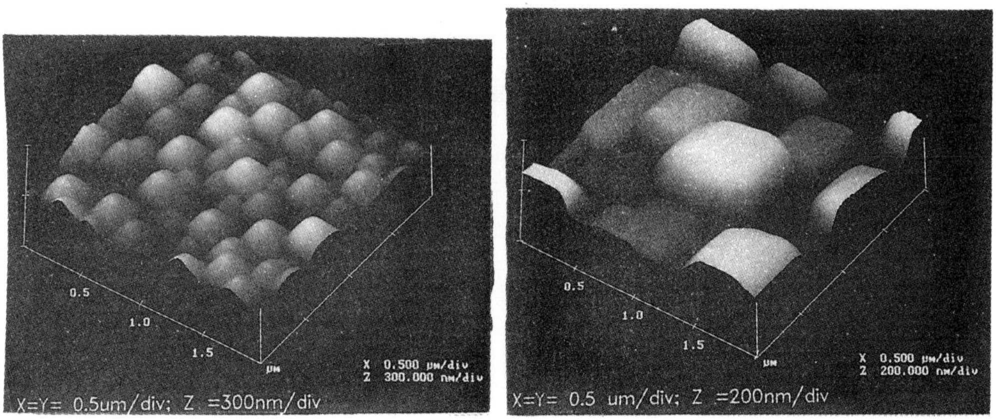

Fig.1 Atomic force micrographs of: a) undoped and b) 5wt% Ag-doped YBCO films

It has been shown by us that the properties YBCO films in normal as well as superconducting states improve dramatically with Ag doping.[12] Au doping too has been found to have a significant effect, especially on the environmental stability of the films.[14] Although X-ray diffraction spectra has not shown any significant structural difference in the undoped and doped films, there is a significant difference in the microstructure of the films as observed with scanning electron microscope or atomic force microscope (AFM). Shown in Fig.1 are AFM micrographs of undoped and 5 wt% Ag-doped YBCO films. It is clear that Ag-doped films not only show enlarged grains but also alignment of grains. Similar is the case with Au-doped YBCO films. The results indicate that 5 wt% Au in the target is about the optimum with largest grain size since any more addition of Au reduces the grain size. This phenomenon was in fact, observed first in the case of Ag-doped YBCO films.

In order to understand the improvements in both Ag- and Au- doped YBCO films, a two-dimensional growth model has been proposed as shown in Fig. 2. The schematics in Fig.2 show a reduction in grain boundary width and alignment of YBCO grains over a specified area within dotted lines. The schematic in Fig.2(b) shows an increase in grain size in Ag-doped films. This 2D model not only explains the observed reduction in the room temperature resistivity with Ag-doping (with a minimum at 5wt% Ag in YBCO target) but also the enhancement in J_c with highest J_c of 1.4×10^7 Acm^{-2} at 77K observed in 5 wt% Ag-doped YBCO films on $SrTiO_3$.

There are primarily two mechanisms which can explain the improved film microstructure with Ag-doping. The first is the enhanced oxidation of YBCO film during growth due to active oxygen generated at the lattice sites by the decomposition of AgO. This is because Ag, which is oxidized in the plume (partial oxidation has been found through emission spectroscopy of laser induced plume of Ag-YBCO target), decomposes at the substrate surface since AgO is unstable at temperatures > 350°C. This releases active oxygen to the growing YBCO grains. Enhanced oxidation of Ag-YBCO film has been verified by quenching experiments by which the oxygen content of YBCO film is frozen immediately after termination of growth.

The second mechanism is the surfactant induced modification of epitaxial grain growth in YBCO films. The modification is induced due to two reasons. The first is the high mobility of Ag atoms at the growth temperature (≥ 700°C). This results in an increase in the surface mobility of constituent materials of the YBCO lattice which in turn results in seeding of nucleation centres at larger distances than those possible in undoped YBCO film. The second reason for the modification is the catalytic effect on the YBCO grain growth in the presence of Ag atoms arising from the well known characteristics of Ag, namely, the wetting between YBCO grains. Due to such characteristics, the addition Ag to YBCO films during in situ formation results in the reduction of surface tension between adjoining grains and between the substrate surface and grains. The wetting of YBCO grains causes enhancement as well as alignment in the growth of grains.

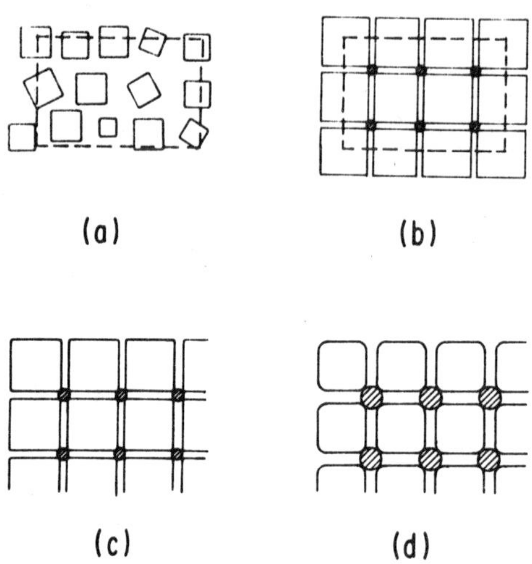

Fig.2. 2D growth model of Ag-doped YBCO films.

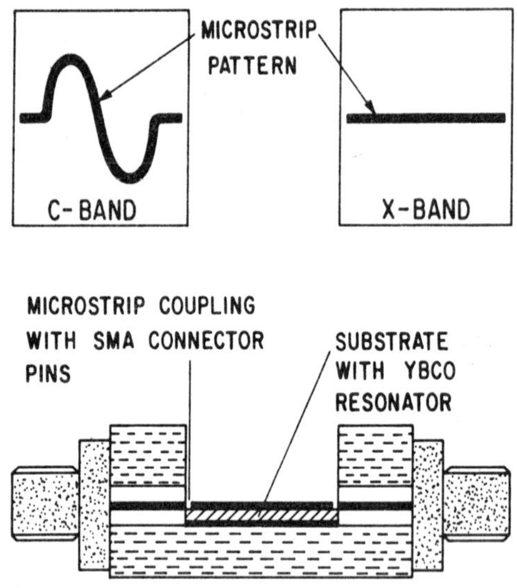

Fig.3. Schematic of c- and x-band microstrip resonators with measurement jig.

IV. MICROWAVE MEASUREMENTS

The surface resistance, R_s of HTS films at microwave frequencies is very sensitive to film quality as well as thickness. Obviously, films with high degree of granularity will have high R_s values. On the other hand, near-epitaxial films with large and well aligned grains have low R_s values. Further, R_s also depends upon applied microwave power, frequency and temperature.

One of the wellknown techniques used for the measurement of R_s of HTS films is the use of planar resonators at appropriate frequency bands. The technique used by us is based on planar resonator having two-port capacitively coupled microstrip configuration. The resonator is designed with a characteristic impedance of 50Ω. The linewidth depends upon the dielectric constant and thickness of substrate. The linewidth for 50Ω impedance is 175μm for 0.5mm thick $LaAlO_3$. Shown in Fig.3 are schematics of c- and x-band microstrip resonators and also of measurement jig. Both the input and output ports are loosely coupled through capacitive gaps between the micro strip line and SMA connector pins such that the measured quality factor, Q_m, is ≈ unloaded Q-factor, Q_o of the resonator. If the resonator is not loosely coupled, the measured Q-factor is the loaded Q-factor, Q_L. Q_o can be calculated from Q_L using $Q_O = Q_L(1 - 10^{L_I/20})$ where L_I is the insertion loss.

The microwave measurements were carried out on YBCO thin films patterned into microstrip resonators. The microwave source used is a HP 83620 A synthesized sweeper and the signals are analyzed using a HP 8757C scalar network analyzer with the associated reflectometry set up. All measurements are carried out at low temperature in the range 15-88K using a closed loop He cryocooler and a temperature controller with a Si diode sensor.

The Q-factor of resonators is measured by calculating the ratio f/Δf, where f is the resonant frequency and Δf is the 3dB bandwidth of resonant curves. The value of Q_o is then calculated using the measured values of Q_m and L_I. The Q-factor due to conductor loss, Q_c, is given by

$$\frac{1}{Q_c} = \frac{1}{Q_o} - \frac{1}{Q_d} - \frac{1}{Q_r} \tag{1}$$

where, Q_d and Q_r are the Q-factors due to dielectric and radiation losses, respectively. Since the losses due to radiation can be minimized by providing effective shielding at half wavelength spacing around the device, and since $1/Q_d \approx \tan\delta < 10^{-4}$ for $LaAlO_3$ at 77K, we can approximately take $1/Q_d + 1/Q_r \approx 10^{-4}$. Hence, Q_c is calculated from Eq.(1). The values of R_s are then calculated from the values of Q_c and other resonator parameters such as dielectric constant, linewidth, impedance etc. using the expression given by Pucel et al.[17].

The penetration depth, λ(T), of YBCO thin films was measured using the technique proposed by Anlage et al[18] which is based on the phase velocity, v_p of a microstrip transmission line

$$\frac{v_p}{c} = \frac{1}{\sqrt{\varepsilon_{eff}}}\left\{1 + \frac{2\lambda \coth t/\lambda}{d}\right\}^{-1/2} \tag{2}$$

where λ and t are penetration depth and thickness of the strip and ground plane films, respectively, and d is the dielectric thickness. Here both films are assumed to be of identical

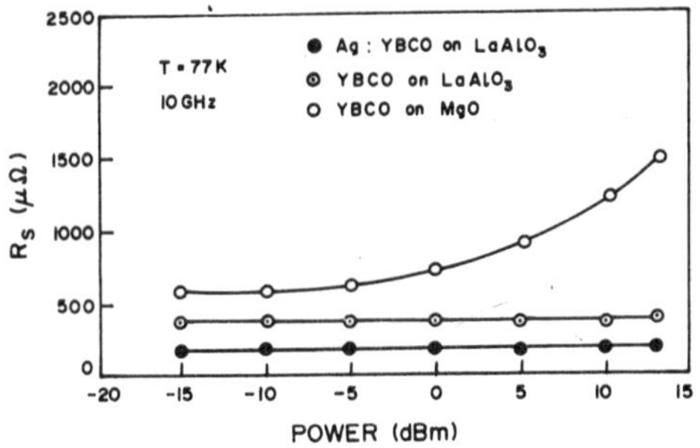

Fig.4. R_s vs microwave power for undoped and 5wt% Ag-deoped YBCO films.

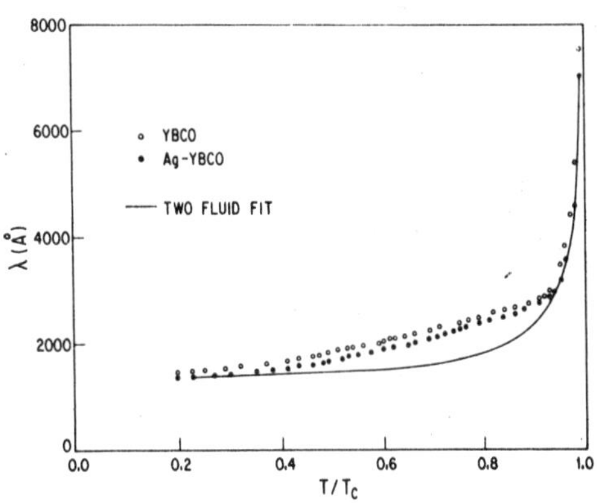

Fig.5. λ(T) vs T/T_c plots for undoped and 5wt% Ag-doped YBCO films.

quality and thickness. ε_{eff} is the effective dielectric constant of the mixed dielectric system and it takes care of the effect of fringing fields. Use of thin dielectric in the range of 10-50 μm thickness increases the sensitivity of λ measurement. Both v_p and ε_{eff} can be determined experimentally from resonant frequency of the microstrip resonator: $v_p = 2Lf_n/n$, where f_n is the nth resonant frequency, n is the mode number and L is the geometric length of the microstrip; and $f = c/2L\sqrt{\varepsilon_{eff}}$, where f is the fundamental resonant frequency. Sensitivity of λ measurement was increased using a 25 μm thick Teflon sheet as a dielectric sandwiched between the YBCO microstrip and the YBCO ground plane films in a flip-chip configuration the same as that used by Anlarge et al.[18] λ(T) values were calculated at various temperatures using phase velocity data and Eq.(2).

V. RESULTS AND DISCUSSION

As mentioned earlier, microstructure and granularity of films determine both power and temperature dependence of R_s. Our results show that Ag-doped YBCO films have a much improved microstructure with significant grain enlargement as well as alignment. This is evident in Fig.1 which shows atomic force micrographs of undoped and 5 wt% Ag-doped YBCO thin films on $LaAlO_3$. The film thickness is 3000A°. The microwave power dependence of R_s of these films is shown in Fig.4. Shown also in the figure for comparison is the microwave power dependence of R_s of YBCO film grown on MgO. Due to poor lattice match and the consequent granularity of YBCO films on MgO, they show a large power dependence.

The measured temperature dependence of λ of undoped and 5wt% Ag-doped YBCO films on $LaAlO_3$ in shown in Fig.5. The solid curve in Fig.5 is a fit to $\lambda(T) = \lambda(0)\left\{1-(T/T_c)^4\right\}^{-1/2}$ two-fluid theory which is not good in the mid-temperature range. The values of λ(0) calculated as per two fluid fit are 1400A° and 1498A° for Ag-YBCO and YBCO films, respectively.

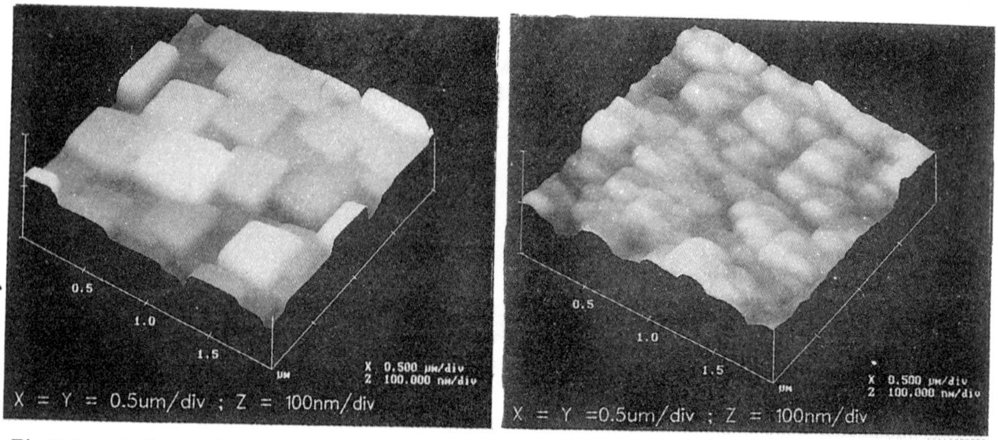

Fig.6. Atomic force micrographs of 5wt% Ag-doped YBCO films with thickness: a) 3000Å and b) 4500Å.

One of the important aspects of YBCO films is the dependence of its microstructure on thickness. Under a given set of growth and substrate conditions the microstructure does not change noticeably upto a certain thickness. This value of thickness has been found to be ~ 3000Å under the growth conditions used by us. As the film thickness is increased beyond 3000Å the microstructure was found to change as observed through AFM. Shown in Fig.6 are atomic force micrographs of 3000Å and 4500Å thick 5wt% Ag-doped YBCO films. The dependence of microwave properties on film thickness is shown in Fig.7 which shows the variation of λ(0) and R_{eff} (thickness dependent value of R_s) with thickness of 5wt% Ag-doped film. It is clear from these two figures that the film granularity increases beyond 3000 A° which in turn deteriorates the film microwave response by increasing both λ and R_s. Shown in Fig. 8 is the variation of R_s (minimum value of R_{eff}) with temperature.

The effect of Au addition in YBCO target has been found to be similar to that of Ag-addition.[14] However, the improvements in transport properties is less than that observed with the best Ag-doped YBCO films. Nevertheless, Au-doped YBCO films have been found to show much improved passivation properties as compared to those of undoped and Ag-doped YBCO films, which is of significant advantage to practical devices.

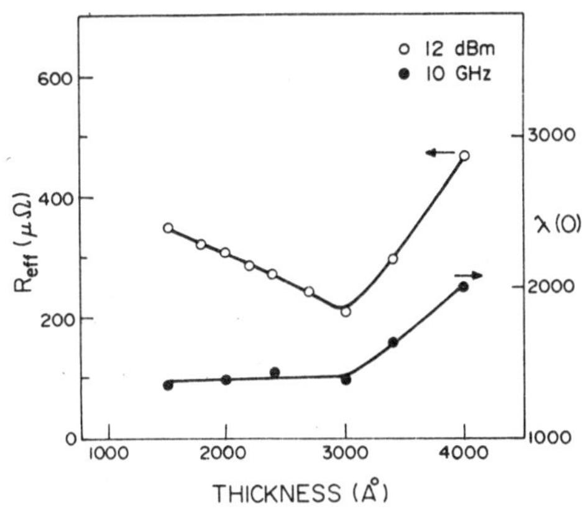

Fig.7. Variation of R_{eff} and λ(0) with thickness of Ag-doped YBCO films.

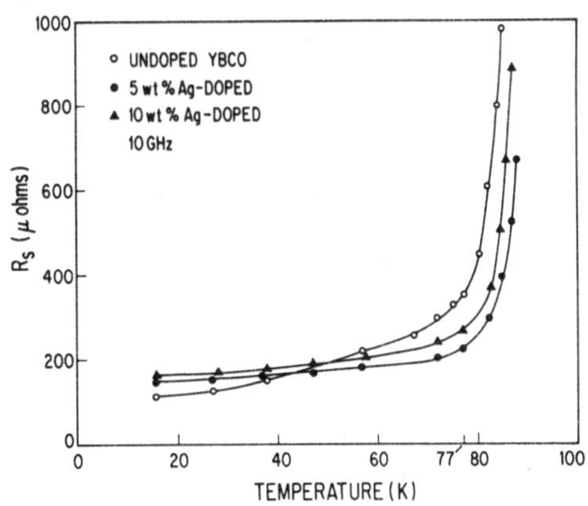

Fig.8. R_s vs T plots for undoped, 5 and 10wt% Ag-doped YBCO films.

VI. CONCLUSION

It is shown that the microstructure of Ag- and Au- added YBCO films improves significantly. The main effect is manifested in enlargement and alignment of grains during film growth by PLAD. The cause of this improvement has been found to be enhanced adatom mobility of film constituents caused by high mobility Ag- or Au- atoms during growth. In the case of Ag addition, enhanced oxidation of YBCO grains has been found to be another mechanism responsible for improved film microstructure.

Improved microstructure of Ag and Au-added YBCO films has been found to significantly improve their microwave transmission properties. Although the reduction in the value of λ(0) is marginal, the reduction in R_s (77K, 10 GHz) has been found to be substantial with Ag and Au addition. Au added films have also shown much improved passivation properties. Finally, the film microstructure has been found to depend upon thickness. Films with thickness >3000A° have more granular microstructure as compared to thinner films which is reflected in the higher values of λ(0) and R_s observed.

ACKNOWLEDGMENT

The author wishes to thank Prof. R. Vijayaraghavan and Prof. L.C. Gupta for discussions and Dr. P.R. Apte and Dr. B.K. Sarkar for work on microwave measurements. Thanks are also due to S.P. Pai, S.C. Purandare, C.P. D'Souza and J. John for experimental assistance.

REFERENCES

1. J. Halbritter, J. Appl. Phys. **68**, 6315 (1990).
2. T.L. Hylton, A. Kapitulnik, M.S. Beasley, J.P. Carini, L. Drabeck and G. Cruner, Appl. Phys. Lett. **53**, 1343 (1988).
3. Y. Matsumoto, J. Hombo, Y. Yamaguchi, M. Nishida and A. Chiba, Appl. Phys. Lett. **56**, 1585 (1990).
4. R. Gross, P. Chaudhari, M. Kawasaki and A. Gupta, Phys. Rev. **B42**, 10735 (1990).
5. N. Savvides, Physica C **165**, 371 (1990).
6. J. Mannhart, in *Earlier and Recent Aspects of Superconductivity*, ed. by J.G. Bednorz and K.A. Muller (Springer, Berlin, 1990).
7. S. Sridhar, D.H. Wu and W. Kennedy, Phys. Rev. Lett. **63**, 1873 (1989).
8. D.E. Oates and A.C. Anderson, IEEE Trans. Magn. **MAG-27**, 867 (1991).
9. R. Pinto, in *Studies of High Temperature Superconductors*, Vol. 17, ed. by Anant Narlikar (Nova Sci. Pub. New York, 1996).
10. R. Kromann, J.B. Bilde-Sorensen, R. deReus, N.H. Andersen, P. Vase and T. Frelotoft, J. Appl. Phys. **71**, 3419 (1992).
11. R. Ramesh, T.S. Ravi, D.M. Hwang, C.C. Chang, A. Inam, T. Venkatesan, X.D. Wu, R.E. Muenchausen, S. Foltyn and N.S. Nogar, Physica C**173**, 163 (1991).
12. D. Kumar, M. Sharon, R. Pinto, P.R. Apte, S.P. Pai, S.C. Purandare, L.C. Gupta and R. Vijayaraghavan, Appl. Phys. Lett. **62**, 3522 (1993).
13. R. Pinto, N. Goyal, S.P. Pai, P.R. Apte, L.C. Gupta and R. Vijayaraghavan, J. Appl. Phys. **73**, 5105 (1993).
14. R. Pinto, P.R. Apte, M.S.R. Rao, R. Chandra, C.P. D'Souza, S.P. Pai, L.C. Gupta, R. Vijayaraghavan, K.I. Gnanasekar and M. Sharon, Appl. Phys. Lett. **68**, 1006 (1996).
15. R. Pinto, D. Kaur, M.S.R. Rao, P.R. Apte, V.V. Srinivasu and R. Vijayaraghavan, Appl. Phys. Lett. **68**, 1720 (1996).
16. R. Pinto, S.P. Pai, C.P. D'Souza, L.C. Gupta , R. Vijayaraghavan, D. Kumar and M. Sharon, Physica C **196**, 264 (1992).
17. R.A. Pucel, D.J. Masse and C.P. Hartwig, IEEE Trans. Microwave Theory Tech. **MTT-16**, 342 (1968).
18. S.M. Anlage, H. Sze, H.J. Shortland, S. Tahara, B. Langley, C. Eom, M.R. Beasley and R. Taber, Appl. Phys. Lett. **54**, 2710 (1989).

DEVELOPMENT OF SQUID BASED ON Nb-AlOx-Nb JOSEPHSON JUNCTIONS

T.S. Radhakrishnan, M.P. Janawadkar, K. Gireesan, R. Baskaran, Rita Saha and L.S. Vaidhyanathan
Materials Science Division, IGCAR, Kalpakkam-603 102, India

Josephson junctions and SQUIDs are the essential components of the superconducting device technology. Recognizing the importance of this technology in the Indian context, a R & D program was initiated a few years ago at Kalpakkam for the development of Nb-AlO_x-Nb SQUID devices. Using electron beam evaporation as the deposition technique high quality Nb-AlO_x-Nb junction have been produced with V_m 60 mV. These junctions have been incorporated in DC SQUID devices of two different designs viz. Ketchen Jaycox design with an on-chip integrated input coil and a multiloop design with the possibility of Additional Positive Feedback (APF). The devices have been operated in the flux locked loop mode with a sensitivity of 75 mV /ϕ_0; spectral density of noise in the white noise regime was 25 $\mu\phi_0 / \sqrt{Hz}$. It is expected that the indigenous availability of SQUID sensors tailored for specific applications will pave way for utilization of these devices in a number of application areas such as biomagnetism, geophysics and nondestructive testing of materials.

I. INTRODUCTION

Josephson junctions and SQUID devices based thereon represent the essential basic components of superconducting device technology. The field of superconducting devices has seen rapid advancements since the introduction of Nb-AlO_x-Nb Josephson junctions in the year 1983 [1]; indeed, their introduction has been the single most important and decisive step towards the establishment of the superconducting device technology, in the global context. Availability of Nb-AlO_x- Nb Josephson junctions, which are mechanically rugged and thermally stable, has triggered intense research and development activity all over the world aimed at utilizing these in a number of potential application areas such as high speed digital circuits, voltage standards, X-ray and millimetre wave detection etc. At the same time DC SQUID sensors based on Nb-AlO_x-Nb Josephson junctions emerged as the worlds most sensitive detectors of magnetic flux and began to be used in a large number of areas such as biomagnetism, geophysical prospecting, nondestructive testing of materials etc., besides being used as laboratory sensors to measure extremely small magnetic signals with unsurpassed sensitivities. However, till the end of the eighties, there was not much activity in India relating to the fabrication and utilization of Nb-AlO_x-Nb Josephson junctions and SQUID devices based thereon. It was in this context that a research and development program was undertaken at our laboratory (Materials Science Division, Indira Gandhi Centre for Atomic Research, Kalpakkam) for developing the technologies related to the fabrication of Nb-AlO_x-Nb Josephson junctions and SQUID sensors. A class 1000 superconducting device fabrication laboratory has been established which houses thin film deposition and photolithographic processing facilities. As a result of this effort, Nb-AlO_x-Nb Josephson junctions comparable in quality to the best junctions reported internationally have been fabricated using the facilities established at this laboratory. DC SQUID sensors, based on both the Ketchen- Jaycox design with an on-chip integrated input coil and the multiloop design with the possibility of additional positive feedback have been fabricated through highly reliable microfabrication techniques.

II. FABRICATION AND CHARACTERIZATION OF JOSEPHSON JUNCTIONS

Nb-AlO$_x$-Nb Josephson junctions are generally fabricated using sputtering as the deposition technique. High deposition rates (60 - 180 nm/min) are used for depositing niobium films to keep possible incorporation of contaminating species like oxygen etc. within the niobium film at a sufficiently low level. An important advantage of the sputtering technique appears to be the ability to control the stress in the deposited Nb films; since the stress can be systematically changed from tensile to compressive as a function of sputtering parameters (gas pressure, cathode bias etc.), it is possible to optimize the sputtering parameters such that the deposited films are in a state of near zero stress. Electron beam evaporated niobium films usually have a high intrinsic tensile stress which is difficult to control. Our preliminary attempts at influencing this intrinsic stress by varying the substrate temperature during deposition did not yield positive results. Nevertheless it was felt that use of electron beam evaporation at moderate deposition rates (18 - 30 nm/min) might be expected to result in smoother niobium films and, consequently, a better quality of Josephson junctions. It may also be expected that during the deposition of the upper niobium electrode by electron beam evaporation, evaporating niobium atoms arrive at the substrate with much less energy when compared to a sputter deposition process and hence cause much less damage to the integrity of the thin and delicate tunnel barrier. Very few reports, however, exist on the fabrication of Nb-AlO$_x$-Nb Josephson junctions by electron beam evaporation. Simon et al have reported fabrication of Nb-AlO$_x$-Nb Josephson junctions by electron beam evaporation, but their junction quality is poor yielding a V_m of only 18-27 mV [2]. Efforts in our laboratory have resulted in the development of a process of fabrication of high quality Josephson junctions using electron beam evaporation; the junctions have near ideal I-V characteristic and exhibit V_m values in excess of 60 mV.

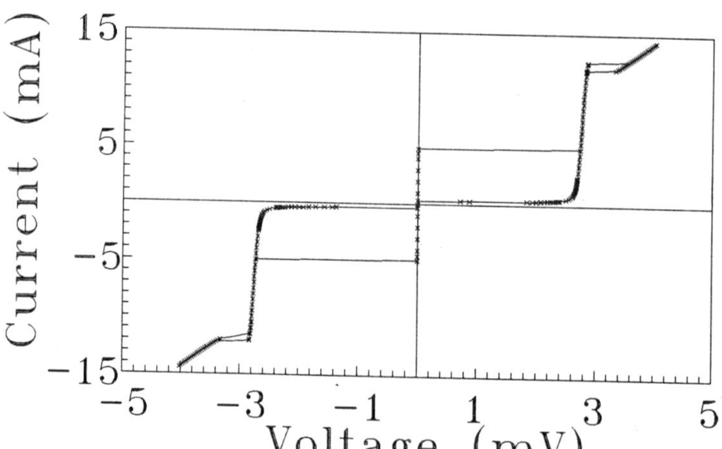

Fig.1. dc I-V characteristic of a typical Nb-AlO$_x$-Nb junction at 4.2K

The process involves deposition of Nb-AlO$_x$-Nb trilayer over the entire wafer without breaking vacuum and the subsequent definition of junction or device geometry via a number of levels of photolithography and etching. The details of the fabrication process have been described in detail elsewhere.[3]

Fig. 1 shows the dc I-V characteristic at 4.2 K of a 100μm × 100μm square junction. The junction exhibited a zero voltage Josephson current exceeding 60% of the value expected on the basis of the Ambegaokar-Baratoff theory. Josephson tunnel current density was 50 A/cm^2; this low value is on account of the fact that the oxidation process was specially optimized to yield a thicker oxide barrier in view of our interest in adopting the fabrication process for DC SQUIDs. The I-V characteristic indicated a clearly resolved energy gap ($\Delta_1 + \Delta_2$) of 2.8 meV and a knee structure attributed to the proximity effect due to the presence of unoxidised aluminum. At 4.2 K, the ratio of resistances R(2 mV)/R(4mV) attained values in excess of 25 and the junction quality parameter V_m, following Adelerhof et al [4], attained values exceeding 60 mV. Barrier uniformity was also investigated by measuring the modulation of the critical current I_c with an in-plane magnetic field. The dependence of critical current on magnetic field is shown in Fig. 2.

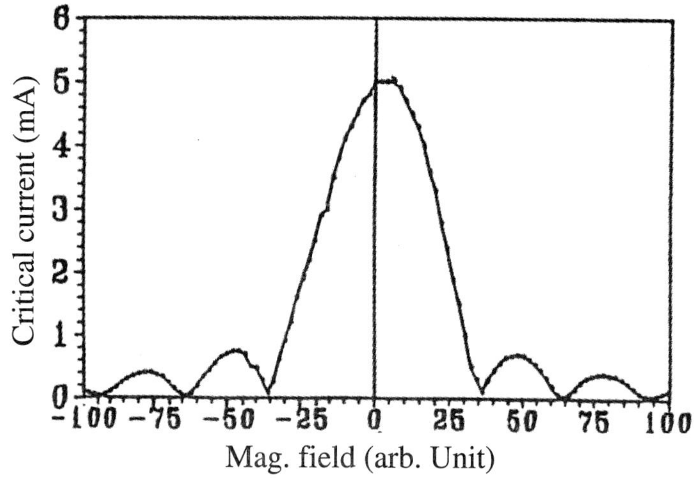

Fig. 2. Variation of I_c with magnetic field

The junction capacitance deduced from the position of the first Fiske step was found to be 260 pF yielding a specific capacitance of 26 fF/μ m^2, in reasonable agreement with values reported in literature[5].

III. FABRICATION AND CHARACTERIZATION OF SQUID DEVICES

The high quality Josephson junctions have been incorporated in DC SQUID sensors of two different designs. One of these is a design first proposed by Ketchen and Jaycox in 1982[6]. This design is based on a superconducting slit washer geometry and involves fabrication of DC SQUID sensors with an integrated spiral input coil which couples the external signal very efficiently into the SQUID loop. The other is a multiloop design proposed by Drung in 1990[7] and has the possibility of providing on-chip additional positive feedback in order to steepen a portion of the I-V characteristic so that the SQUID output can be derived by employing a Direct Offset Integration Technique (DOIT), thus obviating the need for the more conventional electronics involving 100 kHz flux modulation and phase sensitive detection.

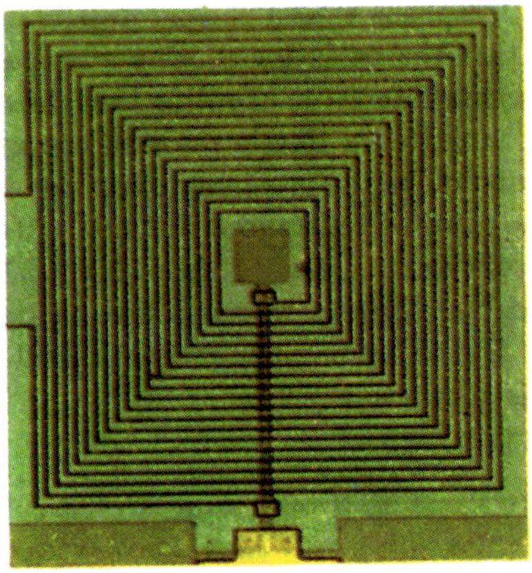

Fig. 3. DC SQUID sensor (Ketchen-Jaycox design)

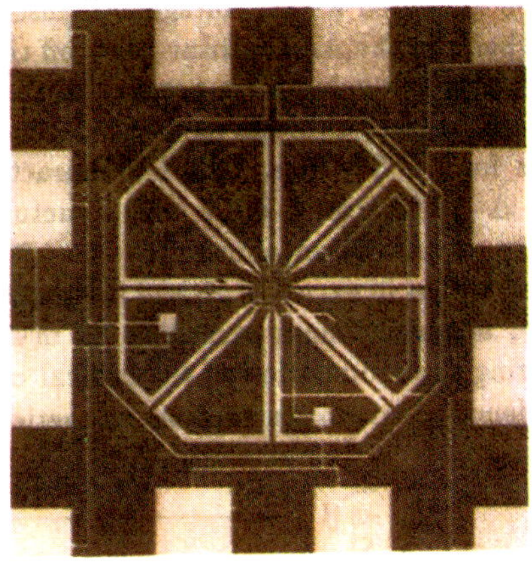

Fig. 4. DC SQUID sensor (Multiloop design)

Fig. 3 shows a micrograph of a Nb-AlO_x-Nb DC SQUID sensor based on the Ketchen - Jaycox design while Fig. 4 shows a DC SQUID sensor based on the multiloop design.

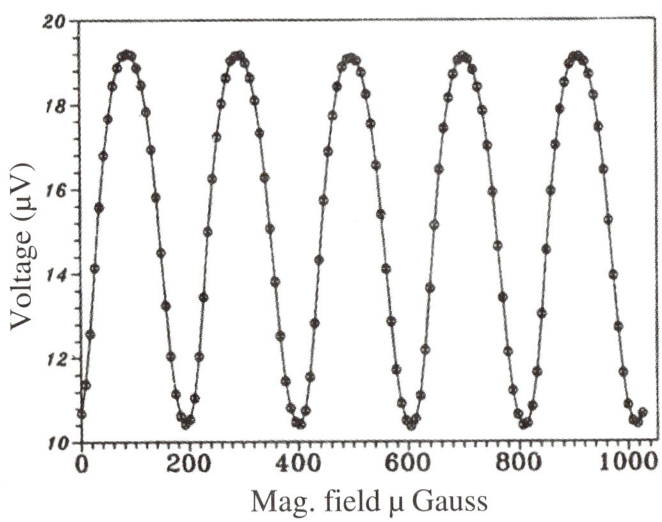

Fig. 5. Modulation of SQUID voltage with applied magnetic flux

Modulation of the SQUID output voltage as the applied magnetic flux is varied is shown in Fig. 5. The voltage modulation depth is 12μ V which conforms to the value expected on the basis of device design. To linearize this response, the devices have been operated in the flux locked loop mode with 100 kHz flux modulation and phase sensitive detection to realize sensitivities as large as 75 mV/Φ_0.

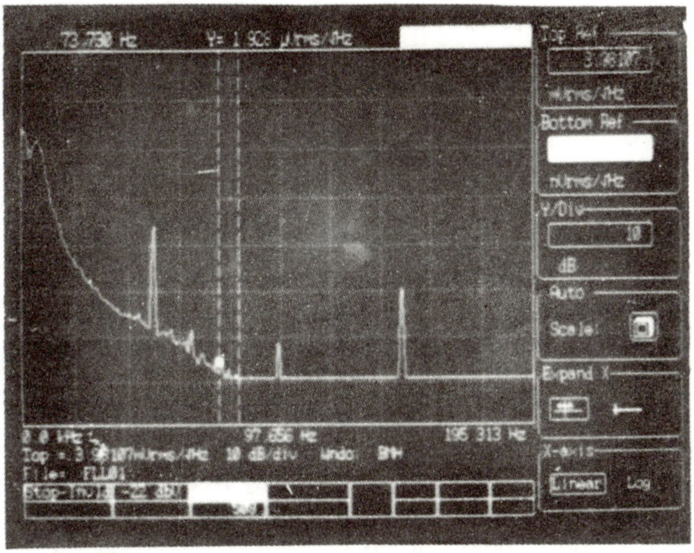

Fig.6. Spectral density of noise voltage

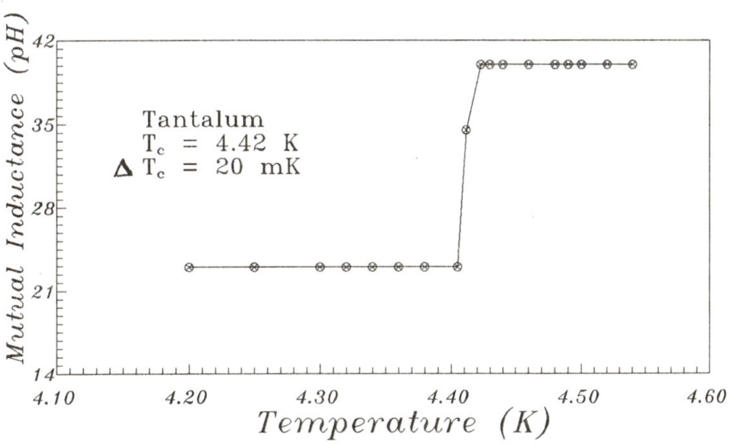

Fig.7. Superconducting transition in a Ta foil measured with the SQUID sensor

The spectral density of voltage noise is shown in Fig.6. White noise is seen to be under 2 $\mu V/\sqrt{Hz}$ which translates to less than $26\mu\Phi_0/\sqrt{Hz}$. Efforts are on to reduce the slightly higher level of 1/f noise as also the knee frequency for the onset of this noise. Elimination of flux trapping in the SQUID structure and the use of improved integrated flux locked loop electronics are expected to lead to significant improvements in both. Further improvements can be achieved by using the bias current reversal techniques. In the flux locked loop mode, it has also been possible to track externally applied low level low frequency magnetic

signals.
As a typical example of measurement using the SQUID, Fig.7 shows a superconducting transition of a Ta foil detected by measuring the change in the mutual inductance between the modulation coil and the SQUID.

IV. CONCLUSIONS

In conclusion, a self contained superconducting device fabrication and characterization laboratory has been set up at Kalpakkam and Nb-AlO$_x$-Nb Josephson junctions and SQUID devices comparable to the ones available on the international scene have been fabricated. This development marks the realisation of an important enabling technology and offers the possibility of embarking on any of a number of applications mentioned earlier. While work is on hand to build a SQUID magnetometer system for use in the laboratory for measurement of weak magnetic signals, it may be possible to realise a wide spectrum of applications through initiatives of interested user groups across the country.

In the global context, despite their development in the eighties, commercial availability of Josephson junctions and SQUIDs has started only recently, and the applications are expected to proliferate. Even after the advent of high temperature superconductors (HTSC), quite a few of the applications still rely on Nb-AlO$_x$-Nb Josephson junctions because of the relatively poor superconducting and other properties of HTSC. The indigenous availability of Nb-AlO$_x$-Nb Josephson junctions and SQUIDs should, therefore, go a long way in establishing beneficial technologies in many important applications areas.

ACKNOWLEDGMENT

The developments reported in this paper have come as a result of intensive work of a project carried during the course of about five years. It is a pleasure to thank Dr.P.K.Iyengar and Shri C.V.Sundaram for responding to our enthusiasm and supporting the establishment of this capital intensive programme. We thank Department of Science and Technology, Government of India for their keen interest in the work in the context of their role in promoting the applications of superconductivity in India. We would also like to thank Dr.Placid Rodriguez and Dr.Baldev Raj for their support and encouragement, Dr.Y.Hariharan for many helpful discussions, Mr.M.K.Ranganathan, Ms.G.Chinnamma, Mr.N.Chinnasamy and Mr.A.Thiruarul for rendering valuable assistance during the various stages of the work.

REFERENCES

1. M. Gurvitch, M. A. Washington and H. A. Huggins: Appl. Phys. Lett. 42, 472 (1983).
2. W. Simon, E. K. Liebmann, M. Simon and E. Bucher: J. Appl. Phys. 72, 4474 (1992).
3. M. P. Janawadkar, R. Baskaran, K. Gireesan, Rita Saha, L. S. Vaidhyanathan and T. S. Radhakrishnan Jpn. J. Appl. Phys. 33, L 1662 (1994).
4. D. J. Adelerhof, M. E. Bijlsma, P. B. M. Fransen, T. Weiman, J. Flokstra and H. Rogalla: Physica C 209, 477 (1993).
5. E. P. Houwman, D. Veldhuis, J. Flokstra and H. Rogalla: J. Appl. Phys. 67, 1992 (1990).
6. M. B. Ketchen and J. M. Jaycox: Appl. Phys. Lett. 40, 736 (1982).
7. D. Drung, R. Cantor, M. Peters, H. J. Scheer and H. Koch: Appl. Phys. Lett. 57, 406 (1990).

DESIGN CONSIDERATIONS FOR HIGH Tc THIN FILM MICROWAVE FILTERS

B. K. Sarkar
Society for Applied Microwave Electronics Engineering and Research
IIT Campus, Powai, Mumbai 400 076, INDIA.

The design considerations for high temperature superconducting microwave filters are presented. The filter designed is at C- band with 0.6% bandwidth. The error analysis, insertion loss calculation, power handling capability, input/output connections and package design are discussed. Theoretical calculations gave insertion loss of 0.8dB at 6 GHz for a conductivity ratio of 50 between high T_c thin film and gold film.

I. INTRODUCTION

Probably the most important application of High Temperature Superconductors (HTS) is the microstripline filters for space applications. The exceptionally high quality factors achievable with HTS films makes the fabrication of HTS filter for narrow bandwidths (< 1% BW) very attractive. The replacement of bulky waveguide filter by such a filter saves the weight and volume by 90%, which is a great relief for satellite designers. Most of the reported works are on broad band filter and without the considerations of substrate and thin film parameters, input/output connectors to thin film bonding and packaging[1-3]. Design of extremely narrow band filters is a challenging task. The design requires weakly coupled lines which can be achieved by widely placed resonators, where slight change in the filter dimensions cause large error in the filter response. These design problems along with the different sources of error are discussed here.

II. FILTER DESIGN CONSIDERATIONS

In developing the narrow-bandpass filter (0.6% BW), we have selected the 500 μm thick $LaAlO_3$ substrate. $LaAlO_3$ has been chosen for filter realization mainly because of its relatively large dielectric constant of 24. This yields effective miniaturization for the filter configuration, since the length of individual elements of the filter, which are in terms of the wavelength of propagation λ_g, will be smaller for larger ε_r. Also single crystal $LaAlO_3$ have low loss microwave properties and also offer very good lattice matching with $YBa_2Cu_3O_{7-\delta}$ (YBCO) films. The filter center frequency is at 6 GHz and we chose a 5 section chebychev bandpass design with 0.01dB ripple.

The microstrip configuration was considered the best choice after it became clear that state of the art HTS deposition had advanced far enough to make double-sided coated substrates. The microstrip configuration is easy to integrate, and the double sided coated substrates used for narrow band filter will yield more than an order of magnitude performance improvement as compared to copper filters.

Since the filters designed have very narrow band width, the traditional end coupled structure results in very long filters. Thus the filter configuration chosen is parallel coupled Microstriplines at the ends along with an end coupled Microstripline in the center. The size of the filter is 23mm X 10mm X 0.5mm. Eesof's Libra package is used for the design optimization. Since the relative dielectric constant of $LaAlO_3$ is 24 and the approximate design equations are normally valid only for ε_r <12, the validity of most of the commercially available software packages (like Eesof, Supercompact etc.) are questionable. Other software packages like IE3D are now available for HTS filter design. For the final optimization small mask changes are necessary to achieve well behaved low ripple band pass characteristics.

1. Error Analysis

a) Effect of variation of temperature

It has been observed that the center frequency of resonance shifts with temperature. The shift is pronounced when T / T_c tends to unity. This is due to the mutual and kinetic inductance effect as the temperature tends towards T_c. The circuit can be modeled for the mutual inductance effect by temperature dependent effective widths and spacing given by,

$$S_{eff} = S + 2\lambda \text{ and } \omega_{eff} = w - 2\lambda$$

where the spacing s and widths w of the lines are much larger than the zero temperature London penetration depth, λ_0 and the temperature dependence of λ, according to the Gorter and Casimir two-fluid model, is given by

$$\lambda = \lambda_0 / \sqrt{1-(T/Tc)^4}$$

Kinetic Inductance effects are more dominant and can be simulated by a temperature dependent length given by

$$l_{eff} = l(1 + \kappa\lambda / h \coth (t / \lambda))^{1/2}$$

where κ is a geometry factor.

b) Effect of change in dielectric constant

The temperature dependence of the dielectric constant gives a frequency shift to the filter response. A 2% change in the dielectric constant gives 1% change in the center frequency, which is unacceptable in the case of narrow band filter design.

In addition, the use of high dielectric constant materials tend to require transmission lines to be extremely narrow where their impedance levels are sensitive to a change of few microns in line width[4]. Narrow lines have higher losses, are very sensitive to pin holes and metalization defects, and are difficult to connect to external circuits.

c) Effect of variation of substrate/ resonator dimensions

Changes in microstrip width due to the photolithography result in a frequency shift in the response. A 10μm change in dimensions gives rise to a frequency shift of 20 to 30 MHz in the center frequency. The change of length of filter's first or last element can result in significant changes in return loss at the band edges. A 10μm change in the substrate thickness gives another 0.5 % shift in center frequency.

2. Insertion Loss

Keeping the band width and skirt response fixed, the insertion loss of a filter is inversely proportional to the unloaded Q of the resonator i.e. the value of surface resistance of the HTS thin film. From the filter's requirement, i.e. an insertion loss of 0.6 dB in the pass band, the unloaded Q should be 7500.[5]

The insertion loss of HTS planar bandpass filters usually comprises of the conductor loss, dielectric loss and radiation loss. The conductor loss is proportional to the surface resistance of HTS film and ground conductor. The dielectric loss is the loss tangent of the substrate. In microwave circuits loss due to the contact resistance or impedance matching also cause further insertion loss.

The contact resistance

$$R_t = R_s + 4R_c/\pi d^2 + R_o$$

where

R_s - Spreading Resistance.
R_c - Specific Contact resistance
d - Diameter of the contact area.
R_o - Interconnect resistance

The connector loss is approximately 0.073 dB for SMA connector and 0.147 dB for K - Type connector.

Radiation loss, substrate mode loss and the loss due to poor ground contact are the other contributions to the insertion loss. Hence, the fabrication techniques, such as the design of the package and the mounting method are extremely important.

3. Power Handling Capability

The power handling capability of the filter is a function of the film quality and is limited by the increased peak current density due to high SWR, which is inversely proportional to the bandwidth. If P_{NL} is the input power and K_{NL} is sheet current density at the onset of non-linearity, then

$$P_{NL} = \frac{\pi n \eta d \omega}{4Q_L}\left(\frac{K_{NL}}{\beta}\right)^2$$

where n - length of resonator expressed in half wavelengths
Q_L - Loaded Q quality factor
ω - microstrip resonator width
d - dielectric thickness
$\beta = K_{max}/K_{av}$
K_{max} - Peak sheet current density
K_{av} - sheet current density averaged across the device cross section

To design a filter with high power handling capability the peak current density needs to be reduced. It is advantageous to design for wide lines on thick dielectric with low effective dielectric constants. Design using overmoded resonators also improves the power handling capability.

4. Packaging

The package design and assembly are key issues in microwave cryogenic devices. It is not only important that the package be electromagnetically sound, designed to minimize radiation losses, but that it be mechanically stable to withstand the thermal cycling without degrading the device characteristics. The package design includes the selection of connectors, ohmic contact deposition, gold wire bonding and the housing design.

a) Connectors

Usually SMA or K - type connectors are used for input and output connections of the filter.

The loss due to these connectors are,

for SMA connector : Insertion loss $= 0.03 \text{ (frequency in GHz)}^{1/2}$

: VSWR $= 1.02 + 0.005 \text{ (frequency in GHz)}^{1/2}$

For K -type connector : Insertion loss $= 0.06 \text{ (frequency in GHz)}^{1/2}$

b) Ohmic contacts and wire bonding

The center pin of the SMA / K -type connector is electrically connected to the ohmic contact of the filter via gap welded Au ribbon. Soldered contacts are avoided because acid based solder flux attack the chemically sensitive YBCO films. The ohmic contact should have the characteristics which allow the formation of a strong wire bond, along with a low contact resistance. A good adhesion of the gold film to the HTS film is also necessary.

Photoresist lift-off techniques are used to form ohmic contact on both surfaces of the substrate. A composite layer Ag/Au is thermally evaporated onto both surfaces of the substrate. Acetone is used to dissolve the photoresist to lift-off the undesired portion of the metal film, before the contacts were annealed in dry O_2 at 500 °C for approximately 0.5 hours to improve adhesion and lower the contact resistance.

c) Package

Packaging reduces the loss due to radiation. Housing is designed in such a way that it acts like waveguide below cut off. Microwave absorbers are added to the package modes and thereby to improve the overall frequency response of the filter.

III. FILTER DEVELOPMENT

Society for Applied Microwave Electronics Engineering and Research (SAMEER) in collaboration with Tata Institute of Fundamental Research (TIFR) has been working on the microwave applications of high temperature superconductors during the last 6 years. Microstripline resonators at S, C and X- band frequencies using YBCO thin film on MgO and

$LaAlO_3$ substrates are developed and reported earlier. Space Application Center (SAC), Ahmedabad has projected the requirement of HTS filters to replace bulky waveguide filters used in the satellite. SAMEER in collaboration with TIFR has taken up the development of HTS filters at C- band, with very narrow band width. The project is funded by Department of Science and Technology, Government of India and the user agency, SAC, Ahmedabad.

CONCLUSION

While it is possible to design high performance microwave filters with very narrow band width using high temperature superconducting thin films, several new design parameters must be evaluated in the design process. Various design considerations for such devices are discussed in this paper. A 0.6% bandwidth band pass filter at 6 GHz is designed. The theoretical calculation showed an insertion loss of 0.8dB at pass band.

REFERENCES

1. G. L. Matthaei , L.Young, and E. M. T. Jones, *Microwave filters, impedance - matching networks, and coupling structures,* Artech House, Inc., Norwood, Mass.
2. S. H. Talisa et al. IEEE Trans. Magn. **27**, 2544 (1991).
3. W. G. Lyons et al. IEEE Trans. Magn. **27**, 2537 (1991).
4. A.K. Ganguly and C. M. Krowne, IEEE Trans. Microwave Theory Tech. **MTT 39**, 1329 (1991).
5. *Microwave Engineer's Handbook, Volume 1,* Horizon House- Microwave, Inc., 1971.

KINETIC INDUCTANCE OF HTS RESONATORS AT VARIOUS MICROWAVE POWER LEVELS

G. P. Srivastava and Mohan V. Jacob

Department of Electronic Science, University of Delhi, South Campus, New Delhi - 110 011, India

Microwave superconducting devices show a drastic deterioration in its performance at high microwave power levels. The flux penetration through the weak links increases the quasiparticle concentration which results in the increase of penetration depth and hence the kinetic inductance. We have modeled an expression to find the kinetic inductance at various rf power levels. The results show that the change in kinetic inductance is proportional to the square of the applied field. This model can explain the reported experimental results at and below the intermediate power levels.

I. INTRODUCTION

The HTS microwave devices show better performance compared to the similar metallic devices at low microwave power levels. The measurements of Q - factor, resonant frequency and insertion loss are effective tools to characterize the device performance[1]. There is a drastic decrease in its performance at high input power levels[1-3]. The less reliability of HTS microwave devices at high input power levels reduces its application and so the analysis of superconducting devices at high power level is an important but a difficult problem to understand.

The electromagnetic behavior of HTS materials was studied[5] by modeling the granular material as an array of weakly coupled, strongly superconducting anisotropic grains. Halbritter[2] explained the nonlinear behaviour of surface resistance by considering the granularity of superconductors and Nguyen et al.[1] obtained an empirical relation, in the low and intermediate power levels. But so far there is no exact models available to study the effect of input power over the kinetic inductance, which changes with the input field and results in the deterioration of the device performance.

In this paper we have modeled the nonlinear behaviour of kinetic inductance. This model can explain the shift in resonant frequency and surface resistance with the variations in input microwave power levels.

II. THEORY

In a superconductor, some ac losses are present due to the presence of normal state electrons which introduces surface resistance R_s. Here we can see a phase shift of 90^0 between I and V as the relaxation time becomes infinite. Since the value of R_s is small, a relation between current I and voltage V can be considered; $I = (jwL)^{-1}V$. This relation introduces an inductance,

L in a superconductor due to the inertial motion of the normal electrons. This is the kinetic inductance which is small at low microwave power levels due to the small concentration of quasi particle. When we expose a superconducting sample to an rf field weak links allow flux penetration as fluxons. The penetration of fluxons at the grain boundaries increases the quasi particle concentration, which introduces the resistive losses and increases the penetration depth. Thus the kinetic inductance increases. The increase in kinetic inductance,

$$\Delta L_k = \mu \Delta \lambda_L \quad (1)$$

where $\Delta \lambda_L$ represents the change in penetration depth.

The surface resistance also changes with the change in penetration depth. The change in surface resistance, ΔR_s, can be calculated using $\Delta \lambda$. Neglecting the higher powers of $\Delta \lambda$, ΔR_s can be expressed by a relation[7],

$$\Delta R_s = \frac{3}{2} \mu^2 \omega^2 \lambda^2 \sigma_1 \Delta \lambda \quad (2)$$

The granular superconductors are characterized by the grain boundary resistance, R_{bn}. This grain boundary resistance and the normal surface resistance, R_s are acting as a serial combination. Inorder to study the device performance we have considered here a superconducting microstrip resonator. By considering all the above factors, we have obtained an equivalent circuit for the microstrip resonator where L and C represents the equivalent inductance and capacitance of the resonator, R_{bn}, the grain boundary resistance, L_k', the kinetic inductance at zero field and ΔL_k the change in kinetic inductance.

It has been observed by several workers[2-4] that the increase in surface resistance shows a quadratic variation of rf field (i.e., $\Delta R_s \propto H^2$) at low and intermediate power levels. We have modeled the nonlinear surface resistance by a generalized way[7] by considering the decrease in current density with the increase in rf field. ie., ΔR_s,

$$\Delta R_s = R_s(0) b_r H_{rf}^q \quad (3)$$

Using eqs. (1) and (3),

$$\Delta \lambda = \frac{2 R_s(0) b_r H_{rf}^q}{3 \mu^2 \omega^2 \lambda^2 \sigma_1} \quad (4)$$

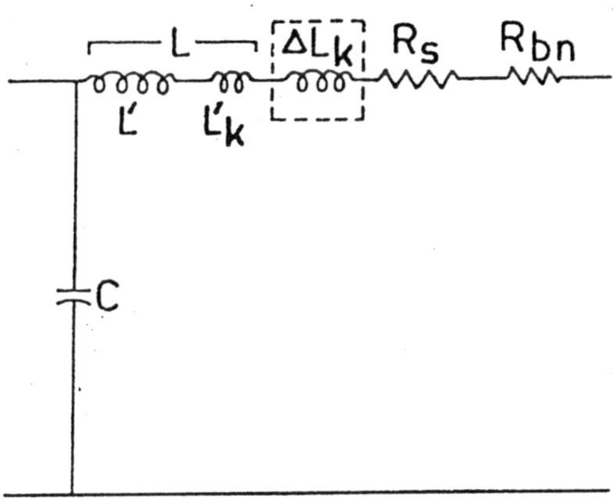

Fig.1. Equivalent circuit for microstrip resonator.

$$\therefore \quad \Delta L_k = \frac{2R_s(0)b_r H_{rf}^q}{3\mu\omega^2\lambda^2\sigma_1} \tag{5}$$

$$\Delta L_k \propto b_1 H_{rf}^q \tag{6}$$

The kinetic inductance at zero rf field,

$$L_k' = \mu\lambda \tag{7}$$

Therefore the total kinetic inductance L_k,

$$L_k = L_k' + \Delta L_k \tag{8}$$

$$= L_k'[1 + \frac{b_r}{3} H_{rf}^q] \tag{9}$$

III. RESULTS and DISCUSSION

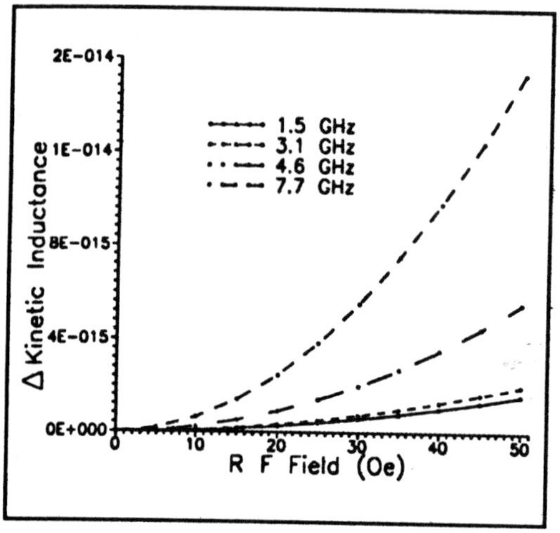

Fig.2. The change in kinetic inductance with rf field.

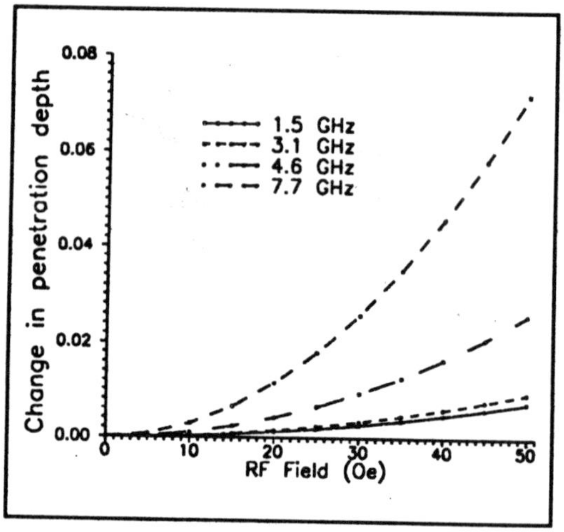

Fig.3. Change in penetration depth with the rf field.

Figure 2 shows the variation of kinetic inductance with the applied input rf field. The parameter b_l is calculated for different frequencies from the reported experimental results. The change in kinetic inductance is estimated using the value of b_l. A quadratic fit to eq.(6) (q = 2) can explain the experimental results of shift in resonant frequency with the field observed by many workers[7,9,10]. At very low power levels the variations are too small to detect. But as the power increases the deviations become higher and hence the device performance shows a drastic decrease. When the applied frequency is less, then the shift in kinetic inductance become negligible even at intermediate power levels. But as the frequency increases the shift become more and more prominent. We have carried out calculations for low and intermediate (<50 Oe) fields. Above this field the increase in kinetic inductance is much faster than H^2. To explain this variation another model has been proposed by us[8]. The change in kinetic inductance can be used for calculating shift in penetration depth at various input microwave power levels. Figure 3 shows the shift in penetration depth with the rf power levels. The nature of variation of both the kinetic inductance and penetration depth are same and shows a quadratic rf field dependence. These results are in very close to the results of Oates et al.[9,1].

It has been observed that $\Delta\lambda \propto \Delta\omega$ [4,7]. Therefore the above results can be used to calculate the change in resonant frequency with the input power levels as observed by Pinto et al[10]. The granularity of high T_c superconductors and the weak coupling between the grains enforces the nonlinear behaviour. A suitable selection of substrate, method of deposition and the conditions of deposition of the superconducting film can effectively reduce the nonlinear behaviour.

ACKNOWLEDGMENT

The authors are thankful to the Department of Science and Technology, Govt. of India for the financial support which they obtained to carry out these investigations.

REFERENCES

1. P.P. Nguyen, D.E.Oates, G.Dresselhaus and M.S. Dresselhaus, Phy. Rev. B**48**, 6400 (1993).
2. J. Halbritter, J. Appl. Phys. (1992).
3. S. Sridhar, Appl. Phys. lett. (1994).
4. P. P. Nguyen, D. E. Oates, G. Dresselhaus, M. S. Dresselhaus and A. C. Anderson, Phy. Rev. B **51**, 6686 (1995).
5. J. R. Clem, Physica C, **153**, 50 (1988).
6. Van Duzer, *Principles Superconducting Devices and Circuits*, Edward Arnold.
7. G. P. Srivastava, Mohan V. Jacob, M. Jayakumar, P. K. Bhatnagar and N. D. Kataria, (to be published).
8. G. P. Srivastava and Mohan V. Jacob (to be published).
9. D. E. Oates, P. P. Nguyen, G. Dresselhaus, M. S. Dresselhaus and C. C. Chin, IEEE Trans. Appl. Supercond. **3**, 1114 (1993).
10. R. Pinto et.al., Solid State Comm. **84**, 1107 (1992).

PENETRATION DEPTH MEASUREMENT NEAR T_c IN $YBa_2Cu_3O_{7-\delta}$ THIN FILMS: ROLE OF CLASSICAL FLUCTUATIONS IN ORDER PARAMETER PHASE

P.R. Apte, V.V. Srinivasu and R. Pinto

Tata Institute of Fundamental Research, Homi Bhabha Road, Mumbai 400 005, India.

We present experimental data and analysis on the microwave penetration depth measured at temperatures close to T_c in a $YBa_2Cu_3O_{7-\delta}$ thin film. The penetration depth varies linearly with temperature, giving evidence for a possible role of classical fluctuations in the order parameter phase.

INTRODUCTION

Microwave studies of High temperature superconductors has become an important area of research[1] because microwaves can probe both the intrinsic and extrinsic properties of these short coherence length superconductors. Extrinsic properties like granularity and percolation, their temperature and magnetic field dependence is very well probed by microwaves[2,3] . Also it was shown that microstructure has significant effect on the penetration depth and also on the shielding properties of Ag-doped and pure $YBa_2Cu_3O_{7-\delta}$ thin films[4,5]. On the other hand, experimental results on both temperature dependence[6] and magnetic field dependence[7] of penetration depth can be named as two outstanding examples illustrating the power of microwave technique to probe the intrinsic properties. The experiments are used to distinguish and also confirm d_{x2-y2} state from the possible symmetries in high-T_c superconductors. These experiments are done at very low temperatures close to 4 K. However at higher temperatures classical fluctuations in the phase of the order parameter can play a role in the temperature dependence of the penetration depth[8,9] . These phase fluctuations are taken into account in a Josephson-coupled uniaxial superconductor model by Coffey[8] giving rise to a linear temperature dependence of the change of penetration depth. However, along with Josephson coupling between the CuO planes, there can be coupling at twin boundaries. In fact, twin boundaries become active weak links in a temperature range of 1K-2K below T_c[10]. Therefore, measurement of penetration depth $\lambda(T)$ in a narrow range of temperature 1K - 2K below T_c gives significant information about the role of classical fluctations in the order parameter phase. In this report, we present experimental data on the microwave penetration depth measured at temperatures very close to T_c in a $YBa_2Cu_3O_{7-\delta}$ thin film and analyse in terms of the role of classical fluctuations in order parameter phase.

EXPERIMENTAL

$YBa_2Cu_3O_{7-\delta}$ thin films are insitu grown by Pulsed Laser Deposition (PLD) technique on $LaAlO_3$ substrates at 720 °C. An excimer laser (KrF, 248 mm) was used for the insitu growth of the films. The films were ~ 3000 A thick. Structural characterization of the films was carried out by XRD and the films are found to be c-axis oriented. The thin films used in this experiment are moderately good with a T_c ~ 88 K. Details of thin film deposition are given elsewhere[11,12]. The temperature controller is programmed so that it gives a sinusoidal temperature variation with time and with a period of say 17 minutes, as shown in Fig.1. The need for this approach is based on the fact that the range of temperature is decided to be 1 K on either side of T_c. And when one wants many data points in 1 K range of temperature, usual method of setting the temperature at a fractional value of 1 K and waiting for hours to get thermal equilibrium takes too long. Also a sinusoidal variation of temperature with time helps in getting many data points in 1 K interval and also good signal averaging is possible.

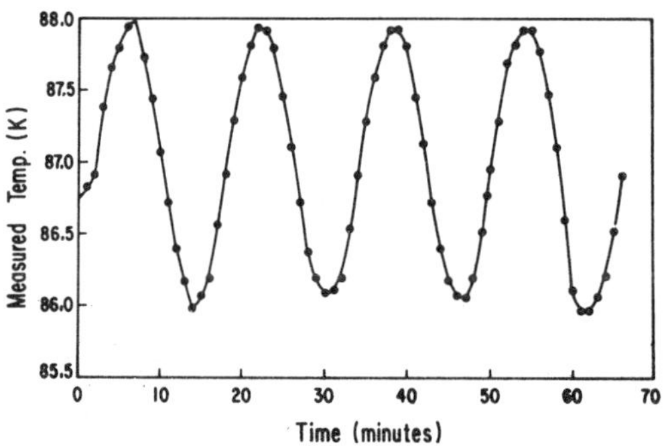

Fig.1. Sinusoidal variation of temperature with time. The solid line is guide to eye.

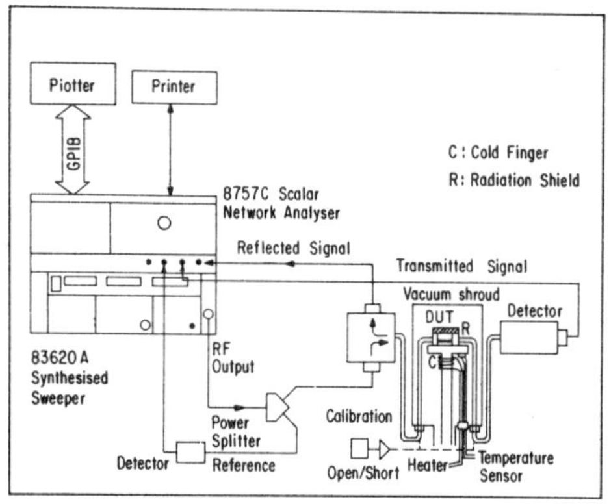

Fig.2 Block diagram of microwave measurement set up.

Microstrip resonator technique, which is based on a finite length of a superconducting microstrip transmission line seperated by capacitive gaps of the input/output ports, is used for the measurement of penetration depth (λ) of superconducting films. The block diagram of microwave measurement set up is shown in Fig.2. Figure.3 shows a schematic of microstrip resonators in C- and X- band frequencies along with a microwave test jig. When temperature is near T_c the resonance characteristics (the bandwidth) is linear to temperature difference from T_c. This results in a truncated sinusoidal Q vs T_c-T response. The extrapolated truncated position by curve fitting gives a nice sinusoidal Q vs T_c-T response. Details of measurement of resonant frequency and the phase velocity and calculation of the penetration depth are reported elsewhere[4].

RESULTS AND DISCUSSION

Figure. 1 shows the very good sinusoidal variation of temperature with time achieved. This enables us to measure the microwave penetration depth in the temperature range ~ 1K near T_c , acurately. Fig.4 shows the normalised change in penetration depth $\Delta\lambda/\lambda(0)$ vs T, over a temperature range of 1 K just below T_c. $\Delta\lambda/\lambda(0)$ is showing a linear depedence on temperature. The linear dependence of $\Delta\lambda/\lambda(0)$ on temperature very close to T_c can be explained in the following.

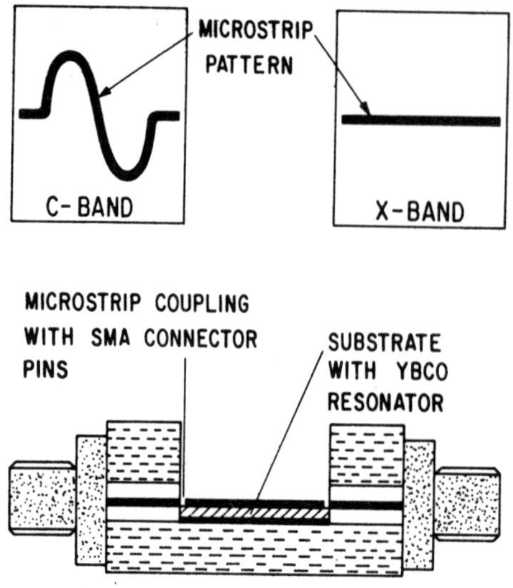

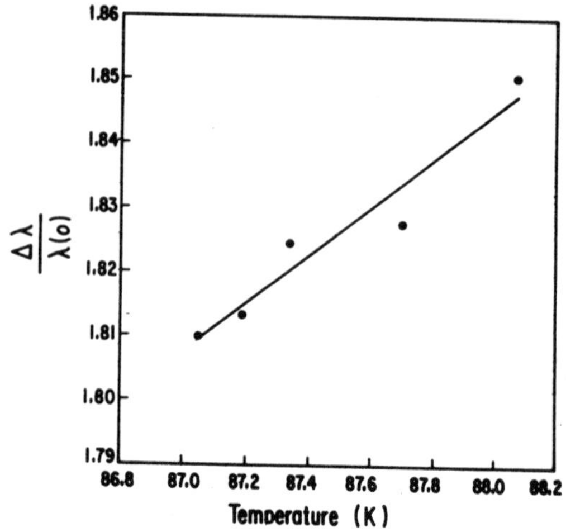

Fig.3 Schematic of microstrip resonators in C- and X- band frequencies along with a microwave test jig.

Fig.4 Linear temperature dependence of $\Delta\lambda/\lambda(0)$ vs Temperature. The solid line is a linear fit.

Coffey[8] has calculated the effect of classical fluctuations in the order parameter phase taking into account the Josephson coupling between CuO planes in these cuprate superconductors and predicted a linear temperature dependence of $\Delta\lambda/\lambda(0)$ at high temperatures in the limit $h\omega_p << 2K_BT$ where classical fluctuations dominate. We extrapolate the argument of Coffey[8] to the twin boundaries as well. It is well established[10] that twin boundaries become active josephson junctions in a temperature range ~ 1K-2K just below T_c. Therefore, classical fluctuations in the order parameter phase can also occur at twin boundaries and should lead to a linear temperature dependence of $\Delta\lambda/\lambda(0)$, measured at ~ 1K-2K just below T_c. Our experimental results of $\Delta\lambda/\lambda(0)$ vs T exactly give the predicted linear temperature dependence giving rise to an evidence for a possible role of classical fluctuations in order parameter phase, as calculated by both groups namely Coffey[8], and Roddick and Stroud[9].

CONCLUSION

Measurement of microwave penetration depth very close to T_c exploiting a programmable temperature controller gives an evidence for a possible role of classical fluctuations in order parameter phase.

REFERENCES

1. Microwave Studies of High Temperature Superconductors, Studies of High Temperature Superconductors, Ed. Anant Narlikar, Vol. **17,** Nova Science Publishers Inc. (1996).
2. S.V. Bhat, V.V. Srinivasu and N. Kumar, Phys. Rev. B **44**, 10121 (1991).
3. V.V. Srinivasu, S.V. Bhat and N. Kumar, Solid State Commun. **89**, 375 (1994).
4. R. Pinto, Davinder Kaur, M.S. Ramachandra Rao, P.R. Apte, V.V. Srinivasu and R. Vijayaraghavan, Appl. Phys. Lett. **68**, 1720 (1996).
5. V.V. Srinivasu, R. Pinto and M.D. Sastry, Applied Superconductivity (in press).
6. W.N. Hardy, D.A. Bonn, D.C. Morgan, R. Liang and K. Zhang, Phys. Rev. Lett. **70**, 3999 (1993).
7. A. Maeda, Y. Ino, T. Hanaguri, N. Motohira, K. Kishio and T. Fukase, Phys. Rev. Lett. **74**, 1202 (1995).
8. M.W. Coffey, Physica C **235-240**, 1961 (1994).
9. Eric Roddick and David Stroud, Physica C **235-240**, 1799 (1994).
10. A. Dulcic, R.H. Crepeau and J.H. Freed, Phys. Rev. B **38**, 5002 (1988).
11. D. Kumar, M. Sharon, R. Pinto, P.R. Apte, S.P. Pai, S.C. Purandare, L.C. Gupta and R. Vijayaraghavan, Appl. Phys. Lett. **62**, 3522 (1993).
12. R. Pinto, S.P. Pai, C.P.D'souza, L.C. Gupta, R. Vijayaraghavan, D. Kumar and M. Sharon, Physica C **196**, 264 (1992).

OBSERVATION OF RF-SQUID EFFECT IN Hg(Tl) BaCaCuO HIGH Tc THIN FILM UPTO 117.5K

Neeraj Khare[1], A.K. Gupta[1], Sangeeta Khare[1], H.K. Singh[2], A.K. Saxena[2] and O.N. Srivastava[2]

[1]*National Physical Laboratory, New Delhi 110 012, India*
[2]*Physics Department, Banaras Hindu University, Varanasi 221 005, India*

Tl-doped Hg based high T_c superconducting film was prepared using two step process. In the first step, the precursor film $Ba_2Ca_{2.2}Cu_{3.3}O_x$ was synthesized using spray pyrolysis technique and in the second step, the precursor film was annealed in Hg-Tl atmosphere. The resulting film exhibits T_c (R=0) at 123K. A rf-SQUID has been fabricated on the film using naturally present grain boundary junctions. The SQUID characteristics have been observed upto 117.5K. This is the highest temperature at which SQUID effect has been reported. The flux noise density, $\sqrt{S_\varnothing}$ at 1 Hz of the rf-SQUID was ~2.5 x 10^{-3} $\varnothing_0 / \sqrt{Hz}$ at 77K and its value is ~4 x 10^{-3} $\varnothing_0 / \sqrt{Hz}$ at 114K.

I. INTRODUCTION

The discovery of superconductivity in Hg-Ba-Ca-Cu-O system with the maximum transition temperature (Tc) of 133K under ambient pressure has aroused a lot of enthusiasm in this field[1,2]. Many efforts have been recently made to synthesize good quality films of Hg-Ba-Ca-Cu-O[3,4]. Since the transition temperatures of these Hg-based superconducting films are much higher than the other available high-Tc systems, it would be very interesting to fabricate and study SQUIDs using these films. The highest temperature upto which SQUID characteristics has been observed in Y-Ba-Cu-O is 87K[5], in Bi(Pb)-Sr-Ca-Cu-O, it is 101K[6] and in Tl-Ba-Ca-Cu-O, the temperature is 101K[7]. Recently dc-SQUID characteristics in Hg-Ba-Ca-Cu-O thin film have been reported[8] upto 111.8K with flux noise density of ~$10^{-3}\Phi_o/\sqrt{Hz}$ at 1Hz and 77K. In this paper, we report the performance of Hg(Tl)-Ba-Ca-Cu-O thin film rf-SQUID. The rf-SQUID characteristics have been observed upto 117.5K. The flux noise density of this SQUID even at 114K is found to be $4\times10^{-3}\Phi_o/\sqrt{Hz}$ at 1Hz.

II. EXPERIMENTAL

Tl doped Hg-1223 ($Hg_{0.8}Tl_{0.2}Ba_2Ca_2Cu_3O_{8+\delta}$) high-$T_c$ superconducting film was synthsized by a two step process. In the first step, a precursor film of $Ba_2Ca_{2.2}Cu_{3.3}O_x$ was deposited on MgO (100) substrate using spray pyrolysis technique. In the second step, the precursor film was annealed in Hg-Tl atmosphere at 860°C for 12 hr and then slowly cooled to room temperature. The experimental details of thin film preparation have been described elsewhere[9]. RF-SQUID pattern consisting of a hole shunted with a microbridge was patterned on the film. The diameter of the hole was 300μm and the dimension of the microbridge was $150\times100\mu m^2$. An eight turn flat spiral coil wound from 38 SWG copper wire was glued on the film. This coil was connected to another sixty turn copper coil to form inductance of the tank circuit whose resonance frequency and Q were 18.9 MHz and 37 respectively at 77K. An rf-oscillator was connected to a tank circuit to provide a rf-bias to the SQUID. The reflected rf-signal from the tank circuit was amplified by a low noise amplifier and was detected using a diode detector. The gain of the amplifier was 60 dB. For observing voltage-flux characteristic of the rf-SQUID, an ac signal was applied to the SQUID through the tank circuit using an audio frequency oscillator. The detector output was connected to the Y-channel of a CRO while the output from the af-oscillator was connected to the X-channel of the CRO. A cylinder consisting of three layers of μ-metal was used to shield the SQUID from external noise. The noise spectrum of the SQUID was recorded using a lock-in-amplifier and a dynamic signal analyzer. A modulation signal of 400Hz was applied to the SQUID through the tank circuit coil. The amplitude of the modulation signal was adjusted so that the current through the coil was less than $I(\Phi_o)/2$, where $I(\Phi_o)$ was the current in the coil required to change the flux in the SQUID loop by one flux quantum. The detector output was connected to the input of the lock-in-amplifier, whose output was connected to the dynamic signal analyzer for recording a voltage noise spectrum, $\sqrt{S_v(f)}$. The flux noise density was calculated using the formula

$$[S_\Phi(f)]^{1/2} = [S_v(f)/(\delta V/\delta \Phi)^2]^{1/2}$$

where ($\delta V/\delta \Phi$) was the transfer function of the SQUID.

III. RESULTS

The XRD pattern of the film (figure 1) shows that the film is almost single phase Hg(Tl)-1223.

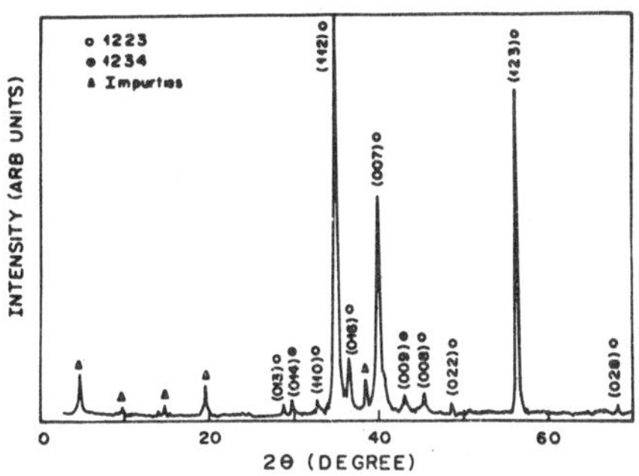

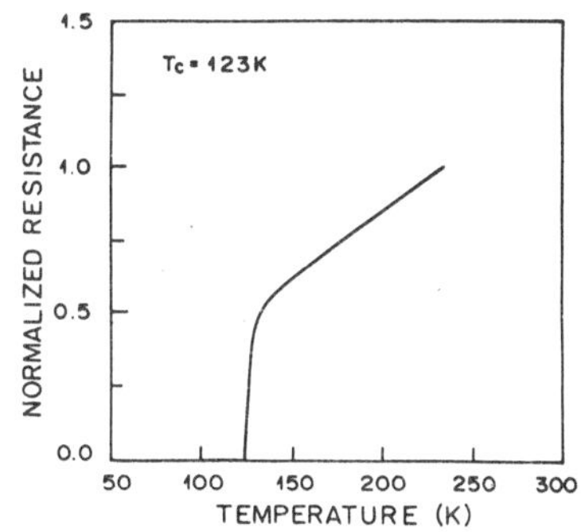

Fig.1. (Left) XRD pattern of Hg(Tl)-Ba-Ca-Cu-O superconducting film (impurity peaks correspond to ($CaHgO_2$).

Fig.2. (Right) Normalized resistance [R(T)/R(235)] versus temperature curve of Hg(Tl)-1223 film.

However, some weak reflections from 1234 phase are also observed. In addition, peaks belonging to insulating phase $CaHgO_2$ are also seen in the XRD pattern. The lattice parameters of the tetragonal unit cell are a=3.857Å and c=15.795Å. The plot of normalized resistance [R(T)/R(235)] versus temperature of the film is shown in figure 2. The onset of transition is around 133K and

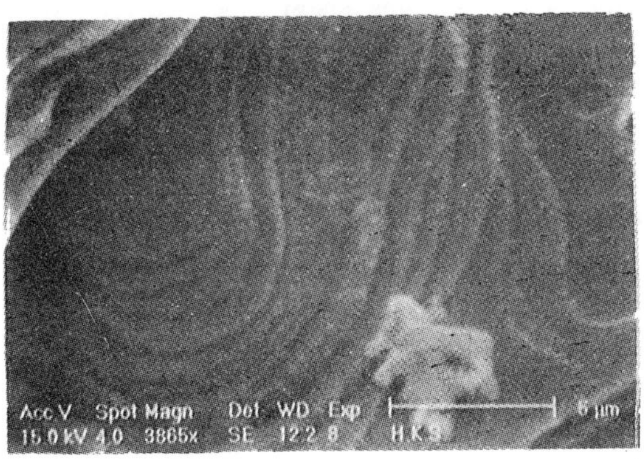

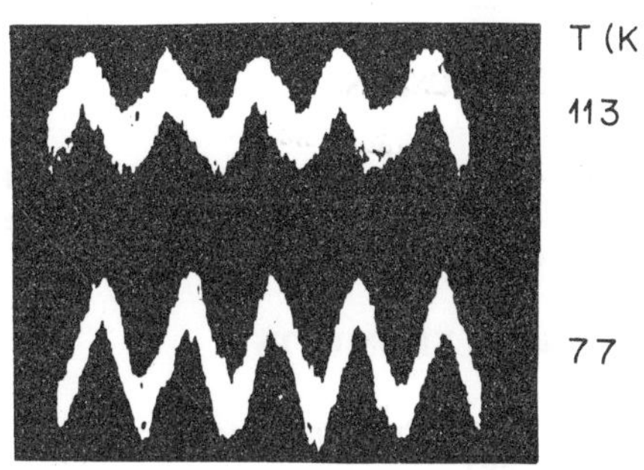

Fig.3. (Left) A SEM micrograph of the Hg(Tl)-1223 film.

Fig.4. (Right) Voltage-flux characteristics of the Hg(Tl)-1223 film rf-SQUID at temperature 77K and 113K. Peak to Peak voltage of the V-Φ modulation is 5mV at 77K and 3.4mV at 113K.

T_c (R=0) occurs at 123 K. The dc transport critical current density (J_c) of the film has been found to be 10^4 A/cm^2 at 77K. The microstructure of the film was obtained using scanning electron microscope. The microstructure (figure 3) shows the presence of large platelike grains indicating melt assisted growth of the superconducting phase. Figure 4 shows V-Φ characteristics of Hg(Tl)-1223 film rf-SQUID at two operating temperatures: 77K and 113K. The peak to peak voltage of V-Φ modulation at the detector output was 5mV at 77K which corresponded to a transfer function of 10 mV/Φ_o. The value

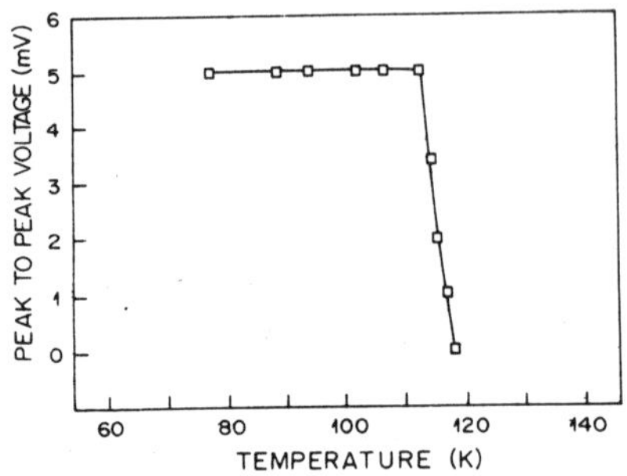

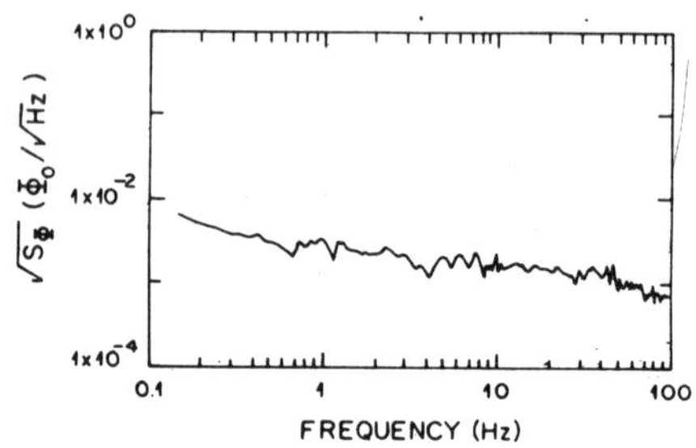

Fig.5. (Left) Variation of the peak to peak amplitude of voltage-flux modulation of Hg(Tl)-1223 film rf-SQUID as a function of temperature.

Fig.6. (Right) The flux noise spectrum of Hg(Tl)-1223 film rf-SQUID at 77K.

of peak to peak amplitude at 113K was 3.4mV. Figure 5 shows variation of peak to peak voltage of V-Φ modulation of the SQUID with temperature. Peak to peak voltage remains almost constant upto 112K, and then it starts decreasing. It was possible to observe the V-Φ modulations upto 117.5K which is higher than the earlier reported temperature at which SQUID behaviour has been observed. Earlier, the dc-SQUID characteristics in Hg-1223 film was observed upto 111.8K [8].

Figure 6 shows flux noise spectrum of Hg(Tl)-1223 film rf-SQUID at 77K. The value of $\sqrt{S_\Phi}$ at 1Hz is found to be $2.5\times10^{-3}\Phi_o/\sqrt{Hz}$. The noise performance of the SQUID has been studied at different operating temperatures. The value of $\sqrt{S_\Phi}$ at 1Hz remains almost same upto 112K and its value increases to $4\times10^{-3}\Phi_o/\sqrt{Hz}$ at 114K.

IV. CONCLUSION

Hg(Tl)-1223 thin film rf-SQUID has been fabricated using natural grain boundary junction. The rf-SQUID characteristics have been observed upto 117.5K. Flux noise density, $\sqrt{S_\Phi}$ (1 Hz) is found to be $4\times10^{-3}\Phi_o/\sqrt{Hz}$ at 114K.

ACKNOWLEDGMENTS

The authors are grateful to Prof. E.S.R. Gopal, Director NPL for encouragement. Two of the authors, HKS and AKS are grateful to the University Grants Commission, New Delhi and Council of Scientific and Industrial Research, New Delhi, for providing research fellowship and research associateship respectively. This work has been partially supported by the Department of Science and Technology, New Delhi.

REFERENCES

1. S.N. Putilin, E.V. Antipov, O. Chmaissem and M. Marezio, Nature **362**, 226 (1993).
2. A. Schilling, M. Cantoni, J.D. Guo and H.R. Ott, Nature **363**, 56 (1993).
3. C.C. Tsuei, A. Gupta, G. Trafas and D. Mitzi, Science **263**, 1259 (1994).
4. S.H. Yun and J.Z. Wu, Appl. Phys. Lett. **68**, 862 (1996); X.S. Wu, Y.M. Cai, L.J. Shen and Z. Wu, Appl. Phys. Lett. **68**, 1723 (1996).
5. R. Gross, P. Chaudhari, M. Kawasaki, M.B. Ketchen and A. Gupta, Appl. Phys. Lett. **57**, 727 (1990).
6. Y. Zhang, Y. Xu and C. Heiden, Appl. Phys. Lett. **56**, 1579 (1990).
7. N. Khare, A.K. Gupta, A.K. Saxena, K.K. Verma and O.N. Srivastava, Supercond. Sci. Technol. **7**, 402 (1994).
8. A. Gupta, J.Z. Sun and C.C. Tsuei, Science **265**, 1075 (1994).
9. H.K. Singh, A.K. Saxena and O.N. Srivastava, Physica C **262**, 7 (1996).

CRITICAL CURRENTS AND MICROSTRUCTURAL DEFECTS IN HIGH T_c THIN FILMS OF $YBa_2Cu_3O_{7-x}$ PREPARED BY DC GLOW DISCHARGE SPUTTERING

P. Kuppusami and V.S. Raghunathan

Physical Metallurgy Section, Metallurgy Division,
Indira Gandhi Centre for Atomic Research, Kalpakkam 603 102, India

The transport properties of thin films of $YBa_2Cu_3O_{7-x}$ grown on (100) MgO substrates by dc glow discharge sputtering is reported. The transport critical current densities and the broadening in the resistive transition as a function of applied field and the anisotropy in the H_{c2} values are discussed. The paper also highlights some of the microstructural defects in thin films.

1 INTRODUCTION

A large critical current density obtained in thin films of $YBa_2Cu_3O_{7-x}$ (YBCO) has been attributed to the epitaxial growth and the imperfections present in the films [1]. However, the defects can act basically in two ways. Distribution of nanometre scale size imperfections provide the necessary flux pinning, while the extended defects such as grain boundaries can limit the flow of supercurrent [2] . For efficient flux pinning, it is advantageous to have a distribution of structural defects whose characteristic lengths of structural and chemical variations may lie in the range of coherence length.

In the following paper, the study of transport properties and the microstructural features in thin films of YBCO is presented.

2 EXPERIMENTAL PROCEDURE

Thin films of YBCO grown on (100) MgO and YSZ substrates were prepared by glow discharge sputtering [3]. The transport critical current densities (J_{ct}) were measured in thin films of YBCO on (100) MgO by standard four probe resistivity technique using microbridges of 50 μm width [4]. The measurements of resistivity against the temperature in a d.c field up to 12 T were carried out in an helium exchange cryostat containing Nb-Sn/Nb-Ti superconducting magnets.

TEM samples were prepared by mechanical polishing down to ~ 50 μm thickness followed by argon ion milling at a voltage of 5kV and at a gun current of 0.5 mA [5]. The planar view of the samples was examined in a Philips EM 400T transmission electron microscope and HREM was carried out at 200kV in JEOL JEM 2000 EX-II transmission electron microscope.

3 RESULTS

3.1 Transport Properties

The values of J_{ct} measured in thin films on (100) MgO at different temperatures in zero magnetic field are shown in Fig.1. It is noticed that J_{ct} at 4.2K and 77K are 6×10^6 A/cm^2 and 5×10^5 A/cm^2 respectively. Since the films contain mostly mosaic grains [5] and therefore the boundaries behave like metallic and show superconductor-normal-superconductor (S-N-S) junctions [6]. Such junctions have been predicted to follow a power law behaviour : $J_c \propto (T_c-T)^n$ with n varying from 1-2 [7]. From the above Figure the value of n is found to be 1.2. Similar results were obtained in the films grown on $LaAlO_3$ prepared by the DC magnetron technique [8].

The measurements of the H_{c2} were carried out by tracing the resistive transitions as a function of temperature under the magnetic fields upto 12T in the films grown on (100) MgO. The temperature dependence of the electrical resistivity was measured with the applied field parallel to c-axis (H||c) as well as perpendicular to the c-axis ($H \perp c$) of the orthorhombic phase. The upper critical fields were plotted against the mid point transition temperatures and were shown in Fig.2. One of the important observation of these plots is that the resistive transition for Ha||c show larger broadening than that for $Ha \perp c$. That is, for the field in the a-b plane, the critical fields are much higher and the depression of T_{co} is correspondingly lower. In the linear regions, the plots show $dH_{c2}/dT = -4.21$ T/K for $H \perp c$ and $dH_{c2}/dT = -1.1$ T/K for H||c field direction.

The following equation given by Werthamer, Helfand, Hohenberg and Maki (WHHM) [9] with extrapolation to low temperature has been used to calculate the $H_{c2}(0)$ using the values of T_c mid-point:

$$H_{c2}(0) = -0.69\, T_{co}\, (dH_{c2}/dT) \qquad (1)$$

This yielded $H_{c2}(0)$ of 230 T for $H \perp c$ and 60 T for H||c directions. Thus the anisotropic ratio in H_{c2} between the two field directions is found to be about 4. This value compares well with the values reported on oriented films [10].

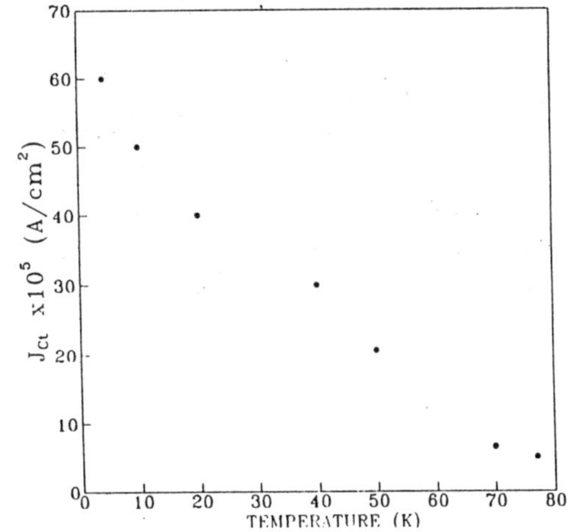

Fig. 1. Temperature dependence of Jct measured on the films grown on (100) MgO substrate.

3.2 Microstructural Features

TEM micrographs of the twins present in thin films grown on (100) MgO is shown in Fig. 3. In contrast to the bulk samples, the width of the twin lamellae are small(≈50-100 nm) but the twin density is rather high. Most of the grains have their c-axis parallel to the electron beam and thus the diffraction pattern from the grains correspond to [001] zone axis. The angular measurements between the grains were found to be about 2° indicating that they consist of low angle boundaries [5].

Fig. 4 is an HREM image of a cross-sectional c-axis oriented YBCO film grown (100) oriented YSZ with the electron beam parallel to the [001] direction. The inset diffraction pattern taken from an area including both film and substrate exhibits a homogeneous epitaxial growth with c-axis parallel to the normal of the substrate. The following cube-to-cube orientation is observed in the present work:

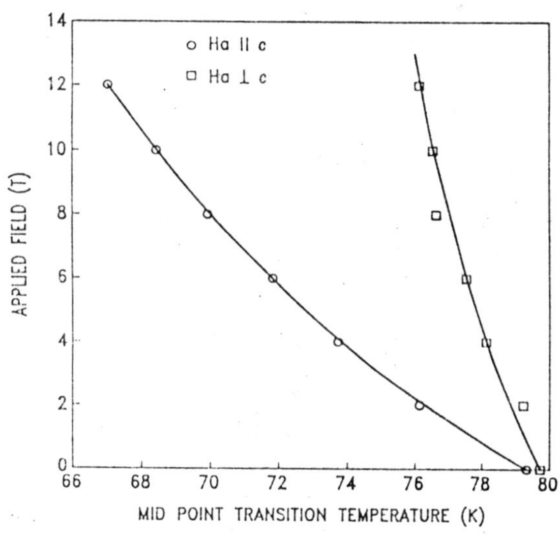

Fig 2. Hc_2 data with Ha||c and $Ha \perp c$ against the mid point transition temperatures.

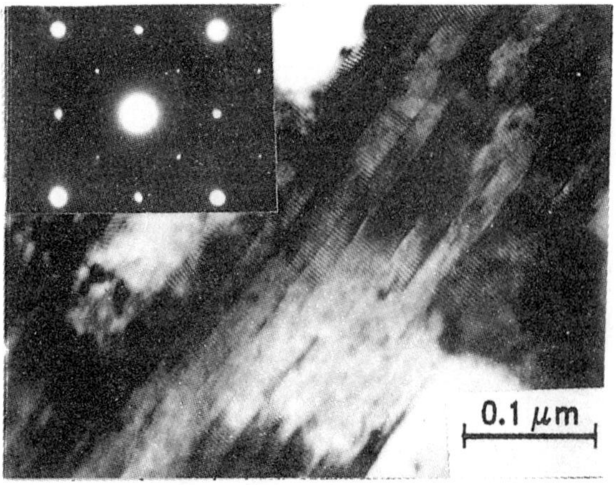

Fig. 3. Twins in c-axis oriented thin films of YBCO grown on MgO; inset is the SAD pattern ; Zone axis : [001].

$$(001)_{YBCO} \parallel (100)_{MgO}$$
$$[100]_{YBCO} \parallel [100]_{YSZ} \quad (2)$$

This orientation relationship tends to dominate at relatively high deposition temperatures.

Another most common defects observed in YBCO thin films are stacking faults and their associated partial dislocations (Fig.5). The faults are parallel to the interface on the (001) plane. It is suggested that they form as a consequence of insertion of extra Cu-O layers [1]. These defects could well provide the necessary flux pinning to explain the high critical current density in these materials.

4 CONCLUSIONS

The following major conclusions are drawn from the study of the transport properties and the microstructural features in thin films of YBCO:

(i) The best J_{ct} values obtained on the films grown on (100) MgO substrate at 4.2K is 6×10^6 A/cm^2 and at 77 K is $\sim 5 \times 10^5$ A/cm^2 and that for the films on $SrTiO_3$ is 2×10^6 A/cm^2 at 77 K, in zero magnetic field. Also, the temperature dependence of J_{ct} without the applied field is found to be proportional to $(T_c - T)^n$ with n = 1.2.

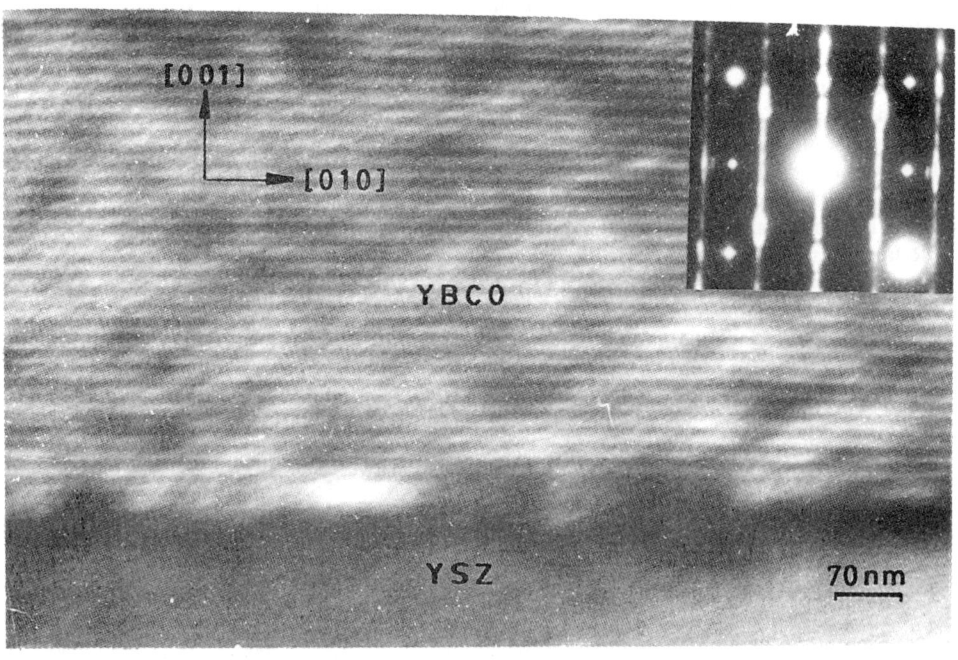

Fig. 4 Cross-sectional HREM of the YBCO film on (100) YSZ. The inset shows a typical SAD pattern taken from a region comprising both film and the substrate; zone axis : [100].

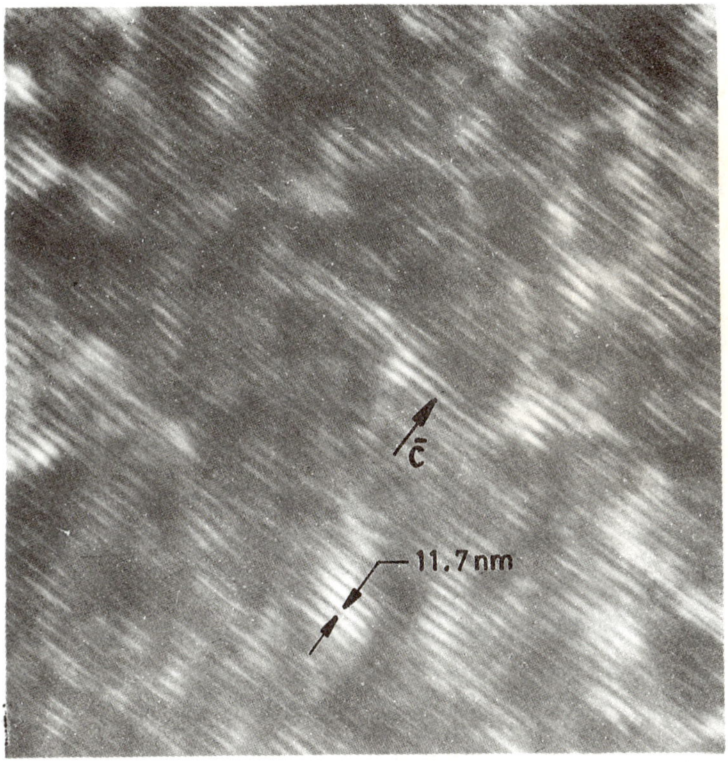

Fig. 5. Stacking faults on (001) planes in YBCO films on (100) YSZ

(ii) The broadening in the resistive transitions as a function of the applied field and the anisotropy in the H_{c2} values strongly support the different pinning mechanisms that are operating in these materials.

(iii) Electron microscopy techniques reveal mostly mosaic grains, twin boundaries and stacking faults. The films show a cube-to-cube orientation with the substrates under present investigation.

Acknowledgements

The authors gratefully acknowledge Dr. Alok Singh and Shri E. Mohandas for electron microscopy work. They also thank Dr. P. Rodriguez, Dr. Baldev Raj, Shri J. B. Gnanamoorthy and Dr. T.S. Radhakrishnan for their constant support and encouragements.

REFERENCES

1. D.Bhatt, S.N. Basu, A.C. Westerheim, A.C. Anderson, Physica C **222**, 283(1994).
2. K. Salama, V. Selvamanickam, L. Gao and K. Sun, Appl. Phys. Lett. **54**,2352 (1989).
3. P.Kuppusami, V.S.Raghunathan, A.G.Vedheswar and Y. Hariharan, Physica C **215**, 213 (1993).
4. P.Kuppusami, V.S.Raghunathan and A.G.Vedheswar, Trans.Indian Inst. Met.**48**, 389 (1995).
5. P.Kuppusami, Ph.D Thesis, Madras University (1995).
6. J.W.C. de Vries, M.A. Gijs, G.M. Stollman, T.S. Baller, and G.N.A. van Veen J. Appl. Phys. **64**, 426 (1988).
7. Michael Tinkham, in Introduction to Superconductivity, McGraw - Hill, Inc. New York, (1975).
8. C.G. Cui, C.C. Lam, B.C. Yang, X.P. Wang, P.C.W. Fung, S.T. Tang, S.L. Li, and X.N. Jing, Solid State Commun. **81**,75 (1992).
9. N.R. Werthamer, E. Helfand and P.C. Hohenberg, Phy. Rev.**147**,295 (1966).
10. P. Chaudhari, R.T. Collins, P. Freitas, R.J. Gambo, J.R. Kirkley, R.H. Koch, R.B. Laibowitz, F.K. LeGoues, T.R. McGuire, T. Penny, Z. Schlesinger, A.P. Segmellar, S. Foner and E.J. Mcnniff, Jr., Phys. Rev. **B 36**, 8903 (1988).

SUBGAP STRUCTURES IN Nb-AlO$_x$-Nb JOSEPHSON TUNNEL JUNCTIONS

Awadhesh Mani, M. P. Janawadkar, Y. Hariharan, R. Baskaran and T. S. Radhakrishnan

Materials Science Division
Indira Gandhi Centre for Atomic Research
Kalpakkam 603 102, India

Subgap structure at a bias voltage corresponding to odd submultiple of gap energy 2Δ is observed in the I-V characteristics of Nb-AlO$_x$-Nb Josephson tunnel junction at 1.2 K. It is argued that the observation of this structure at a voltage bias of 2Δ/3e in our junctions cannot be explained by mechanisms such as multiple particle tunneling and multiple Andreev reflections. A qualitative explanation for the occurrence of this structure has been offered within the framework of Josephson self coupling mechanism.

Introduction

Subgap structures (SGS) in the I-V characteristics of Josephson tunnel junctions (JTJ) have been the subject of intensive investigation for last several years. These structures occur in the I-V characteristics of JTJ at threshold voltages which correspond to gap energy 2Δ divided by an integer n. The presence of such SGS in the I-V characteristics of high quality Nb-AlO$_x$-Nb JTJ has attracted the attention of several groups in recent times. In careful measurements at fairly low temperature SGS are found to occur at one half, one third and in some cases at one fifth of the measured gap voltage 2Δ/e[1-6] and their observation has been attributed by different research groups to various mechanisms such as multiple particle tunneling (MPT)[7,8], multiple Andreev reflections (MAR)[9,11] and Josephson self coupling (JSC)[12,13]. In MPT two or more quasi-particles simultaneously and incoherently tunnel through the barrier and hence the tunneling probability for a n-particle tunneling process is proportional to $|T|^{2n}$, where $|T|^2$ is tunneling probability for single quasi-particle. The MAR process involves the tunneling of particles across the barrier via a reflection mechanism in which an electron like incident quasi-particle is partially reflected and partially transmitted as a hole like one under the influence of a discontinuity in superconducting pair potential at the two interfaces between the superconducting electrodes and the tunnel barrier. A process involving n Andreev reflections and n+1 barrier traversals results in a SGS at a bias voltage of 2Δ/ne (n even), while a process involving (n-1) Andreev reflections and n barrier traversals results in a SGS at a bias voltage of 2Δ/ne (n odd). Hence the SGS at a bias voltage of both 2Δ/2e and 2Δ/3e scale as $|T|^4$. In JSC process the photons emitted at Josephson frequency due to ac Josephson effect assist the tunneling of quasi-particles either directly or by breaking the copper pairs in the superconducting electrodes.

Tunnelling probability in this case is proportional to $|T|^2$. JSC mechanism involving tunnelling assisted by self emitted photons causes odd series of SGS[12] at threshold voltages corresponding to gap energy $2\Delta/(2n+1)e$.

A number of investigations on the SGS in Nb-AlO_x-Nb Josephson junctions exist in literature; however the question as to which mechanism is really responsible for the occurrence of these structures is far from being unambiguously resolved and the interpretation of the experimental results is still controversial. According to one view[4] the experimental SGS are generated by MAR, while another point of view[1,2] interprets these structures as a manifestation of MPT. In this paper we present the results of our investigation of SGS at $2\Delta/3e$ in Nb-AlO_x-Nb JTJ. Though the critical current density of our junctions is fairly low, our investigations reveal a pronounced SGS at $2\Delta/3e$. We further present a qualitative explanation for the occurrence of this odd subharmonic structure invoking JSC as the main operative mechanism in this case.

Experimental details

A JTJ has been fabricated using electron beam evaporation as the deposition technique. Sandwiches of Nb/Al-AlO_x-Al/Nb were deposited on the whole wafer by sequential electron beam evaporation of the metal films onto glass substrates. Over the base layer of niobium of thickness 250 nm aluminium film of total thickness 5 nm was deposited. The tunnel barrier was formed by exposing the freshly deposited aluminium layer to pure oxygen at a pressure of 133 Pa for 1 hr. A second layer of aluminium having a thickness of 4 nm was deposited over this tunnel barrier and finally a 150 nm thick Nb layer was deposited to form the trilayer. Using Selective Niobium Etching Process (SNEP) junctions having an area of 100 μm x 100 μm were fabricated. The fabrication procedure is described in detail elsewhere[14].

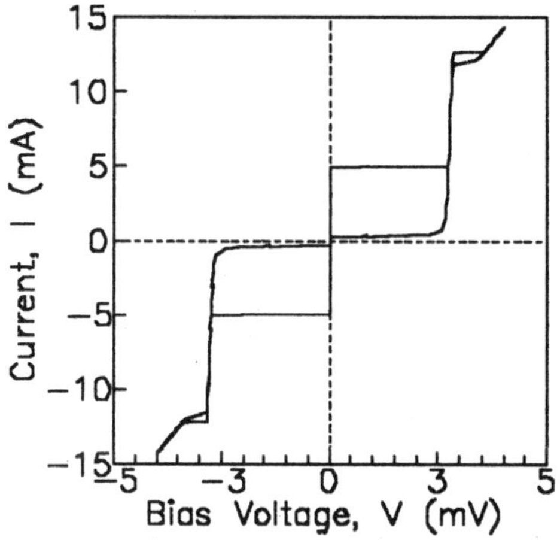

Fig.1. The I- V characteristics of the JTJ measured at 4.2 K

Measurements of the I-V characteristics of the junction were performed in a home made 1K cryostat insertable in a 50 mm neck commercial liquid helium dewar having a capacity of 100 ltr. Liquid Helium from the main dewar is drawn via a small capillary to fill a 1K pot ,which in turn is pumped by a mechanical pump of pumping capacity 25 lit/sec to produce a stable temperature of 1.2K. A superconducting solenoid is incorporated inside the cryostat in close proximity of 1K pot to provide a small magnetic field in order to suppress the Josephson current. The first and second derivatives of the I-V characteristics were simultaneously measured by using harmonic detection technique using an ac modulation voltage of 50 μV at a frequency of 571 Hz.

Results

Fig.1 shows the dc I-V characteristics of junction measured at 4.2 K. The quality parameter V_m of the junction, determined by taking product of critical current and the subgap resistance R_{sg} measured at a bias voltage of 2 mV, was found to have value of 60 mV. The product of R_n (normal resistance at a bias of 4 mV) and area A of junction, taken to provide a measure of the transparency of the tunnel barrier, was found to have values of 4.23×10^{-9} $\Omega.m^2$. The junction capacitance was deduced from the position of the first Fiske step[15] and found to be 260 pF. We estimate thickness of the oxide barrier to be 30Å from the junction capacitance using formula $C = \varepsilon_0\varepsilon_r A/d$. We have used ε_r=9 for the relative dielectric constant of AlO_x barrier[16]. Thus our junction has a relatively thick insulating layer which results in a low critical current density J_c of 50 A/cm^2. Fig.2 shows I-V characteristics of the junction along with the plots of dV/dI and d^2V/dI^2 vs bias voltage measured at 1.2 K in the absence of an external magnetic field. These characteristics exhibit high dynamic resistance in subgap regime. After applying a small magnetic field parallel to the plane of the junction , a pronounced SGS at 2Δ/3e emerges clearly in the I-V characteristics as shown in fig. 3. Plots of dV/dI and d^2V/dI^2 have also been shown. The position of the structure is determined by locating the minimum in dV/dI.

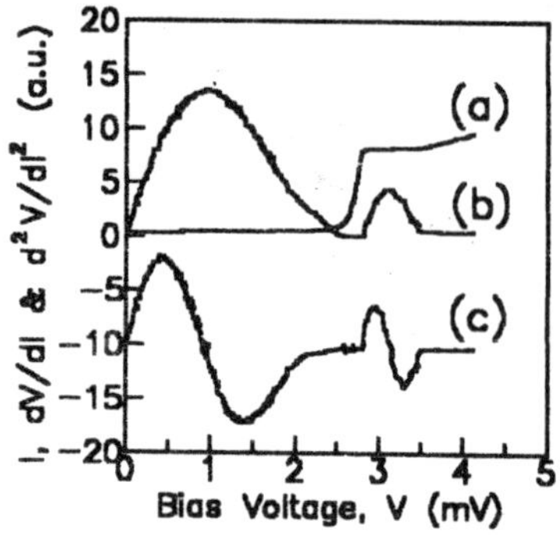

Fig. 2 (a) I-V characteristics, (b) dV/dI- vs. V and (c) d^2V/dI^2 vs. V measured at 1.2 K without external magnetic field. The curve (c) has been shifted arbitrarily along Y - axis for the sake of clarity

Discussion and Conclusion

Fig. 3 shows the appearance of a structure at a bias voltage of 0.91 mV which corresponds to a value of 2Δ/3e (2Δ/e= 2.72 mV). As mentioned earlier the effective thickness of the insulating barrier of our junction is quite high (30 Å) and the transparency is low. With such low barrier transparency, it is very unlikely that MPT is responsible for the observed structure since the tunnelling of three quasi-particles through a relatively thick insulating barrier will have a negligibly small tunneling probability $p \sim |T|^6$. Moreover there exists an appreciable amount of subgap current 300 nA near the 2Δ/3e structure. As reported by Foden et al[2]. the

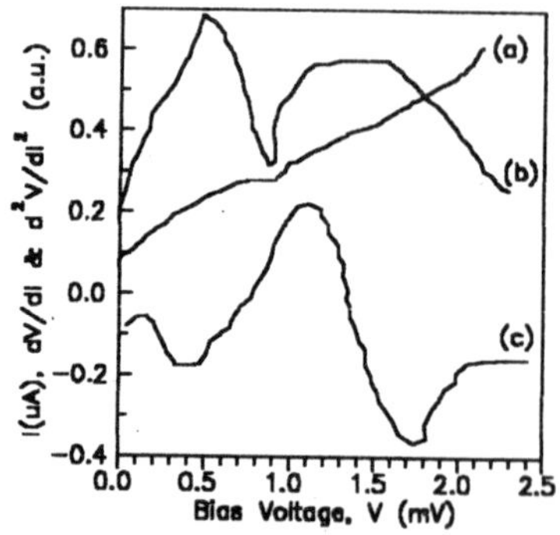

Fig. 3 (a) I-V characteristics; (b) dV/dI vs. Vand (c) d^2V/dI^2 vs. V measured at 1.2 K in presence of a small magnetic field . The curve (c) has been shifted along the Y - axis for the sake of clarity.

magnitudes of the current steps corresponding to $2\Delta/3e$ structure in junctions with tunnel barriers thinner than ours range between only 2 nA to 10 nA. In such a situation it would be almost impossible to detect the small excess current due to incoherent tunnelling of three independent particles as envisaged in the MPT mechanism. Further the absence of a SGS at a bias voltage of $2\Delta/2e$ argues against the MPT or MAR mechanisms. The mechanism involving JSC thus emerges as the only mechanism consistent with our experimental observations. Our results are in general agreement with the observations of Maezawa et al[6]. whose data has shown reasonable agreement with the predictions of Werthamer's microscopic theory[12] invoking JSC.

In conclusion it appears that the mechanism responsible for the observed SGS in our junctions is Josephson self coupling. Further investigations are presently underway to unequivocally establish the origin of the SGS by carrying out a detailed quantitative analysis.

Acknowledgement

We are grateful to Dr. H. K. Sahu for his useful suggestions and discussions at various stages of this work.

References

[1]C. L. Foden, N. Rando, A van Dordrecht, and A. Peacock, Phys. Rev. **B47**, 3316(1993)

[2]R. Cristiano, L. Frunzio, R. Monaco, C. Nappi, and S. Pagano, Phys. Rev. **B49**, 427(1993).

[3]A. W. Kleinsasser, R. E. Miller, W. H. Mallison, and G. B. Arnold, Phys. Rev. Lett. **72**, 1738(1994)

[4]E. C. O. Kirk, M. G. Blamire, R. E. Somekh, J. E. Evetts, D. Van Vechten and M. N. Lovellette IEEE Trans. Magn. **MAG-27**, 3137(1991).

[5]D. J. Adelerhof, E. P. Houwman, P. B. M. Fransen, D. Veldhuis, J. Fokstra, and H. Rogalla, IEEE Trans. Magn. **MAG-27**, 3153(1991).

[6]M. Maezawa , M. Aoyagi, H. Nakagawa, I. Kurosawa, and S. Takada IEEE Trans. Appl. Superconductivity **5**, 3073(1995).

[7]J. R. Schrieffer, and J. W. Wilkins, Phys. Rev. Lett. **10**, 17(1963).

[8]J. W. Wilkins, *Tunneling Phenomena in Solids* (Plenum, New York, 1969).

[9]A. F. Andreev, Sov. Phys. JETP **19** 1228(1964)

[10]G. B. Arnold, J. Low Temp. Phys. **68,** 1(1987)

[11]T.M. Klapwijk, G. E. Blonder, and M. Tinkham, Physica **109&110B**, 1657(1982)

[12]M. R. Werthamer, Phys. Rev. **147**, 255(1966).

[13]A. B. Zorin, K. K. Likharev, and S. I. Turovets, IEEE Trans. Magn. **MAG-19**, 629(1983).

[14]M. P. Janawadkar, R. Baskaran, K. Girieesan, Rita Saha, L. S. Vaidhyanathan, and T. S. Radhakrishnan, Japn. J. Appl. Phys. **33**, L1662(1994).

[15]K. Gireesan, M. P. Janawadkar, R. Baskaran, Rita Saha, L. S. Vaidhyanathan, and T. S. Radhakrishnan, Proc. DAE Symp. Solid State Physics Bombay, India (1993)p363.

[16]S. Morohashi, and S. Hasuo, J. Appl. Phys. **61**, 4835(1987)

ANALYSIS OF MICROWAVE SURFACE RESISTANCE OF YBCO THIN FILMS BY STRUCTURAL DEPENDENT TWO FLUID MODEL

Mukul Misra, N.D. Kataria, M. Jayakumar and A.K. Gupta
Superconductivity Division, National Physical Laboratory, New Delhi-110012

An analysis of microwave surface resistance of high T_c thin film based on modified two fluid model is presented. It incorporates the structural anomalies due to grains and grain boundaries in the material. The experimental surface resistance values of the films, having different structural morphology are fitting well with this theory.

I. INTRODUCTION

Characterization of high T_c superconducting (HTS) materials at microwave frequencies are necessary to investigate and verify the usefulness of its application with ultimate performance compared to that of non superconducting materials. This type of measurements are also a good probe for understanding the basic mechanism of superconductivity. Information concerning the value and anisotropy of the energy gap, the symmetry of superconducting order parameters and the mechanism of energy dissipation, for low T_c superconducting materials, could be deduced[1]. But, in granular HTS materials behaviour is masked by grain boundary weak links and grain misorientation effects[2]. Surface resistance, R_s , of HTS thin films is most important parameter for microwave device applications. Large losses occur when the microwave enters the intergranular region which initiates breakdown of Josephson coupling between weakly coupled grains. The orderliness and orientation of the films are of great concern for microwave device applications due to different values of surface resistance from film to film.

The original Two Fluid Model (TFM), proposed by Gorter and Casimir[3], assumes a homogeneous superconductor and analysis of R_s for mixed oriented films and polycrystalline become difficult as the theory does not account for anisotropic properties of HTS materials. For instance the value of penetration depth for a c-axis oriented film is about 140 nm compared to that of 750 nm measured for ab-plane film[4].

In TFM it is assumed that superconductors have two kind of current carriers, superconducting charge carriers and normal charge carriers with n_s and n_n being their carrier densities respectively. the sum of these carrier densities is the total carrier density $n = n_s + n_n$. Both n_s and n_n are temperature dependent and given by

$$n_s(t) = 0 \quad \text{for } t > 1 \quad \text{and} \quad n_s(t) = n\,(1 - t^4) \quad \text{for } t < 1$$
$$n_n(t) = n \quad \text{for } t > 1 \quad \text{and} \quad n_s(t) = n\,t^4 \quad \text{for } t < 1$$

where $t = (T/T_c)$ is the reduced temperature.

The London penetration depth is an important parameter because it has strong influence (λ^3) on the microwave surface resistance R_s. The value of λ is inversely proportional to the square root of the superconducting charge carrier density and hence it is temperature dependent. By considering n_s as proposed in TFM the variation of λ with temperature could not be fitted well with experimental results specially in case of HTS materials. Many workers proposed modifications in the TFM regarding to the variation of n_s with temperature for better explanation of experimental results[4,7].

Dionne et al. modified the Two Fluid Model incorporating the structural anomalies and developed covalent electron transfer theory (CET) using hopping electron activation energy[8]. In this paper, a treatment of the surface resistance of YBCO superconductors based on CET has been performed. The theory agrees well with the experimental data.

II. THEORY

The surface resistance R_s of HTS at an angular frequency ω in the temperature below T_c is given by

$$R_s = 0.5\,\omega^2\,\mu^2\,\lambda^3(t)\,\sigma \tag{1}$$

where λ is temperature dependent penetration depth, σ is the real part of complex conductivity. Dionne et.al.,[9] modified the Two-Fluid model incorporating the structural anomalies and developed covalent transfer theory (CET) which assumes the Boltzmann type exponential partition function for the super carrier density as a function of temperature using hopping electron activation energy given by

$$n_n = n\exp(-W/t)$$

$$\frac{n^e_s(t)}{n^e_s(0)} = \{1-\exp(W-W/t)\} \tag{2}$$

where $W = (E_{hop}/kT_c)$ and $t = T/T_c$, E_{hop} is the hopping electron activation energy. $n^e_s(t)$ and $n^e_s(0)$ are the effective supercarrier densities at temperatures T and 0 °K respectively. T_c is the transition temperature from normal to superconducting state and k is the Boltzmann constant. The above supercarrier density will give the temperature dependence of λ as

$$\frac{\lambda(t)}{\lambda(0)} = \left[\frac{n^e_s(0)}{n^r_s(t)}\right]^{1/2} = \{1-\exp(W-W/t)\}^{-1/2} \tag{3}$$

where $\lambda(t)$ and $\lambda(0)$ are the penetration depth at the temperatures T and 0 °K respectively.

For metal oxides with mixed valanced cations the normal conductivity of superconductors is expressed as

$$\sigma_n = \frac{1}{C}\frac{x}{1-x}\frac{1}{k\,T_c\,t}\,(\eta P + (1-\eta P)\exp(W/t))^{-1} \tag{4}$$

where C is the parameter that contains the diffusion constant and its value strongly depends on the grain orientations and discontinuities at grain boundaries. x is the carrier concentration for chemical formula unit predicted as, x = 0.25y - 1.5, for $YBa_2Cu_3O_y$, yielding x = 0.25 for y =7. $0<\eta P<1$ is the product of charge transfer probability (η) and carrier dispersal parameter (P) related to microscopic homogeneity of the material. Now substituting Eq. (2) and Eq. (3) in Eq. (1) we have

$$Rs^{cet} = \frac{0.5\omega^2\,\mu^2\,\lambda^3(0)}{C\left(1-\exp\left(W-\frac{W}{t}\right)\right)^{3/2}}\,\frac{x}{(1-x)}\,\frac{1}{kTc}\,\frac{1}{t}\left(\eta P+(1-\eta P)\exp\left(\frac{W}{t}\right)\right)^{-1} \tag{5}$$

Quantitative agreement of this model with experimental results for YBCO thin films deposited on different substrates has been found reasonably good by taking different values of W for different films.

III. EXPERIMENT

Thin films of YBCO were prepared in-situ by dc magnetron sputtering using stoichiometric 123 target. During deposition the substrates were held at 730 C in 800 mTorr of Ar-O_2 mixture in the ratio of 2:1[10]. The depositing condition vary while depositing on different substrates to obtain best superconducting characteristics and morphology. in terms of highest J_c, sharpest resistive transition and well oriented texture. The deposition of YBCO on $LaAlO_3$(LAO), $SrTiO_3$(STO) and MgO were made simultaneously at a condition that suits best for LAO and are not far from the optimized conditions on STO and MgO. Thickness of all the films were approximately 400 nm and had dc resistivity $T_c(0)$ of 90 K, 88 K and 83 K on LAO, STO and MgO, respectively. The film deposited on LAO and STO had a high degree of texture as compared to that of MgO. The films were preferentially oriented c-axis normal to the substrates. The SEM micrographs shows the surface of the film on LAO is much smoother than those on STO and MgO. The measured J_c at 77 K on patterned film on LAO was 8×10^6 A/cm^2 and that on STO and MgO was 2×10^5A/cm^2 and 8×10^5 A/cm^2 respectively at zero magnetic field.

The surface resistance of YBCO films were measured using a cylindrical cavity in the TE_{011} mode with the film embedded in the center of the bottom end plate. The details of the measurement method are discussed elsewhere[11]. The temperature dependence of the cavity characteristics with YBCO film and with copper sample were recorded for the determination of surface resistance of the YBCO.

IV. DISCUSSION

The structural and superconducting properties of the YBCO films depend strongly on the substrate material and its surface on which these are grown. Unfortunately those substrates on which the films are relatively easy to deposit are not the best from the microwave point of view. In order to study the microwave properties of YBCO, the films were two to three times thicker than the microwave skin depth at 10 GHz to avoid microwave leakage through the film. Fig.1 shows the measured surface resistance of YBCO grown on LAO, STO and MgO fabricated simultaneously by taking the optimization conditions for LAO. The variation in the surface resistance with respect to substrate are due to two reasons; (i) fabrication process and other optimization conditions (ii) structural orientation related to the homogeneity of the grains. The change in the value of T_c is also attributed to these reasons which can be exploded by analyzing these with normal state conductivity as in the Covalent transfer theory (CET).Table 1 shows the parameter values optimized for our experimental surface resistance data at 10 GHz for YBCO thin films on different substrates.

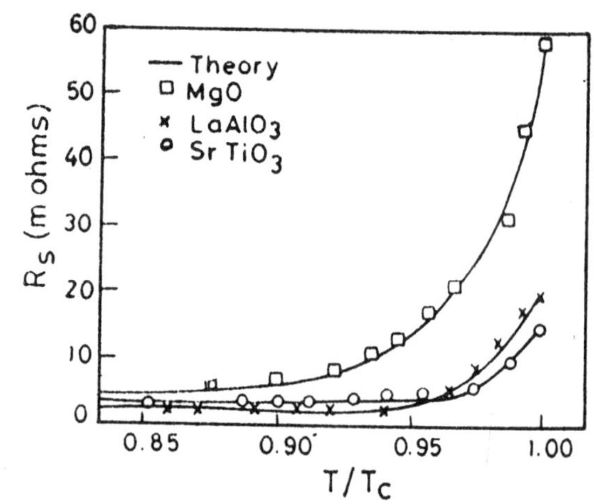

Fig.1. Theoretical fitting using CET with experimental values of R_s of YBCO thin films on different substrates measured at 10 GHz.

Table. 1

Substrate	$T_c(0)$ °K	$\lambda(0)$ nm	$C\times10^{-5}$ Ω/eV	ηP –	W (E_{hop}/kT_c)	x –
LAO	90	180	2.7	0.34	1.5	0.25
STO	87	180	2.7	0.34	1.2	0.25
MgO	83	180	2.7	0.34	0.65	0.25

It is known that the major contribution to the microwave losses occur at the high angle grain boundaries between the two mis-aligned grains in YBCO. These grain boundaries exhibit Josephson junction behavior that controls the macroscopic electromagnetic behaviour of all superconducting copper oxides. The grain boundaries limits the transport critical current density and at low temperatures the inherent temperature independent residual resistance observed depends inversely on the critical current density. Mechanism controlling the physical properties varies from boundary to boundary within the high angle regime and even from boundary facet to boundary facet due to anisotropic nature of the coherence length and the fact that the grain boundaries in YBCO are often non-stiochiometric with regard to both cation composition and oxygen concentration. These heterogeneity in the grain boundary structure could contribute the microwave losses on different samples of YBCO depending on the micro structure and also coupling strength between the grains.

Figure 1 shows the theoretical fitting using CET analysis with the experimental surface resistance values measured on the three substrates. To fit the experimental data with the CET the parameter W was varied. The value of the parameter W is related to the structural property of the material. The value of W= 1.5, 1.2 and 0.65 for films on LAO, STO and Mgo respectively can qualitatively describe the surface quality of the film. The lowest value of measured R_s of YBCO on LAO is 2.4 mΩ at 77 K can be interpreted due to the reduction in the number of weak links in the improved in-plane epitaxy of the film followed closely by the deposition on STO. The higher value of surface resistance in the case of film on MgO is due to the contribution of the surface impedance offered by the copper enriched grains distributed over the epitaxial surface of the film. It is to be pointed out that the lowest value of R_s on LAO as compared to that of STO and MgO is due to the fact that the films were prepared at optimized conditions for LAO.

V. CONCLUSION

The difference in the surface resistance of YBCO films on different substrates attributed to the structural orientation of the grains and its homogeneity. This has been analyzed using covalent transfer theory which is based on the hopping energy concept of carriers participating in superconductivity. The penetration depth and the surface resistance of YBCO thin films has been optimized using this theory which are useful for designing the microwave devices.

ACKNOWLEDGMENT

The authors are thankful to Prof. E.S.R.Gopal, Director of National Physical Laboratory for encouragement. Mukul Misra would like to acknowledge the award of Research Fellowship from CSIR.

REFERENCES

1. J.P. Carini, L. Drebeck and G. Gruner, Mod. Phys. Lett. **B3**, 5 (1989).
2. T.L. Hylton, A. Kapitulnik, M.R. Beasely, J.P. Carini, L. Drebeck, and G. Gruner, Mod. Phys. Lett., **B3**, 5 (1989).
3. G. Gorter and H.B.G. Casimir, Physica **1**, 306 (1934).
4. B. Pumpin et al., Physica **C**, 162, 151 (1989).
5. J. G. Ma and I. Wolff, IEEE Trans. Microwave Theory Tech. **MTT 43**, 1053 (1995).
6. D.A. Bonn, P. Dosanjh, R. Liang and W.N. Hardy, Phys. Rev. Lett., **68**, 2390 (1992).
7. E.J. Pakulis, R.L. Sandstorm, P. Chaudhary and R.B. Laibowitz, Appl. Phys. Lett., **57**, 940 (1990).
8. G.F.Dionne, IEEE Trans. Appl. Supercond. **3** (1993).
9. G.F.Dionne, IEEE Trans. Magn. **27** (1990).
10. R. Chandra, V. Kumar and A.K.Gupta, Ind. J. Pure Appl. Phys. **32** (1994).
11. M.JayaKumar, N.D.Kataria and G.P.Srivastava, IEEE Trans. Magn. (1994).

140 MeV Si-ION IRRADIATION EFFECT IN YBCO THIN FILMS

T. Mohanty[1], B.N. Dash[1], N.C. Mishra[1], K.Patnaik[1], L. Senapati[2], O.G. Singh[2], D. Kanjilal[2], G.K. Mehta[2], R. Pinto[3]

[1]*Physics Department, Utkal University, Bhubaneswar 751 004, India*
[2]*Nuclear Science Centre, P.B. 10502, New Delhi 110 067, India*
[3]*Tata Institute of Fundamental Research, Mumbai 400 005, India*

On-line resistance measurement in a temperature interval of 77K to 124K was performed on epitaxial thin films of YBCO irradiated with 140 MeV Si ions. Irradiation was done at 77K. The superconducting transition temperature(T_c), resistance at 124K (R_{124K}) and transition width (T_c) show peak at a fluence of $2x10^{13}$ ions/cm^2. Beyond this fluence, these parameters show their usual expected trend. It is expected that the appearance of peak at low fluence may be due to induction of point defects at such sites which have a direct effect on the superconducting coupling.

I. INTRODUCTION

Ion irradiation in crystalline material gives rise to atomic displacements and structural disorder[1]. In the MeV and GeV ranges of the ion beam, most of the initial ion energy is lost to the electronic excitation of the target ions and a small fraction results in the creation of defects through elastic nuclear recoil. In high T_c cuprate superconductors, the defect production through the former process has been observed to depend sensitively on the quality of the sample [2]. While, heavy ions in the MeV range produce columnar defects, the lighter ions in the same energy range has been shown to produce localised atomic size defects [3]. Effect of such point defects on the superconducting and normal state properties are expected to be different from that due to extended defects. While the extended defects have been shown to reduce T_c, increase J_c and normal state resistivity in general [4] there has not been unanimous agreement on the effect of localized point defects on these parameters. Sample quality is one of the many factors that contributes to the different effects of lighter ion irradiation inducing mostly point defects on the physical properties of cuprates.

In this report we present the effect of 140 MeV Si ion irradiation on YBCO thin films. While a T_c enhancement of about 3K is observed at a fluence of $2x10^{13}$ ions/cm^2, the fluence dependence of T_c, normal state resistance and transition width is shown to have a correlation which we explain based on localised point defects affecting the mechanism of superconductivity in this system.

II. EXPERIMENTAL

$YBa_2Cu_3O_{7-x}$ thin films on $LaAlO_3$ substrate were prepared by laser ablation technique. The films were epitaxial in nature with well ordered grains as revealed from AFM study. The thickness of the films were around 3000A. 140 MeV ^{28}Si ion irradiation was performed using the 16MV tandem pelletron accelerator at the NSC, New Delhi. Temperature of the films mounted on a sample holder could be varied from 77K to 300K inside the beam chamber. This permitted on-line four probe d.c. resistivity measurement as a function of temperature. Data acquisition was done using a computer controlled data acquisition system interfaced through a GPIB card. Accurate ion beam dosimetry was achieved by surrounding the sample holder with a Faraday Cup. The films were irradiated for varying fluences at 80K. After each irradiation, the beam was switched off and resistivity data was acquired with increasing temperature.

III. RESULT AND DISCUSSION

The electronic energy loss S_e, nuclear energy loss S_n and the range of ion beam for 140 MeV Si ion beam in YBCO as obtained from TRIM calculation are of the order of 0.44 KeV/A, 0.3 eV/A and 26 μm respectively. The film thickness being of the order of 0.3 μm, the projectile ions are expected to traverse the entire film interacting with the target ion only through electronic energy loss and come to rest on the substrate. The ratio of S_e to S_n being of the order of 1900, suggests that electronic stopping of the beam can cause dynamic annealing of defects intrinsically present in the system as well as the defects created during irradiation [3]. In addition, electronic energy loss also create defects and a threshold S_e of 2KeV/A is believed to create continuous amorphous tracks in YBCO[5]. However, many authors [6] have shown that defects can still be produced at a S_e lower than the threshold value. The S_e of 0.44 KeV/A in the present experiment being much less than the threshold S_e cannot lead to production of continuous amorphous tracks but can result in atomic size point defects as has been seen by Liu et al [3]. Coupled with the dynamic annealing of defects, the very low concentration of defects that can result during Si ion irradiation is expected to have different effects on the superconducting and normal state properties as compared to extended defects observed in irradiation by higher energy heavier ions like Pb [7], Xe [8] and Kr [9].

The normal state resistivity has been shown[10] to be a sensitive diagonistic probe of defect formation, defect density, defect mobility and the annealing behaviour of the irradiated superconductors. Depending on the ion energy and sample quality, exponential dependence of room temperature resistivity on fluence[11], deviation from exponential beaviour [12] have been observed. In most of the cases monotonic increase of resistivity is seen to occur with increasing fluence, while a few reports[13] have indicated a saturation behaviour in the low fluence regime and involving lighter ions. Our experiment involving Si ion falls in the later regime. The resistance at 124K (R_{124K}) as a function of fluence [Fig.1] not only shows a plateau at low fluences like that observed [3,13], but a peak in this plateau. At higher fluences the resistance increase occurs following the usual trend. The superconducting transition temperature instead of showing a monotonic decrease with fluence, also shows a peak [Fig. 2] at the fluence coinciding with the resistance peak.

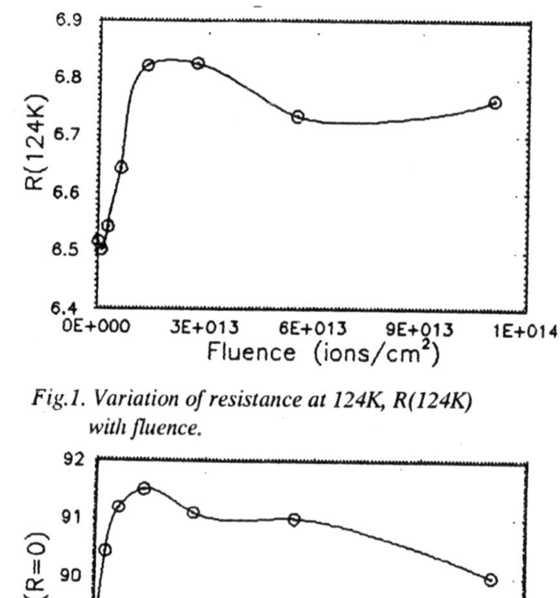

Fig.1. Variation of resistance at 124K, R(124K) with fluence.

Fig.2. Variation of T_c (r=0) with fluence.

Except a few reports [14], where T_c enhancement has been observed under irradiation, most of the reports indicate T_c suppression with irradiation. Albenque et al [15] have shown a correlation of the electronic energy loss induced T_c and normal state resistance variation with fluence. It should be noted here that Xe ion in GeV energy range was used for irradiation where amorphous track formation occurs. Such extended defects are expected to suppress subtle features arising due to variation in sample quality. At lower fluences and involving lighter ions the T_c variation has been shown to be crucially dependent on sample quality [12]. In the later kind of experiments intrinsic defects play a crucial role in deciding the changes induced by irradiation induced defects. However, a correlation between T_c and resistance variation is still to be expected as observed by Albenque et al [15]. The T_c decrease at higher fluence [Fig. 2] thus correlates well with the resistance increase [Fig. 1]. In the low fluence regime on the other hand T_c was seen to increase instead of showing the expected decrease; showing a peak coinciding with the peak of the resistance. A possible explanation for this unusual behaviour of T_c dependence on fluence could be the formation of a kind of ordering being induced by ion irradiation. The transition width [T_c], which is a measure of inhomogeneity and disorder in the sample, however shows an increase with increasing fluence, also having a peak coinciding with the peak in T_c [Fig. 3]. This rules out any kind of order being induced by irradiation and being responsible for the rise in T_c. The resistance increase also supports this view. Therefore, we believe that this unusual behaviour arises due to localised point defects which have a direct influence on the mechanism of superconductivity.

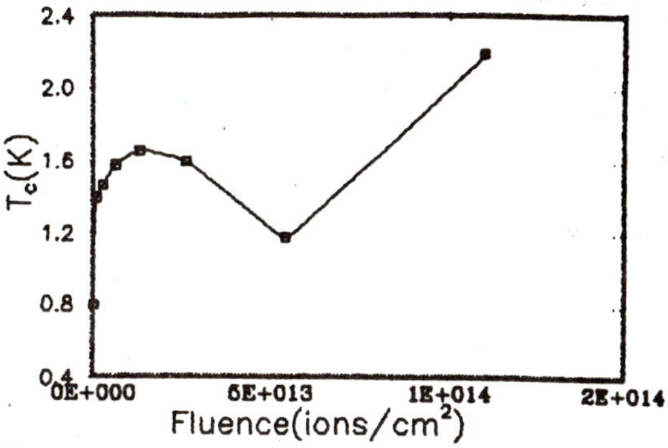

Fig. 3. Variation of transition width (90%-10%) of the T_c onset with fluence.

Comparing our observation with that of ref. 16, we note that the room temperature resistance vs oxygen content shows a peak not at a oxygen content of 7 (O_7), where well ordered chains are established but at ~ $O_{6.9}$ [Fig. 4] with randomly distributed oxygen vacancies in the chains. The T_c variation with oxygen content also shows a peak at the same oxygen content where room temperature resistivity peaks.

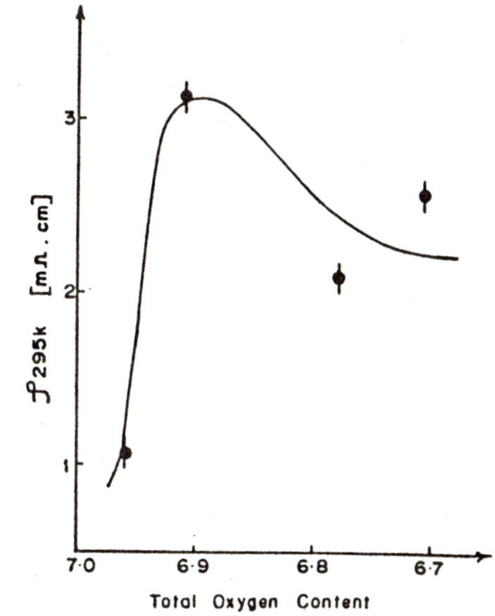

Fig.4. Variation of room temperature resistivity (295K) with total oxygen content of YBCO showing a maximum near 6.9.

A crystallochemical analysis by Mohapatra et al [17] have explained this correlation of T_c with room temperature resistivity [Fig. 5] based on metastability induced charge and spin fluctuations. According to the later approach if isolated oxygen vacancies are created in well ordered CuO chains of YBCO due to ion irradiation, the Cu ions near the vacancies would find themselves in three fold oxygen coordination. While, Cu^{+1} can have only a two fold oxygen coordination, at least a four fold oxygen coordination is admissible to Cu^{+2} and Cu^{+3} [18,19]. Therefore the three fold oxygen coordinated Cus near the vacancies remain in an unstable charge state. At lower concentration of these unstable sites, spatial equilibration is prevented, and these Cus fluctuate between +2 and +1 charge states. Such charge fluctuation behaviour has been shown to induce antiferromagnetically coupled spin fluctuations through the process of creation and annihilation of holes [17].

Though the mechanism of superconductivity is still obscure, the spin fluctuations playing a decisive role in the mechanism of superconductivity has been speculated by many[20]. Therefore, the scattering of charge carriers in the normal state which results in a peak in resistivity at a particular defect concentration, also results in the enhanced T_c by providing the coupling between the charge carriers in the superconducting state at this defect concentration. At higher fluences with larger defect concentrations, disorder sets in reducing T_c and increasing resistivity as expected.

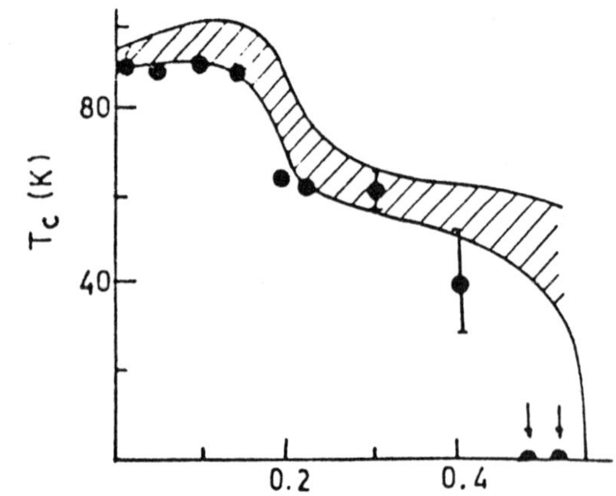

Fig.5. Variation of T_c of YBCO with x. Hatched region: magnetic susceptibility data; Closed circles: resistivity data.

REFERENCES

1. Chr. Lehmann, *Defects in Crystalline Solids*, (North-Holland, Amsterdam, 1968) vol.**10**, P 153.
2. Y. Katoh and A. Kohyama, Nucl. Instrum. Methods B **102**, 12 (1995).
3. J.R. Liu, Z.H. Zhang, J. Wu, K. Ma, Y.J. Zhao, N. Yu, P. Hsieh and Wei Kan Chu, Nucl. Instrum. Methods B **62**, 74 (1991).
4. M.A. Kirk, M.C. Frischherz, J.Z. Liu, L.R. Greenwood and H.W. weber, Philos. Mag. Lett. **62**, 41 (1990).
5. B. Hensel, B. Roas, S. Henke, R. Hoptengartner, M. Lippert, J.P. Strobel, M. Vildic, G. Saemann-Ischenko, S. Klaumunzer, Phy. Rev. B **42**, 4135 (1990).
6. S.A. agnihotry, R.H. Bhawalkar, S. Chandra, L. Senapati, D. Kanjilal and G.K. Mehta, Supercond. Sci. Tech. **5**, 582 (1992).
7. V. Hardy, D. Groult, M. Hervieu, J. Provost, B. Raveau, S. Bouffard, Nucl. Instrum. and Methods **B 54**, 472 (1991).
8. V. Hardy, D. Groult, J. Provost, B. Raveau, S. Bouffard, Physica C **178**, 255 (1991).
9. F. Studer, S. Provost, D. Groult, B. raveau, S. Bouffard, M. Toulemonde, J.C. Jousset and Rullier-Albenque, Mater. Res. Soc. Symp. Proc. **99**, 877 (1988).
10. S.M. Meyer, Nucl. Instrum. and Methods, **168**, 265 (1980)
11. D. Bourgault, D. Groult, S. Bouffard, J. Provost, F. Studer, N. Nguyen, P. Raveau and M. Toulmonde, Phys. Rev. B **39**, 6549 (1989).
12. A.D. Marwick, G.J. Clark, D.S Yee, R.B. Laibowitz, G. Coleman and J.J. Cuomo, Phys. Rev. B **39**, 9061 (1989).
13. Nirupa Sen, A.K. Bandyopadhay, P. Sen, Umesh Tiwari, Deepak Varandani, V.P.S. Awana, and A.V. Narlikar, Sol. St. Comm. **82**, 555 (1992).
14. M. Okada and T. Kawa kubo, Rad. Eff. and Defects in Solids **38**, 137 (1989).
15. F. Rullier- Albenque, A. Legris, S. Bouffard, E. Paumier and P. Lejay, Physica C **175**, 111 (1991).
16. S. Ruseicki, B. Bucher, E. Kaldis, E. Jilek, J. Karpinski, C. Rossel, B. Pumpin, H. Keller, W. Kundig, T. Krekels and G. Van Tenddoo, J. Less-Common Met. **164-165**, 31 (1990).
17. H.P. Mohapatra, D. Behera, S. Misra, N.C. Mishra and K. Patnaik, J. of Supercond. **6**, 359(1993).
18. A.F. Wells, Structural Inorganic Chemistry, **4**th edn. (Clarendon Press, Oxford, 1975).
19. B. Raveau, C. Michel, M. Hervieu and J. Provost, Physica C **153-155**, 3 (1988).
20. G. De Gennes, C.R. Acad, Sci (Paris) **305**, 345 (1987).

MAGNETO-CONDUCTIVITY OF $NdBa_2Cu_3O_{7-\delta}$ THIN FILM IN THE PARACONDUCTIVITY AND PARACOHERENCE REGIMES

Narayani Senthilkumaran, S.Ravi*, K.Sethupathi, V.Sankaranarayanan and G.Rangarajan

Department of Physics, Indian Institute of Technology, Madras 600 036 India

**Department of Physics, NERIST, Nirjuli 791 109*

Temperature variation of in-plane electrical resistivity in the absence and in the presence of magnetic fields 1T to 4T applied both parallel and perpendicular to the ab-plane has been measured on a c-axis oriented superconducting thin film of $NdBa_2Cu_3O_{7-\delta}$. The magneto-conductivity data were analyzed using Aslamazov-Larkin(AL) and Maki-Thompson(MT) models by considering both orbital and Zeeman effects. Although the contribution of MT term to the zero field case was found to be insignificant, the magneto-conductivity data could be explained only by taking into account the MT term. The estimated values of the average phase breaking time τ_φ(100 K) and coherence lengths parallel and perpendicular to the ab- plane $\xi_{ab}(0)$,$\xi_c(0)$ are 4.5×10^{-13} s, 21Å and 2.4Å respectively. The data has been analyzed in the paracoherence region in the absence as well as in the presence of magnetic fields. We observed a cross-over from disordered phase fluctuations to ordered phase fluctuations. The disordered phase fluctuations close to transition temperature T_c are attributed to the weak links due to suppression of some low J_c grains into normal state by the applied magnetic field.

I. INTRODUCTION

The high temperature oxide superconductors are weakly coupled in nature. The rounding of electrical resistivity above the superconducting transition temperature, T_c (mid-point of transition) can be explained in terms of excess conductivity due to the formation and destruction of cooper pairs. In this region the magnitude of the order parameter becomes identical within the grains. Below the transition temperature the rounding-off of the electrical resistivity is due to the fact that the phase of the order parameter between the individual grains is not equal and at $T = T_{co}$ when the resistivity of the material is zero, complete phase coherence is achieved. In the present work, we report the temperature variation of in-plane electrical conductivity in zero magnetic field and also in the presence of the field from 1T to 4T applied both parallel and perpendicular to the ab-plane on a c-axis oriented thin film of $NdBa_2Cu_3O_{7-\delta}$ with a T_c of 84.4K. The Aslamazov-Larkin (AL) magneto-conductivity data were analyzed using AL and Maki-Thompson (MT) models considering both the orbital and Zeeman effects. The excess conductivity data has been analyzed in the paracoherence regions both in the absence and presence of magnetic field parallel and perpendicular to ab-plane to extract the critical exponent in this region. The superconducting transition in the paracoherence region is analogous to the paramagnetic to ferromagnetic transitions in a 3D XY ferromagnet. Such studies have already been carried out on other high temperature superconductors like $YBa_2Cu_3O_{7-\delta}$[1], $Tl_2CaBa_2Cu_2O_8$, $CaLaBaCu_3O_{7-\delta}$, $CaSmBaCu_3O_{7-\delta}$[2] and BiSrCaCuO system[3]. The value of critical exponents obtained are found to be either close to -1.33 or -2.7. In the case of 3D-XY ferromagnet system, the critical exponent of -1.33 corresponds to a pure ferromagnet and the value -2.7 to a disordered ferromagnetic system[4].

II. EXPERIMENTAL DETAILS

Thin film of $NdBa_2Cu_3O_{7-\delta}$ is grown on $LaAlO_3$ substrate using pulsed laser deposition and the stoichiometric ceramic targets. The temperature of the substrate was always maintained at 700°C. The temperature variation of electrical resistivity in zero field and in the presence of the applied field was measured using a four probe technique employing a commercial liquid helium cryostat. The sample exhibited superconducting transition at 84.4K. The magnetic field was applied both parallel and perpendicular to the ab-plane using a superconducting magnet supplied by Quantum Design and the temperature was measured to an accuracy better than 0.05K.

III. RESULTS AND DISCUSSIONS

The variation of electrical resistivity ρ with temperature T in the absence of magnetic field and for different values of applied magnetic field along the ab-plane ($H_{\parallel}$) and perpendicular to the ab-plane ($H_{\perp}$) are depicted in Fig.1(a) and (b). In the absence of the applied field the $d\rho/dT$ vs T plot exhibits a single sharp peak at 84.4K and this is taken as the mean field transition temperature, T_c for further analysis.

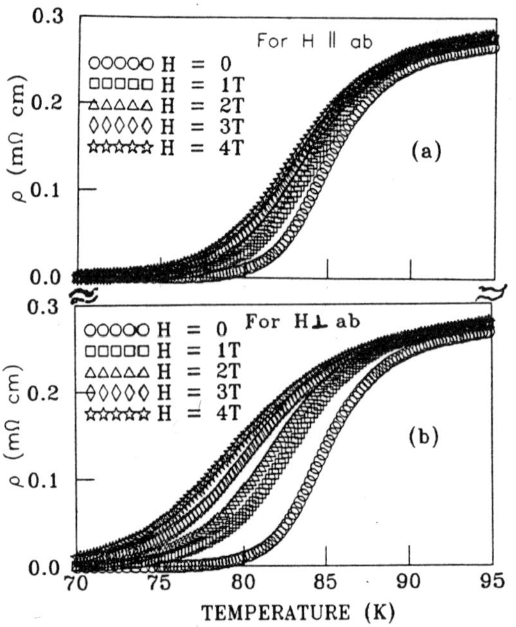

Fig.1. Resistivity as a function of temperature for different fields (a) along ab-plane and (b) perpendicular to ab-plane.

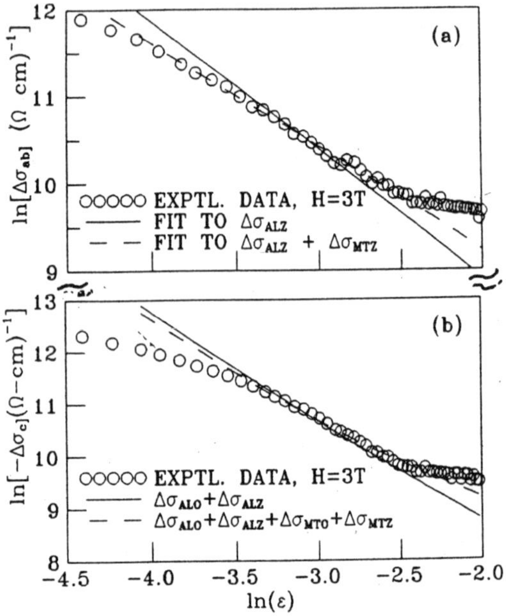

Fig.2. Plot of $\ln(-\Delta\sigma)$ versus $\ln(\epsilon)$ for an applied field of 3T (a) along ab-plane and (b) perpendicular to ab-plane.

A. Paraconductivity region

The fluctuations-induced magneto- conductivity, $\Delta\sigma_H$, calculated from the above experimental data was analyzed in terms of Aslamazov-Larkin (AL) and Maki-Thompson (MT) models by considering both orbital and Zeeman effects. $\Delta\sigma_H$ is a sum of four different contributions resulting from the interaction of the magnetic field with the carrier orbital and spin moments[5-7]. The contributions from orbital [ALO, MTO] and Zeeman effects [ALZ, MTZ] to the AL and MT terms are given by[6,7].

$$\Delta\sigma_{ALO} = \frac{-e^2}{64\hbar d\epsilon^3}\,\frac{2+4\alpha+3\alpha^2}{(1+2\alpha)^{5/2}}h^2 \tag{1}$$

$$\Delta\sigma_{MTO} = \frac{-e^2}{48\hbar d(1-\alpha/\delta)\epsilon^3}\left[\frac{\delta^2}{\alpha^2}\,\frac{1+\delta}{(1+2\delta)^{3/2}} - \frac{1+\alpha}{(1+2\alpha)^{3/2}}\right]h^2 \tag{2}$$

B. Paracoherence region

The experimental excess conductivity $\Delta\sigma$ in the paracoherence region was calculated as a function of reduced temperature $\epsilon = (T-T_{co})/T_{co}$ using the following Eq.

$$\Delta\sigma = \frac{1}{\rho(T)} - \frac{1}{\rho(T_c)} \tag{6}$$

where $\rho(T_c)$ is the resistivity at T_c. Fig 3 shows the plot of $\ln(\Delta\sigma/\sigma(T_c))$ vs $\ln(\epsilon)$ when no magnetic field is applied. Here $\sigma(T_c)$ is the measured conductivity at T_c. In the paracoherence region one would expect $\Delta\sigma$ to vary as ϵ^b where b is the critical exponent. Then

$$\ln\left(\frac{\Delta\sigma}{\sigma T_c}\right) = a + b\ln(\epsilon) \tag{7}$$

where a is a constant. The experimental $\Delta\sigma$ data were fitted to Eq. 7 by varying the parameters a and b. The fitted data are shown as solid line in Fig.3 for H = 0. The value obtained for the critical exponent b from the fit is 1.83 which is closer to -1.33 as predicted for pure ferromagnets.

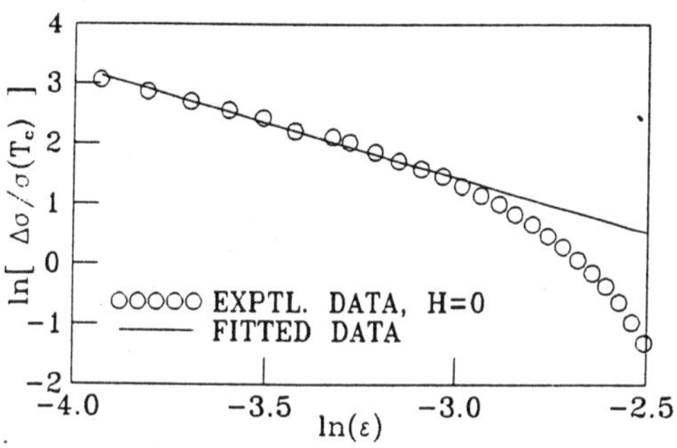

Fig.3. Plot of $\ln(\Delta\sigma/\sigma(T_c))$ versus $\ln(\epsilon)$ for when H = 0T.

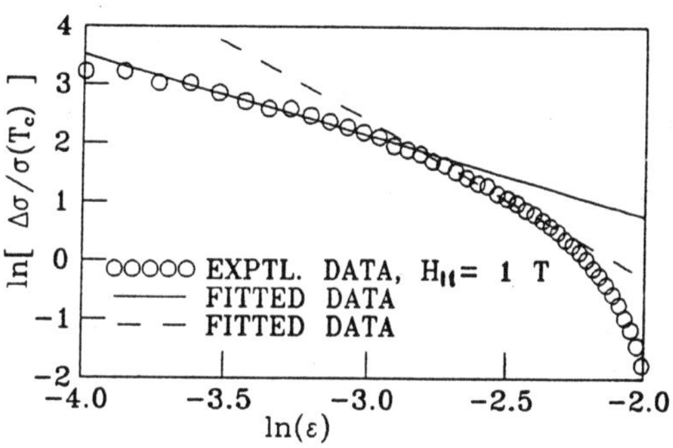

Fig. 4. Plot of $\ln(\Delta\sigma/\sigma(T_c))$ versus $\ln(\epsilon)$ for when $H_{\parallel} = 1$ T.

Similar fitting procedure was also carried out for the excess conductivity data obtained from ρ vs T measurements at different applied magnetic fields both parallel and perpendicular to ab- plane. It is assumed that even in the presence of magnetic field the transition from T_c to T_{co} is mainly due to phase fluctuations.

As can be seen in Fig.4, there are two different slopes for $H_{\parallel}$ = 1T. The critical exponent obtained from the fit in the low temperature region (LTR, $\ln(\epsilon)$ = -4 to -3, close to T_{co}) is -1.29. In the high temperature region (HTR $\ln(\epsilon)$ = -3 to -2, close to T_c) b is found to be -2.13. Here the former one is close to -1.33 and the later one is close to -2.7. The critical exponent obtained in the HTR is comparable to that of disordered ferromagnetic materials and this could be because of the suppression of some of the low J_c grains into normal state by the applied magnetic field and normal state grains might act as weak links, thereby affecting the critical exponent. In the LTR, all the grains may be in the superconducting state and since there is no weak link, the exponent is comparable to that of pure ferromagnetic materials.

The cross - over temperature (T_{od}), i.e the temperature at which cross over from disordered phase fluctuations (-2.7) to ordered phase fluctuations (-1.33) occurs is estimated from the plot of experimental $\ln(\Delta\sigma/\sigma(T_c))$ vs $\ln(\epsilon)$. The T_{od} values are listed in the Table II for different fields. We can see that as the field increases the cross over temperature decreases, i.e more and more grains become normal, thereby extending the disordered phase fluctuations regime to much lower temperature.

In order to check the observed critical exponents, we have plotted $(\Delta\sigma/\sigma(T_c))^{-1/1.33}$ versus temperature presuming that the critical exponent is -1.33. A typical graph for $H_\perp$ = 1T is shown in Fig.5. The zero resistivity temperature T_{co}' was estimated by extrapolating the curve to $(\Delta\sigma/\sigma(T_c))^{-1/1.33} = 0$. This estimated values of T_{co}' are also given in Table II for different fields and they are comparable to the respective T_{co} values obtained from the experimental ρ vs T data, thereby indicating that the critical exponent is close to -1.33 in the low temperature region.

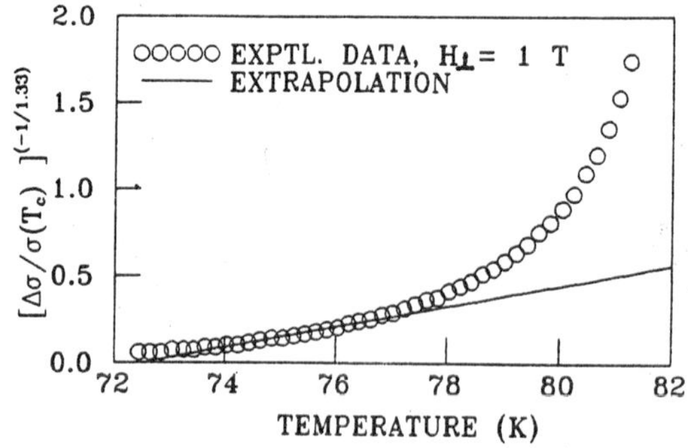

Fig.5. Plot of $(\Delta\sigma/\sigma(T_c))^{-1/1.33}$ as a function of temperature for $H_\perp$ = 1 T.

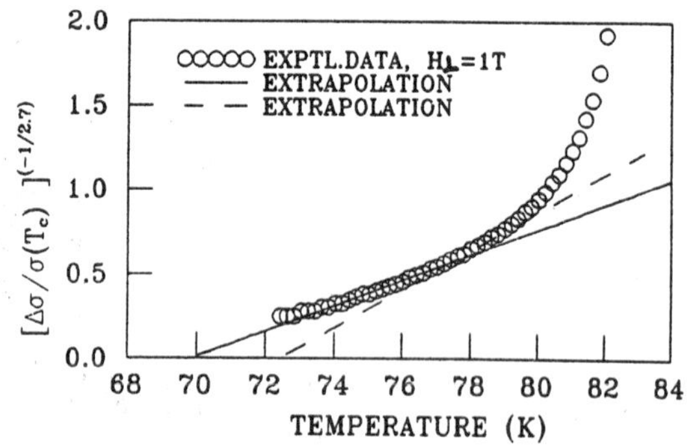

Fig.6. Plot of $\ln(\Delta\sigma/\sigma(T_c))^{-1/2.7}$ as a function of temperature for $H_\perp$=1 T.

By presuming that the critical exponent is -2.7 plots of $(\Delta\sigma/\sigma(T_c))^{-1/2.7}$ vs T are drawn for various fields. Fig.6 depicts such a graph for $H_\perp$=1T. The zero resistivity temperature was estimated by extrapolating the curve in the low temperature region (T_{co}'') and the high temperature region (T_{co}'''). The estimated T_{co}'' values are quite different from T_{co} values and it indicates that the critical exponent is not close to -2.7. However the T_{co}''' values are comparable to the respective T_{co} values indicating that the critical exponent in the HTR is close to -2.7.

Table II:Parameters obtained in the paracoherence region.

H T	T_{co} K	T_c K	$\rho(T_{co})$ mΩ-cm	b LTR	 HTR	T_{co}' K	T_{co}'' K	T_{co}''' K	T_{od} K
0	77.5	84.4	0.112	-1.83	-	77.9	76.4	-	-
$H_{\parallel}1$	75.3	83.8	0.124	-1.29	-2.13	75.0	72.6	76.1	79
$H_{\parallel}2$	74.1	83.3	0.126	-1.44	-2.71	74.2	71.6	74.2	78
$H_{\parallel}3$	73.2	82.8	0.122	-1.14	-2.64	72.7	70.1	73.6	77
$H_{\parallel}4$	72.3	82.7	0.133	-1.10	-2.82	71.8	68.4	72.1	76.3
$H_\perp 1$	72.3	82.6	0.130	-1.37	-2.67	72.7	69.8	72.4	76.7
$H_\perp 2$	70.4	81.5	0.128	-1.41	-2.52	70.6	66.3	70.1	75.1
$H_\perp 3$	68.5	80.2	0.127	-1.37	-2.53	68.7	65.2	68.5	73.3
$H_\perp 4$	66.8	78.7	0.119	-1.34	-2.31	66.8	63.1	66.2	71.5

$$\Delta\sigma_{ALZ} = -0.526\,\frac{e^2}{\hbar d\epsilon^3}\frac{1+\alpha}{(1+2\alpha)^{3/2}} \tag{3}$$

$$\Delta\sigma_{MTZ} = \frac{-e^2}{16\hbar d\epsilon}\left[\frac{1+\delta}{(1+2\delta)^{3/2}} - \frac{1+\delta+\delta/\alpha}{[(1+\delta/\alpha)(1+2\delta+\delta/\alpha)]^{3/2}}\right]\left[\frac{\omega_s\tau_\phi}{\hbar}\right]^2 \tag{4}$$

where $\alpha = (2\xi_c^2(0))/(d^2\epsilon)$, $h = 2e\xi^2_{ab}(0)H/\hbar$, $\delta=(16\xi_c^2(0)k_BT\tau_\varphi)/(\pi d^2\hbar)$, $\omega_s = g\mu_BH$ is the Zeeman energy, $\epsilon=(T-T_c)/T_c$ is the reduced temperature, d is the interlayer spacing, $\xi_{ab}(0)$ and $\xi_c(0)$ are the coherence lengths parallel and perpendicular to the ab plane, μ_B is the Bohr magneton, g is the 'g' factor, τ_φ is the phase breaking time, k_B is the Boltzmann constant and e is the charge of the electron.

Typical plots of $\ln(-\Delta\sigma_{H\parallel})$ vs $\ln(\epsilon)$ for $H_\parallel$ = 3T and $H_\perp$ = 3T are shown in Fig. 2(a) and 2(b). The data for $\Delta\sigma_{H\parallel}$ could be fitted well to the theory over a large range of $\ln(\epsilon)$ only if contribution to excess conductivity both from AL and MT terms, $\Delta\sigma_{ALZ}$, $\Delta\sigma_{MTZ}$ are considered. The fit of the experimental data is shown as a dashed line in Fig 2(a). The details of the fitting procedure are described elsewhere[8].

From the fitted value of α_o, the dimensionality cross-over temperature T_o is estimated using the relation,

$$T_o = T_c\exp\left(\frac{2\xi_c(0)}{d}\right)^2 \tag{5}$$

The value of T_o is found to be (101.5±1)K and is higher than Tc by about 17K. By taking d as the c-axis length of the unit cell, $\xi c(0)$, To and $\tau\varphi(100\text{ K})$ for H|| = 1,2,3 and 4T are estimated and are given in Table I. The above analysis demonstrates the significance of MT term in $\Delta\sigma_{H\parallel}$.

In the case of $H_\perp$ to ab-plane, both orbital and Zeeman effects contribute to excess conductivity. The Dashed line in Fig. 2 (b) shows the fit to the experimental data when all the four terms $\Delta\sigma_{ALO}$, $\Delta\sigma_{ALZ}$, $\Delta\sigma_{MTO}$, $\Delta\sigma_{MTZ}$ are taken into account. Reasonable fit is obtained for $\ln(\epsilon)$ from -3.5 to -2.5. The fitted values for $\xi_{ab}(0)$, $\tau_\varphi(100\text{ K})$ are also given in Table I for $H_\perp$ = 1,2,3 and 4T.

Table I. List of parameters obtained from the fit of magneto-conductivity data to Eq.1-4.

H T	For H ∥ ab			For H ⊥ ab	
	$\xi_c(0)$ Å	T_o K	τ_φ(100K) 10^{-13} s	$\xi_{ab}(0)$ Å	τ_φ(100K) 10^{-13} s
1	2.3±0.04	97.2±2.7	9.7±0.1	27.4±0.6	3.0±1.6
2	2.4±0.1	100.1±3.1	2.4±0.5	21.7±1.1	0.9±0.4
3	2.6±0.11	103.1±3.5	3.8±2.4	19.3±0.7	0.5±0.2
4	2.1±0.09	96.6±2.9	15.5±5.8	15.7±0.5	0.3±0.1

In conclusion the magnetoconductivity data for H || ab and H ⊥ ab could be explained only by taking into account the MT terms. The average values of $\xi_c(0)$, $\xi_{ab}(0)$ and $\tau_\varphi(100\ K)$ estimated from the above analysis are 2.4 Å, 21.0 Å and $4.5x10^{-13}$ s and are comparable to those of $YBa_2Cu_3O_{7-\delta}$ films reported in the literature[9,10]. It may be noted that the presence of Zeeman effect in this material indicates that the superconductivity here is "BCS-like"[9].

The paracoherence analysis show a cross-over from disordered phase fluctuations to ordered phase fluctuations. The disordered phase fluctuations in the high temperature region (close to T_c) is attributed to the weak lines due to suppression of some low J_c grains into normal state by the applied magnetic field. This argument can be substantiated from the observed decrease in cross-over temperature when the field is increased.

ACKNOWLEDGMENTS

The authors acknowledg K.V.Rao and K.Sreenivas of Royal Institute of Technology, Sweden for providing the thin films. This work is supported by the Department of Science and Technology (NSTB-DST), New Delhi.

REFERENCES

1. J.Rosenblatt, A.Raboutou, P.Peyral and C.Lebeau, Rev. Phys. Appl. **25**, 73 (1990).
2. H.Krishnan, V.Sankaranarayanan, R.Srinivasan, Physica B **165 & 166**, 1403 (1990).
3. E.Issac Samuel, S.Ravi and V.Seshu Bai, Solid State Comm. **96**, 441 (1995).
4. A.Raboutou, P.Peyral, C.Lebeau and J.Rosenblatt, Physica A **157**, 159 (1989).
5. K. Maki and R.S. Thompson, Phys. Rev. B, **39**, 276, (1989).
6. S. Hikami and A.I. Larkin, Mod. Phys. Lett. B, **2**, 693 (1988).
7. A.G. Aronov, S. Hikami and A.I. Larkin, Phys. Rev. Lett., **62**, 965 (1989).
8. Narayani Senthilkumaran, S.Ravi, K.Sethupathi, G.Rangarajan and K.Sreenivas, IEEE Trans. Magn. (in press).
9. Y. Matsuda, T. Hirai, S. Komiyama, T. Terashima, Y.Bando,K.Iijima, K. Yamamoto and K. Hirata, Phys. Rev. B, **40**, 5176 (1989).
10. W. Holm, Ö. Rapp, C.N.L. Johnson and U. Helmersson, Phys. Rev. B, **52**, 3748 (1995).

MICROWAVE SURFACE RESISTANCE IN $Lu_{1-x}Pr_xBa_2Cu_3O_{7-\delta}$ THIN FILMS

V.V.Srinivasu[+], Pratap Raychaudhuri[+], C.P.D'Souza[+], R.Pinto[+] and R.Vijayaraghavan[#]
[+]Tata Institute of Fundamental Research, Homi Bhabha Road, Mumbai-400 005, India.
[#] SAMEER, IIT Campus, Powai, Mumbai-400 076, India.

We report microwave surface resistance $R_s(T)$ measurements on $Lu_{1-x}Pr_xBa_2Cu_3O_{7-\delta}$ thin films insitu grown by Pulsed Laser Deposition (PLD) technique, focussing on the effect of Pr doping in this system. Anomalous peaks in $R_S(T)$ plots are observed for both dilute and strong doping concentrations of Pr ($x \sim 0.07$ and $x \sim 0.2$). Also aging effects are observed in these films, with a decrease in residual surface resistance and vanishing of the peaks with aging of these films. We interpret the peaks in $R_S(T)$ vs T curve due to the different competing processes involving the temperature dependences of quasi particle scattering time, pair condensation , penetration depth.

INTRODUCTION

Microwave loss measurements in High-T_c superconductors (HTSC) continue to be of great interest as they throw new insights into both intrinsic and extrinsic properties of these materials[1] . Intrinsic loss[2,3] depends upon the quasi particle number density, scattering time, impurities etc . In clean single crystals surface resistance $R_S(T)$ has a non monotonic dependence on temperature and infact, there is a broad peak in $R_S(T)$ vs T plot[2,3] . Recently, Lofland et al[4] ., have observed a large peak in the temperature variation of c-axis microwave loss at a few K below T_c, in a-axis oriented thin films of $YBa_2Cu_3O_{7-\delta}$. In view of the interesting experimental data on anomalous behaviour of microwave loss dependence on temperature, we report here our results on microwave surface resistance measurements on $Lu_{1-x}Pr_xBa_2Cu_3O_{7-\delta}$ thin films. We notice that for dilute as well as strong doping of Pr, ($x \sim 0.07$ and $x \sim 0.2$), $R_S(T)$ vs T curves exhibit anomalous peaks. These peaks are absent in the $Lu_{1-x}Pr_xBa_2Cu_3O_{7-\delta}$ with $x = 0$. Also strong aging effects are observed in these films.

EXPERIMENTAL

Thin films of $Lu_{1-x}Pr_xBa_2Cu_3O_{7-\delta}$ were grown in situ on <100> $LaAlO_3$ substrates by Pulsed Laser Deposition (PLD) technique. Details of target preparation, growth conditions and the characterization are described elsewhere[5,6] . These films are single phasic and c-axis oriented. The microwave measurements were carried out at X-band frequencies on microstrip resonators fabricated on 10mmX10mmX0.5mm $LaAlO_3$ substrates using 2500 A thin $Lu_{1-x}Pr_xBa_2Cu_3O_{7-\delta}$ films. The design and fabrication of the X-band microstrip resonator and the method of calculating surface resistance is described elsewhere[7,8]. The aging effects are measured, by leaving the sample undisturbed in the microwave jig, placed in the low temperature chamber.

RESULTS AND DISCUSSION

Figure 1. shows the surface resistance $R_S(T)$ measured at 9.1 GHz as a function of temperature for the thin film $Lu_{1-x}Pr_xBa_2Cu_3O_{7-\delta}$ with $x \sim 0.07$. The $R_S(T)$ vs T plot shows two peaks. One peak is at ~ 62 K (peak A) and another at ~ 35 K (peak B). Figure .2 shows the $R_S(T)$ vs T plot for the thin film $Lu_{1-x}Pr_xBa_2Cu_3O_{7-\delta}$ with $x \sim 0.2$, measured at 9.1 GHz. The peaks have broadened and moved closer to each other. Infact, peak A is shifted towards lower temperature while Peak B shifted towards higher temperature, indicating that both arise out of different reason. Figure 3(a) and 3(b) show the aging effects (after 24 hours and 100 hours of aging, respectively) of the thin film $Lu_{1-x}Pr_xBa_2Cu_3O_{7-\delta}$ with x~0.07 on $R_S(T)$ vs T plot. Here it is clearly observed that the peaks vanished and the residual resistance also decreased when compared with that of fresh film. Also the peak heights are much smaller compared to those observed by Hardy et al[3] . It is possible that our measurements are on samples with more impurity (Pr) concentrations and also our thin films are not completely free of grain boundaries like single crystals. As it is well established, these grain boundaries can partially mask the intrinsic effects and here in our case the height of peaks.

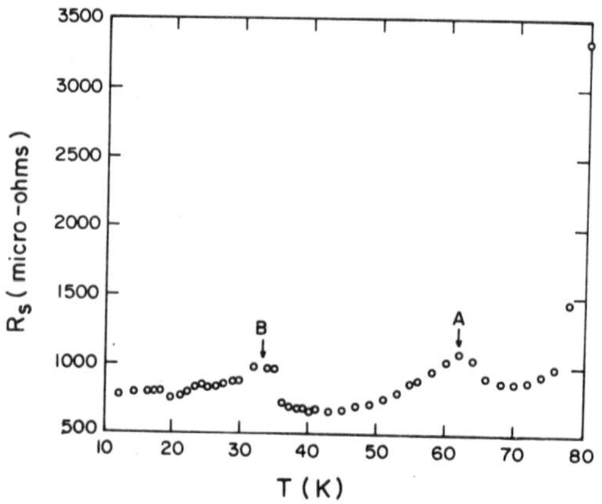

Fig.1 Microwave Surface Resistance vs Temperature measured at 9.1 GHz, for as-deposited $Lu_{1-x}Pr_xBa_2Cu_3O_{7-\delta}$ thin film with x~0.07.

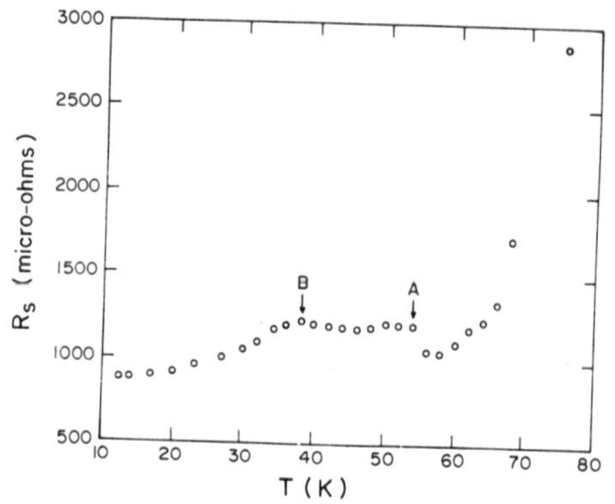

Fig.2 Microwave Surface Resistance vs Temperature measured at 9.1 GHz, for as-deposited $Lu_{1-x}Pr_xBa_2Cu_3O_{7-\delta}$ thin film with x ~ 0.2.

These peaks can be understood in the following physical picture given by Hardy et al[3] In the local limit, surface resistance $R_S(\omega,T)$ can be written as

$$R_S(\omega,T) = (\mu_o^2/2)\, \omega^2\lambda^3(T)\sigma_1(\omega,T) \tag{1}$$

Where

$$\sigma_1(\omega,T) = (1/(\mu_o\lambda^2(0)))\, x_n(T)(\tau(T)/(1+\omega^2\tau^2(T))) \tag{2}$$

with $x_n(T) = (1-x_s) = 1-(\lambda(0)/\ \lambda(T))^2$ representing the normal fluid fraction. Below T_c , scattering time increases rapidly and reaches a saturation value, limited by defects or impurities. Also x_n decreases below T_c as the sample is cooled. So there can be a competition between x_n (T) and τ (T) leading to a peak in $\sigma_1(\omega,T)$ itself, which inturn shows up in $R_S(T)$. We believe that the peak at 64 K (peak 'A') is coming from this origin.This view is supported by the fact that the peak position shifts to lower temperature and also broadens as shown in Fig.2 as x (impurity concentration) is increased from x~0.07 to x~0.2, Hardy et al[3] have observed similar effects (Fig.6 in ref.[3]). The fact that there are no peaks in $R_S(T)$ vs T plot for x = 0 and that they are observed only for x $\neq$ 0 indicates that Pr doping breaks pairs and enhances the competition between x_n (T) and τ (T), leading to the observation of peaks in $R_S(T)$ vs T plots.

As shown in Fig.1 and 2, the x dependence of the position of peak 'B' is opposite to that of peak 'A'. Peak 'B' shifts to higher temperature as x is increased from x~0.07 to x~0.2. This shows that the origin of peak 'B' is different from that of peak 'A'. By examining equation (1) a peak is also possible, from the competition between $\lambda^3(T)$ and $\sigma_1(\omega,T)$ itself. Distinguishing this competition process from the competing process between x_n and $\tau(T)$ is relatively easy. Since the variation of x_n (~ $1-(\lambda(0)/\lambda(T))^2$) is much slower than that of $\lambda^3(T)$, the peak due to competition between x_n and $\tau(T)$ (peak 'A') should occur at much higher temperature than the peak arising out of competition between $\lambda^3(T)$ and $\sigma_1(\omega,T)$ (peak 'B'). As Pr concentration (x) increases, T_c decreases, penetration depth increases, and therefore the peak 'B' shifts to higher temperatures. The increase in penetration depth with decrease of T_c is justified at least in the two fluid picture with the relation $\lambda(T) = \lambda(0)\{1-(T/T_c)^4\}^{-1/2}$.

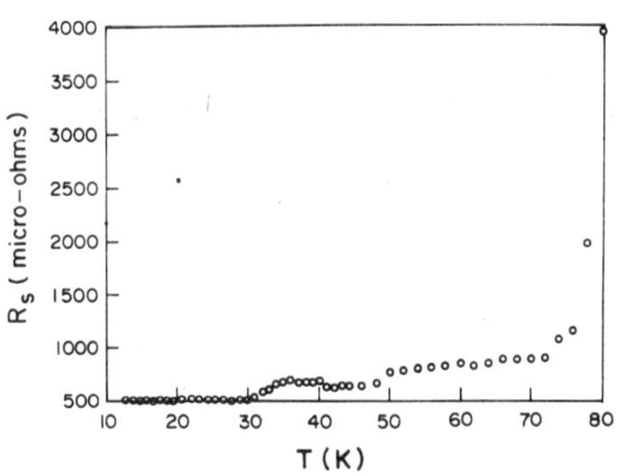

Fig.3(a) Microwave Surface Resistance vs Temperature measured at 9.1 GHz, for a $Lu_{1-x}Pr_xBa_2Cu_3O_{7-\delta}$ thin film with x ~ 0.07, after 24 hours.

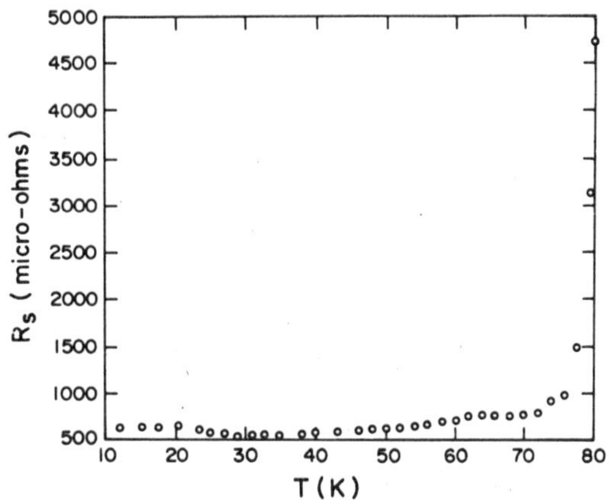

Fig.3(b) Microwave Surface Resistance vs Temperature measured at 9.1 GHz, for a $Lu_{1-x}Pr_xBa_2Cu_3O_{7-\delta}$ thin film with x ~ 0.07, after 100 hours.

The aging effects shown in Fig 3. can be understood in the same picture as above. The $Lu_{1-x}Pr_xBa_2Cu_3O_{7-\delta}$ system in the thin film form is a strained lattice[8,9], and relaxes with time. The relaxation of lattice strains, can in turn drastically affect the quasi particle scattering time, which in turn affects the different competing processes described above, leading to vanishing of the peaks in $R_S(\omega,T)$ vs T. It was shown by Lippert et al[9]., that high residual surface resistance is due to the lattice imperfections . The aging effects in our case show a large decrease of residual resistance within a few days of aging. This also supports our explanation that lattice strains are relaxing with time. A fresh $Lu_{1-x}Pr_xBa_2Cu_3O_{7-\delta}$ film with lot of lattice strains (lattice imperfections) shows larger residual surface resistance than an aged film (see Fig.1 and Fig.3), which is to be expected.

CONCLUSION

Microwave surface resistance in $Lu_{1-x}Pr_xBa_2Cu_3O_{7-\delta}$ thin films varies non-monotonically with temperature and exhibits two peaks in $R_S(\omega,T)$ vs T plot for both dilute and strong doping of Pr in the system. This is explained in terms of the different competitive processes involving the temperature dependences of quasiparticle scattering time, pair condensation, and penetration depth. Strong aging effects are observed and are attributed to the relaxation of lattice strains with time in the system.

REFERENCES

1. A.M. Portis, "Electrodynamics of High-Temperature Superconductorrs", Lecture notes in Physics Vol **48**, World Scientific, (1992).
2. Kuan Zhang, D.A. Bonn , Ruixing Liang, D.J. Baar and W.N. Hardy, Appl. Phys. Lett. **62**, 3020, (1993).
3. W.N. Hardy, S. Kamal, D.A. Bonn, Kuan Zhang, Ruixing Liang, E. Klein, D.C. Morgan and D.J. Baar, Physica B **197**, 609, (1994).
4. S.E. Lofland, B. Ratnam, S.M. Bhagat, I. Takeuchi, Z. Trajanovic, P. Warbuston and T. Venkatesan, Solid State Commun.**94**, 471 (1995).
5. R. Pinto, K.I. Gnanasekar, H.V. Keer, A.S. Tamhane, L.C. Gupta and R. Vijayaraghavan, Physica C **227**, 120 (1994).
6. K.I. Gnanasekar, R. Pinto, A.S. Tamhane, L.C. Gupta, S.P. Pai, P.R. Apte, M. Sharon and R. Vijayaraghavan, Phys.Rev. B **52**, 1362 (1995).
7. R. Pinto, P.R. Apte, L.C. Gupta, R. Vijayaraghavan, K. Easwar and B.K. Sarkar, Supercond.Sci.Technol.**4**,577 (1991).
8. R. Pinto, Navdeep Goyal, S.P. Pai, P.R. Apte, L.C. Gupta and R. Vijayaraghavan, J. Appl. Phys.**73**, 5105 (1993).
9. M. Lippert, J.P. Strobel, G. Saemann-Ischenko, S. Orbach, S. Hensen, G. Muller, H. Piel, J. Schutzmann, K.F. Renk, B. Roas, W. Gieres, Physica C **185-189**, 1041 (1991).

DEPENDENCE OF SUPERCONDUCTING RESONATOR Q ON TEMPERATURE

G.P. Srivastava[1], Vincent Mathew[1] and Agnikumar G. Vedeshwar[2]

[1]*Department of Electronic Science, University of Delhi South Campus, New Delhi 110 021, India*
[2]*Department of Physics and Astrophysics, University of Delhi, Delhi 110 007, India*

The dependence of Q of a superconducting microstrip resonator on temperature has been analysed. The Fourier transformed dyadic Green's functions are derived and the resonant frequency is calculated in a Galerkin procedure for different temperatures. The attenuation constant and propagation constant are evaluated at the resonant frequency and the unloaded Q of the resonator is estimated.It has been observed that the unloaded Q is very sensitive to the operating temperature.

I. INTRODUCTION

High temperature superconducting (HTS) materials are used in many microwave and millimeter wave devices such as filters, resonators, delay lines etc. The stability in performance of these devices with respect to variation in the different physical conditions is a factor to be examined for a better design of such circuits. Experimental results[1] show considerable influence of temperature on HTS circuit performance. In this paper we present an analysis of this effect using the rigorous spectral domain formulation, in order to obtain an exact view of such dependence.

The temperature dependance of the resonator performance can be due to two mechanisms: First, the complex conductivity of the superconducting material is dependent on temperature. Second, the thermal expansion of the substrate material will cause the resonator patch to elongate as a function of temperature. These two mechanisms are included in this study.

II. METHOD OF ANALYSIS

The temperature dependance of the complex conductivity will enter the analysis through the surface impedance of the HTS strip,

$$Z_s = \left[\frac{j\omega\mu}{\sigma(\omega,T)}\right]^{1/2} \tag{1}$$

Here the complex conductivity $\sigma = \sigma(\omega,T)$ of the superconductor is calculated by the phenomenological two fluid model[2]. The effect of the thermal expansion of the substrate is to elongate the superconducting strip as can be represented by the standard relation,

$$L(T) = L(0)\,[1 + \alpha' T] \tag{2}$$

where α' is the coefficient of thermal expansion of the substrate and T, the temperature in Kelvin.

The method of analysis employed here follows the method used for normal resonator[4] to obtain transverse field components E_x and E_z in spectral domain. Dyadic Green's functions are derived by incorporating the complex resistive boundary conditions for the superconducting strip[3] in the spectral domain formulation. The resulting equation is given by,

$$\begin{bmatrix}\tilde{E}_z\\ \tilde{E}_x\end{bmatrix} = \begin{bmatrix}\tilde{G}_{xx} & \tilde{G}_{xz} - Z_s\\ \tilde{G}_{zx} - Z_s & \tilde{G}_{zz}\end{bmatrix}\begin{bmatrix}\tilde{J}_x\\ \tilde{J}_z\end{bmatrix} \tag{3}$$

where,

$$\tilde{G}_{xx} = k_{x}k_{z}\,(\gamma_0 + \mu_r\gamma_1\tanh\gamma_1 d)\,\Delta^{-1} \tag{4a}$$

$$\tilde{G}_{xz} = (\epsilon_r\mu_r k_0^2 - k_z^2)\,\gamma_0 + \mu_r(k_0^2 - k_z^2)\,\gamma_1\tanh(\gamma_1 d)\,\Delta^{-1} \tag{4b}$$

$$\tilde{G}_{zx} = (\epsilon_r\mu_r k_0^2 - k_x^2)\,\gamma_0 + \mu_r(k_0^2 - k_x^2)\,\gamma_1\tanh(\gamma_1 d)\,\Delta^{-1} \tag{4c}$$

$$\tilde{G}_{zz} = \tilde{G}_{xx} \tag{4d}$$

$$\Delta = [\gamma_1\tanh(\gamma_1 d) + \epsilon_r\gamma_0]\,[\gamma_1\coth(\gamma_1 d) + \mu_r\gamma_0] \tag{4e}$$

$$\gamma_0^2 = k_x^2 + k_z^2 - k_0^2 \tag{4f}$$

$$\gamma_1^2 = k_x^2 + k_z^2 - \epsilon_r\mu_r k_0^2 \tag{4g}$$

$$k_0 = \omega\sqrt{\mu_0\epsilon_0} \tag{5}$$

where G_{ij} (i,j = x or z) represent the dyadic Green's functions and $J_{i,j}$ corresponding to the normal microstrip resonator[4].

The resonant frequency for the fundamental mode is calculated from Eq. (3) in a Galerkin procedure, using a set of basis functions[4]. Once the resonant frequency is estimated, the spectral domain method can again be used to find out the attenuation constant α and propagation constant β at the same resonant frequency for an infinitely long microstrip line with the same cross sectional geometry. Now the unloaded Q can be estimated from the relation[5],

$$Q_u = 2\pi[1 - \exp(-4\pi\alpha/\beta)]^{-1} \tag{6}$$

RESULTS AND DISCUSSIONS

In the present calculation, we have used the YBCO thin film resonator with $LaAlO_3$ substrate (ϵ_r = 24 and coefficient of thermal expansion = 8 x $10^{-6}K^{-1}$). YBCO is having transition temperature T_c = 92 K and London penetration depth λ_L = 0.15 μm at T = 0 K. The dimensions used are a = 5 mm, w = 1 mm, L = 4 mm and d = 2 mm. The resonant frequency obtained is 4.4392 GHz at T = 0 K.

The operating temperature is increased from 0 K to 90 K and the resonant frequency is estimated at different temperatures. Fig.2 shows the shift in the normalised resonant frequency with temperature. Curve (1) shows the frequency shift due to thermal expansion, curve (2) due to change in kinetic inductance and curve (3) due to both mechanisms. It can be seen that the total shift in the resonant frequency is about 10 MHz which is not negligible.

Fig.3 shows the variation of Q_u with temperature. Our analysis shows that the resonator Q_u is very sensitive to the changes in the operating temperature, even below 77K (liquid nitrogen boiling point). The relative difference of Q_u can be estimated to be about 20 % between 60K and 80K. These calculated results are in good agreement with the experimental results[1].

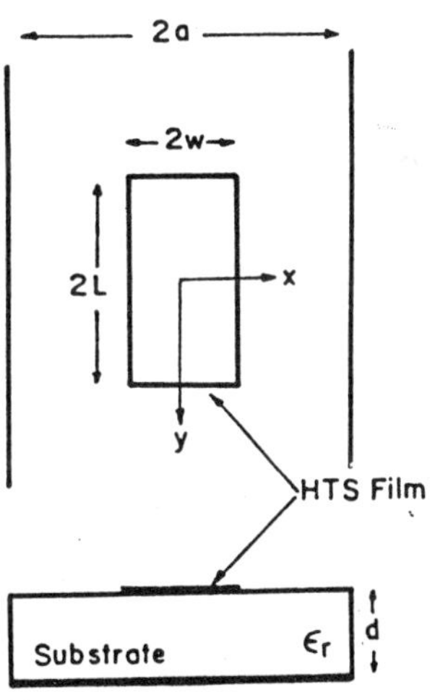

Fig. 1. Schematic of YBCO thin film resonator

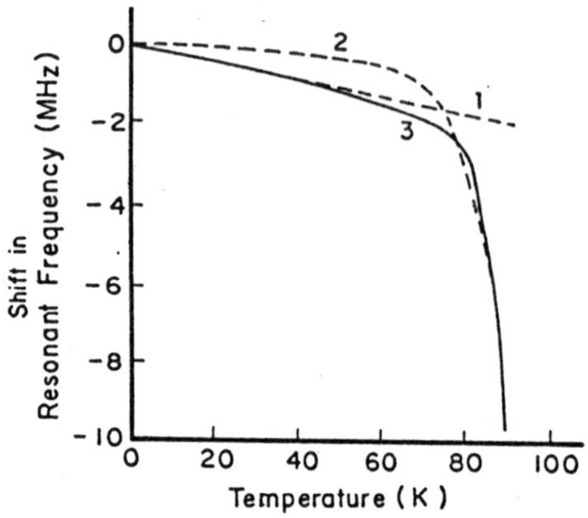

Fig. 2. Shift in normalized resonant frequency with temperature

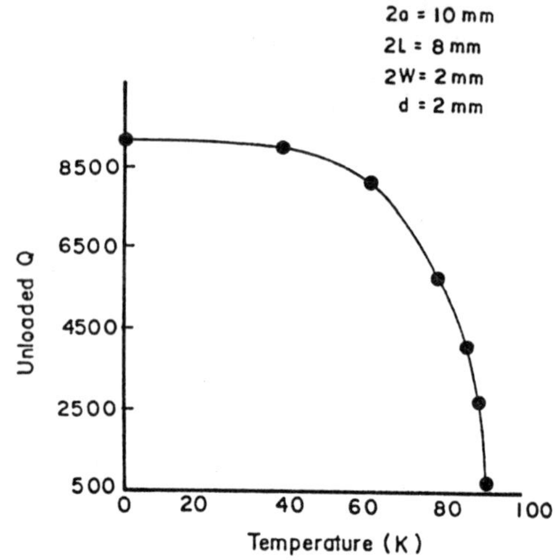

Fig. 3. Variation of unloaded Q (Qu) with temperature

The presented analysis has shown how the performance of the superconducting microstrip resonator is affected by changes in the operating temperature. This analysis can be modified to deal with many more complicated problems associated with present day research in the field. Inclusion of substrate anisotropy, consideration of thick superconducting strips etc. are examples.

ACKNOWLEDGMENTS

One of the authors (VM) is thankful to UGC for the financial assistance.

REFERENCES

1. C. Wilker, Z.Y. Shen , P. Pang , W.L. Face , A.L. Mathews, and D.E. Laubacher, IEEE Trans. MTT-39, No. 9, PP. 1462 (1991).
2. M. Tinkham, *Introduction to superconductivity*, McGraw Hill, Inc., New York, 1996.
3. J.M. Pond, C.M. Crowne, and W.L. Carrter, IEEE Trans. MTT-37, No.1, pp.181 (1989).
4. T. Itoh, IEEE Trans. MTT-22, No.11, pp.946-951 (1974).
5. J. Kessler, R. Dill, and P. Russer, IEEE Trans. MTT-39, No. 9, pp. 1566 (1991).

GROWTH OF $DyBa_2Cu_3O_x$ THIN FILMS BY MOLECULAR BEAM EPITAXY

J. C. Vyas, K. P. Muthe, D. K. Aswal, D. P. Gandhi, S. K. Gupta, R. Pinto*, G. P. Kothiyal, S. C. Sabharwal and V. C. Sahni

Technical Physics and Prototype Engineering Division, Bhabha Atomic Research Centre, Trombay, Mumbai 400 085, India
**Tata Institute of Fundamental Research, Colaba, Mumbai 400 005, India*

In-situ growth of superconducting $DyBa_2Cu_3O_x$ thin films on single crystal $SrTiO_3$ substrates has been investigated under molecular beam epitaxial conditions and using a dc glow discharge source to provide activated oxygen. The effect of the substrate temperature on phase formation and texture of the films has been investigated. Epitaxial films with c-axis orientation have been prepared at 700°C. The films were found to undergo superconducting transition at ~ 90 K and exhibited a critical current density of ~$7x10^5$ Amp/cm^2 at 77 K.

I. INTRODUCTION

For device applications, the fabrication of S-I-S and S-N-S junctions or multilayer structures based on high temperature superconductors (HTSCs) requires the preparation of a sharp interface with precise composition control. These requirements arise due to the fact that the HTSC materials have very short coherence length which necessitates the preparation of a very thin intermediate layer (~10 Å). Molecular beam epitaxy (MBE) has the inherent capability of yielding films with surface smoothness much better than that obtained with any other known thin film deposition technique.[1] Consequently, efforts have been made to prepare films of HTSC and related materials under MBE conditions.[2-7] In order to obtain a molecular beam of the species being evaporated, the mean free path has to be maintained larger than the source to substrate distance (about 20 cm). Therefore, the oxygen over pressure maintained during the deposition should be less than $2X10^{-4}$ Torr.[8] However, under this condition, the high T_c phase is unstable at temperatures higher than 500°C that are required for formation of high T_c phase. In order to meet both the above conditions, several better oxidants like atomic oxygen, ozone and NO_2 etc. have been attempted.[9] In the study reported here, we investigate the conditions suitable for the preparation of epitaxial thin films of $DyBa_2Cu_3O_x$ (DBCO) under MBE conditions and in presence of atomic oxygen.

The growth of DBCO phase is known to occur at oxygen pressures and temperatures that govern the stability of CuO.[10] The deposition of DBCO was carried out at different temperatures between 500 and 700°C and the resultant films were characterized using Reflected high energy electron diffraction (RHEED), X-ray diffraction (XRD), Scanning electron microscopy (SEM) and four probe resistivity techniques. Based on the feedback obtained, the conditions conductive for the preparation of epitaxial films and exhibiting good superconducting properties were found out.

II. EXPERIMENTAL

The deposition of composite films of DBCO was carried out using effusion cells for the evaporation of Ba and Dy and electron gun for Cu_2O of the Riber, EVA-32 MBE system. The substrates, prior to deposition, were cleaned first using an organic solvent and then by heating them at 750°C for an hour under 10^{-8} Torr vacuum. A streaky RHEED pattern for the cleaned substrates was obtained, which ensured its good surface condition. A dc discharge tube, operating in the current regulation mode, was employed to produce active oxygen. The flow of oxygen was regulated to maintain a pressure of 10^{-5} Torr in the growth chamber, which corresponds to the flow of $3.5X10^{17}$ gas species/sec. The fraction of atomic oxygen generated here depends upon the discharge current. In the present case, a discharge current of 15 mA was maintained, which according to the data available in the literature[11] yielded atomic oxygen flux equal to ~ $2X10^{15}$ atoms/cm^{2}sec. An ionization gauge for the measurement of Dy and Ba fluxes and a quartz crystal thickness monitor for Cu flux were used. The source materials used were the metal chunks of Ba and Dy and Cu_2O pellets. All of them were of 4N purity. The films were characterized using RHEED, SEM, XRD and four probe resistivity techniques.

III. RESULTS AND DISCUSSION

Earlier studies carried out on the growth and oxidation behavior of Ba and Dy films have shown that both the materials could be easily oxidized at temperatures that are required for the preparation of DBCO phase. Barium was seen to deposit as BaO under molecular oxygen pressure of 10^{-6} Torr and substrate temperatures higher than 600°C.[12] Also, Dy films were seen to have unusually high affinity for oxygen. These films when grown under UHV conditions, and at temperatures as low as 200°C were found to get oxidized by picking up oxygen from the oxide substrate itself.[13] Thus, the requirement of oxygen for both these materials was found to be compatible with MBE conditions. However, for the preparation of CuO films, atomic oxygen flux much higher than the metal flux is found necessary.[14] The total metal flux maintained in present study was about $3.6X10^{14}$ atoms/cm^{2}sec.

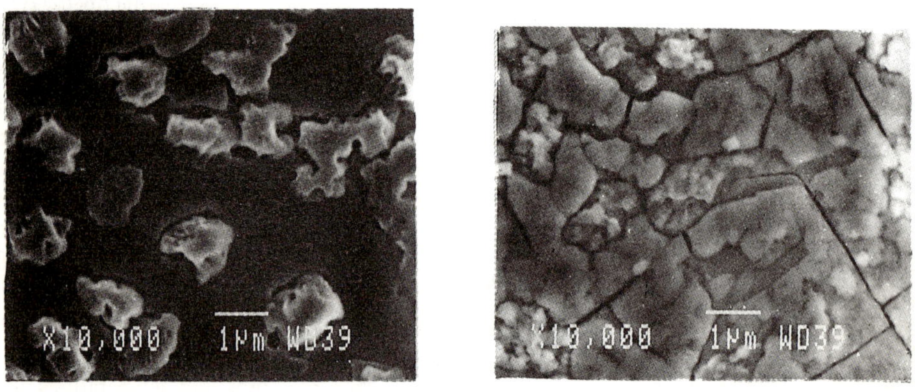

(a) (b)

Fig.1. SEM photographs for films deposited at a) 650°C and b) 700°C.

The films grown at substrate temperature of ~ 520°C turned out to be amorphous in nature. On annealing at 850°C under flowing oxygen, crystallization and superconductivity could be induced in these films. The annealed films were found to have a T_c of 85K and the critical current density (J_c) measured was lower than 10^4 Amp./cm^2 at 77 K. The reason for poor J_c is attributed to the polycrystalline nature of the films, which was revealed by the XRD measurements. For substrate temperature of 650°C, island growth with average grain size of 1 μm was observed to take place, as shown in Fig. 1(a). The grains were disjointed and hence the films did not show any superconductivity even after annealing at 850°C. The XRD pattern obtained in this case, however, showed the formation of DBCO phase. On increasing the substrate temperature to 700°C, the growth of films with a streaky RHEED pattern and having line spacing similar to a c-axis oriented film was observed. A typical RHEED pattern recorded in this case is shown in Fig. 2.

Fig.2. RHEED pattern for a film deposited at 700°C.

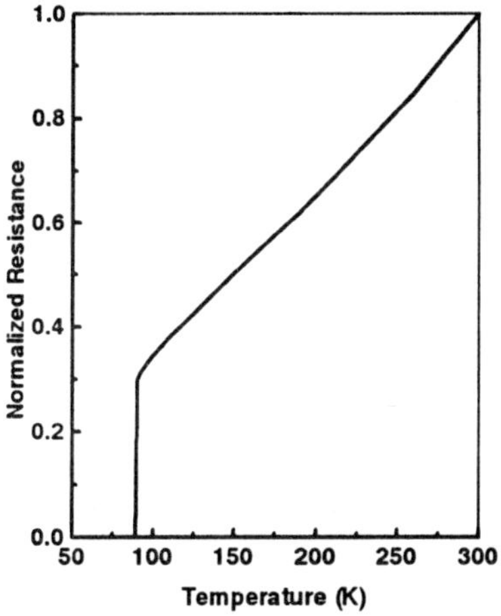

Fig.3. Resistance vs Temperature plot for a typical film deposited at 700°C.

This result suggested the onset of layer by layer (LBL) growth mechanism. The XRD studies carried-out on these films confirmed their c-axis orientation. However, the SEM photograph shown in Fig. 1(b), indicated the growth mechanism to be of mixed type namely layer by layer (LBL) and island growth.[15] Thus the results suggested that the growth mechanism changes from island to LBL on raising the growth temperature from 650°C to 700°C. Further these results indicate that a properly optimized substrate temperature should yield pure LBL growth.

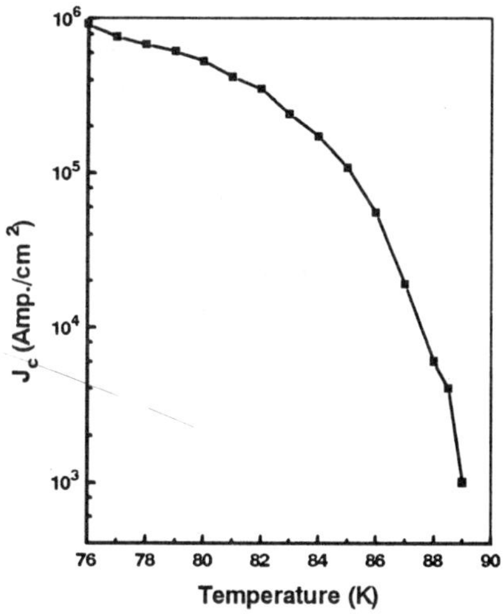

Fig.4. Variation of J_c Vs T for a film deposited at 700°C.

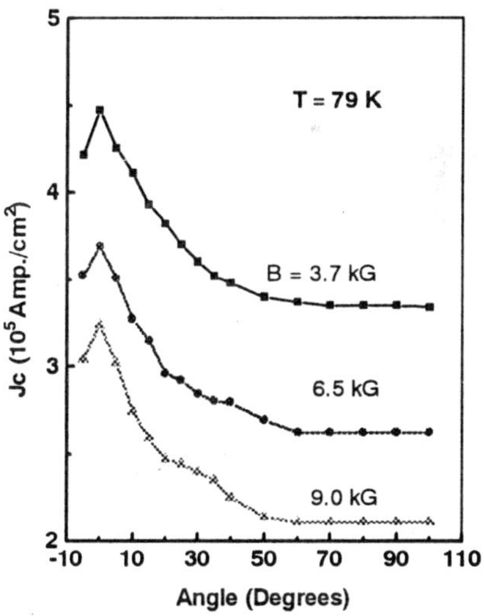

Fig.5. Angular dependence of J_c at different magnetic fields.

For performing four probe resistivity measurement, the films were patterned with an excimer laser to form a bridge of ~100 μm line width and 1mm length. The R vs T plot for a typical film deposited at 700°C is shown in Fig. 3. The superconducting transition in this case occurred at 90 K. The critical current density as a function of temperature was measured and the results obtained are shown in Fig. 4. The films were found have a critical current density of ~7×10^5 Amp/cm^2 at 77 K. The variation of critical current density as a function of angle between the substrate surface and an external magnetic field was studied. The results obtained for different magnetic fields have been shown in Fig. 5. At all the angles the dependence of J_c on magnetic field is found to be weak. This result showed that the films were free from large angle grain boundaries. The angular dependence of critical current shown in Fig. 5 matches with that expected for a c-axis oriented film.[16]

IV. CONCLUSION

The formation of DBCO phase is observed to take place for substrate temperatures higher than 650°C and under sufficient atomic oxygen flux. The growth mechanism below 700°C is found to be of island type and on increasing the growth temperature it progressively becomes layer by layer type. The films prepared at 700°C are found to be epitaxial with c-axis orientation and show a T_c of 90 K and J_c equal to $7X10^5$ Acm^{-2} at 77 K.

REFERENCES

1. M. A. Hermann, and H. Sitter, *Molecular Beam Epitaxy: Fundamentals and Current Status* (Springer, Berlin, 1989).
2. R. J. Shah, H. F. Hess, H. L. Stomer, A. F. White and K. T. Short, Appl. Phys. Lett. **53**, 441 (1988).
3. J. Kwo, M. Hong, D.J.Travor, R.H. Fleming, A.E.White, R. C. Ferrow, A. R. Kortan, and K. T. Short, Appl. Phys. Lett. **53**, 2683 (1988).
4. J. S. Harris, J.N. Ecstein, E. S. Hellman, and D. G. Schlom , J. Cryst. Growth **95**, 607 (1989).
5. J. Kwo, J. Cryst. Growth **111**, 965 (1991).
6. Y. Ishizuka, Y. Terashima, and T. Miura, Japan. J. Appl. Phys. **30**, L1006 (1991).
7. J. Bae, P. Seshadri, and K. M. Choudhary, J. Vac. Sci. Technol. A **10**, 281 (1992).
8. D. G. Schlom, A. F. Marshall, J. T. Sizemore, Z. J. Chen, N. Eckstein, I. Bozovic, K. E. Von Dessonneck, S. Harris and J. C. Bravman, J. Cryst. Growth **102**, 361 (1990).
9. H. Nonaka, S. Ichimura, T. Shimizu and K. Arai, *Critical Reviews in Solid State and Material Sciences* **20**, 285 (1995).
10. D. K. Aswal, Ph. D. Thesis, Bombay University (1995).
11. L. Laska, K. Masek and T. Ruzicka, Czech. J. Phys. B **29**, 498 (1979).
12. K. P. Muthe, J. C. Vyas, G. P. Kothiyal, D. P. Gandhi, A. K. Debnath, S. K. Gupta, S. C. Sabharwal and M.K. Gupta, J. Cryst. Growth **118**, 213 (1992).
13. M. K. Gupta, G. P. Kothiyal, V. C. Sahni, J. C. Vyas, D. P. Gandhi, K. P. Muthe, S. C. Sabharwal, J. Appl. Phys. **78**, 927 (1995).
14. K.P. Muthe, Savita N. Narang, J. C. Vyas, Debarti Bhattacharya, S. K. Gupta, G. P. Kothiyal, V. C. Sahni and S. C. Sabharwal, DAE Solid St. Phys. Symp., Mumbai (1996).
15. T. Frey, C. C. Chi, C. C. Tsuei, T. Shaw and F. Bozso, Phys. Rev. B **49**, 3483 (1994).
16. M. Tachiki and S. Takahashi, Solid St. Comm. **72**, 1083 (1989).

EFFECT OF DISORDER ON THE PEAK EFFECT AND STRUCTURE OF FLUX LINE LATTICE IN 2H-$NbSe_2$

S. S. Banerjee [1], N. G. Patil [1], Subir Saha [1], S. Ramakrishnan [1], A.K. Grover [1], S. Bhattacharya[1], G. Ravikumar [2], P.K. Mishra [2], T.V. Chandrasekhar Rao [2], V.C. Sahni [2], C.V. Tomy[3], G. Balakrishnan[3], D.Mck Paul [3] and M. J. Higgins [4]

[1]Tata Institute of Fundamental Research, Mumbai - 400005, India
[2]TPPED, Bhabha Atomic Research Centre, Mumbai - 400085, India
[3]Department of Physics, University of Warwick, Coventry, CV4 7AL, U.K.
[4]NEC Research Institute, 4 Independence Way, Princeton, New Jersey 08540, USA

The effect of disorder on flux line lattice (FLL) melting is studied via the Peak Effect phenomenon. On the upper branch of the Peak Effect curve, where this effect is more robust, we observe a stepwise disordering of the FLL, suggesting two first order transitions spanning the Peak Effect region. The lower, reentrant part of the curve is found to be strongly affected by disorder. Both the branch and the "nose", i.e., the turnaround of the curve, disappear in systems with stronger disorder and the observed results are consistent with the line separating a disentangled liquid or a glass from an entangled flux liquid.

I. INTRODUCTION

The effects of disorder on the phase boundaries in the flux line lattice (FLL) provide a way to measure the competition between inter-vortex interaction and both thermal and quenched disorder[1]. Due to the variety of materials with a large range of quenched disorder and transition temperatures, and the easy tunability of the interaction using the magnetic field, the FLL provides a rich testing ground of the theoretical concepts[2] and of the phase behaviour without and, more notably, with disorder[3]. In the case of the much-studied cuprates, quenched disorder is known to destroy[4] a sharp melting phase boundary of FLL; the residual effect is the presence of the so-called irreversibility line separating a less mobile state of FLL from a more mobile one for small driving forces. Whether the latter is indeed a phase transition line or a mere crossover of dynamic response remains uncertain[5]. A second question is concerned with the Peak Effect (PE), a sharp rise in the apparent critical current density in close proximity of the melting line. In some descriptions, this effect itself is associated with melting, while in some other, it represents a premelting softening of the FLL. The precise relationship between PE and FLL melting and whether it depends on specifics of systems under study continue to remain subjects of current work[6,7]. It was shown recently[8] that in the case of weak pinning single crystal samples of hexagonal 2H-$NbSe_2$, the PE curve bears a remarkable similarity to the theoretical predictions of Nelson[9] of the melting phase boundary of pure FLL, including the lower reentrant branch[2]. The location of this reentrant line, including the turnaround, often referred to as the "nose", is also in good quantitative agreement with simulations[10].

The disappearance of peak effect at field values below the nose was argued [8] to be consequence of dominance of pinning on the interaction. In order to further explore these ideas, we undertook investigations on the effects of increasing disorder on the magnetic signatures of the PE curve in 2H-$NbSe_2$. Specifically, we find that the entire low field part of this curve is extremely susceptible to disorder; neither the "nose" nor the lower branch of the PE curve survives stronger disorder. The upper branch is relatively more rugged; but even for this branch, over a wider parameter space, the full disordering of the FLL occurs via steps, suggestive of first order transitions and ending, remarkably, with the exact location for the disappearance of thermomagnetic history effect in the magnetization screening response.

II. EXPERIMENTAL RESULTS

The experiments were carried out on a single crystal of Fe-doped $NbSe_2$, similar to samples where transport studies of metastability were reported recently[12]. Fig.1(a) shows a dc magnetization hysterisis loop for H||c at 5.1 K recorded with a standard Quantum Design Inc. Squid magnetometer. Note that the magnetization loop opens up anomalously at the onset of the PE at a field marked H_m, reaching a peak at H_p. The resulting ``magnetization bubble" closes at a higher field H_{irr} below the nominal H_{c2}. Fig.1(b) shows the H-dependence of the real part of the ac susceptibility χ'. In this experiment, the PE is manifested as an anomalous enhancement of the diamagnetic response that appears at H_m and reaches a maximum value at H_p. Together, these results (Figs 1(a) and 1(b)) show the self-consistency between the anomalous dc magnetization hysterisis and the ac susceptibility behaviour in detecting the PE.

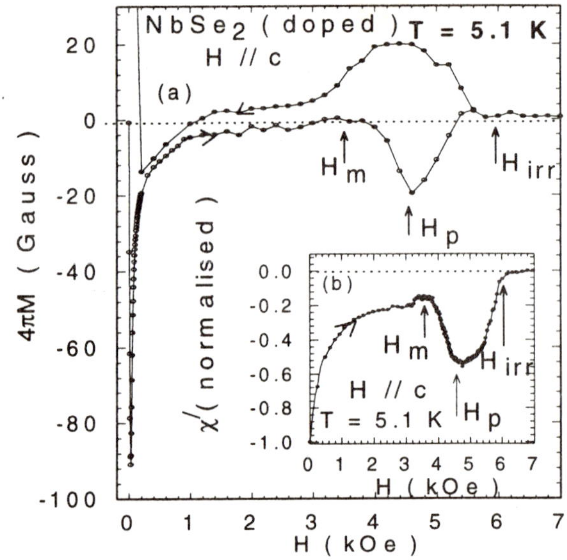

Figure 1: The main panel (a) shows dc magnetisation hysterisis curve for H//c in doped crystal of 2H-$NbSe_2$ ($T_c(0) \approx 6$ K) at 5.1 K. The inset (b) depicts the in-phase ac [f=211 Hz, h_{ac}=0.5 Oe (r.m.s)] susceptibility (χ') vs H at nearly the same temperature. The fields H_p and H_{irr} in the inset are marked as per dc data in the main panel.

Fig.2 shows the temperature dependence of the real part of the ac susceptibility χ' at a fixed H in the doped sample for two different thermomagnetic histories : field cooled (FC) and zero-field-cooled (ZFC). For ZFC, we observe a sharp drop in χ' at the onset of the PE, marked T_m, following which χ' continues to decrease until at a higher temperature marked T_p it shows yet another sharp drop, after which it rises rapidly. A comparison with the field cooled (FC) data shows a distinct difference with the ZFC data. The χ' response in FC state at low temperatures ($<T_m$) is more diamagnetic, implying a more strongly pinned (i.e., more disordered) lattice, in agreement with recent transport measurements[12]. Importantly, the thermomagnetic memory effect disappears at T_p above which the FC and ZFC curves merge into one. The peak effect extends over a region spanned by two temperatures, T_m and T_p. For ZFC, the more weakly pinned and hence the more ordered state, we see significant structure in χ', namely, the sudden drops, while the more disordered FC state shows much less structure. Since thermomagnetic history effect disappears above T_p, the lattice is disordered in equilibrium and hence a candidate for a pinned liquid or an amorphous FLL. Field cooling through this regime freezes FLL in a disordered structure of the vortices and thus the FC state is more akin to a supercooled liquid or a glass. Transport data[12] showed that the FC state, metastable in the peak regime and below, when "annealed" with a greater than critical current passed through it, produces a state with a critical current nearly identical to the ZFC state, i.e., the ZFC state is closer to the equilibrium structure of the system. Thus, the structures observed in the ZFC state closely approximate the equilibrium features of the system. In what follows, we present a tentative explanation of the observations in Fig. 2. First we note that both the drop features are extremely sharp and narrower than 5mK, whereas the normal to superconducting transition itself (see zero

field curve in Fig.3(b)) is nearly fifty times as large. The two sudden drop features are thus signatures of well-defined first order phase transitions. The sudden drop in χ' for ZFC at T_m is analogous to the sharp increase of J_c at H_m in Fig. 3 of Ref [12].

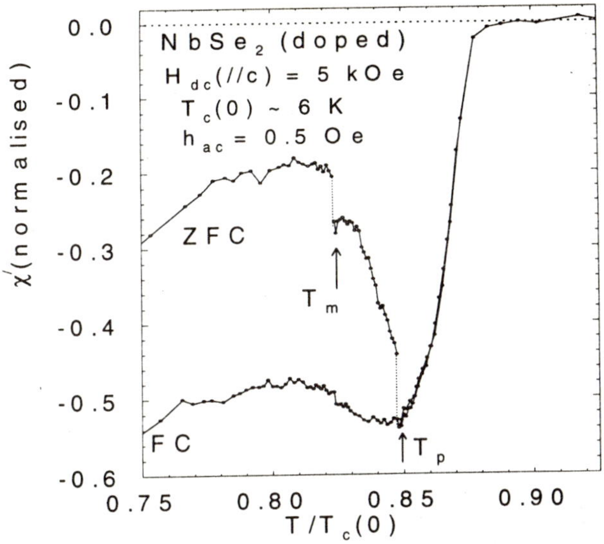

Figure 2 : Temperature variation of the in-phase ac susceptibility χ' at H(//c)=5 kOe in the doped crystal of 2H-NbSe$_2$ for two thermomagnetic histories, zero field cooled (ZFC) and field cooled (FC). The arrow mark the temperatures T_m and T_p at which discontinuous jumps are observed in ZFC case. The χ' data are normalised such that the extremum (diamagnetic) screening response corresponds to the value 1.

Within the Larkin - Ovchinnikov[14] model of collective pinning, the pinning force F_p is given by $F_p = (n_p f^2/V_c)^{1/2}$, where f is the elementary pinning interaction, n_p is the density of pins and V_c is the correlation volume (= $R_c^2.L_c$, where R_c and L_c are collective pinning radial and longitudinal lengths[2,3,14]). Since the pinning interaction can only decrease with increasing T, a sharp increase in F_p at T_m can only occur via the sudden drop in V_c and it must come from a sudden drop in one (or more) of the elastic moduli of FLL. At a higher T, we observe a second drop in V_c at T_p, which we attribute to the final disordering of the FLL through the collapse of another elastic constant. A plausible implication is that the two elastic constants are the shear (c_{66}) and tilt (c_{44}) moduli. One can envisage two different scenarios : the first, proposed by Kes and coworkers[15], suggests that the first drop in V_c corresponds to the collapse of the tilt modulus and the crossover of the effective dimensionality of the FLL from two to three. Since it was not observed before, the second anomalous drop in V_c is not explained in this case. A second

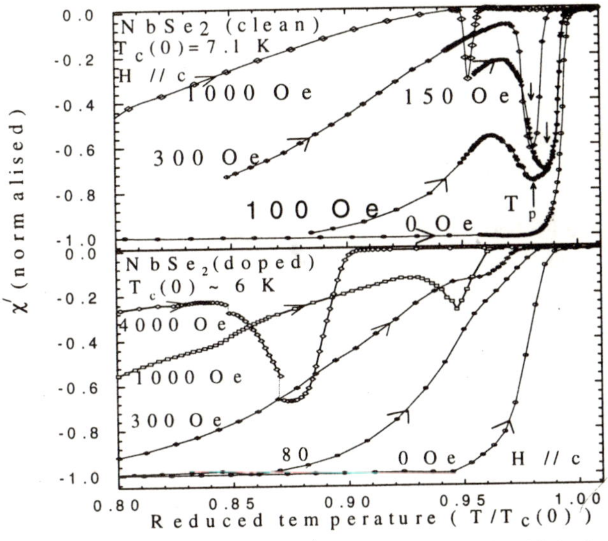

Figure 3: Temperature variation of χ' at different H (//c and ZFC) in the low field region for a clean and a doped crystal of 2H-NbSe$_2$. The doped crystal is more disordered than the undoped clean specimen as it has lower value of $T_c(0)$ and a broader transition in zero field. The vertical arrows in Fig. 3(a) mark the peak temperatures $T_p(H)$.

scenario would suggest that the onset (near T_m) of the PE is caused by the collapse of c_{66}; above T_m, one has a line liquid which is still correlated along H. In that case, the second anomaly at T_p could be the collapse of the tilt modulus c_{44}, which completes the disordering of the FLL. Above T_p, one has a fully disordered state which could explain the disappearance of the thermomagnetic history effect above T_p. Which (if any) of the scenarios is correct is unclear at present, but it is nevertheless obvious that the full disordering of the FLL does not happen in one step.

Figs.3(a) and 3(b) show isofield measurements of χ' in an earlier[8,16] clean crystal specimen and the present doped crystal, respectively. For the cleaner system (in Fig. 3(a)), the data span fields greater than and less than the "nose" location (cf. Fig. 4(b)); one observes a sharper peak effect and the nonmonotonic variation of T_p (marked by arrows) with H leading to the reentrance characteristic in PE curve as shown in Fig. 4(b). In the doped system (see Fig. 3(b)), however, the peak effect starts to broaden at significantly higher fields (than the "nose" location), it manifests as a shoulder in χ' around 150 gauss and for further lower values, it can no longer be located precisely. It is important to note that the gradual broadening of the PE transition and its ultimate disappearance are precisely what would be expected for the increasing effects of disorder in destroying the phase boundary defining the melting of FLL[5,17]. Also, there is no nonmonotonicity in the field range where the PE is observed in the doped sample. In other words, neither the lower branch, nor the nose can be detected unambiguously for the more disordered system.

III. DISCUSSION

The (H,T) phase boundaries emerging from ac and dc magnetization experiments for both doped and clean samples are shown in Fig. 4. The two PE lines in Fig. 4(a) for doped sample correspond to the two jumps in χ' at T_m and T_p at fixed H (cf. Fig.2). Clearly, a simple Lindemann criterion of a one step melting process cannot be applied to the two lines. Nevertheless, we attempt a fit to the centre of the two lines, for lower T and higher H, i.e., the so called upper branch of the PE curve, for an order of magnitude estimate of Lindemann parameter (c_l) by using the relation 5.5 of Ref. 2 , $B_m(T) = \beta_m(c_l^4 / Gi) H_{c2}(0)(1-T/T_c)^2$. A fit to this expression with values of various parameters kept the same as those in Ref. 8, yields a reasonable value for the Lindemann parameter ($c_l \sim 0.1$).

The PE becomes immeasurably weak at around 150 gauss for the doped case (see Fig. 3(b)), while in the clean system it disappears below the nose around 30 gauss. We note (Fig. 4(b)) that the upper branch of the PE curve is also affected by disorder: the curve shifts down with increasing disorder (cf. χ' curves at H = 300 Oe in clean and doped samples in Fig. 3(a) and Fig. 3(b), respectively). This can be understood on general grounds since increased disorder reduces the correlation length of the vortex solid which can thus melt with weaker thermal fluctuations. The lower branch of PE curve is more profoundly altered. As stated earlier neither the nose, nor the lower branch is detectable in the more disordered doped case, which implies that the solid phase is squeezed upwards to higher H. In other words, larger intervortex interaction is needed to stabilise the solid phase. The disappearance of the PE curve at fields lower than 30 Oe even for the clean case was attributed [8] to the dominance of the disorder for the dilute array of vortices, as per arguments given for the Nelson-Le Doussal line[2,17]. The relevant boundary is governed by the entanglement length L_E becoming equal to the pinning length L_c, following the relationship 6.47 of Ref. 2 :

$$\frac{L_E}{L_C} = \frac{\pi \kappa^2 \ln\kappa}{\sqrt{2}} \left[\frac{a_o}{2\pi\kappa(0)}\right]^2 \left[\frac{j_c}{G_i j_o}\right]^{1/2} \frac{(1-t)^{4/3}}{t}$$

where $a_0^2 \propto 1/B$ and the rest of the symbols have their usual meaning[2]. In the clean sample, it was found that $L_c / L_E \approx 1$ at $B \approx 30$ Oe[8].The relation (1) implies that the crossover field (corresponding to $L_c / L_E = 1$) is proportional to the square root of the critical current density j_c, which is larger for the doped case by nearly a factor of 50. This coupled with the nominal changes in the values of κ and λ(0) of the doped sample as compared to those in a clean sample yields a crossover field of ~ 4×10^2 Oe at t ≈ 0.96, in reasonable agreement with the observed value of 150 Oe (cf. Fig 3(b)). Thus, the loss of the PE curve is consistent with the transformation of a thermally entangled flux liquid at low fields to a pinning dominated disentangled liquid or a glass[18]. Whether this crossover is accompanied by changes in the dynamic response of the system remains unknown.

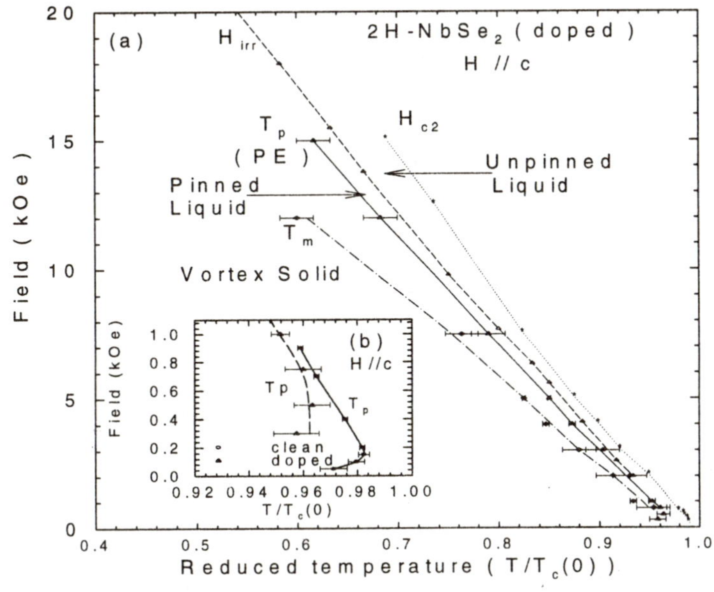

Figure 4 : Magnetic phase diagram for H//c in doped crystal of $2H\text{-}NbSe_2$ having $T_c(0)$~6 K. The two PE lines, $T_p(H)$ and $T_m(H)$, and the irreversiblity line $H_{irr}(T)$ can be identified from the data such as in Fig. 2 and Fig.1, respectively. The inset (b) shows a comparision between $T_p(H)$ lines in the doped and a clean crystal. The $T_p(H)$ line in the latter sample displays the nose and the reentrant characteristics at H < 150 Oe.

The irreversibility line (H_{irr}) in Fig.4(a), has been obtained from data such as in Fig.1. Below this line (and above T_p line), the FLL is disordered but pinned and is thus very likely a candidate for pinned liquid state as mentioned above. The irreversibility line thus corresponds to a crossover between a pinned and an unpinned liquid, as the correlation time of a very viscous flux liquid (plastic time t_{pl} as mentioned in [2]) becomes comparable to the pinning time t_{pin}.

IV. CONCLUSION

In conclusion, we have observed a stepwise disordering of the FLL in the upper branch of the peak effect curve in a more disordered sample of $2H\text{-}NbSe_2$. In a cleaner sample these steps, if present, are presumably too close to be resolved. We also find that the role of disorder is to shrink the solid phase[19] in the H-T space for both branches . For the lower reentrant branch in particular, the effect of disorder is very dramatic in the form of a disappearance of the "nose" and the entire branch, due, presumably, to the intersection of the Nelson Le Doussal line with the melting line of FLL at higher field values[2,17], below which the entangled flux liquid phase is destroyed by disorder.

REFERENCES

1. D. S. Fisher, Phys. Rev. B **31**, 1396 (1985).
2. G. Blatter *et al*, Rev. Mod. Phys. **66**, 1125 (1994) and references therein.
3. M. J. Higgins and S. Bhattacharya Physica C**257**, 232(1996) and references therein.
4. J. A. Fendrich, W. K. Kwok *et al*, Phys. Rev. Lett. **74**, 1210 (1995).
5. M. I. J. Probert and A. I. M. Rae, *ibid*, **75**, 1835 (1995) and E. Zeldov *et al*, *ibid* 73, 1428 (1994).
6. A. I. Larkin, M. C. Marchetti, and V. M. Vinokur, *ibid* **75**, 2992 (1995).
7. C. Tang *et al*, Europhys. Lett. **35**, 597 (1996).
8. K. Ghosh *et al*, Phys. Rev. Lett. **76**, 4600 (1996).
9. D. R. Nelson, *ibid*, 60, 1973 (1988); J. Stat. Phys. **57**, 511 (1989).
10. G. Blatter, V. Geshkenbien , Phys. Rev. Lett.**77**, 4958 (1996).
11. P. Kes, Nature **376**, 729 (1995).
12. W. Henderson *et al*, Phys. Rev. Lett. **77**, 2077 (1996).
13. Complications with measurements on superconductors with Quantum Design Inc. SQUID magnetometer sometimes give rise to spurious effects in the PE region due to the magnetic field inhomogeneities along the scan length. This aspect is discussed elsewhere, see, G. Ravikumar *el al*, Physica C **276**, 9 (1997). The data presented in Fig. 1(a) are free of possible artifactual effects.
14. A. I. Larkin and Yu. N. Ovchinnikov J. Low. Temp. Phys.**34**, 409 (1979).
15. R. Wordenweber and P. H. Kes, Phys. Rev. B **34**, 494 (1986).
16. S. Ramakrishnan *et al*, Physica C **256**, 119, (1996).
17. D. R. Nelson and P. Le Doussal, Phys. Rev. B **42**, 10113 (1990).
18. See Fig. 1. of D. R. Nelson, in *The Vortex State* pp 41-61, N. Bontemps *et al.* (eds.), Kluwer Academic Publishers, The Netherlands (1994). Also see Fig. 25 of Ref.2.
19. Gautam I. Menon and Chandan Dasgupta, Phys. Rev. Lett.**73**, 1023(1994).

NON-RESONANT MICROWAVE AND RF ABSORPTION INSIGHTS INTO SUPERCONDUCTING STATE OF HTSC

S.V. Bhat[1], Y.S. Sudershan[1], Amit Rastogi[1], A.K. Grover[2]
K. Oka[3], Y. Yamaguchi[3] and Y.Nishihara[3],

[1]*Department of Physics, Indian Institute of Science, Bangalore 560 012, India*
[2]*Tata Institute of Fundamental Research, Mumbai 400 005, India*
[3]*Electrotechnical Laboratory, Tsukuba, Ibaraki 305, Japan*

The magnetic field dependent dissipative processes in the mixed state of HTSC give rise to intense, low field signals when studied using continuous wave electron paramagnetic resonance (EPR) and nuclear magnetic resonance (NMR) spectrometers. The combination of field modulation and phase sensitive detection commonly used in these spectrometers provides a highly sensitive technique of measuring the field derivative of the power absorbed, dP/dH, by the sample. The technique can be used to detect superconductivity in miniscule amounts of a sample, to determine the quality of a given superconducting material and also to obtain useful insights into its mixed state. We shall illustrate the applications of the technique through our recent results on (1) non-resonant microwave absorption by intrinsic Josephson junctions and (2) non-resonant rf absorption signatures of reentrant melting line, in single crystals of $Bi_2Sr_2CaCu_2O_8$.

I. INTRODUCTION

The study of microwave and rf response of superconductors, especially in the presence of an externally applied magnetic field, is of important fundamental as well as technological interest. Not only can it yield information on the material parameters crucial to many of the device applications but can also provide a test for any plausible microscopic theory of superconductivity. After the advent of the high-T_C superconductors (HTSC)[1,2] a very sensitive and contactless technique of studying the response of superconductors to electromagnetic radiation in microwave and rf regions has come into vogue. It is based on the observation [3,4] that when examined with conventional CW EPR and NMR spectrometers, the new superconductors exhibit, at low magnetic fields, intense signals in their superconducting state, even with miniscule amounts of sample. The sensitivity of the technique results from the magnetic field modulation and lock-in detection commonly used in CW EPR and NMR spectrometers. This mode of detection yields the derivative dP/dH of the field dependent power absorption, in the absence of any hysteresis.

The microwave and rf power dissipation in HTSC is strongly dependent upon even low values of the applied magnetic field due to their granular nature. The ceramic samples (powders and sintered pellets) of course contain an abundance of intergrain contacts which act as weak-links between the superconducting grains. In addition, even nominally single phase, good single crystals and single crystalline thin films are known to consist of various defects such as oxygen vacancies, twin boundaries, stacking faults, dislocations and intergrowths. The very short coherence lengths ($\xi(0) \sim 1 - 15$ Å) and the relatively high transition temperatures of HTSC lead to weak-link structures originating in the intrinsic layered structure as well as in local non-stoichiometry[5] and disorder. This hierarchy of weak-links[6,7] forming Josephson junctions (Jjs) leads to the high sensitivity of the dissipation to the applied field. Magnetic field can induce losses in Jjs mainly through two different mechanisms depending upon the size of the junction. In the so called 'short junction' limit, when the Josephson penetration depth λ_j is larger than the junction dimensions, the magnetic field leads to the 'diffraction reduction' of the critical current density J_c of the junction according to the relation

$J_c = J_c(0)\sin(\pi H/H_o)/(\pi H/H_o)$, corresponding to the entry of one flux quantum ϕ_o (= 2.07×10^{-7} Gcm^2) into the junction. The dissipation in such a situation can be treated using the approach of Dulcic et al.[8] applicable to the classic case of resistively shunted junctions (RSJ). Alternatively if the junction is large enough for the quantized flux in the form of Josephson vortices to be nucleated, the Lorentz force driven damped motion of these fluxons causes dissipation. The situation was first treated by Portis et al.[9]. The RSJ mechanism coupled with the diffraction reduction of J_c when averaged over Jjs of varying strengths and orientations leads to a monotonically increasing dissipation with field. So does the loss arising out of damped fluxons since with increasing field , the number of fluxons contributing to the loss inside the junction would increase. This monotonic increase of the loss with the field gives rise, in the derivative mode, to a signal which is opposite in phase to that of an EPR or an NMR signal. This is the "normal" behaviour and over the last decade it has been reported in a large number of publications[10].

In this paper, we shall first briefly discuss various features of the non-resonant microwave and rf response which can be used to characterize the superconducting samples and then illustrate through our recent results, the power of the technique in illucidating the low-field high-temperature phase diagram of the highly anisotropic $Bi_2Sr_2CaCu_2O_8$ (Bi2212) single crystals.

The rf and microwave responses of the HTSC in the normal state is mostly independent of the applied magnetic field. However, just below T_c, there is a strong, magnetic field dependent dissipation for the reasons mentioned earlier. This fact can obviously be used as a marker of the normal ↔ superconducting transition. This method is highly sensitive because of the magnetic field modulation and phase sensitive detection commonly used in CW EPR and NMR spectrometers. The contactless nature of the technique is an added advantage especially in case of miniscule samples. In case of mixed phase samples so commonly obtained in the standard 'mix-grind-heat-pelletize-sinter' procedure of making them, if there is only a minority superconducting phase present, NRMA would be the technique to give a clear indication of its existence. In such a situation R-T experiment

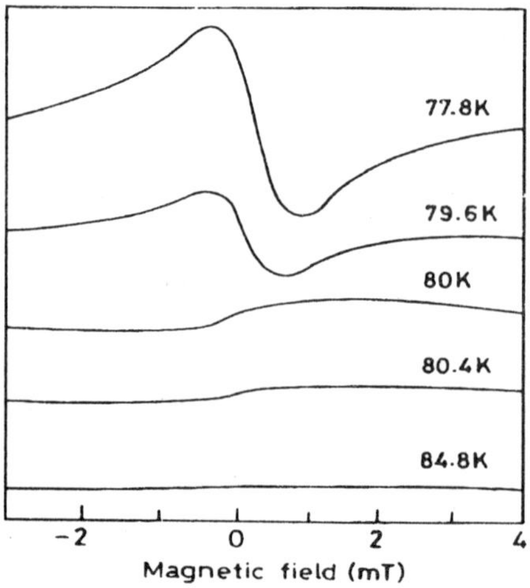

Fig.1. Non-resonant microwave absorption derivative signal of $YBa_2Cu_3O_{7-x}$ as a function of temperature.

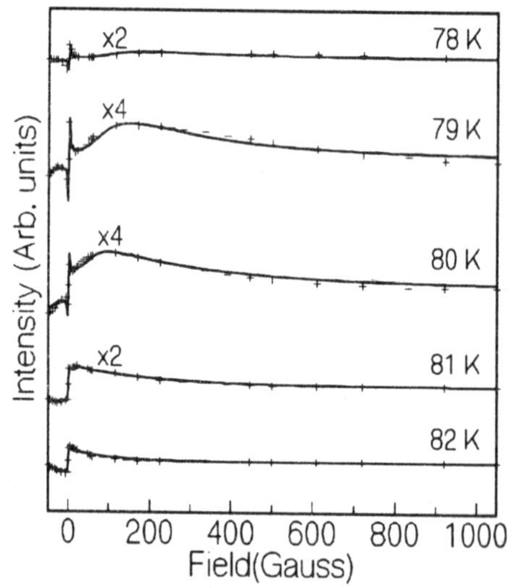

Fig.2.NRMA signals recorded from an air-annealed single crystal of Bi2212 at various temperatures for H ∥ c axis and H_{mw} ∥ ab plane. The (+) marks represent observed and continuous lines represent the calculated signals. [Ref. 13].

might show a small decrease in resistivity, ambiguous at best. Figure 1 shows the first-ever NRMA result on a HTSC where a normal to superconducting transition is clearly seen around 84 K.

II. INTRINSIC JOSEPHSON COUPLING IN Bi2212 SINGLE CRYSTALS

The highly anisotropic Bi2212 superconductor consists of stacks of superconducting CuO_2 bilayers separated by non-superconducting medium of BiO and SrO layers. This intervening medium behaves as a Josephson junction. The demonstration of AC and DC Josephson effects in Bi2212 by Kleiner et al.[11] provided a direct evidence for the existence of intrinsic Josephson coupling. The vortices in such highly anisotropic systems have very different structures for in-plane (H ∥ ab plane) and out-of-plane (H ∥ c-axis) orientations of the magnetic field (H). In the former, the field penetrates as Josephson vortices whereas in the latter there is stacking of discrete two dimensional vortices called 'pancakes'. A vertically straight stack of pancakes is in a phase-locked state i,e. the c-axis critical current is maximum. In case of a tilted field with respect to c-axis or a disruption of the vertical stack of pancakes due to a shear along the ab plane, the pancakes are coupled across the junction medium via Josephson strings[12]. As mentioned above, the vortex motion is one of the main sources of magnetic field dependent electromagnetic absorption in HTSC. In the following, we study the vortex dynamics in Bi2212 single crystals using the technique of non-resonant microwave absorption (NRMA)[13]. In the process we illustrate the evolution of intrinsic Josephson coupling as the temperature approaches the transition value. Single crystals of Bi2212 grown by travelling flux zone method and subsequently air- annealed ($T_c \approx 83$ K) were subjected to NRMA studies. For these studies an X band Bruker ER 200D CW EPR spectrometer equipped with low temperature assembly was used. The details of the measurements are described in ref. [13].

Figure 2 shows the NRMA signals recorded at various temperatures for H ∥ c axis (H_{mw} ∥ ab plane) orientation. The signal at 77K shows that NRMA is dominated by a narrow component. Such narrow signals are usually observed in ceramic samples of HTSC[14], where they originate from intergranular weaklinks. In case of single crystals of Bi2212, such signals arise from inevitably present interunit cell weaklinks e.g. intergrowths. As the temperature increases, we start to see another component of absorption as a broad wing like signal (see the NRMA signal at 79K). This component grows in intensity across 80-82 K and diminishes as the temperature approaches its nominal T_c value. We associate this particular component with the evolving Josephson coupling between the CuO_2 bilayers in this temperature interval. As mentioned earlier, at sufficiently low temperatures, for H ∥ c axis, straight stacks of pancake vortices are formed. Thermal motion of pancakes sets in at higher temperatures disrupting the stacks. For H ∥ c axis and H_{mw} ∥ ab plane, the induced microwave currents exert a Lorentz force on the vortices and tend to move them along ab plane. When the pancakes move away from the core axis, a phase change between the pancake vortices associated with the Josephson current across the layers is created, giving rise to Josephson vortices connecting the pancakes. Since the vortices are mobile, the coupling of the pancakes through the Josephson strings can be understood to be in a state of dynamic equilibrium[15] for a given field and temperature. We believe that the dominant contribution to the microwave power loss comes from the viscous motion of these Josephson strings which terminate at either end on the pancakes driven by induced microwave currents. Another feature of the broad component of signals in Fig. 2 is the saturation after a temperature dependent field value. This means that the junctions are decoupled (no Josephson strings) so that the power loss ceases to be a function of the field.

We have analysed these observations in terms of a surface impedance model due to Portis et al[16]. This model takes into account the effects of vortex motion on the surface impedance of a HTSC. The viscosity of medium plays a crucial role. The expression for complex surface impedance is given by $Z = R_s + iX_s$ where,

$$R_s = X_o[(-1 + (1 + 4f^2H^2/B_o^2)^{1/2})/2]^{1/2} \qquad (1)$$

$$\text{and } X_s = X_o[(1 + (1 + 4f^2H^2/B_o^2)^{1/2})/2]^{1/2}$$

with $X_0 = 4\pi\omega\lambda\mu^{3/2}/c^2$ and $B_o = 8\pi\omega\lambda^2\mu\eta/\phi_o$ where λ is the London penetration depth, ω the microwave frequency, η the coefficient of viscosity experienced by the moving vortices, μ the magnetic permeability and ϕ_o the unit flux. Fraction of free vortices is given by f. In the NRMA context, the induced microwave currents cause dissipation by driving the vortices and real part is the loss. We therefore, compute the derivative (dR_s/dH) of the real part of the surface impedance as a function of static field H. To identify the losses due to pancake and Josephson vortices, we have used different viscosities ($\eta \sim 10^{-7}$ cgs units for Josephson vortices and $\sim 10^{-6}$ cgs units for pancakes). To account for the narrow component of the signal, we have used the RSJ model of Dulcic et al[8]. The decoupling of the junctions is explained by assuming a distribution of junction strengths[17]. The microwave power loss due to decoupling of the junctions is given by

$$P_J = \Delta P_o\,(1 - \exp(-H/H_D)) \qquad (2)$$

where ΔP_o is the power absorbed when all the junctions are decoupled and H_D is the decoupling field. Finally we account for the microwave power loss due to thermal activation of the pinned vortices. The fraction of pinned vortices is (1-f). Therefore, the fraction of thermally activated vortices[18] is given by

$$F \propto (1\text{-}f)\,\exp[(U(H,T,\omega)/kT] \qquad (3)$$

where $U = U_o[(1-\ T/T_c)^{3/2}/H]\ln(\omega/\omega_p)$ and ω_p is the depinning frequency[18]. We incorporate this fraction F into Eq. (1) to calculate the loss in irreversible region. For reversible region f = 1. On the basis of the above analysis we have calculated the lineshapes (represented by continuous lines in Fig. 2) which conform to the experimentally observed signals.

To verify that we are in the flux flow regime, NRMA signals were recorded at different microwave powers. It was observed that the intensity of the signals was proportional to the square root of the power. The signals were also recorded for H ‖ ab and H_{mw} ‖ c axis. There was no change in the signal shapes as a function of temperature. We expect that in this orientation the loss is due to the Josephson vortices in a lock-in state[11]. Since there are no pancakes warranted in this orientation, there is no evolution of the signal shapes.

III. REENTRANT MELTING OF THE VORTEX LATTICE IN Bi2212 SINGLE CRYSTALS

Experiments probing the vortex structure in the layered superconducting compounds like Bi2212 have unearthed many interesting phenomena. Depending on the technique, the interpretations have been diverse. Whether it is melting[19] (first or second order ?) of the vortex lattice or a decoupling transition[20] is still not established. While the decoupling field has (1/t - 1) dependence, the melting based on Lindemann criterion is expected to follow the $(1 - t)^2$ dependence. In addition, the presence of another vortex liquid regime is predicted[21] at fields close to H_{C1} with an essentially $1/\lambda^2_{(ab)}(T)$ dependence [21]. We discuss our results in the light of the predictions by Blatter et al.[21] on the reentrant melting of the vortex lattice.

We also have studied the isothermal magnetic field (H//c) dependent rf (20 MHz, H_{rf}//a) losses in Bi2212 single crystals using a continuous wave NMR spectrometer equipped with an Oxford instruments CF200 continuous He flow cryostat. Figures 3a & 3b, show dP/dH *vs.* H at various temperatures. The signals are stacked vertically to illustrate the temperature evolution of the features. For instance, the dP/dH at 72 K on the positive field side first decreases with the increase in field, then increases again to cross the field axis at H_M as indicated by the arrows. The signals are symmetric about zero field and similar features are observed for negative fields as well.

The behavior at lower temperatures is shown in Fig. 3b for positive fields only. Note that dP/dH is independent of field till H_P. For $H>H_P$, the behavior is similar to those at higher temperatures.

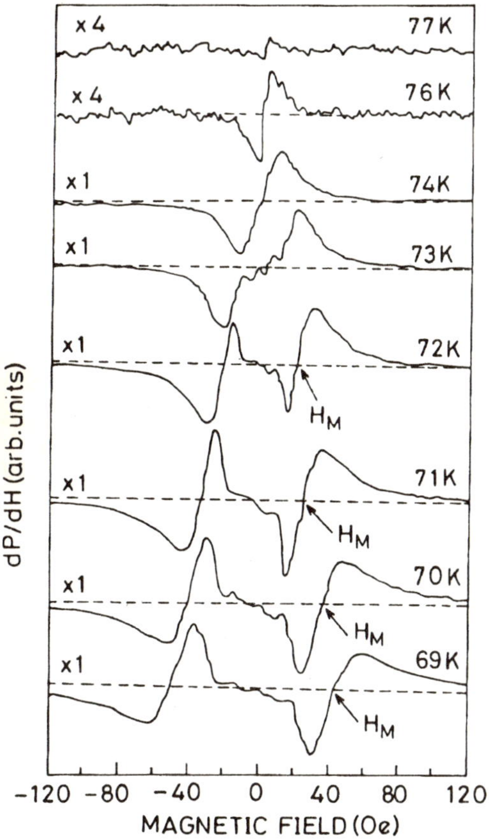

Fig. 3a. dP/dH vs. H at various temperatures. T_C=78 K.

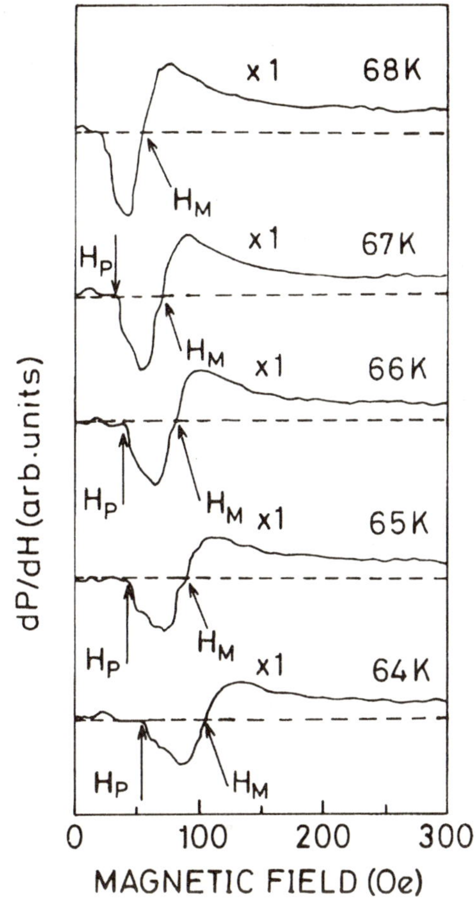

Fig. 3b. dP/dH at lower temperatures. Note that it is field independent till H_P.

To illustrate how the absorbed rf power P(H) behaves as a function of field we have numerically integrated the derivative signals at 72 K and 68 K and these are shown in Figs. 4a and 4b, respectively. Features corresponding to those in Figs. 3a and 3b are marked by the vertical arrows that also serve to indicate the field position of the features. Close to T_C (=78 K), P(H) initially decreases as the field is increased, reaches a minimum at H_M, then increases monotonically for $H>H_M$. At lower T (Fig. 4b), P(H) is independent of field till H_P, then decreases to a minimum, followed by the monotonic increase. The plateau in P(H) becomes apparent only at lower temperatures while at higher T, H_P is lower in magnitude and hence, rf penetrates much early in field. In the low field regime (at $H \approx H_{C1} \approx H_P$) the intervortex separation R is very large, and as the field is increased R decreases causing the intervortex repulsion to grow stronger, thus suppressing the motion in response to induced rf currents. For instance, the vortex diameter at 68 K is $\Lambda \approx 2\lambda_{ab} \approx 6155$ Å where $\lambda_{ab}(T) = \lambda_{ab}(0)/(1-t^4)^{1/2}$, $\lambda_{ab}(0) = 2000$ Å and $t = T/T_C$. The observed $H_M = 57$ Oe at 68 K corresponds to an intervortex separation of $R \approx (2\phi_0/\sqrt{3}H_M)^{1/2} \approx 6460$ Å which is comparable to Λ. Therefore the intervortex repulsive interactions are significant even at these low fields and as the field is increased from zero onwards, the intervortex separation $R > \Lambda$ for $H < H_M(T)$ and $R < \Lambda$ for $H > H_M(T)$. At $H \approx H_M$, $R \approx \Lambda$ and in this regime of separation, in addition to the repulsive forces between the flux lines, the attractive electromagnetic interaction[21,22] between the pancake vortices (comprising the line vortex) in neighboring layers also becomes dominant. These two effects together serve to confine the 2D vortices to the stack thereby enhancing its stiffness which inhibits their motion in response to rf currents. Therefore, the decrease in loss marks the stiffening of the flux lines and hence is a transformation from a liquid like state near H_P to a lattice at $H \approx H_M(T)$. Further increase in field causes the suppression of the c-axis superconducting order parameter (i.e. Josephson coupling between CuO_2 bilayers) causing[23] the lattice to melt again. The suppression of the coherence across the bilayers (in which the order parameter is still non-zero in view of their higher critical fields) can also lead to the vortex currents being induced in them but not in the intermediate BiO-SrO layers. At fields well above H_M, the vortex lattice that had begun to melt can now decouple into discrete 2D vortices residing in the bilayers. In Fig. 5 we show the $H_P(T)$ and $H_M(T)$ *vs.* t and to substantiate the

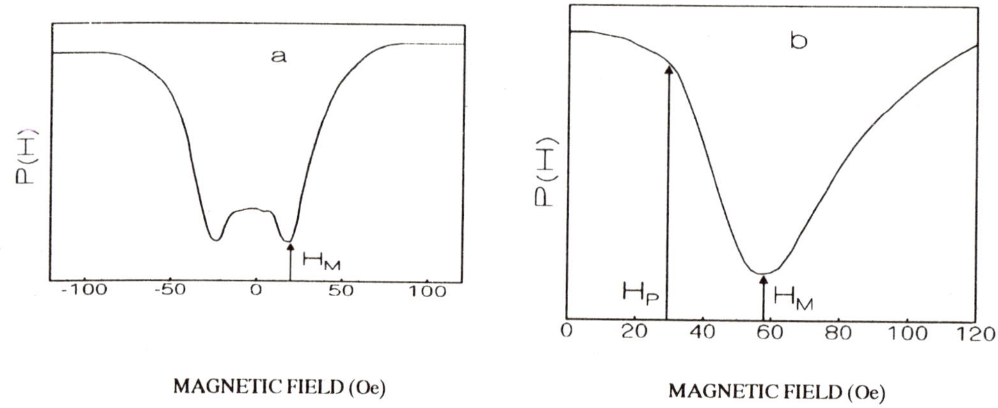

Fig. 4. The losses P(H) at (a) 72 K and (b) 68 K (vertical scale in arbitrary units).

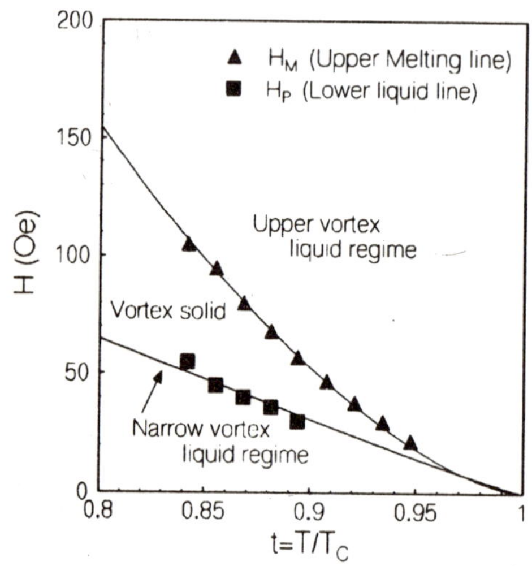

Fig. 5. H_P & H_M vs. T. The lines are fits of the data to Eqs. 4 and 5 discussed in the text.

above interpretation with regard to the reentrance into the liquid phase (from the lower liquid phase near H_P to the lattice at H_M and then into the upper liquid regime above H_M) we have fit $H_P(T)$ data to the lower melting line

$$H_P \approx (\phi/4\lambda^2_{ab}(T))[\ln(4\pi c_L \varepsilon_o \lambda_{ab}/(3\pi)^{1/4}T)]^{-2} \quad (4)$$

and $H_M(T)$ data to the upper melting line

$$H_M \approx (1 - t^2)^{3/2} \quad (5)$$

as predicted in ref. 3. It is seen that the fits are reasonably good lending support to our identification of $H_P(T)$ and $H_M(T)$ with the lower and upper melting lines, respectively.

IV. CONCLUSIONS

We have reviewed the use of non-resonant rf and microwave absorption technique to detect and characterize HTSC. Through the dynamics of the vortices in response to induced microwave currents, the evolution of Josephson coupling between the CuO bilayers in Bi2212 single crystals is illustrated. In addition, at rf frequencies the signatures of stiffening of the vortex lattice and the resulting reentrant vortex lattice melting as a function of field is also demonstrated in single crystals of Bi2212.

ACKNOWLEDGMENTS

YSS and AR would like to thank UGC and CSIR, India for financial support.

REFERENCES

1. J.G. Bednorz and K. A. Muller, Z. Phys. **B 64,** 189 (1986).
2. M.K. Wu, J.R. Ashburn, C.J. Tong, P.H. Hor, R.L. Ming, L. Gao, Z.J. Huang, Y.Q. Wang and C.W. Chu, Phys. Rev. Lett. **58,** (1987) 1908.
3. S.V. Bhat, P. Ganguly and C.N.R. Rao, Pramana J. Phys. **28,** L425 (1987).
4. S.V. Bhat, P. Ganguly, T.V. Ramakrishnan and C.N.R. Rao, J. Phys. **C 20,** L559 (1987).
5. A. Dulcic, R.H. Crepeau, J.H. Freed, L.F. Scheemeyer and J.V. Waszczak, Phys. Rev. **B 42,** 2155 (1990).
6. V.V. Srinivasu, S.V. Bhat and N.Kumar, Solid State Commun. **89,** 375 (1994).
7. F. Iga, A.K. Grover, Y. Yamaguchi, Y. Nishihara, N. Goyal and S.V. Bhat, Phys. Rev. **B 51,** (1995).
8. A. Dulcic, B. Rakvin and M. Pozek, Europhys. Lett. **10,** 593 (1989).
9. A.M. Portis, K.W. Blazey, K.A. Muller and J.G. bednorz, Europhys. Lett. **5,** 467(1988).
10. K.W. Blazey in Springer series in Solid State Science, *Earlier and Recent Aspects of Superconductivity,* eds. J.G. Bednorz and K.A. Muller, vol. **90,** 282 (1990).
11. G. Blatter, M.V. Feigl'man,V.B. Geshkenbein, A.I. Larkin and V.M. Vinokur, Rev. Mod. Phys. **66,** 1147 (1994).
12. R. Kleiner, f. Steinmeyer, G. Kunkel and P. Muller, Phys. Rev. Lett. **68,** 2394 (1992).
13. Amit Rastogi, Y.S. Sudershan, S.V. Bhat, A.K. Grover, K. Oka, Y. Yamaguchi and Y. Nishihara, Phys. Rev. **B 53,** 9366 (1996).
14. V.V. Srinivasu, Boben Thomas, M.S. Hegde and S.V. Bhat, J. Appl. Phys. **75,** 4131 (1994).

15. R. Busch, G. Ries, H. Werthner, G. Kreiselmeyer and G. Saemann-Ischenko, Phys. Rev. Lett. **69,** 522 (1992).
16. A.M. Portis, K.W. Blazey, K.A. Muller and J.G. Bednorz, Europhys. Lett. **5,** 493 (1988).
17. E. Silva, M. Giura, R. Fastampa, G. Balestrino, M. Marinelli and E. Milani, Phys. Rev. **B 45,** 12566 (1992).
18. F. Zuo, M.B. Salamon, E.D. Bukowski, J.P. Rice and D.M.Ginsberg, Phys. Rev. **B 41,** 6600 (1990).
19. E. Zeldov, D. Majer, M. Konczykowski, V.B. Geshkenbein, V.M. Vinokur and H. Shtrikman, Nature **375,** 373 (1995).
20. R.A. Doyle, D. Liney, W.S. Seow, A.M. Campbell and K. Kadowaki, Phys. Rev. Lett. **75,** 4520 (1995).
21. G. Blatter, V.B. Geshkenbein, A.I. Larkin and H. Nordborg, Phys. Rev. **B 54,** 72(1996).
22. G. Blatter, M.V. Feigel'man, V.B. Geshkenbein, A.I. Larkin and V.M. Vinokur , Rev. Mod. Phys. **66,** 1125 (1994).
23. D.S. Fisher, M.P.A. Fisher and D.A. Huse, Phys. Rev. **B 43,** 130 (1991).

DYNAMIC PHASES OF A DISORDERED VORTEX LATTICE

S. Bhattacharya and M.J. Higgins
NEC Research Institute, Princeton, NJ 08540, USA

A vortex lattice in the mixed state of a type-II superconductor, interacting repulsively and pinned by random impurities require a threshold force in order to be depinned. The nature of the moving state of the flux assembly is determined by a competition among three components : the intervortex interaction, randomness provided by the quenched disorder(pins), and in addition, the external drive. Depending upon the relative strengths of these competing components, one observes three distinct types of motion, termed elastic flow, plastic flow and fluid flow. In the elastic flow regime, for the most rigid lattice, all vortices participate in motion and move with the same time-average speed. In the plastic regime, for a softer lattice, the motion is spatially inhomogeneous with some regions("chunks" whose transverse dimension is much larger than the lattice constant) of vortices moving and others pinned. This regime provides highly anomalous dynamics with distinct I-V curves, fingerprint phenomenon, large broad-band noise, strong thermal instability and frequency dependent response at very low frequencies. All these phenomena are best understood through the dynamics of flux line defects(dislocations). In the fluid flow regime, the motion of vortices is extremely inhomogeneous with individual rows of vortices participating in the motion, i.e., the transverse "chunk" size becomes of the order of the nearest neighbor spacing. One expects these types of dynamics to be characteristic of collective driven transport of disordered many body systems in general.

Since the advent of high-T_c superconductivity, the physics of vortices in type-II superconductors has been a subject of unprecedented activity. Foremost and most popular among these new areas of investigation has been the existence of new phases of the vortex assembly. In the absence of quenched pinning disorder, thermal fluctuations alone can melt the lattice and result in slivers of flux liquid states on either side of the flux solid phase. A search for the flux liquid states on the upper as well as the lower side of the flux solid phase has been among the dominant experimental activities. A second area of activity is centered around the effect of quenched disorder on the solid phase. Following the pioneering work of Larkin and coworkers[2], quenched random pinning was known to destroy the long range order of the lattice. New glassy phases of the vortex solid caused by different types of disorder, random and correlated, were predicted[3] and much experimental work is concerned with the identification of these phases of the vortex assembly.

A third area, now gaining in prominence, is inspired by the possibility that inter-vortex correlation can also be affected in the moving state by the driving force and may result in new dynamic phases[4] and may induce phase transitions, or crossovers, among these moving states. Guided by early simulations[5] that led to an experimental determination of a dynamic phase diagram and characterization of different moving phases[4], considerable theoretical as well as experimental work now seek to address these questions. Specifically, true phase transitions have been postulated theoretically[6]. Of particular significance is the possibility that the phase behavior of the vortex assembly in the moving states may be quite general in driven disordered systems. In this summary we describe briefly some experimental results for the low-T_c anisotropic superconductor 2H-$NbSe_2$; for details, see a recent review[7,8].

The vortex lattice is an elastic medium consisting of elastic strings interacting repulsively. For very dilute vortices the interaction is exponentially weak, but for dense vortices the interaction is logarithmic. As the magnetic field is increased, the lattice becomes more rigid, but as the field approaches the upper critical field H_{c2}, the lattice begins to soften again as the induction becomes nearly uniform with small modulations at the lattice sites. As a result, the lattice can be melted by thermal fluctuations both at small B and at large B.

Thus one expects that as the magnetic field is increased towards the upper critical field, a change of the nature of the phases will be observed accompanied by different types of motion. In the process one encounters a strange result : The resistance of the system, measured at a fixed current shows a pronounced minimum near H_{c2}, as shown in Fig.1. This is caused by an anomalous peak (the "peak effect"), between field values marked H_{pl} and H_p (which we call the peak regime,) in the critical current J_c. Above Hp the critical current depletes rapidly to zero. This anomalous behavior is thought[2] to result from a precipitous softening of the lattice. Within the collective pinning scenario, the critical current is given by

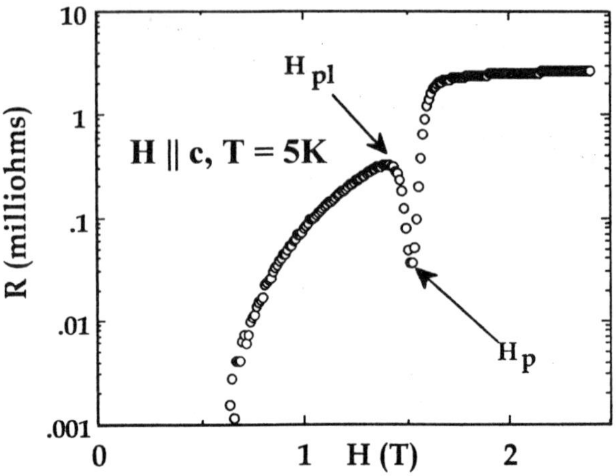

Fig.1. Peak effect is seen in the anomalous drop in resistance between H_{pl} and H_p, slightly below the upper critical field.

$$J_cB = (n_p f^2/V_c)^{1/2}$$

where n_p is the density of pins and f is the pinning interaction parameter. The correlation volume V_c of the lattice is directly related to the elasticity of the lattice and thus has to decrease faster than the pinning interaction which also goes to zero as H_{c2} is reached in order for the critical current to show a maximum. Despite decades of work, a specific mechanism that can yield a quantitative understanding of the peak effect is still lacking.

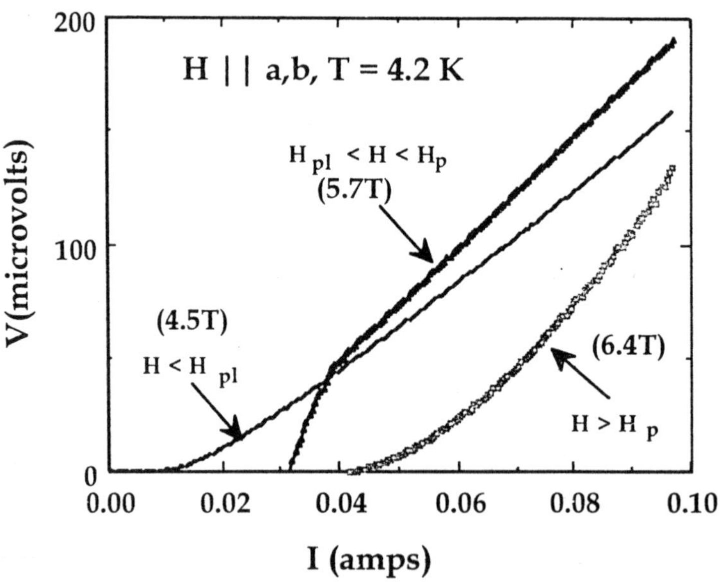

Fig.2. I-V curves for fields below, within and above the peak regime. Note the change in curvature in the peak regime and a sharper onset of voltage.

Figure 2 shows that the increase in J_c is simply not due to a softer lattice but also a fundamentally different depinning process that leads to qualitatively different I-V curves. In the peak regime, the curves are convex upwards as opposed to the typical concave upwards curves usually seen in much of the H,T space, both above and below the peak regime. Comparison with numerical simulations suggest that this depinning process is spatially inhomogeneous : instead of the entire system depinning, the onset of motion is through depinning of distinct pieces (chunks) while the rest of the system remains pinned. This phenomenon gives rise to the "fingerprint" effect[9], easily observed when the chunks themselves are sizable fractions of the entire system and their number is small.

Over a range of current values the depinning process continues in this manner, but as the current increases, the underlying pinning potential becomes less important and the elasticity of the lattice allows the moving pieces to join and heal. At large enough forces, the system recovers the coherent motion of the elastic medium. This is characterized in the experiments by the fact that at large currents a quasilinear regime is reached where the differential resistance reaches the Bardeen-Stephen value, unaffected by disorder as shown by I_{cr} in Fig.3. This crossover current thus marks the entry of dynamics back into the elastic flow regime.

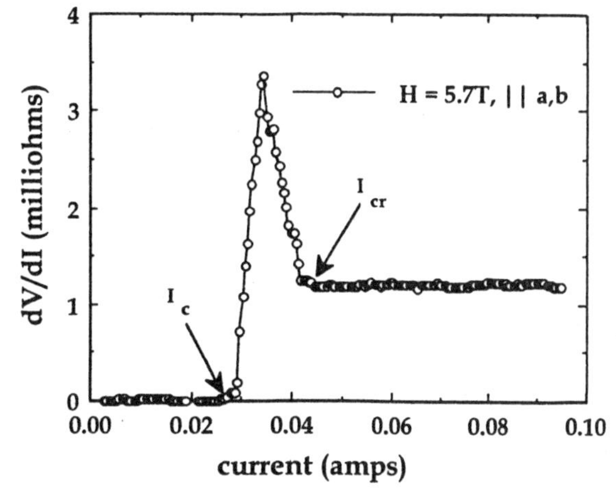

Fig.3. The variation of differential resistance with the driving current. I_c marks the onset of motion and I_{cr} marks a crossover to the quasilinear regime with the Bardeen-Stephen slope, i.e., a nearly pinning free region.

It has been suggested theoretically that this crossover process is actually a true phase transition[6]; a dynamic recrystallization of the broken up lattice. Whether and under what circumstances this is the correct theoretical scenario is a subject of active debate.

At even higher fields above the peak of the peak effect, the system reverts back to a smooth I-V curve, once again concave upwards, as shown in Fig.2, reminiscent of the elastic regime below the peak effect region. One expects this regime to represent an amorphous vortex assembly, either a glass or a pinned liquid. Here the onset of motion is entirely inhomogeneous, termed fluid flow, reminiscent of directed percolation. But the elastic coherent dynamics is not achieved at large currents since the system is thermally melted.

For both the elastic regime below the peak region and the fluid flow region above the peak effect, the I-V curves can be fitted well by a power-law scaling form :

$$V \sim (I - I_c)^{\beta}$$

with different exponents b that appear to depend on the effective dimensionality of the lattice. What physical significance these exponents have is unclear at this time. More theoretical work is needed to resolve this question.

Combining these results we show in Fig.4 the main result obtained in this study which is a generic dynamic phase diagram of the moving and pinned phases of the vortex lattice, described above. Considerable work is needed to understand the low field region where the critical currents increase and joule heating effects prevent measurements at high currents. Specifically, one expects that a dilute vortex assembly will also be strongly affected by quenched disorder and glassy phases may result in this regime as well.

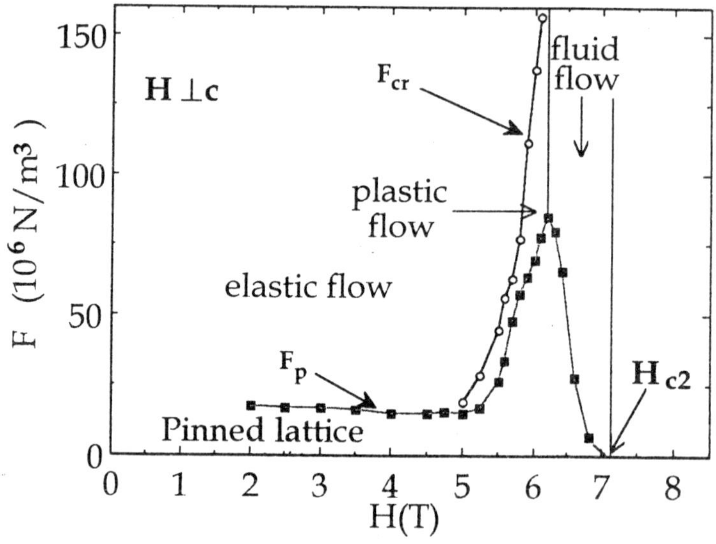

Fig.4. Dynamic phase diagram describing different flow regimes. $F_p = J_c B$ and $F_{cr} = J_{cr} B$; H_p agrees with estimates of the melting field of the lattice, above which the pinned state is amorphous. See text for discussions.

These results suggest several directions for future study. First, the theoretical understanding of the peak effect must consider its proximity to, or coincidence with, vortex lattice melting. The sudden onset of plastic flow underlies the peak effect and thus an elastic medium approximation is unlikely to produce the right result. Even if one interprets the peak effect as a precursor softening of the lattice before a melting transition, as is often done, a quantitative understanding is likely to require an explicit treatment of the breaking up process of the lattice caused by pinning and the concomitant shrinkage of the correlation volume. How to incorporate plasticity into the general framework remains a challenge.

A second question is related to the existence of an equilibrium broken-up state of the pinned lattice in the peak effect region from which the moving plastic flow phase originates. An experimental resolution of question is complicated by the fact that the drive itself can generate defects in the otherwise defect-free pinned state prior to depinning and before it eventually heals them at larger forces. Recent noise measurements[9] point to a persistence of metastability in the pinned state which strongly suggests that, in some cases, the broken-up lattice preexists in the pinned state as well.

Finally, the studies described above focuses on a single sample with a fixed realization of quenched disorder and varies interaction by changing the magnetic field. It would be particularly fruitful to elaborate on this question with samples with systematically varied levels of pinning, both random and correlated, to investigate further the fascinating problem of statics and dynamics of disordered many-body systems of which the mixed state of the type-II superconductor is a specially important example.

REFERENCES

1. D.R. Nelson, Phys.Rev.Lett. **60**, 1973 (1988).
2. A.I. Larkin and Y.N. Ovchinnikov, J. Low Temp. Phys. **34**, 409 (1974).
3. M.P.A. Fisher, Phys.Rev.Lett. **62**, 1511 (1989); D.R. Nelson and V.M. Vinokur, ibid, **68**, 2398 (1992).
4. S. Bhattacharya and M.J. Higgins, Phys. Rev. Lett. **70**, 2617 (1993); Phys.Rev.B **49**, 10005 (1994).
5. H.J. Jensen et al, Phys Rev.Lett. **60**, 1676 (1988); A.C. Shi and A.J. Berlinsky, ibid, **67**, 1926 (1991).
6. A.E. Koshelev and V.M. Vinokur, Phys. Rev. Lett. **73**, 3580 (1994).
7. M.J. Higgins and S. Bhattacharya, Physica C**257**, 232 (1996) and references therein.
8. S. Bhattacharya and M.J. Higgins, Phys. Rev. B**52**, 64 (1995).
9. R. Merithew et al., Phys. Rev. Lett. 77, 3197 (1996) and references therein.

CORRELATION LENGTHS IN THE MAGNETIC FLUX LATTICE OF NIOBIUM IN THE VICINITY OF THE PEAK EFFECT

P.L. Gammel[1], U.Yaron[1], D.J. Bishop[1], D.A. Huse[1], E. Bucher[1], M.R. Eskildsen[2], K. Mortensen[2]

[1]*Bell Laboratories, Lucent Technologies, Murray Hill, NJ, USA*
[2]*Riso National Laboratory, 4000 Roskilde, Denmark*

Preliminary analysis of Small Angle Neutron Scattering (SANS) data from the Flux-Line Lattice (FLL) in a Niobium crystal reveals structural disordering near the peak effect in the transport critical current. The disordered FLL seen in field cooled experiments is a structure that has long been used to understand transport data in this regime. However, these SANS data represent the first direct measurements of all three correlation lengths-one longitudinal and two transverse-through the peak effect. In the vicinity of the peak effect, first positional, then orientational order disappear. The longitudinal correlations, which are much longer, have a weak maximum in the vicinity of the peak effect. These static structural data give an insight into the highly disordered flux line structures near the peak effect, but do not distinguish between vortex glass and pinned liquid phases. Dynamic SANS data could, in principle, be used to define the precise nature of the vortex phases in this regime.

I. INTRODUCTION

The past ten years have seen an enormous increase in the understanding of the equilibrium phase diagram of the flux line lattice (FLL). High temperature superconductivity in the cuprates, combined with spectacular success in growth of clean crystals in this system led to an explosion of experiments probing the interplay between thermal fluctuations, dimensionality and inter-vortex interactions. The resulting phase diagram is now relatively well understood, and has been extensively reviewed[1].

The response, both static and dynamic, of the flux line lattice to quenched, static disorder has become an increasingly important topic for research[2]. One of the most dramatic consequences of this interaction is an old, well studied, increase of the transport critical current just below H_{c2}-- the peak effect. While a general understanding of this phenomenon has been around for years, the resurgence of interest in the flux line lattice has prompted new and exciting studies of the detailed responses and history dependencies.

The qualitative argument for the peak effect stems from Pippard[3]. If the elastic constants of the flux line lattice soften more rapidly than the pinning interaction in the vicinity of H_{c2}, then the flux lines are more able to distort and accommodate the distribution of pinning centers. The resulting increase in the critical current J_c is thought to occur when the FLL correlation length drops to one lattice constant a_o. Recently, peak effects have been seen near other phase transitions in the FLL, specifically melting, and are thought to be related to the vanishing of the shear elastic modulus in this case.

II. EXPERIMENTAL

In this paper, we report small angle scattering investigations (SANS) of the FLL in a Niobium crystal which exhibits a pronounced peak effect in transport and magnetization data. The crystal was a cylinder with the [1,1,0] crystal axis aligned nearly with the cylinder axis. The transport properties of this crystal were good, but not exceptional, which are the conditions for a pronounced peak effect in Niobium. The RRR of about 1000, the transition width of .04K, and the critical current in the 10^3 A/cm^2 range are all examples of this quality. What was unusual about this crystal, from our experience, was the small crystal mosaic of 0.2° as measured with neutrons, with no sign of additional domains.

For the SANS data, the magnetic field and neutron beam were both applied parallel to the cylinder axis. Neutron wavelengths of 9.9A and 10.9A were used with an 18% energy resolution. This energy resolution was always a small part of the overall experimental resolution, which was dominated by the beam divergence (set between 0.13° and 0.18° FWHM). The crystal was mounted with a Cadmium mask exposing a circular hole 3.7mm in diameter, reducing problems from specular reflection and the distorted FLL near the sample surface.

Fig. 1. *The strong scattering arising from the short penetration depth and the well ordered flux line lattice in clean Niobium allows the observation of three orders of diffraction peaks at 1.1kOe and 4.6K. The flux line lattice is hexagonal and oriented by the Niobium crystal lattice in the [110] direction.*

III. RESULTS AND DISCUSSION

Shown in Fig. 1 is the a map of the scattered intensity detected by an area detector 6 m from the sample at 1.1kOe and 4.6K. At this field, near the maximum of the scattering intensity, almost 10% of the incident beam is scattered into the first order FLL Bragg peaks. Three orders of diffraction peaks are easily seen in this linear intensity plot, showing a well ordered, perfectly hexagonal lattice. At this field, the peak widths are not resolution-limited, and so represent accurately the positional and orientational ordering of the FLL. The scattered intensity drops quite rapidly as H_{c2} is approached, as $(H-H_{c2})^{2-3}$, making experiments near the peak effect time consuming even for this strong scattering material. While much more extensive and controlled transport experiments have been conducted in $NbSe_2$, there is little hope of studying the peak effect with SANS in that material.

Unfortunately, the field range over which Niobium can be used to study the FLL is limited by its being nearly type-I. In Fig.2, the scattering wave vector, q for field cooled experiments is shown as a function of applied field. The field independent q at low fields is a signature of the intermediate mixed state[4], consisting of patches of FLL in an otherwise superconducting background. Only for fields between 1kOe and H_{c2} = 2.35kOe is the sample in the mixed state. Hence, detailed studies of the order in the FLL are limited to $h=H/H_{c2}>0.5$.

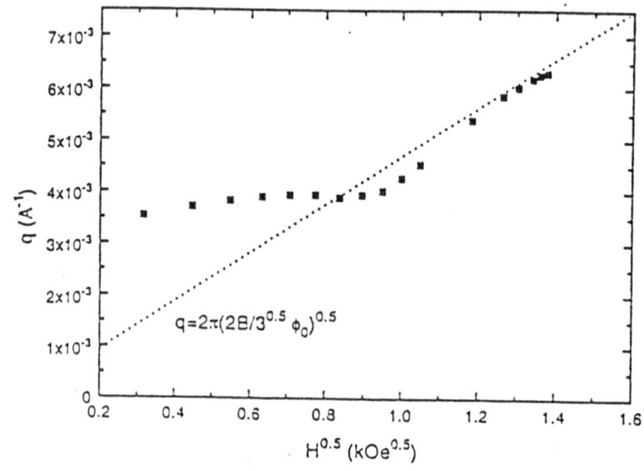

Fig. 2. The reciprocal lattice vector, q for the flux line lattice as a function of applied field at 4.6K. For fields below about 1kOe, the sample is in the intermediate mixed state. This state, stable in marginally type-II materials, consists of patches of flux line lattice with fixed lattice constant. Only for fields between 1kOe and H_{c2}=2.35kOe is the sample in the mixed state, where the flux line lattice evolves smoothly with increasing field.

The positional and orientational correlation lengths can be extracted from the widths in the plane of the detector[5]. To perform a detailed set of experiments on these lengths, we took two-dimensional scattering data with the neutron beam precisely aligned with the magnetic field. In this case, none of the flux line lattice peaks are in the exact Bragg condition, but the rocking curves were sufficiently wide for the neutron wavelengths used, that the intensity was only reduced by about a factor of 2. This scattering geometry was chosen to avoid experimental bias based on different cuts through reciprocal space used in the full rocking curve analysis. In the intermediate mixed state, the correlation lengths grow with increasing field, reaching a broad maximum near h=0.5. Data near this maximum are difficult to interpret due to the experimental resolution. Certainly, a gradual broadening of the peaks in the plane of the detector from their resolution limited values is seen for h>0.7. Figures 3 and 4 show the scattering pattern for h=0.88 and h=0.96. For h=0.88, additional broadening is especially obvious in the azimuthal direction. For h=0.96, only an isotropic ring of scattering is seen. This is in qualitative agreement with other SANS data[6] which interpreted this broadening in terms of a melting transition. We have extended the data to values of h much closer to one than that case, allowing observation of the ring of scattering.

More extensive data sets have been taken and used to calculate the correlation lengths from the width in the plane of the detector at both T=6.4K and T=4.6K. Fits to the radial width show

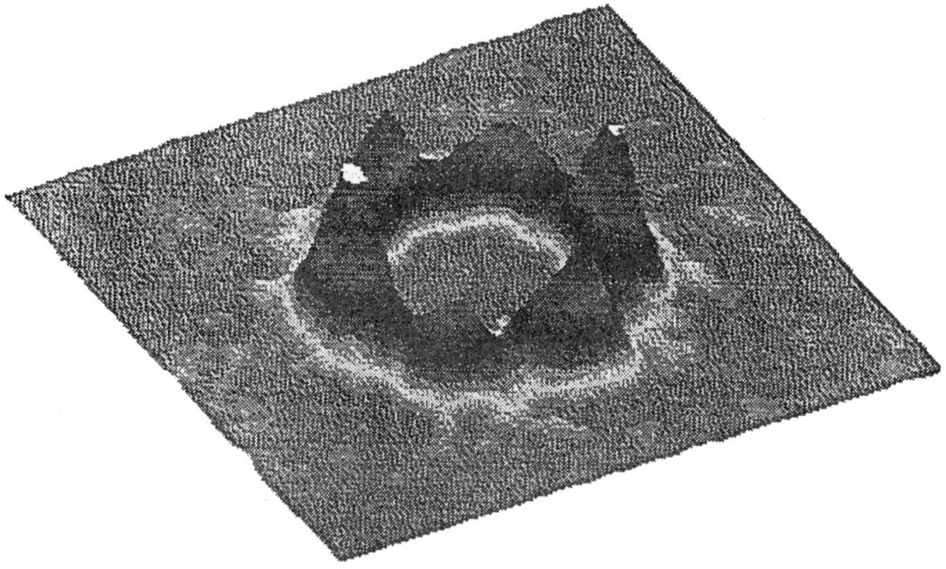

Fig. 3. The scattering from the flux line lattice at 2.15kOe (h=0.88) and T=4.6K. The positional correlation length, determined from the radial width, of the flux line lattice has dropped to one lattice constant at this field. Substantial loss of orientational ordering is also seen from the azimuthal broadening.

Fig. 4. The scattering from the flux line lattice at 2.25kOe (h=0.96) and T =4.6K. The scattering pattern is now an isotropic ring, showing complete loss of orientational and positional ordering. The rocking curves have begun to broaden at this point, but the longitudinal correlation length remains high at more than 50 lattice constants.

that the positional correlation length drops to $1a_o$ between h=0.80 and h=0.85. Near the point where the positional correlation length drops to $1a_o$ the azimuthal width starts a relatively much more rapid increase, leading to the isotropic ring of scattering by h=0.96.

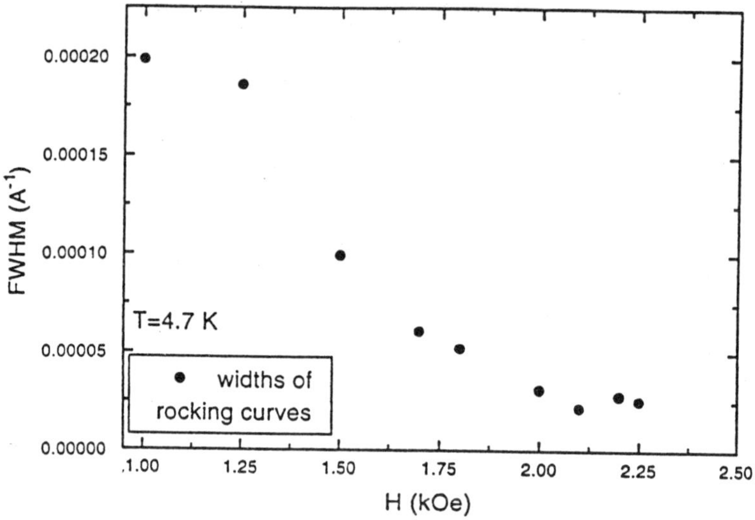

Fig. 5. The rocking curve widths versus applied field at T=4.6K. The broad rocking curves at low field show the flux lines have relatively short longitudinal correlation lengths. The rocking curves are never resolution limited, but reach a minimum width near 2kOe corresponding to a correlation length near 10μm. Above this field, the rocking curves broaden slowly, with the longitudinal correlation length dropping to 5μm at 2.25kOe.

Throughout this regime, we also performed rocking curves, which can be used to extract the longitudinal correlation length, which gives the effective straightness of the flux lines. The raw data for the rocking curve widths is shown in Fig. 5. Clearly, throughout most of the mixed state, the flux lines become increasingly straight. The minimum just about 2kOe is within a factor of two of the resolution limit of our experiment. Changing these data in lengths gives a longitudinal correlation length of about 1μm at 1kOe, growing to 10μm at the peak near 2kOe (h=0.82). Above this value, while not obvious from the scale in Fig. 5, the longitudinal correlation length starts to drop. The value at h=0.92, the highest field where we were able to obtain good fits to the rocking curves was only 5μm. This is still $50a_0$, and much longer than any other correlation length. So, throughout the entire regime of the peak effect, the flux line lattice can be thought of as a highly disordered system of relatively long lines.

In order to locate these features relative to the signatures of the peak effect, we have performed a variety of measures of the critical current. These are shown in Fig. 6, combined with arrows indicating different structural regimes. The first experiment we performed consisted of M-H loops on the same sample used in the SANS data. The upper part of such a loop, showing the classic 'fish-tail' signature of the peak effect is shown in the figure, together with the critical current extracted from the critical state model. The peak effect, about a 50% increase in the extracted value of J_c is quite distinct at h=0.97. Direct transport experiments were more problematic, because surface (or impurity) superconductivity in these Niobium samples gives substantial critical currents up to $\sim 2H_{c2}$. This value, much higher than the traditional value for surface superconductivity, is well documented in the literature[7] on Niobium. To suppress this we tried a variety of annealing and surface preparation techniques. The general trend of the transport data,

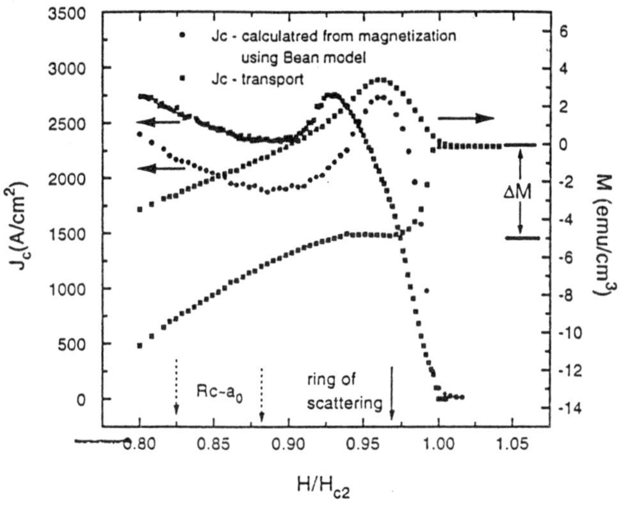

Fig. 6. Critical current data extracted from transport and magnetization data. While these are slightly shifted, the general trend is quite readily apparent. The flux line lattice forms a disordered system of lines (correlation length about one lattice constant) near the onset of the peak in the critical current. Near the maximum critical current, all order has vanished and only an isotropic ring of scattering is seen. This is in agreement with the generally accepted scenario for the flux line lattice order near the peak effect. These static data do not distinguish between a ring of scattering due to a pinned liquid or an amorphous vortex glass.

shown in Fig. 6 for a bar cut from one end of the SANS sample, was always the same. There is a distinct peak at h=0.93 and a shoulder at h=0.97, the location of the peak in the magnetization data. Transport data on the full SANS sample were similar, but more noisy, and limited to very close to H_{c2} because of the large currents involved. The precise details of the transport data were strongly dependent on surface preparation.

The qualitative comparison between the SANS data and the various measures of the peak effect is quite striking. The onset of the peak correspond to complete loss of positional order and the maximum in the longitudinal correlation length. The peak in the critical current is close to the point where the orientational order is also lost, and one is left with an isotropic ring of scattering.

It is tempting to interpret this isotropic ring of scattering as the signature of a pinned liquid of flux lines. Static structure factors alone are not sufficient to determine whether the FLL is a pinned liquid or highly disordered solid. The flux liquid would not be expected to occur over a measurable interval in Niobium based on the Lindemann criterion[8]. Two additional experiments, which we have not yet performed, could help in the final elucidation of the distinction. The first would be attempts to quench in the disordered system close to H_{c2}. This would give an idea of the timescales involved. The second would be to look at the disordered system in the presence of transport currents[9]. Roughly, one might expect a pinned liquid to remain disordered for current above J_c, whereas an amorphous solid would order into a lattice as we saw in SANS experiments at low fields in $NbSe_2$. Recently, Henderson et al.[2] have made rather specific predications of the expected ordering in the vicinity of the peak effect based on a variety of transport data in $NbSe_2$.

IV. CONCLUSIONS

In conclusion, we have performed SANS studies to measure directly the detailed disordering of the flux line lattice through the peak effect in Niobium. We have found a strong disordering as expected. The positional correlation length drops to a_o at the onset of the peak, and the longitudinal correlation length, while always high, exhibits a weak maximum also near the peak. These static data are not sufficient to constrain our knowledge of the phase diagram at the peak effect. However, Niobium provides an excellent opportunity for unique and detailed understanding of the peak effect. In this system one can grow large, relatively prefect single crystals. The short penetration depth allows for easy imaging of the flux lattice. Through annealing, the strength of the peak in the critical current can be controlled, and even eliminated. Finally, imaging experiments of the FLL in the flowing state would allow one to separate the pinned liquid and vortex glass phases, either one of which can explain our static data.

REFERENCES

1. G. Blatter, *et al.*, Rev. Mod. Phys. **66**, 1125 (1994).
2. W. Henderson, *et al.*, Phys. Rev. Lett. **77**, 2077 (1996) and references therein. U. Yaron, et al., Nature **376**, 753 (1995).
3. A.B. Pippard, Phil. Mag. 19, 217 (1969). A.I. larkin and Yu. J. Ovchinnikov, J. Low temp. Phys. **34**, 409 (1979).
4. R.P. Huebner, *Magnetic Flux Structures in Superconductors,* Springer-Verlag, N.Y. (1979).
5. U. Yaron, et al., Phys. Rev. Lett. **73**, 2748 (1994).
6. J.W. Lynn, et al., Phys. Rev. Lett. **72**, 3413 (1994).
7. B. Slettenmark and H.U. Astom, J. Appl. Phys. **40**, 3985 (1969). J. Good and E. Kramer, Phil. Mag. **24**, 339 (1971).
8. A. Houghton, R.A. Pelcovitz and A. Sudbo, Phys. Rev. B**40**, 6763 (1989).
9. P. Thorel, et al., J. Phys. (Paris) **34**, 447 (1973).

SECOND PEAK EFFECT OF $La_{2-x}Sr_xCuO_4$ SINGLE CRYSTALS

Y. Kodama[1,2], Y. Yamaguchi[1], K. Oka[1], K. Kajimura[1,2], K. Ghosh[3],
A. K. Grover[3] and Y. Nishihara[1,4]

[1]*Electrotechnical Laboratory, Tsukuba, Ibaraki 305, Japan*
[2]*University of Tsukuba, Tsukuba, Ibaraki 305, Japan*
[3]*Tata Institute of Fundamental Research, Mumbai, India*
[4]*Ibaraki University, Mito, Ibaraki 310, Japan*

Magnetization measurement data in the superconducting region are reported in single crystals of $La_{2-x}Sr_xCuO_4$. The second peak of the magnetization changes closely correlated with the change in the superconducting state from underdoped to overdoped regions. It is pointed out that the collective pining state which gives the strong second peak effect is formed by a distribution of normal metallic states which fluctuatingly appear in the crossover region from superconducting to normal metallic state.

1. Introduction

The $La_{2-x}Sr_xCuO_4$ (LSCO) system has a simple crystal structure consisting of a single CuO_2 plane and covers the full range of compositions from under- to overdoped superconductivity. Using these advantages, many experimental studies have been performed as a function of hole doping. It has been known that many physical properties changes by this carrier doping. With increasing doping, the temperature dependence of spin susceptibility changes from that of a system with strong singlet correlation among carriers to the one where they are relatively weak [1]. In the underdoped regime the c-axis resistivity is semiconducting and the c-axis optical conductivity also shows a pseudogap effect [2,3]. The metallic conductivity and a Drude peak develop only in the overdoped regime.

We have made measurements of magnetization curves and relaxation effect of magnetization in single crystals of LSCO system [4] and found that the superconducting magnetic properties also change from under- to overdoped superconductivity. In the underdoped regime, a weak second peak appears in the magnetization curve at the field less than 1 kOe at 10 K. On the other hand, in the overdoped regime, a strong second peak is observed at the field above 10 kOe at 10 K. The peak below x=0.14 shows little concentration dependence and that above it increases rapidly with x up to 40 kOe at x=0.18. The results show that there are two kinds of pinning centers in the LSCO system and the change of the electronic structure from under- to overdoped regime gives a strong influence on the formation of the pinning center in this system. This paper gives details of magnetization results, a correlation of the second peak to c-axis resistivity and discussion on an origin of the second peak.

2. Experimental Results

2.1 Samples

Single crystals of $La_{2-x}Sr_xCuO_4$ with the Sr concentration between x=0.1 and x=0.18 are prepared by the traveling solvent floating zone method [5]. After the cutting of the crystals, sample with x=0.115 and x=0.15 are annealed in O_2 at 900 C° for 24 hours, 700 C° for 24 hours then 400 C° for 50 hours.

The c-axis resistivity of these samples varies as shown in Fig. 1. The samples with Sr concentration $x \leq 0.15$ (underdoped regime) show a semiconducting behavior. The ab-plain resistivity of all the samples shows a metallic behavior. The resistivity at low temperature changes from 2D like one to 3D like one at $x \approx 0.15$. The results are consistent with published data [6]. The value of resistivity decreases with increasing Sr concentration. The temperature width of superconducting transition is less than 1 K between 10 to 90 % of resistivity transition.

The superconducting magnetic property in the underdoped regime changes from 3D to 2D like one with increasing temperature near T_C. Figure 2 shows the temperature

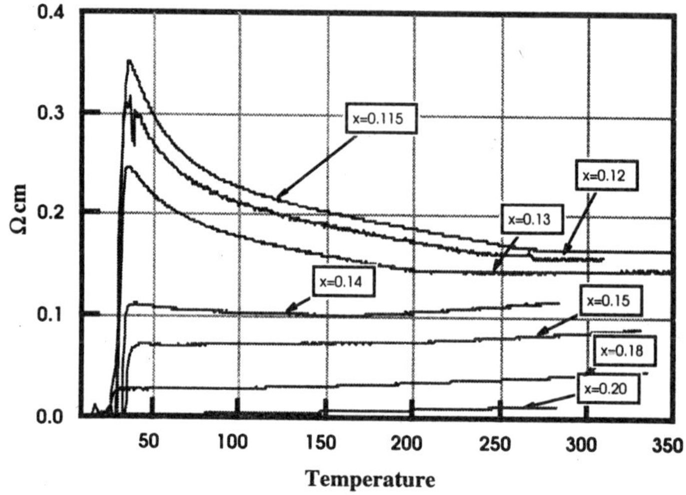

Fig. 1: Temperature dependence of resistivity for LSCO with various Sr concentrations.

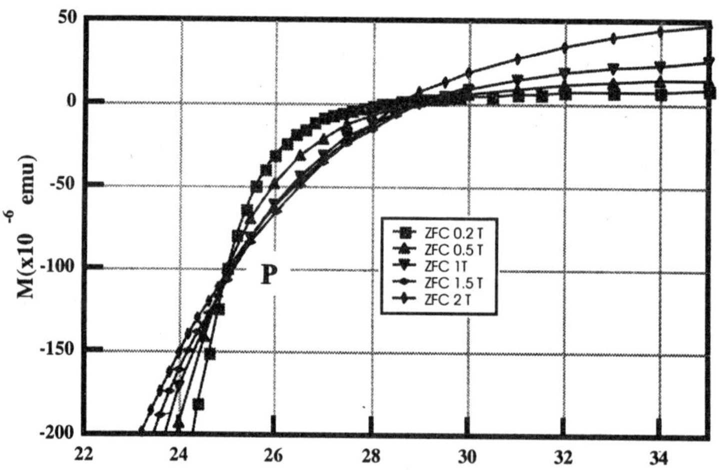

Fig. 2 Temperature dependence of magnetization at various applied fields for LSCO (x=0.1). *P* indicates a crossover point between 3D and 2D.

dependence of zero-field cooled magnetization of the sample with x=0.10 in magnetic field between 0.2 and 2 T near T_C. All curves cross at 25 K which is lower by about 2 degrees than T_C (28.5 K).

Above results show that these samples has high quality of homogeneity in magnetic and transport properties.

2.2 Second-peak of magnetization

Magnetization curves for the magnetic field parallel to the c axis were measured. Measurements were made by step-by-step method with a waiting time of 30 sec at several temperatures using the SQUID magnetometer. The results of x=0.115 and 0.15 are shown in Fig. 3. For x=0.115, a little increase of magnetization was observed at low field region. A distinct increase of magnetization was observed in the samples with x=0.15. The peak position at 10 K changes between underdoped and overdoped regions as shown in Fig. 4. From these results, it is found that a small second peak is observed in the underdope region and a distinct large second peak appears in the overdope region. The temperature dependence of the second peak position is shown in Fig. 5. The field of the peak position H_{SP} is nearly proportional to the square of $(1\text{-}T/T_C)$ as shown in Fig. 5. It is clear that there are two kinds of H_{SP} values in this system.

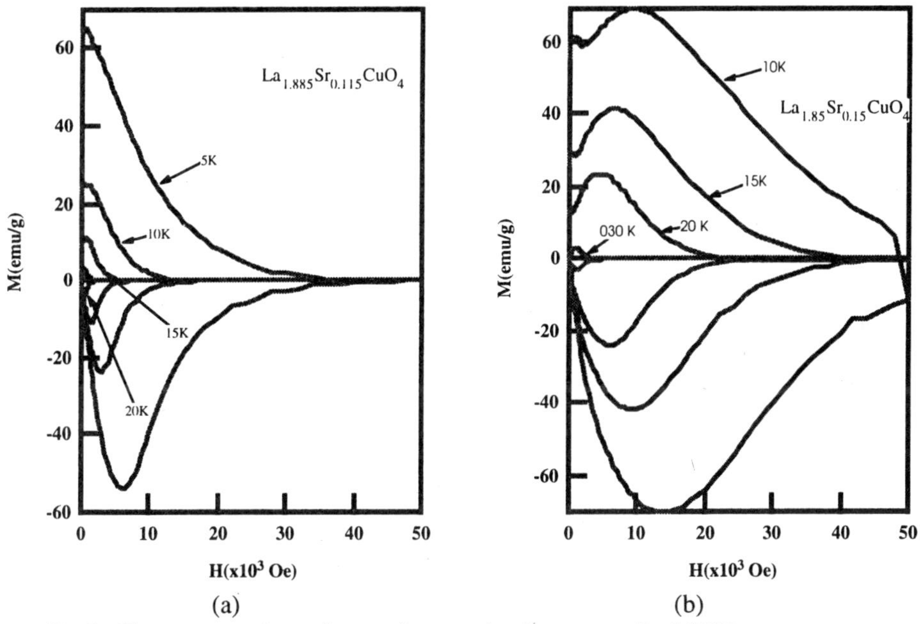

Fig.3 Temperature dependence of magnetization curves for LSCO.
(a) A small peak is observed at a low field. (b) A distinct second peak is observed at an intermediate field.

2.3 Magnetic relaxation

The magnetic relaxation was measured every 200 sec at a fixed magnetic field in zero-field cooled (ZFC) and in field cooled (FC) states at a fixed temperature. The result in FC state agrees with that in ZFC one. ZFC results were shown in Fig. 6 for x=0.15.

In the high temperature superconducutor, the magnetic relaxation shows nonlogarithmic behavior due to a short coherence length and a large thermal fluctuation.

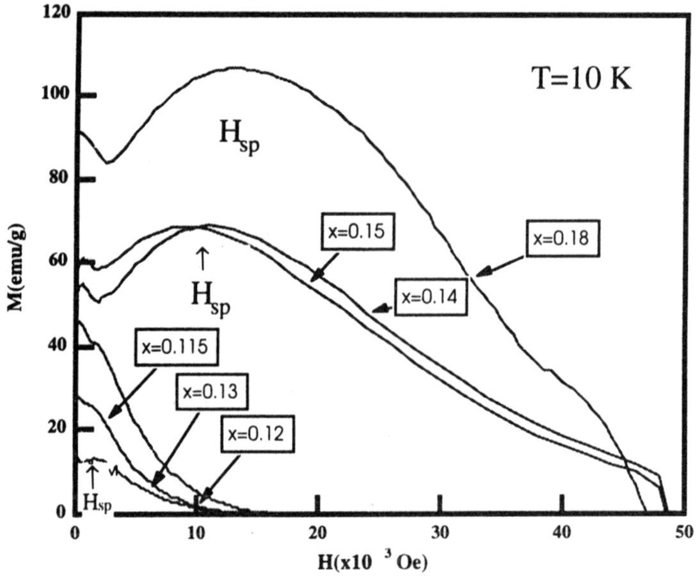

Fig. 4 Sr concentration dependence of magnetization curves for LSCO with various Sr concentration. The peak position of overdoped region is different from that of underdoped region.

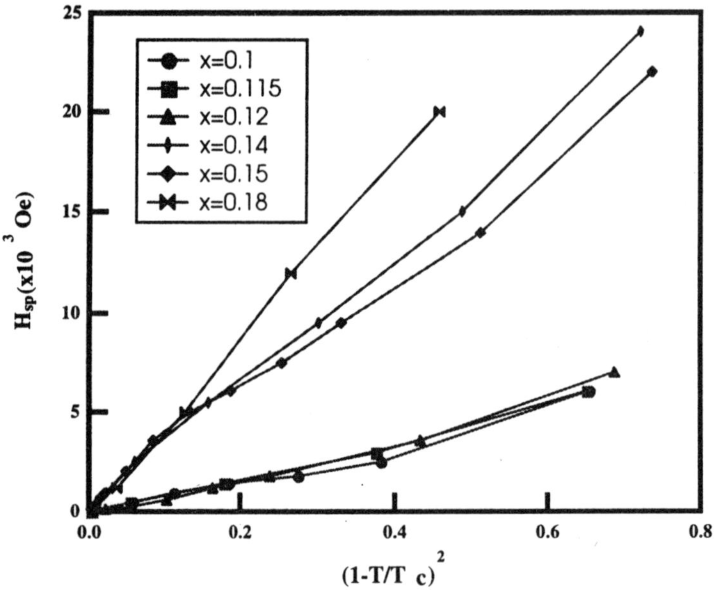

Fig. 5 Temperature dependence of H_{sp} plotted against $(1\text{-}T/T_c)^2$.

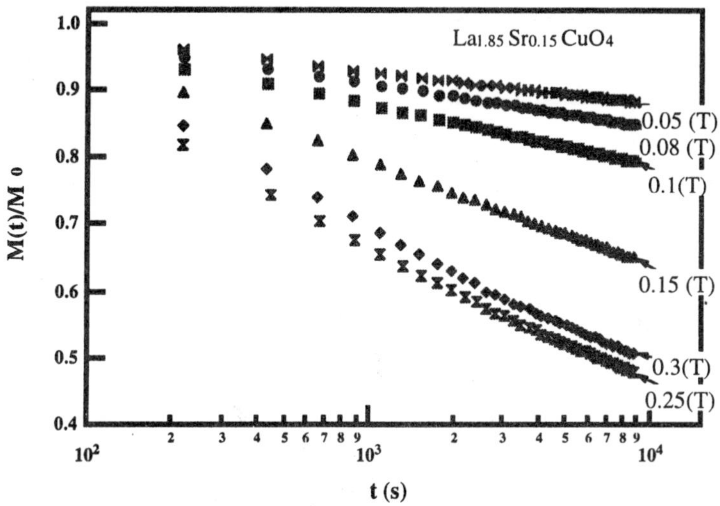

Fig. 6 Time dependence of magnetization for LSCO (x=0.15) at 30 K. Magnetization decreases almost linarly with ln(t).

This phenomenon is explained by the collective creep model concerned with elastic deformation of short-range-ordered volume of the flux line lattice in the state of randomly distributied weak pinning centers [7]. In the collective creep model, the time dependence of magnetization can be expressed by

$$M(t)=M_0\left[1+\frac{\mu k_B T}{U_c} ln(\frac{t}{\tau})\right]^{-1/\mu} \tag{1}$$

where the glassy exponent μ is a function of magnetic field and temperature. U_c is a pinning energy and its current dependence is expressed by $U_c \propto J^{-\mu}$. The glassy exponent μ is estimated by normalized relaxation rate;

$$S=-d\ln M(t)/d\ln t=\frac{T}{U_c+\mu\ln(t/\tau)} \tag{2}$$

The μ value changes from 1/7 to 3/2 to 7/9 with increasing temperature and magnetic field and these values correspond to single vortex, small boundle and large boundle states, respectively.

Figure 7 shows the field dependence of μ value derived using eq. (2). In the sample with x=0.115, μ value indicates a large negative value in the low field region and then approaches zero with increasing applied field at 20 K. In the sample with x= 0.15, μ value is ~1/7 at ~200 Oe, which corresponds to the single vortex state. At ~350 Oe it reaches ~3/2 which corresponds to the value of the small boundle state. At ~450 Oe it shows 7/9 which corresponds to the value of the large boundle state. Finally, it becomes a negative value which can not be explained by the collective creep theory. In the negative value region, it is thought that a catastrophic flux flow occurs. According to Matsushita *et al.*[8], the field dependence of pinning force F_p in this region can be expressed by $F_p \propto b^{1/2}(1-b)^2$, where b is H/H_{irr}. The measured results of F_p agree well with the calculated curves of F_p as shown in Fig. 8. It is concluded that the flux state changes from the collective creep state to the catastrophic flux flow one with increasing applied field.

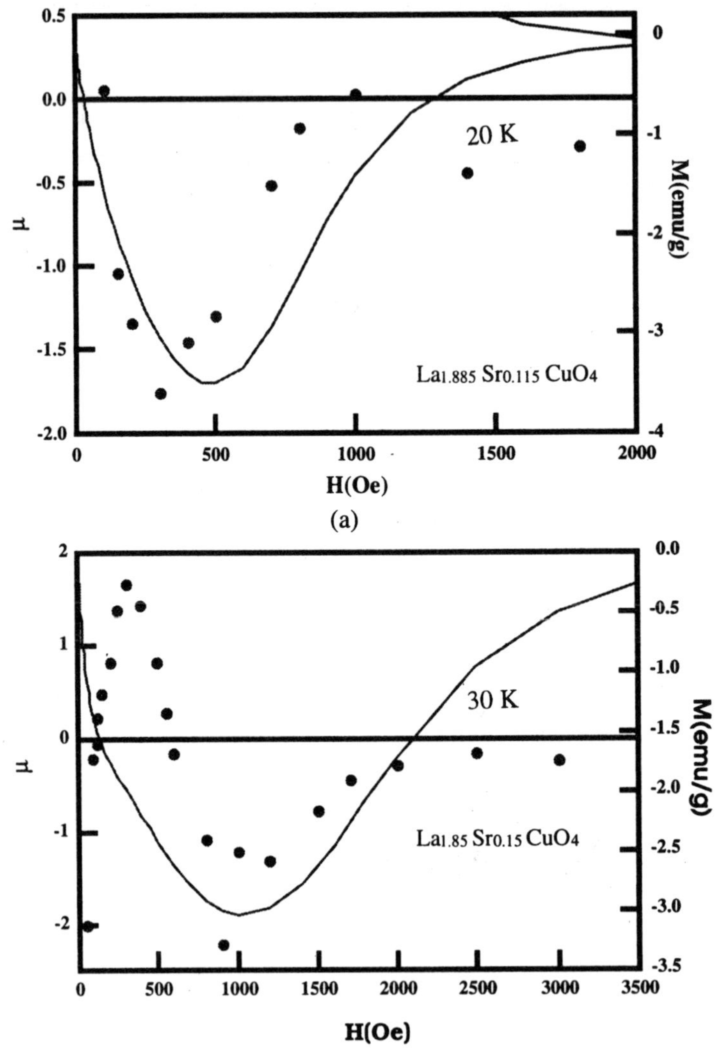

(b)

Fig. 7 Field dependence of glassy exponent μ for LOCO. Circles indicate μ and broken lines indicate magnetization. (a) x=0.115 at 20 K, (b) x=0.15 at 30 K.

3. Discussion - *Origin of the second peak*

The position of the second peak shows a clear Sr concentration dependence as shown in Fig. 4. Figure 9 shows the difference in c-axis resisitivity between 40 K and 250 K and the position of the second peak of the magnetization curves at 10 K. We can see a clear correlation between them. In the semiconductive region the value of the pinning field is ~1 kOe and shows little concentration dependence. In the metallic region, the pinning field is over 10 kOe and increases with increasing Sr concentration.

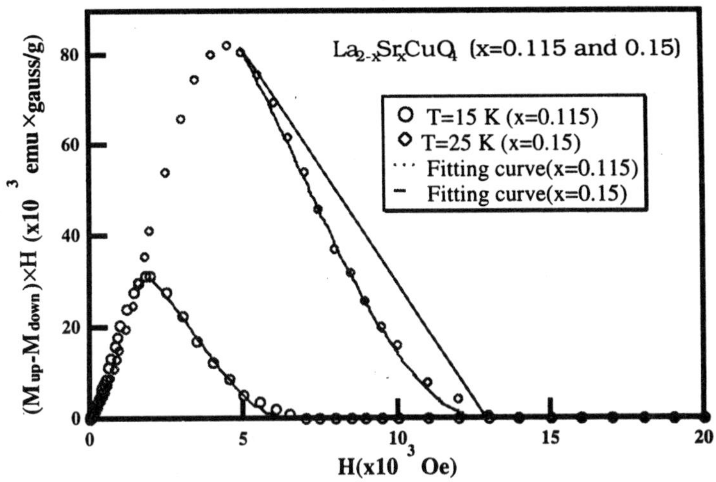

Fig. 8 Field dependence of pinning force ($(M_{up}-M_{down})\times H$ for LSCO (x=0.115 and 0.15). Broken curves show calculated curves by the catastrophic flux flow model.

The semiconducting c-axis resisitivity is connected with a pseudogap behavior in optical reflectivity spectra [3]. No gap feature is seen in the in-plain resistivity and optical spectrum. In this region the spin gap behavior is observed in NMR [9] and magnetoresistance [6]. The negative magnetoresistance is observed only in the underdopoped region and

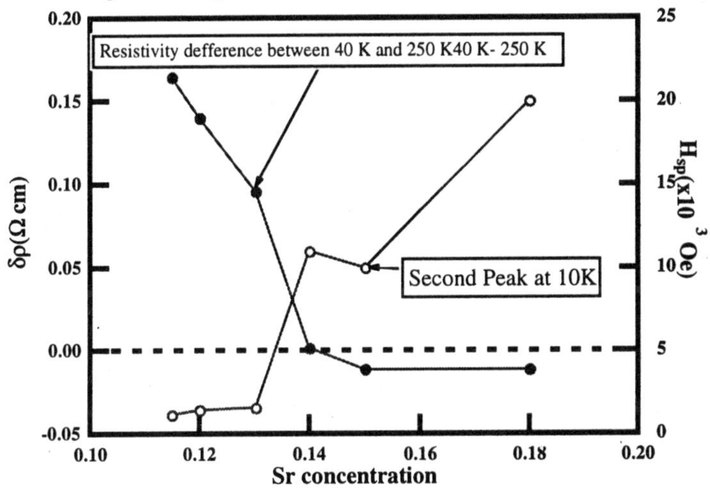

Fig. 9 Relation between temperature dependence of c-axis resistivity and the second peak. $\delta\rho$ is a resistivity diference between 40 K and 250 K. The strong second peak appears in the negative $\delta\rho$ region.

origin is ascribed to an effect of spin degrees of freedom. The charge confinement (divergence of c-axis resisitivity) and a formation of the spin gap have been discussed recently in terms of the non-Fermi-liquid model as a sinario of high-T_c superconductivity [10].

In the overdoped normal metallic region both in-plane and c-axis resistivities show almost same metallic behaviors. The magnetoresistance of normal state component is observed even in the optimum to the overdoped superconduction region ($0.15 \leq x \leq 0.20$) [6]. Therefore, the optimum to overdoped region is regarded as a crossover region from superconducting to normal metallic phases.

As the origin of the second peak of $La_{2-x}Sr_xCuO_4$ system, an effect of oxygen vacancies is discussed [11]. However, according to the results of neutron diffraction [12], oxygen occupancy in CuO_2 plane starts to decrease to over $x=0.2$. Our experiment of the well oxygen-annealed sample shows that the second peak becomes clear in the optimum to the overdoped region below $x<0.2$ which corresponds to the crossover region between superconducting and normal metallic phases. Therefore, we would like to point out that one origin of the second peak appeared in the field strength above 10 kOe is attributed to a distribution of weak pinning centers formed by a normal metallic state which fluctuatingly appears in the crossover region.

The peak in the underdoped region may be originate from atomic inhomogeneities in the sample, such as fluctuation of Sr concentration or oxygen vacancies. Since the tetragonal to orthorhomibic phase boundary lies over $x>0.2$, we cannot attribute this phase transition to the origin of these second peaks.

4. Conclusion

The field strength of the second peak changes with the Sr concentration. In the optimum and the overdoped region the second peak appears above 10 kOe at low temperatures, while in underdoped region it appears below 1 kOe. This change correlates well with the change in the c-axis resisitivity and not so sensitive to the oxygen annealing. The results suggest that the collective pinning state in the optimum and the overdoped region is formed by the normal metallic state which fluctuatingly appears in the crossover region from superconducting to normal metallic state.

References

1 Y. -Q. Song *et al.*, Phys. Rev. Lett. **70**, 3131 (1993).
2 Y. Nakamura and S. Uchida, Phys. Rev. B **47**, 8369 (1993).
3 S. Uchida *et al.*, Phys. Rev. B **53**, 14558 (1996).
4 Y. Kodama *et al.*, *Advances in Superconductivity VIII* eds. H. Hayakawa and Y. Enomoto (Springer, Tokyo 1996) p.197.
5 K. Oka *et al.*, J. Crystal. Growth. **137**, 479 (1994).
6 T. Kimura *et al.*, Phys. Rev. B **53**, 8733 (1996).
7 M. V. Feigel'man *et al.*, Phys. Rev. Lett. **63**, 2303 (1989).
8 T. Matsushita and H. Küper, J. Appl. Phys. **63**, 5048 (1988).
9 M. Matsumura *et al.*, J. Phys. Soc. Jpn. **65**, 699 (1996).
10 H. Fukuyama et al., J. Low Tem. Phys. **99**, 429 (1995).
11 T. Kobayashi et al., Appl. Phys. Lett. **62**, 1830 (1993).
12 J. T. Jorgensen *et al.*, *Advances in Superconductivity III* eds. K. Kajimura and H. Hayakawa (Springer, Tokyo 1990) p.337.

PEAK EFFECT OF THE CURRENT IN $YBa_2Cu_3O_{7-\delta}$ SINGLE CRYSTALS

H. Küpfer[1], Th. Wolf[1], A.A. Zhukov[2], C. Lessing[1], W. Jahn[1], R. Meier-Hirmer[1],

1 *Forschungszentrum Karlsruhe, Institut für Technische Physik Postfach 3640, D-76021 Karlsruhe, Germany*
2 *Physics Department, Moscow State University, Moscow 117234, Russia*

The current density j in $YBa_2Cu_3O_{7-\delta}$ shows frequently a maximum at fields B_p far above the selffield with very high values of up to $1.2 \cdot 10^5$ A/cm² (77 K, B_p = 2 T). This peak effect (PE) originates from uncorrelated pointlike defects or small clusters and is present for currents within the a,b plane and along the c axis of the crystal. The peak is observed in twinned and twin-free crystals but its current and relaxation behavior is strongly influenced in the presence of twins. For sufficiently pure, twin-free crystals j decreases monotonously with B and has very low values, j (77 K, 2 T) < 100 A/cm². Increasing oxygen deficiency δ causes a small PE which rises after thermal diffusion of the oxygen vacancies pointing to small clusters being responsible for the PE. In the region between $0.01 \leq \delta \leq 0.06$ where the PE develops, the peak appears first immediately below B_{irr}. The further evolution of the peak with increasing δ is, below B_p, in qualitative agreement with the static collective pinning theory. Above B_p the current shows a saturation behavior in accordance with plastic interaction between vortices. Thermal influence may be responsible for a decrease of the effective defect concentration N with increasing temperature T. This dependence between N and T prevents the scaling behavior of j which is not observed in the region where the peak develops.

I. INTRODUCTION

For possible applications of high-T_c superconductors necessary conditions are high current densities at high magnetic fields and tolerable low resistivities. Investigations of the current flowing within the a,b plane of the crystal do not reveal a granular behavior and therefore allow to study flux pinning exclusively. Such a separation from the influence of granularity on the current behavior is important for the characterization and optimization of the current.

In bulk $YBa_2Cu_3O_{7-\delta}$ high currents at low fields are achieved for instance by 211 precipitates whereas in single crystals frequently the maximum of the current appears at much larger fields. To introduce such a peak effect (PE) of the current into a melt textured material is - besides the fundamental aspect of flux pinning - the reason why the PE is of great actual interest. In this paper we characterize the PE by its superconducting properties, the responsible defect structure and the influence of the twins. The evolution of the peak is investigated starting from a crystal with a very low current decreasing monotonously with field. A comparison with the development of the peak in low-T_c superconductors is the basis for the discussion about the relevant mechanism. Finally we relate the observed non-scaling behavior of the current, in the regime where the peak develops, to thermal influence on the defect concentration N. The discussion of all these topics shall be restricted to the higher temperature region.

II. EXPERIMENTAL

The magnetic moment m of the crystals due to an induced shielding current was measured with a vibrating sample magnetometer. Measurements were made as a function of magnetic field B, temperature T and angle ϕ between the magnetic field direction and the c axis of the crystals at a constant sweep rate of the magnetic field dB/dt = 10^{-2} T/s. This value corresponds to an electric field E at the sample surface between 10^{-2} and 10^{-1} μV/cm. The current density j was determined from the hysteresis width Δm, the difference of m between increasing and decreasing magnetic field and the extended Bean model. Resistive measurements were used for the determination of the transition from the vortex liquid into the vortex solid state. The $RBa_2Cu_3O_{7-\delta}$ (R = Y, Nd) crystals were grown from CuO-BaO flux in ZrO_2 crucibles by the slow-cooling method[1]. Some crystals were investigated at different oxidation states. The oxygen deficiency δ was determined from Ref. 2 and 3 for Y and Nd, respectively. The twin spacing in some crystals were observed by TEM[4].

III. CHARACTERIZATION OF THE PEAK

In $RBa_2Cu_3O_{7-\delta}$ maxima of the current far above the self-field may result from correlated and/or uncorrelated pinning structures. They can be discriminated by the dependence of m(ϕ). Isotropic uncorrelated defects produce a current density with a smooth angular dependence from the intrinsic anisotropy of the superconducting properties. But a very sharp variation is obtained from correlated structures like twin boundaries or CuO planes. From both a peak of j(B) may be observed which is however different from the PE or "fishtail" effect.

In the geometry B parallel to the a,b plane of the crystal (ϕ = 90°) intrinsic pinning between CuO layers results within the trapping angle $\phi_T \approx 1°$ in an increase of j (ϕ) for the current flowing along a,b. Current maxima vs. B appear within the lock-in angle $\phi_L \approx 0.1°$ at fields of commensurate vortex-vortex and interlayer CuO distances [5,6]. These lock-in oscillations occur for different temperatures at the same matching fields.

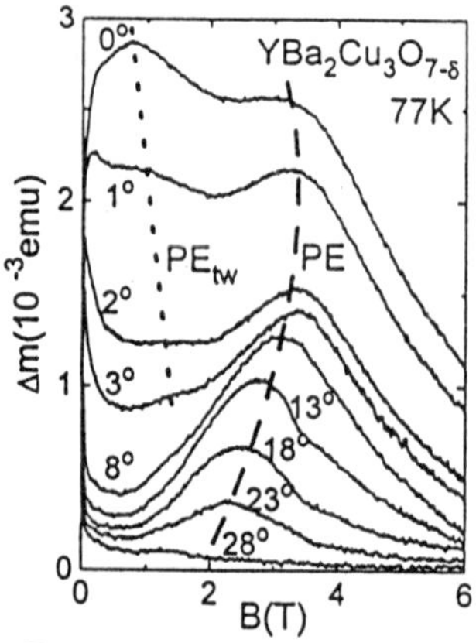

Fig. 1: Hysteresis width $\Delta m(B)$ at different angles, $\phi = 0°$ corresponds to B||c.

In contradiction to this phenomenon the peak from twin boundaries (PE_{tw}) may easily be mixed up with the peak effect from an uncorrelated structure (PE). For a brief clarification we discuss the measurements from crystals showing both effects. In Fig. 1 the angular dependence of Δm around B||c demonstrates that the PE_{tw} disappears if the misalignement between field and c axis is above 8° in agreement with an interaction from a correlated structure. The mean twin spacing of the crystal in Fig. 1 is roughly 180 nm which corresponds to the distance between vortices at a magnetic field of 0.07 T.

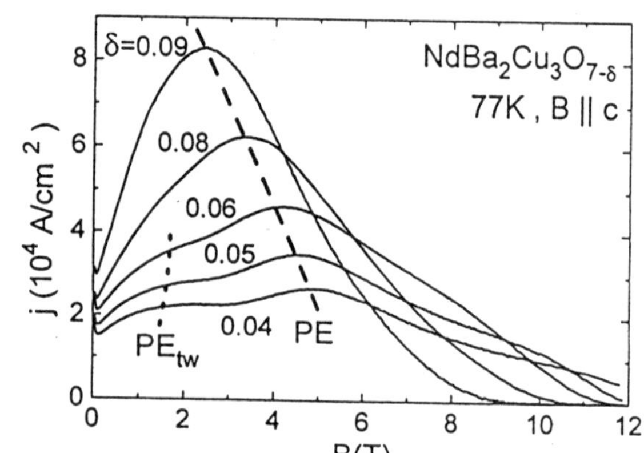

Fig. 2: Current density j(B) at different oxygen deficiency δ in a twinned Nd 1,2,3 crystal.

However, the PE_{tw} occurs at much larger fields which excludes a matching interaction between twin walls and vortices. The peak from the twin structure is only observable if the interaction of the competing, uncorrelated defects is sufficiently low. Therefore PE_{tw} becomes more pronounced in the higher temperature region where twin boundary pinning dominates. Increasing defect concentration N from increasing oxygen deficiency δ results in the development of the PE and in a depression of PE_{tw} as shown in Fig. 2 for a Nd 1,2,3 crystal from Ref. 7. A quantitative correlation between the twin spacing and the value of the current is very difficult to find because besides pinning from the twin walls for the Lorentz force perpendicular to the walls, also easy plastic shear of vortices may occur[8,9] for the Lorentz force directed parallel. The closed current loops of the magnetization measurements have always both components. Their ratio depends on the sample geometry and on the pattern of twin complexes of both {110} and {111} types. It should be emphasized that this peak from the twin structure is caused by a different mechanism than the peak observed from two twin planes just below the vortex melting line which is related to the softening of the vortex lattice shear modulus[10].

A. Properties and responsible defects

The following discussion is related only to the PE in the temperature region above 60 K where it becomes well pronounced. Angular dependent measurements from Ref. 6 in Fig. 3 show that the PE is present in the whole regime from $\phi = 0°$ (B||c) up to 90° (B||a,b). This observation is in agreement with an uncorrelated statistically distributed defect structure from which the PE originates. Fig. 3 shows a maximum of Δm for B||c which decreases with ϕ mainly due to the observation[11] that the shielding current flows only within the a,b plane up to very large ϕ values. In the geometry B||a,b, both shielding currents, flowing along a,b and along c have a PE (inset of Fig. 3). The PE for the current along c moreover supports a randomly distributed defect structure. Both currents show the same field dependence and the ratio (j||a,b)/(j||c) is close to the anisotropy parameter.

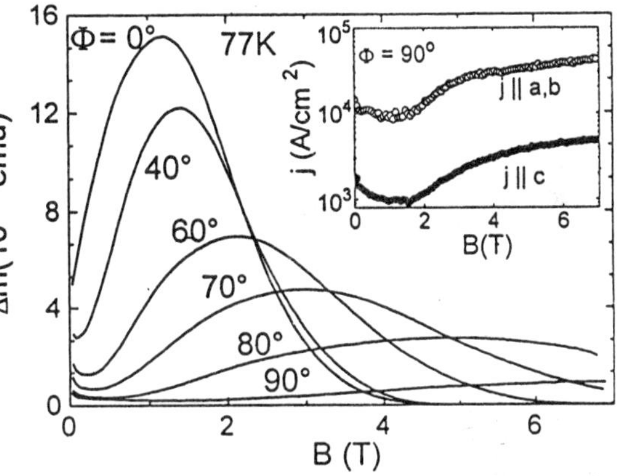

Fig. 3: Hysteresis width $\Delta m(B)$ at different angles ϕ. The inset shows both currents flowing at B||a,b.

A well known property of j with a PE is the scaling behavior: $j(B,T) / j(B_p,T) = f(B/B_p)$. The temperature range within which this relation holds depends on the influence of twins and strongly interacting defects like 211 precipitates and on the oxygen content[12]. In twin-free, fully oxidized crystals the scaling is valid in the whole temperature region. Deviations occur usually below 60 K where the peak broadens and may even disappear[9] and at temperatures close to T_c. The ratio $B_{irr}(T) / B_p(T)$ is about constant but sample dependent with values between 2 and 4.

The decrease of the current with B above B_p was related to granularity produced by the interaction of normal[13] or reversible[14] regions which approach B_{c2} or B_{irr} of a second phase at B_p. This possibility was rejected by transport measurements of the PE[15,16] which showed the same j(B) dependence as magnetic measurements up to B_{irr}. This confirms that the length scale within which the current flows does not change as required for a granular behavior above B_p.

From a number of investigations on quite differently prepared single crystals it was established that the PE results from weakly interacting pointlike defects or small clusters[6,7,12]. These pinning centers stem from impurities of the starting material or from the crucible, from doping or substitution[17] or from irradiation with protons or electrons. For a sufficiently large N of such defects the PE is present for all oxygen contents $0 \le \delta \le 0.55$, especially also in the fully oxygenated state[12]. Oxygen deficiency changes the superconducting parameters but the introduced oxygen vacancies are not the dominating pinning centers. This is in contrast to very pure crystals which have for $\delta \le 0.01$ a low and monotonously decreasing current (Fig. 4). With increasing δ the peak develops due to the increasing oxygen defect concentration as shown in Fig. 4 and for Nd 1,2,3 in Fig. 2. Higher impurity content leads to increasing concentration of other than oxygen defects, consequently the PE partly remains for $\delta = 0$ and the variation of j with δ in such crystals is much smaller[12]. At 77 K the maximum current which can be achieved from oxygen deficiency is close to 10^5 A/cm^2 at B_p in Y and Nd 1,2,3. Larger oxygen deficiencies than 0.1 result in lower current densities at 77 K from the degradation of T_c and especially from decreasing irreversibility field B_{irr}. For instance, in the Nd 1,2,3 crystal from Fig. 2 B_{irr} at 77 K was between 13 and 14 T at the highest oxygen content[7]. This extremely large value decreased to about 10 T at $\delta = 0.09$ where the maximum current was obtained.

In pure twin-free crystals the peak develops from oxygen reduction in the region $0.01 \le \delta \le 0.06$. Such crystals were deoxidized and then rapidly cooled down to room temperature. Before the reduction j decreased monotonously with B. After the thermal treatment j(B) reveals a very weak and undeveloped peak as shown by the dashed line in Fig. 5. Then the crystal was stored about 5 weeks at room temperature and measured again. The current at B_p increased by about one order in magnitude. We relate this to small scale diffusion of oxygen vacancies which results in the formation of small clusters. This observation demonstrates that the responsible pinning centers are not point defects but slightly larger agglomerates in full agreement with Ref. 18. There is further support for this observation from γ-irradiation of crystals without a PE. Here the introduced point defects increase the current below 20 K but do not produce a peak.

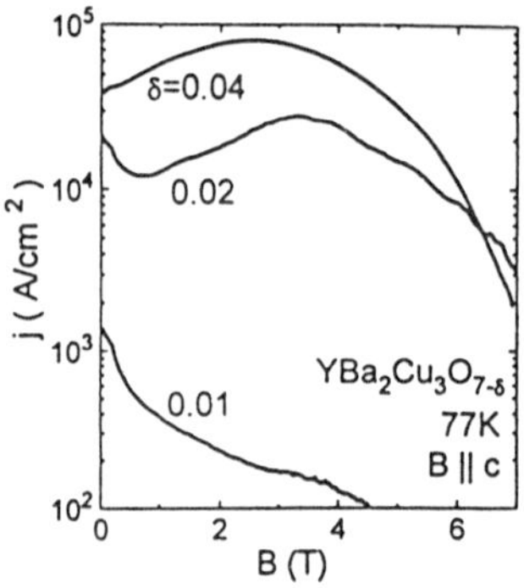

Fig. 4: Current density j(B) of a twinned crystal with different oxygen levels δ.

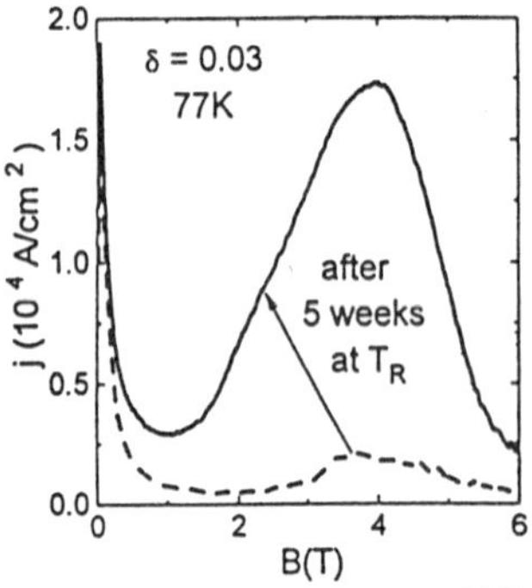

Fig. 5: Current density j(B) of a twin-free crystal before and after it was stored at room temperature T_R.

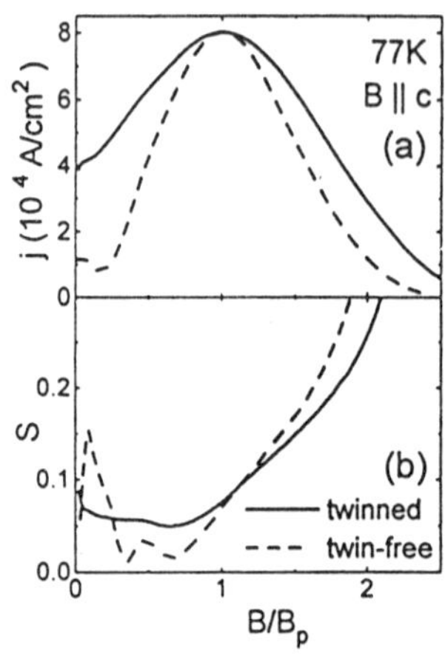

Fig. 6a,b: Current density $j(B/B_p)$ and normalized relaxation rate $S(B/B_p)$ for a twinned and another twin-free crystal.

B. Influence of the twin structure

The PE is present in twinned and twin-free crystals. But the B, T and δ dependencies of the current, the normalized relaxation rate S and $B_{irr}(T)$ are strongly influenced by the twin structure. Fig. 6a,b show j and S of a twinned and another detwinned crystal which have the same current density at about the same B_p. However, in the field region below and above the peak, the current is lower in the twin-free crystal. In angular measurements $\Delta m(\phi)$ passes in the detwinned crystal a shallow minimum at B||c[6]. Further a calculation for $\phi=0°$ based on the measurement $\Delta m(B,\phi)$ and neglecting the current along c results in a perfect agreement with the measured Δm at $\phi=0$. Both results prove an uncorrelated defect structure being responsible for the irreversible behavior. But in a twinned crystal with a PE such a comparison results in large differences. The solid line in Fig. 7a represents the measured Δm of the twinned crystal from Fig. 6a whereas the dashed line corresponds to the calculation using Δm $(B,\phi=40°)$ far outside the trapping angle of the twins ($\phi_T \approx 15°$). The same features as in Fig. 6a are observed: below and above the peak region Δm is increased by pinning from the twin boundaries. However, in the peak region the current value becomes much lower than expected from the measurement outside the trapping angle. This interference between uncorrelated disorder from pointlike defects and correlated disorder from twins is further proved by $\Delta m(\phi)$ measurements shown in Fig. 7 b,c,d. Below and above the peak (Fig. 7 b,d) Δm has a cusplike maximum at $\phi = 0°$. The dip of Δm in the peak region (Fig. 7c) may result from easy channeling of vortices along twin walls[8,9]. The depression of the peak becomes even more pronounced at lower temperatures where instead of a maximum a plateau in j(B) occurs [8,9].

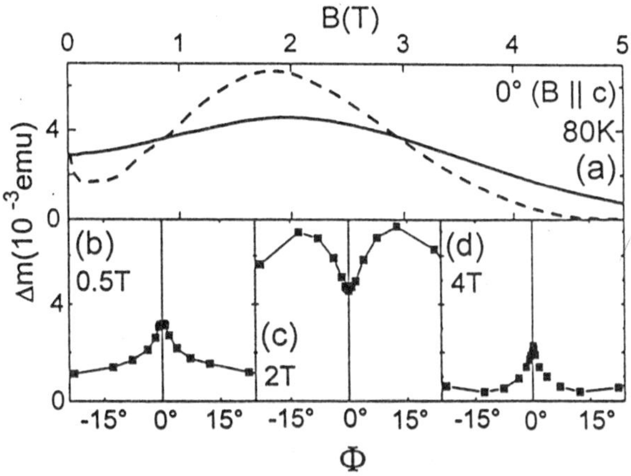

Fig. 7a-d: Hysteresis width $\Delta m(B)$ in a twinned crystal measured and calculated (dashed line) (a) and angular dependence $\Delta m(\phi)$ at different B (b-d).

At higher temperatures the field dependence of the current with a PE in twin-free crystals shows in contrast to a twinned crystal a pronounced minimum below Bp (Fig. 6a). This is in accordance with measurements of twinned crystals outside the trapping angle or at B||a,b (Fig. 3 and 7) or with the PE in Y 2,4,8[18]. The larger S values in the peak region of the twinned crystal (Fig. 5b) are related to channeling of vortices. In contrast to one shallow minimum of S(B) at about $B_p/2$ in twinned crystals, the twin-free crystals reveal two minima and a much larger relaxation rate at low fields.

The influence of the twins on j and S demonstrates that the mechanism responsible for the peak can only be studied in twin-free crystals.

IV. EVOLUTION OF THE PEAK FROM OXYGEN REDUCTION

Several mechanisms were proposed for the explanation of the PE in high- and low-T_c superconductors[13,20,21,22]. From weak elementary pinning forces like pointlike defects or small agglomerates one expects an elastic collective interaction of the vortex lattice characterized by a correlation volume V_c. Within $V_c = R_c^2 L_c$ the order of the vortex lattice is still established, R_c is the transverse and L_c the longitudinal size. The rise of the current is related to: (i) a lower relaxation with increasing field[20] due to a transition from the single collective vortex - ($R_c < a$) into the small bundle regime ($\lambda > R_c > a$) where a is the mean vortex distance; (ii) an increase of the elementary pinning interaction if second phase particles become normal conducting or reversibel at about B_p[13,14], (iii) a synchronization or softening of the vortex lattice[21]; or (iv) in the static collective pinning model, to the transition from a large bundle - ($R_c > \lambda$) into a single vortex regime ($R_c = a$) at a field B_o. In explanation (i), based on the collective creep model, the unrelaxed current is expected to decrease monotonously with B, the PE occurs then from relaxation which is predicted to be larger in the single collective vortex regime. Whereas in explanation (iv) relaxation is neglected and the PE occurs from the change of V_c which is predicted to decrease above B_o. The decrease of j above B_p is related to granularity[13,14] or plastic interaction of vortices[12,23,24] which was already discussed in connection with the saturation of the current in low T_c superconductors[25].

The concentration of the defects is in none of the models a parameter important for the j(B) dependence except in the static collective pinning theory in which the transition at B_o depends strongly on N. This parameter allows a more rigorous check of the proposed explanations which do usually not consider the development of the PE. In the following this evolution of the peak with increasing N from oxygen reduction is discussed in two naturally twin-free, pure Y 1,2,3 single crystals[26]. Both crystals (#A and #B) are from the same batch and about reversible above 70 K. They show a transition from the liquid into the vortex solid state which is in agreement with a melting of the vortex lattice[27]. Crystal #B was subsequently deoxidized. The peak for $\delta = 0.007$ (Fig. 8) appears first immediately below B_{irr} which is close to the melting line. This observation makes the relaxation model (i) less probable because it would require a single collective vortex regime extending close up to the melting line. With increasing oxygen deficiency the peak broadens and increases in height. The peakfield Bp and the field Bo at which the current starts to increase shift to lower values with larger N.

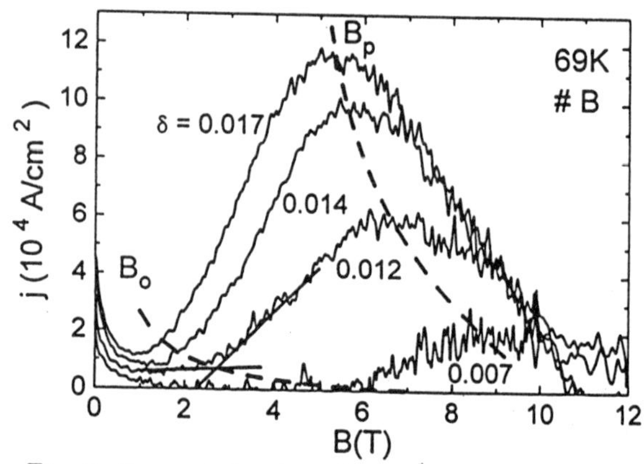

Fig. 8: Current density j(B) in a twin-free crystal at different oxygen deficiencies δ.

In the field region above B_p the current becomes about independent from concentration and j approaches a saturation behavior. There is some qualitative agreement with explanation (iii): the peak seems to be a precursor of melting as discussed in Ref. 10. But there are much more similarities with the evolution of the peak in low-T_c superconductors. For instance in V_3Si with defects from fast neutron irradiation the development of the PE shows the same behavior[28]. With increasing N the peak broadens and becomes higher, B_p and B_o shift to lower values (Fig. 9). At B_o the transverse length R_c becomes equal to the effective penetration depth in rough accordance with the measurements. From B_o to B_p R_c shrinks to the mean vortex distance correlated with an increase of j. Above B_p the saturation behavior gives experimental evidence for plastic interaction in contradiction to the collective model. Between B_o and B_p there is a continuous transition from elastic to

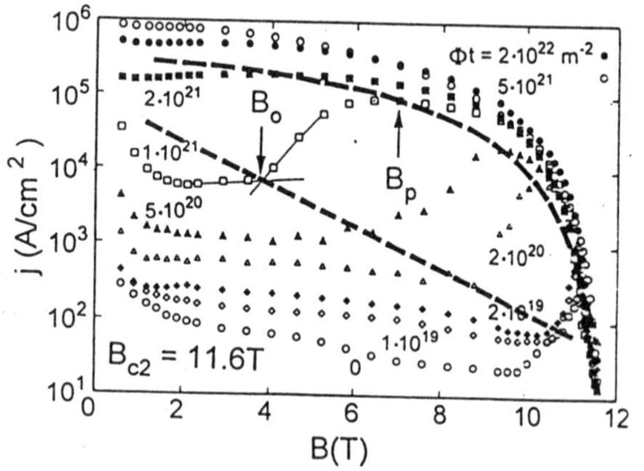

Fig. 9: Current density j(B) in V_3Si irradiated with different fast neutron fluences ϕt and measured at a temperature where the upper critical field B_{c2} is 11.6 T (from Ref. 28).

plastic interaction. The collective pinning model which describes the PE of V_3Si up to B_p is based on the dispersion of the tilt modulus of the vortex lattice and on a melting field B_m very close to B_{c2}. A very rough transformation in order to adopt this model for Y 1,2,3 would be to use an effective shear modulus which approaches zero at B_m and to replace B_{c2} by B_m. The model is then in qualitative agreement also with the PE in Y 1,2,3 up to B_p. The influence of relaxation must be considered further but this effect does not seem to be responsible for the PE. The change of the pinning mechanism between B_o and B_p was also indicated by the change of the E(j) characteristics[15,24,29] in this field region. The negative curvature of log E vs. log j transforms into a power law or exponential E(j) dependence at a field slightly below B_p. A more rigorous check of $B_o(N)$ and the elastic interaction regime below is prevented from the unknown N and a peculiar phenomenon which is finally discussed.

Low T_c superconductors with a PE show a j(B,T) scaling behavior also in the concentration regime where the peak develops[28]. However, in Y1,2,3 with a PE from oxygen deficiency the current does not follow a scaling in this regime. Fig. 10 shows j(B) at different T for a constant δ. The peakfield at high T is close to B_{irr}, then, with decreasing T, B_p shifts to lower values (inset of Fig. 10) opposite to $B_{irr}(T)$. No scaling behavior is observed in this temperature range down to about 55 K from where B_p increases again with B_{irr} as usually observed. The non-scaling region where B_p and B_{irr} have a different temperature dependence reminds on the development of the PE with increasing defect concentration from oxygen deficiency. A comparison shows that the evolution of the PE at a constant temperature with increasing δ (Fig. 8) is equivalent to the development of the PE with decreasing temperature at a constant δ (Fig. 10). From this similarity it is obvious to assume an effective defect concentration which increases with decreasing T. The pinning potentials from the oxygen defects are expected to have a distribution. Thermal oscillations of vortices reduce the gradient of the potentials and broaden the distribution. If kT is within the distribution as indicated from relaxation measurements, only the potentials above kT contribute to the current. The corresponding effective N increases with decreasing T as well as with increasing δ. Such a possibility of a temperature dependent defect concentration prevents a scaling behavior because j(N) is different in different field regimes. This non-scaling is restricted to the temperature region where B_p shifts to larger values with higher temperatures. Whereas for $T < 55$ K beyond the minimum of B_p (inset of Fig. 10), kT is expected to be below the distribution. Now all pinning sites contribute, N remains constant and scaling is observed.

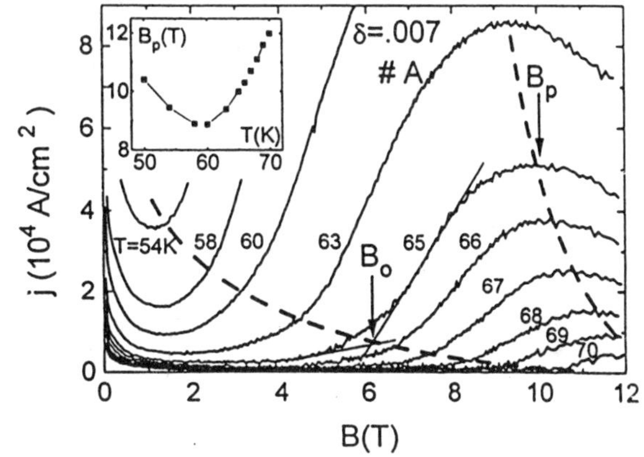

Fig. 10: Current density j(B) in a twin-free crystal at different temperatures T. The inset shows the peakfield B_p vs. T.

V. SUMMARY

The peak effect (PE) in $RBa_2Cu_3O_{7-\delta}$ single crystals (R = Y, Nd) is present in twinned and twin-free samples and at all angles from B||c to B||a,b. It originates from an uncorrelated pinning structure like randomly distributed pointlike defects or small agglomerates of oxygen vacancies, impurities or substitutions. The interference between the dominating pointlike disorder in the vortex lattice and the correlated disorder from twins results below and above the peak in a cusplike increase of the current at B||c from twin wall pinning. However, in the field region of the peak a pronounced depression of the current is observed.

Starting from a twin-free crystal without a PE the peak develops first immediately below the irreversibility field. With further increasing defect concentrations from oxygen deficiency the peak broadens and increases in height. This evolution of the PE is very similar to that one in low-T_c superconductors and in qualitative agreement with the static collective pinning model. Above the peakfield the saturation of the current is an experimental evidence for a plastic interaction of vortices.

REFERENCES

1 Th. Wolf et al., J. Cryst. Growth **96**, 1010 (1989).
2 D.J.L. Hong and D.M. Smith, J. Am. Ceram. Soc. **74**, 1751 (1991).
3 K. Kishio et al., Mat. Res. Soc. Symp. Proc. **156**, 91 (1989).
4 K. Saito et al., to be published in the Proc. of LT 21 Conf., Prague Aug. 1996.
5 M. Oussena et al., Phys. Rev. Lett. **72**, 3606 (1994).
6 H. Küpfer et al., Phys. Rev. B **54**, 644 (1996).
7 Th. Wolf et al., to be published in the Proc. of the 8th IWCC, Kitakyushu May 1996.
8 M. Oussena et al., Phys. Rev. B **51**, 1389 (1995).
9 A.A. Zhukov et al., Phys. Rev. B. **52**, R9871 (1995).
10 W.K. Kwok et al., Phys. Rev. Lett. **73**, 2614 (1994).
11 A.A. Zhukov et al., unpublished.
12 A.A. Zhukov et al., Phys. Rev. B **51**, 12704 (1995).
13 M. Daeumling et al., Nature (London) **346**, 332 (1990).
14 L. Klein et al., Phys. Rev. B **49**, 4403 (1994).
15 S.N. Gordeev et al., Phys. Rev. B **49**, 15 420 (1994).
16 V.F. Solovjov et al., Phys. Rev. B **50**, 13 724 (1994).
17 M. Murakami et al., to be published in the Proc. of the MOS Conf., Karlsruhe Aug. 1996.
18 A. Erb et al., to be published in the Proc. of the MOS Conf., Karlsruhe Aug. 1996.
19 Ming Xu et al., Phys. Rev. B **48**, 10 630 (1993).
20 L. Krusin-Elbaum et al., Phys. Rev. Lett. **69**, 2280 (1992).
21 A.B. Pippard, Philos. Mag. **19**, 217 (1969).
22 A.I. Larkin and Yu. N. Ovchinnikov, J. Low. Temp. Phys. **34**, 409 (1979).
23 H. Küpfer et al., Phys. Rev. B **52**, 7689 (1995).
24 Y. Abulafia et al., Phys. Rev. Lett. **77**, 1596 (1996).
25 E.J. Kramer, J. Appl. Phys. **44**, 1360 (1973).
26 H. Küpfer et al., to be published in the Proc. of the 8th IWCC, Kitakyushu May 1996.
27 C. Lessing, diploma thesis, University Karlsruhe Sept. 1996.
28 R. Meier-Hirmer et al., Phys. Rev. B **31**, 183 (1985).
29 H. Küpfer et al., Phys. Rev. B **50**, 7016 (1994).

NOVEL VORTEX PHASES IN SUPERCONDUCTING THIN FILMS WITH LATTICES OF ANTIDOT PINNING CENTERS

V. V. Metlushko[1&2], L. E. DeLong[1], M. Baert[2], E. Rosseel[2], M. J. Van Bael[2], K. Temst[2], R. Jonckheere[3], V. V. Moshchalkov[2], and Y. Bruynseraede[2]

[1] *University of Kentucky, Lexington, KY 40506, USA,*

[2] *Laboratorium voor Vaste-Stoffysica en Magnetisme, Katholieke Universiteit Leuven, B-3001 Leuven, Belgium,*

[3] *IMEC, B-3001 Leuven, Belgium*

We report AC magnetic susceptibility and magnetic relaxation measurements on a superconducting W_xGe_{1-x} ($x \approx 0.67$) film with an antidot lattice used as an ordered array of artificial pinning centers. We find new interstitial vortex phases stablized by collective flux line interactions in a "supermatching regime" extending well above the saturation field defined by maximal filling of the antidot lattice by flux lines. A nonmonotonic field dependence of matching anomalies in susceptibility and the existence of a critical depinning angle in the relaxation reveal the existence of a frustrated flux line lattice phase of relatively low stability.

I. INTRODUCTION

Irradiation of high-T_c superconductors (HTS) with relativistic heavy ions creates columnar damage tracks that increase the critical current density j_c by up to several orders of magnitude, and is therefore of great technological interest [1-2]. Huge enhancements of j_c also can be realized in superconducting $[Pb/Ge]_n$ multilayers with submicron, square or triangular antidot lattices (ADL), precisely created by e-beam lithography [3-7]. Such *ordered arrays* of artificial pinning centers substantially simplify studies of "matching", which generally occurs at a magnetic field $H = H_\Phi = H_1$ when the number of flux lines (FL) coincides with that of the pinning centers, and at its multiples $H_m = mH_1$, where m is an integer.

The number n of FL pinned by an individual antidot (AD) in an ADL can be changed either by varying the AD diameter D ($>> \xi$), the temperature T (due to the strong temperature dependence of the coherence length $\xi(T)$ near T_c), or by using different materials characterized by different $\xi(0)$. When n reaches a "saturation number" [8] $n_s = D/4\xi(T)$, additional FL can no longer be trapped by AD, and they will be repelled into the interstitial region between the AD [3-7], where they will be "caged" by the repulsions of nearby FL [9-10]. As a result, the system is composed of two different types of FL: (i) FL strongly pinned by AD, and (ii) FL weakly pinned at the interstices (IFL). The clear differences in behavior between FL strongly pinned by AD and IFL are reflected in an observed transition [4,7] between a completely localized state ("insulator") with all FL localized at AD for $n \leq n_s$, and a delocalized state ("metal") when a number $\Delta n \equiv n - n_s \propto (B - B_m)/\Phi_o$ of the FL are pushed into the interstices. Similar results were reported for irradiated $YBa_2Cu_3O_7$ single crystals [11,12]. This transition is analogous to the "Mott transition" predicted by Nelson and Vinokur [13] for the case of correlated disorder among *randomly distributed* pinning defects, where the matching condition $H = H_\Phi$ is attained when the *average* densities of FL and defects are equal.

Details of the dynamics and thermodynamic phase stability of the IFL have not been adequately studied at fields above the Mott transition. In particular, certain matching fields corresponding to $n > n_s$ were found to be absent while others were prominent in magnetization and transport data for $[Pb/Ge]_n$ multilayers with ADL [14], and in perforated Al-films [15]. Further, these experiments are probably closely related to observations of a pronounced matching effect in a damaged $Tl_2Ba_2CuO_6$ single-crystal at $H = 3H_\Phi$ [16], and radiation damage effects that persist to $H > 40H_\Phi$ in high-T_c films [17]. The elevated fields involved in the latter studies indicate that collective interactions among FL weakly pinned between columnar defects (i.e., the random-damage analogues of IFL) are responsible for these interesting effects.

In principle, the static FL arrangement can be determined directly from Bitter decoration experiments [18,19], or from electron holography [20,21]; however, a decoration can be performed more than once on a given sample only with difficulty, and holography requires specially prepared thin specimens. Therefore, additional methods to identify novel vortex arrangements are desirable.

The strong pinning of FL by the ADL was previously known to stablize FLL having no IFL for $n \leq n_s$. In this work, we report the existence of *new "supermatching" IFL phases* ($n > n_s$) in a film with an ADL, as inferred from analyses of anomalies in the DC magnetization, AC magnetic susceptibility and magnetic relaxation observed at various fields, angular orientation, and AC drives and frequencies.

II. EXPERIMENTAL DETAILS

Measurements of the AC susceptibility $\chi = \chi' + i\chi''$ sensitively reflect the behavior of both the current (χ') and losses (χ'') within the mixed state, and have the additional advantage of permitting detailed studies of FL dynamics using amplitude (h_o) and frequency (ω) variations of the driving AC field. Since the weakly pinned IFL are mainly responsible for dissipation [4-7], a comparison of the real and imaginary parts of χ will reveal differences in behavior between FL pinned by AD and IFL. We have performed accurate measurements of the drive and frequency dependences of χ using a Quantum Design MPMS SQUID magnetometer, whose gain is frequency-independent [22] over extremely wide ranges 10^{-4} G $\le h_o \le 3$ G, and 10^{-2} Hz $\le \omega \le 10^{3}$ Hz.

It is well known [23] that DC magnetization measurements using the MPMS SQUID can be altered by the applied field inhomogeneity encountered by a sample moving along the magnet axis. We minimized this problem by performing DC measurements with a scan length of only 2 cm, corresponding to a field homogeneity of better than 0.005%, and by checking these data with AC measurements. We concentrated our AC and DC experiments on the temperature interval near T_c in order to minimize field gradients and maintain uniform matching conditions $H \approx H_n$ throughout the sample [24-26].

Amorphous W_xGe_{1-x} ($x \approx 0.67$) films of thickness t = 600 Å were prepared on Si substrates with a SiO_2 sublayer using electron beam coevaporation and MBE. A square lattice of AD of diameter D = 0.35 μm and lattice period d = 1 μm was created using e-beam lithography and lift-off techniques [7]. The resulting amorphous W_xGe_{1-x} film is a type-II superconductor with $T_c \approx 4.50$ K, $\xi_0 = 60$ Å, $\kappa \approx 82$ and $\Lambda(0) = 2\lambda(0)^2/t \approx 8$ μm [7]. For T = 4.4 K $\approx 0.98T_c$, the temperature-dependent penetration depth Λ(T=4.4 K) $\approx$ 54 μm is much higher than the separation d between AD at temperatures of interest, indicating a strong overlapping of FL and their collective behavior.

III. EXPERIMENTAL RESULTS

DC magnetization and AC susceptibility data are shown in Fig. 1, which demonstrates that low frequency susceptibility measurements (ω = 0.1 Hz, corresponding to a characteristic measurement time $\tau^M = 1/\omega = 10$ s) are consistent with DC SQUID magnetization measurements ($\tau^M \approx 60$ s). Sharp anomalies at matching fields $H_n = nH_1 = n\Phi_o/d^2 = n \times 20.7$ G are observed with both techniques, although the DC data lack the sensitivity of the AC method at higher fields. The saturation number can be calculated [8] using $n_s = D/4\xi(T) \approx 2.2$, which means that at T = 4.4 K, up to 2 FL (i.e., a biquantum vortex) may be trapped by an AD, and a third FL will be repelled into the interstice. We therefore associate the anomalies at H_1, H_2 and H_3 with distinct vortex phases having a one-Φ_o fluxoid at an AD, a two-Φ_o fluxoid at an AD (Fig.1, n_s=2), and a two-Φ_o fluxoid at an AD and one IFL (Fig.1, (a)), respectively. At fields $H > H_3 = 62.1$ G, when a second FL is pushed into the interstice, M(H) and χ' (which reflect currents) drop nearly to zero. This means that the FL configuration for $H \approx H_4$ (Fig.1, (b)) is much less stable than for $H \approx H_3$, as corroborated below. On the other hand, for $H \approx H_5$ (Fig.1, (c)) there is a pronounced minimum in χ' reflecting a "reentrant screening current", which is similar to a "peak effect" (a nonmonotonic decrease of magnetization hysteresis and j_c with increasing field [27]).

Measurement of the normalized magnetization relaxation rate S(H) is a useful probe of the FL pinning potential U(H). DC relaxation measurements were first performed using the MPMS, which operates with an initial time delay of 60-100 s. This delay and the field inhomogeneity along the scan length complicate SQUID relaxation measurements, especially near T_c. Alternatively, Van der Beek, Geshkenbein and Vinokur have shown [28] that variable frequency measurements of $\chi(\omega)$ in a "strongly nonlinear regime" are equivalent to a measurement of S, and $\chi \propto \{(kT/U)\ln(1/\omega\tau)\}^{1/\beta}$, where the relaxation time τ is a constant. Since the AC gain of the Quantum Design SQUID is frequency-independent [23] we can accurately determine $S = \partial(\ln(\chi))/\partial(\ln(\omega)) \propto \partial(\ln(M))/\partial(\ln(t)) \propto 1/U$, where U is the pinning potential. Theory predicts [9] a sharp increase in S(H) when FL first enter the interstices, so we can use AC susceptibility data as a powerful tool to explore the nucleation and dynamics of IFL.

We performed detailed studies of χ vs. AC field amplitude h_o at T = 4.4 K, H = 0-200 G, and ω = 0.01 Hz - 1 kHz, in order to demonstrate that our data describe a strongly nonlinear regime. We found that a linear regime $\chi' \propto h_o$ exists only for a very limited range of h_o below a threshold amplitude [29] $h_o^{\dagger} \approx 0.02$ G, where also $\chi'' \rightarrow 0$. The drive level h_o = 1 G used in our S(H) measurements is therefore two orders of magnitude higher than the nonlinear threshold amplitude, and one order of magnitude greater than the amplitude $h_o = h^* \approx 0.1$ G necessary for the AC field to penetrate to the center of the sample. We found strongly nonlinear response $\chi(h_o)$ for h_o = 1 G, with $\chi(\omega) \propto \ln(\omega)$ (see insert in Fig.2), as predicted in Ref. 28.

The relaxation rate is essentially field-independent for $H \le H_2$, as shown in Fig. 2. The slight increase at $H = H_1$ can be explained [3] by the appearence of a small potential barrier surrounding the AD for n = 1. At $H = H_2$, $n = n_s(T = 4.4K) = 2$, and a third FL per unit ADL cell is repelled into the interstice for $H > H_2$. As a result, a substantial jump in S(H) is observed for $H > H_2$ for both components of χ. This behavior is consistent with previous (but less sensitive) DC measurements [4,7]. At $H = H_3$, n =

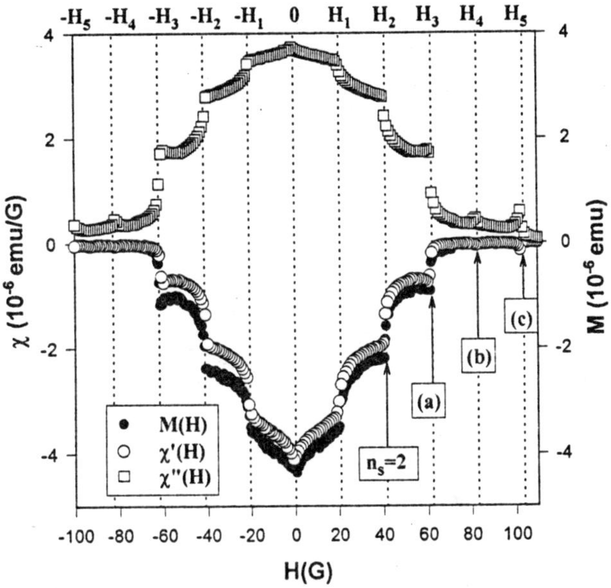

Fig. 1. AC susceptibility $\chi(H) = \chi' + i\chi''$ for AC field amplitude $h_o = 0.2$ G and frequency $\omega = 0.1$ Hz, and a portion of the DC magnetization loop M(H), for a W_xGe_{1-x} ($x \approx 0.67$) film with a square ADL (with AD separation d =1 μm and AD diameter D = 0.35 μm) for T = 4.4 K ≈ $0.98T_c$. The vertical dotted lines denote the expected matching fields H_m = m x (20.7 G). The arrows denote matching vortex configurations with zero, one, two and three IFL (n_s = 2, (a), (b) and (c), respectively), as discussed in the text.

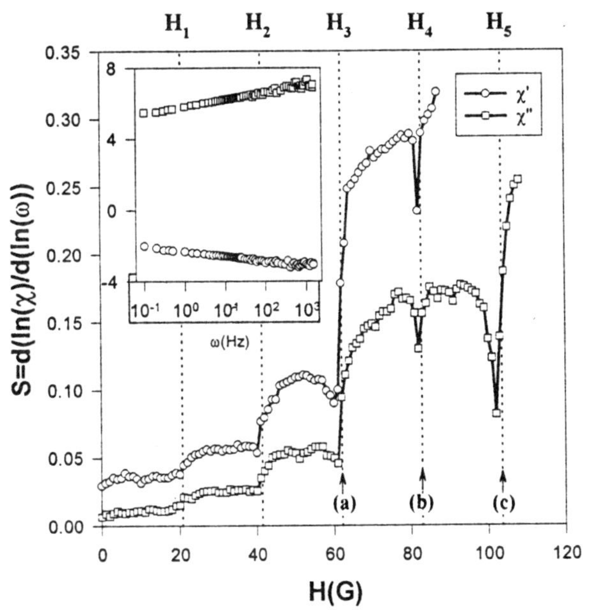

Fig. 2. Normalized relaxation rate $S = \partial(\ln(\chi))/\partial(\ln(\omega))$ for the W_xGe_{1-x} film of Fig. 1, for T = 4.4 K and AC drive h_o = 1 G. The insert shows the frequency dependence of the AC susceptibility χ. The vertical dotted lines denote the expected matching fields H_m = m x (20.7 G). Arrows denote matching vortex configurations with one, two and three IFL ((a), (b) and (c), respectively), as discussed in the text.

n_s+1 (two-Φ_o fluxoid at AD, one IFL; see Fig. 3-(a)), and the data of Fig. 2 indicate that this is a quite stable configuration, even if we take into account that the IFL are pinned much more weakly than fluxoids pinned by AD. An enormous jump in S(H) is apparent for fields H > H_3, when a second FL appears in the interstice (Fig.3-(b)), and the current relaxation (χ`) becomes strongly nonlogarithmic.

We can associate the data of Figs. 1 and 2 with probable FLL configurations based upon recent Lorentz microscopy studies of Nb films with regular arrays of pinning centers [30], as shown in Fig. 3. It is clear that phases with *two* IFL are much less stable than the case of *one* IFL, because *pairs of IFL* must form a frustrated FLL, as shown in Fig.3-(b), which illustrates two possible FLL arrangements. Therefore, FL are expected to switch between unstable FLL domains or superstructures, thereby explaining the remarkable jump in S(H) and the drop of the current to approximately zero at H_3 (Fig. 2). A weak matching effect in S(H) was found at H = H_4, although evidence for this effect is apparently unresolved in DC M(T,H) (see Fig. 1) and transport measurements [7]. Three FL occupy the interstice (see Fig.3-(c)) and S(H) exhibits a very deep minimum at H = H_5 (comparable to that at H = H_3), which implies that a FLL phase similar to case (c) of Fig. 3 generates a clear minimum in free energy. The relative size of the S(H) anomalies at H_2 and H_3 indicates that the change in pinning potential at the Mott transition at H_2 has much less effect on S(H) than the interstitial FLL rearrangement at H_3.

Nelson and Vinokur [13], and Radzihovsky [10] have proposed that weakly pinned IFL will behave differently from FL strongly pinned by columnar defects as a function of the angle between the applied field and the defect axis; and the IFL are expected to depin at a relatively low characteristic angle $\alpha^{depinning}$ [10,13]. We have generalized this idea to anticipate that differences in stability of various vortex phases (such as in Fig. 3) can be distinguished in the dependence of χ on the angle α between the DC field and the AD axis, and our results from such experiments are shown in Fig. 4.

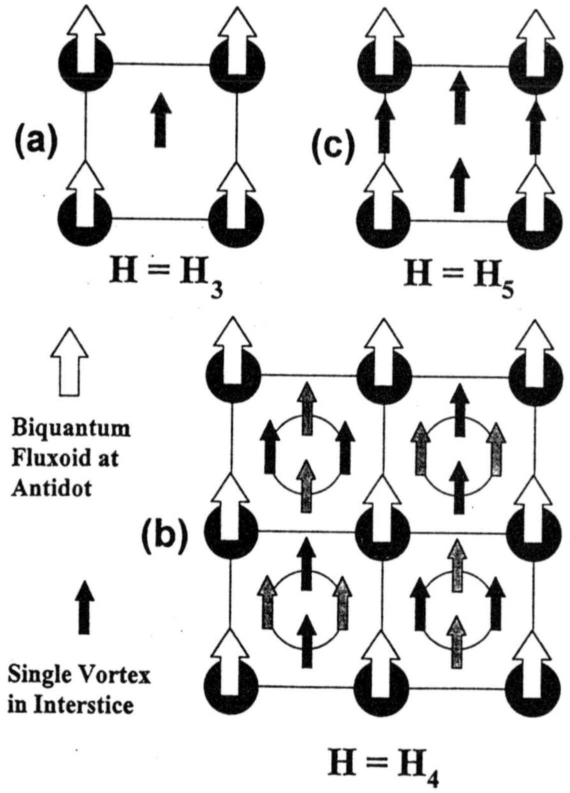

Fig. 3. Proposed vortex configurations for n > n_s at matching fields H = H_n: (a) The number of FL per unit cell n = n_s + 1 corresponds to a two-Φ_o fluxoid at the AD (n_s = 2, large arrow) and one FL in the interstice. (b) A "frustrated" configuration with n = n_s + 2, corresponding to a two-Φ_o fluxoid at the AD and two FL in the interstice (small black arrows). An alternative, rotated IFL configuration (small shaded arrows) illustrates potential competing FLL domain structures. (c) A relatively stable configuration with n = n_s + 3, corresponding to a two-Φ_o fluxoid at the AD and three FL in the interstices.

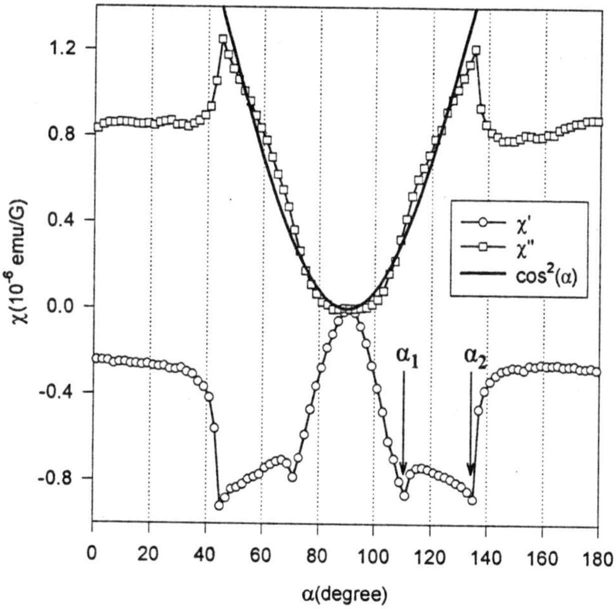

Fig. 4. The angular dependence of χ of a W_xGe_{1-x} film of Fig. 1 for external field H = 60 G, T = 4.40 K, h_o = 1G, and ω = 0.1 Hz. The arrows indicate anomalies in $\chi(\alpha)$ and the solid line illustrates the dependence $\chi \propto \cos^2(\alpha)$. See the inset of Fig. 6 for a definition of α.

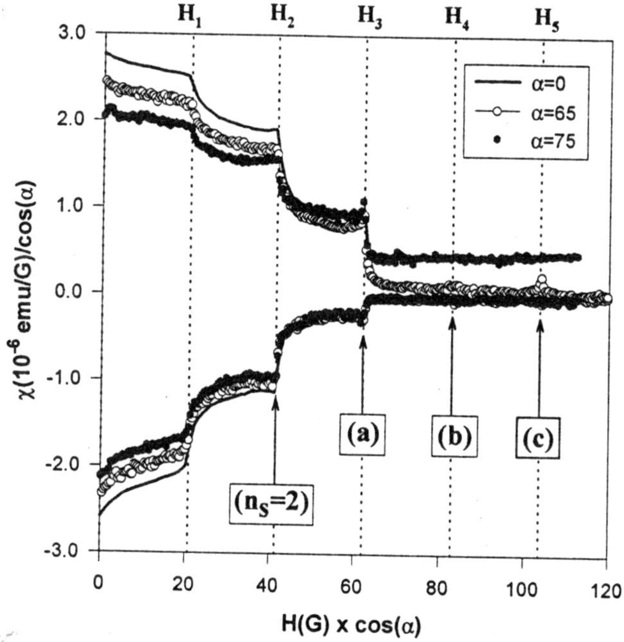

Fig. 5. Scaled susceptibility $\chi/\cos(\alpha)$ vs. perpendicular component of DC field $H_\perp = H\cos(\alpha)$ for the W_xGe_{1-x} film of Fig. 1 at fixed angles α shown, T = 4.40 K, reset AC drives $h_o = 1G/\cos(\alpha)$ and frequency ω = 0.1 Hz. The arrows indicate matching vortex phases defined in Figs. 1 and 3. See the inset of Fig. 6 for a definition of α.

A $\cos^2(\alpha)$ dependence of the response of the FLL [10,13] (solid line in Fig. 4) follows from assuming the sample is sensitive only to the component of magnetic field perpendicular to the sample surface, $H_\perp = H\cos(\alpha)$, and that the measuring geometry only allows the determination of a longitudinal component of magnetic response $\propto \chi\cos(\alpha)$. We have found an approximate $\cos^2(\alpha)$ dependence for fixed external fields $H < H_1 = 20.7$ G (data not shown). However, the situation changes for $H > H_1$. For example, for $H = 60$ G $> H_2$, two very sharp anomalies appear at $\alpha_1 \approx 44°$ and $\alpha_2 \approx 72°$, as shown in Fig. 4. The origin of these anomalies becomes clear if we calculate the component of magnetic field perpendicular to the film surface: $H_\perp(\alpha_2) = 60$ G x $\cos(\alpha_2) \approx 43$ G $= H_2$, and $H_\perp(\alpha_1) = 60$ G x $\cos(\alpha_1) \approx 20$ G $= H_1$. This shows that a film ($t \approx 600$ Å) with a ADL is sensitive only to $H_\perp$, and we should expect anomalies in χ whenever $H_\perp = H_m$; moreover, these data do not simply reflect differences in the angular dependence of the stability of various FL configurations.

We therefore performed additional measurements of $\chi(H)$ at various fixed angles in order to better assess the stability of different FLL, as shown in Fig.5. The field dependences of $\chi(H,\alpha)$ indicate that the matching fields $H_m(\alpha)$ and the absolute values of χ are invariant under the scaling $H_m(0) = H_m(\alpha)\cos(\alpha)$ and $\chi_m(0) = \chi_m(\alpha)/\cos(\alpha)$, respectively, at least for $\alpha \leq 65°$ (Fig. 5). It is important to note that in order to maintain the $\cos(\alpha)$ scaling of the absolute value of χ, the amplitude of AC field must be reset to $h_o^\alpha = h_o/\cos(\alpha)$ for each measurement. Although the data approximately scale for fixed $\alpha = 0°$ and $65°$, and exhibit a matching effect at $H = H_5$, the scaling breaks down and we observe a strong increase in dissipation (χ'') for $\alpha \geq 75°$, and $H > H_3$; the matching effect at $H = H_4$ is also lost.

The angular dependences of the dissipation can also be checked while maintaining the perpendicular DC field at particular matching conditions $H_\perp = H\cos(\alpha) = H_3$ (one IFL) and $H_\perp = H_4$ (two IFL) as shown in Fig. 6. The absolute values of dissipation χ'' approximately follow a $\cos(\alpha)$ dependence for $\alpha \leq 65°$ at these matching fields, but for the frustrated vortex configuration (b) with $n = n_s+2$, a sharp increase in dissipation appears at a stability limit $\alpha = \alpha^{depinning} = 72°$, implying that two IFL can no longer be pinned for $\alpha > \alpha^{depinning}$.

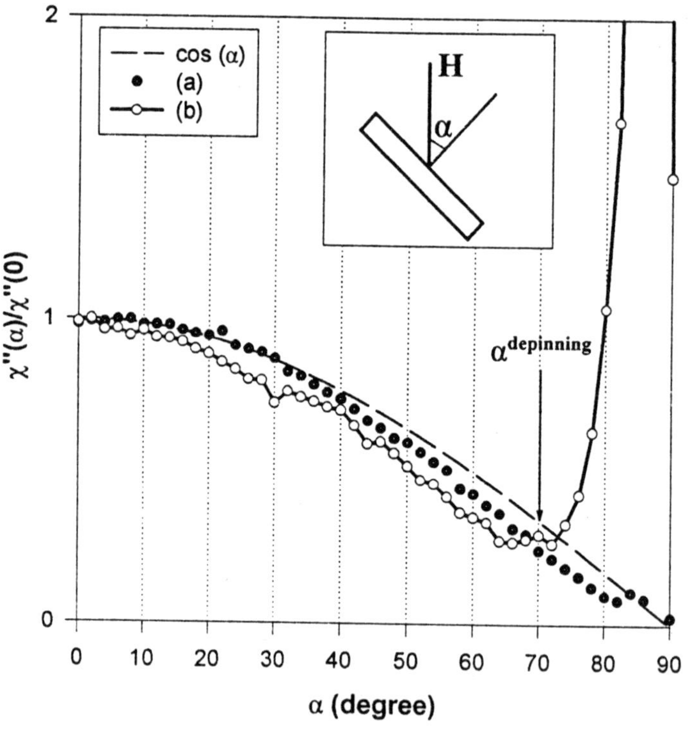

Fig. 6. The angular dependences of normalized dissipation $\chi'' = \chi''(\alpha)/\chi''(0)$ at the third ($H_3(\alpha) = H_3\cos(\alpha)$, $n = n_s+1$; (a)) and fourth ($H_4(\alpha) = H_4\cos(\alpha)$, $n = n_s+2$; (b)) matching fields for a W_xGe_{1-x} film with a square ADL. The arrows indicate the depinning angle $\alpha^{depinning}$ where an anomalous increase in dissipation for less-stable configuration (b) with $n = n_s+2$ is observed. The inset indicates the sample orientation with respect to the applied field **H**.

IV. CONCLUSION

In conclusion, we have detected anomalies in the DC magnetization, and the angular, field, AC drive and frequency dependences of the AC susceptibility and magnetic relaxation that reflect the dynamics and stability of novel vortex phases formed by two different types of FL in a superconducting thin film with an ADL. One subgroup of FL is strongly pinned by AD and the other is weakly pinned in the interstitial region between AD. The strong pinning epitaxy of the ADL was previously known to stablize FLL having no IFL for matching fields corresponding to a number of FL per ADL unit cell $n \leq n_s = D/4\xi(T)$. We find novel, stable vortex phases of IFL exist at surprisingly high fields in a "supermatching" regime where $n > n_s \approx 2$, and that the relative stability of these phases is nonmonotonic in field (similar to a peak effect) due to frustration. The reduced stability of the frustrated phase is consistent with our observation of a theoretically predicted [10,13] critical depinning angle. Similar supermatching regimes and interesting angular and FL dynamical effects should be accessible in future studies of other ADL systems with n_s =3, 4, etc. These results are relevant to the problems of FLL phase transitions, dissipation and peak effects [27] under investigation in a variety of superconductors with ion damage or other types of correlated disorder.

ACKNOWLEDGMENTS

Research at the University of Kentucky was supported by U.S. NSF Grant #DMR-9512589. Research in Katholieke Universiteit Leuven was supported by the Flemish Fund for Scientific Research (FWO) and the Concerted Action (GOA) Programs. We thank Mr. Byron Watkins for help in preparing the manuscript.

REFERENCES

1. L. Civale, et al., Phys. Rev. Lett. **67**, 648 (1991); M. Konczykowski et al., Phys. Rev. B **44**, 7167 (1991-I).
2. W. Gerhauser, et al., Phys. Rev. Lett. **68**, 879 (1992); M. Leghissa, et al., Phys. Rev. B **48**, 1341 (1993).
3. V. V. Metlushko et al., Solid State Commun. **91**, 331 (1994).
4. M. Baert et al., Phys. Rev. Lett. **74**, 3269 (1995).
5. M. Baert et al., Europhys. Lett. **29**, 157 (1995).
6. V. V. Moshchalkov et al., Jpn. J. Appl. Phys. **34**, 4559 (1995).
7. E. Rosseel et al., Phys. Rev. B **53**, R2983 (1996).
8. G. S. Mkrtchyan and V. V. Shmidt, Sov. Phys.: JETP **34**, 195 (1972).
9. I. B. Khalfin and B. A. Shapiro, Physica C **207**, 359 (1993).
10. L. Radzihovsky, Phys. Rev. Lett. **74**, 4923 (1995).
11. K. M. Beauchamp et al., Phys. Rev. Lett. **75**, 3942 (1995).
12. L. Krusin-Elbaum, L. Civale, J. R. Thompson and C. Field, Phys. Rev. **53**, 11744 (1996).
13. D. R. Nelson and V. M. Vinokur, Phys.Rev. B **48**, 13060 (1993).
14. V. V. Moshchalkov et al., Phys. Rev. B **54**, 7385 (1996).
15. A. T. Fiory, A. F. Hebard and S. Somekh, Appl. Phys. Lett. **32**, 73 (1978).
16. E. R. Nowak, S. Anders, H. M. Jaeger, J. A. Fendrich, W. K. Kwok, R. Mogilevsky and D. G. Hinks, submitted to Phys. Rev. B.
17. D. G. Steel, J. D. Hettinger, J. T. Parkman, M. E. Yoder, K. E. Gray, B. G. Glagola, M. M. Eddy, Physica C, to be published (1996).
18. C. P. Herring , Phys. Rev. Lett. **47A**, 105 (1974).
19. A. Bezryadin and B. Pannetier, J. Low Temp. Phys. **102**, 73 (1996).
20. K. Harada et al., Nature **360**, 51 (1992).
21. J. E. Bonevich et al., Phys. Rev. Lett. **70**, 2952 (1993).
22. R. E. Sager, A. D. Hibbs and S. Kumar, IEEE Trans. Magn. **28**, 3072 (1992).
23. MPMS Users Manual, Quantum Design 1995; A. D. Huxley, et al., J. Phys.: Condens. Matter **5**, 7709 (1993).
24. V. V. Moshchalkov et al., J. de Physique IV **6**, C3-335 (1996); Y. Bruynseraede et al., SPIE Proc. **2697**, 328 (1996).
25. L. D. Cooley and A. M. Grishin, Phys. Rev. Lett. **74**, 2788 (1995).
26. D. Dominguez and J. V. Jose, Phys. Rev. B **53**, 11692 (1996).
27. M. J. Higgins and S. Bhattacharya, Physica C **257**, 232 (1996).
28. C. J. van der Beek, V. B. Geshkenbein and V. M. Vinokur, Phys. Rev. B **48**, 3393 (1993).
29. A. M. Campbell, J. Phys. C **2**, 1492 (1969).
30. K. Harada et al., Science **274**, 1167 (1996).

FLUX PINNING AND APPLICATIONS OF HIGH TEMPERATURE SUPERCONDUCTORS

Masato Murakami

ISTEC, Superconductivity Research Laboratory, 1-16-25 Shibaura, Minato-ku, Tokyo 105, Japan

Next to zero resistivity, the attractive property of superconductors is their response to magnetic fields. High temperature superconductors are classified as type II superconductors, which can be penetrated by magnetic flux lines. When type II superconductors contain pinning centers, they behave like paramagnetic, diamagnetic and ferromagnetic materials, depending on the temperature and the magnetization process.

I. INTRODUCTION

The most attractive property of superconductors is zero resistivity, but their response to magnetic field is also highly attractive. In particular, the pinning effect cause their magnetic properties to change in various ways [1].

Materials can be classified according to their magnetic properties as diamagnetic, paramagnetic, ferromagnetic, antiferromagnetic and ferrimagnetic. The interest in superconductors lies in the fact that they can exhibit diamagnetic, paramagnetic, or ferromagnetic properties, depending on the temperature and magnetization process.

Superconductors are generally known to behave as perfect diamagnets. This property of complete field exclusion is known as the Meissner effect, after its discoverer. The phenomenon of a small magnet floating above a superconductor is due to the Meissner effect. Of all the materials in nature, only the superconductor can exhibit perfect diamagnetism. However, complete diamagnetism can only be observed for weak magnetic fields, and superconductors are divided into two categories according to their behavior when perfect diamagnetism is destroyed.

Type I superconductors transform to a normal state (or paramagnetism), as sketched in the magnetization curve of Fig. 1(a). In contrast, the external field partially penetrates into the type II superconductor. As the external field is increased, the penetration of the field is enhanced, and finally at the upper critical field (H_{c2}), the superconductors become normal. The first critical field at which the Meissner state is destroyed is called the lower critical field (H_{c1}). The magnetization curve of type II superconductors is illustrated in Fig. 1 (b). Between H_{c1} and H_{c2}, the superconducting and normal states co-exist, and therefore this state is called a mixed state.

II. QUANTIZATION OF MAGNETIC FIELD

It is very interesting to note that the magnetic field which enters the superconductor is quantized, and the result is called fluxoid or vortex[2]. The mechanism of flux quantization is identical to the quantization of the electron orbital in atoms. The screening current flowing around the fluxoid is persistent, therefore, the phase of the superelectron wave of such current must be an integral multiples of 2π, otherwise the currents are attenuated. It is known that the magnitude of a single fluxoid is $\phi_0=2.07\times10^{-15}$Wb. Figure 2 shows a schematic diagram of the mixed state and quantized fluxoids. The centers of fluxoids are non-superconducting, and there is a change from normal conduction to superconduction over a distance of the coherence length (ξ). The radius of screening supercurrent or the magnetic

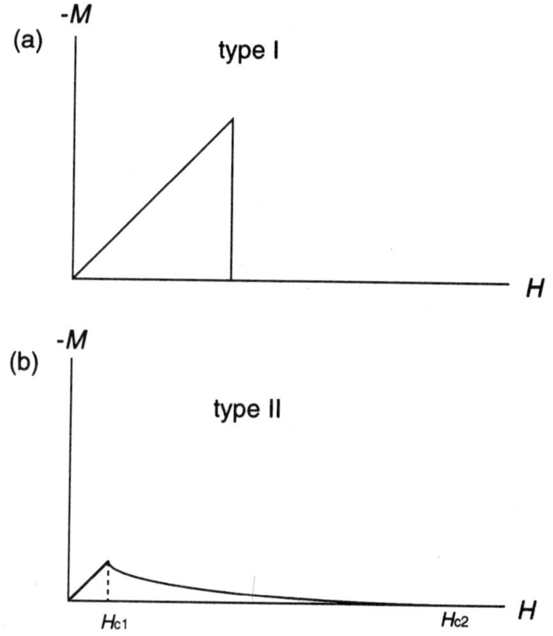

Fig. 1. Sketch of magnetization curves for (a) type I and (b) type II superconductors.

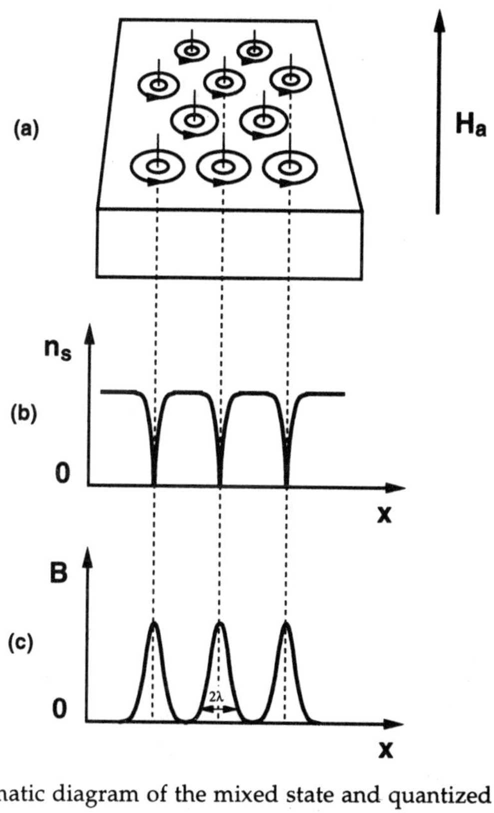

Fig. 2. Schematic diagram of the mixed state and quantized fluxoids.

field is equal to the field penetration depth (λ). One can consider that the the screening currents flowing over a distance of λ is producing the flux quantum ϕ_0.

III. PINNING EFFECT

In type II superconductors, the diamagnetism is weakened by the penetration of magnetic field when the field exceeds H_{c1} (Fig. 1(b)). However, when they contain some structural defects, their magnetic properties are dramatically changed.

Normal to superconducting transition

The transition from normal to superconducting state occurs at T_c, because the free energy of the superconducting state becomes lower than that of the normal conducting state, as illustrated in Fig. 3. This transition is an ordinary second-order phase transition.

In the presence of magnetic field, the free energy of the superconducting state is increased. As mentioned above, the superconductor can exclude magnetic field completely, which is known as the Meissner effect. However, the exclusion of magnetic field adds the excess energy of $(1/2)\mu_0 H^2$ to the superconducting state.

In type I superconductors, at the point where the energy required to exclude a magnetic field becomes equal to the energy gained on entering the superconducting state (the condensation energy: $f_s - f_n$), the material will transform to the normal conducting state. The corresponding magnetic field is called the thermodynamic critical field (H_c).

On the other hand, in the case of type II superconductor, before reaching H_c, the field starts to penetrate into the superconductor at H_{c1}, thus lowering the energy for field exclusion.

The difference between type I and type II superconductors can be summarized in terms of the surface energy of the boundary between the superconducting and normal conducting regions (S/N boundary). In type I superconductors, S/N boundary has a positive surface energy, and therefore such boundary will be difficult to be present. In contrast, in type II superconductors S/N boundary has a negative surface energy, thus the system can lower its free energy by creating S/N boundaries, and it can go into the mixed state.

The surface energy is determined by the ratio of λ to ξ. When λ/ξ is larger than $1/\sqrt{2}$, the interfacial energy is negative, and the superconductors behave like type II.

Here, it is important to note that in the mixed state of type II superconductor, although the overall energy is lowered by allowing the penetration of magnetic field, at the region where the fluxoid has penetrated, there is a local energy loss equal to the amount of the condensation energy. This loss is called the penalty energy and is the source of flux pinning.

Pinning mechanism

Let us assume that a superconductor contains a nonsuperconducting particle as shown in Fig. 4. The fluxoid can lower its energy by interacting with such nonsuperconducting particles.

As mentioned before, magnetic field enters a type II superconductor to compensate the energy needed to repel magnetic field when the external field exceeds H_{c1}. The system can lower its total free energy by allowing the magnetic field to enter the superconductor, although the region where the fluxoid has penetrated locally loses the condensation energy. The loss of energy due to the penetration of fluxoid is thus called the penalty energy and is $1/2\mu_0 H_c^2$ (or $B_c^2/2\mu_0$) per unit volume.

When the fluxoid interacts with nonsuperconducting region, the penalty energy can be saved per interacted volume, since such region is normal conducting. The saved energy is $U_p = (B_c^2/2\mu_0)\pi\xi^2 d$ where d is the size of a nonsuperconducting particle. On the other hand, the penalty energy must be paid to move the fluxoid from normal to superconducting region, whereby the fluxoids are pinned at normal regions. It is also clear that the fluxoid can be pinned even by a superconducting region where H_c is small compared to the surroundings, since the penalty energy becomes smaller, although the flux pinning is less effective than normal regions.

The pinning force f_p is given by $f_p = dU_p/dx$ and in a simple case

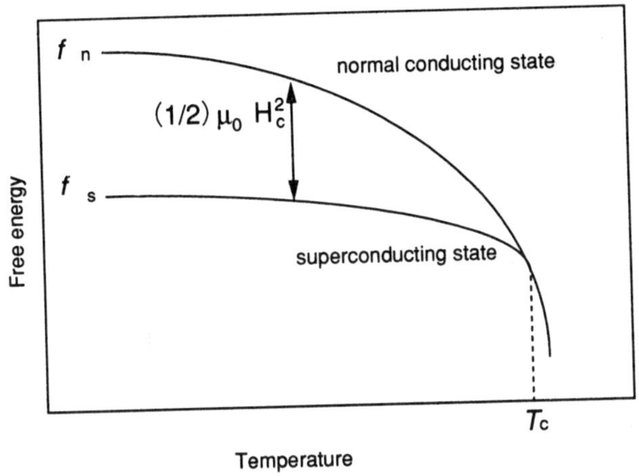

Fig. 3. Temperature dependence of free energy for the normal (f_n) and the superconducting state (f_s).

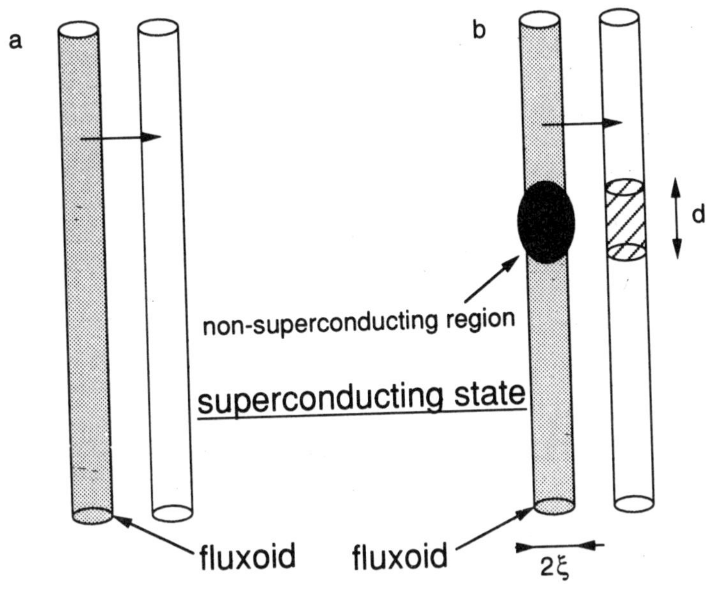

Fig. 4. Schematic illustration of the mechanism of flux pinning by non-superconducting particle.

$$fp = U_p/\xi = (Bc^2/2\mu_0)\pi\xi d.$$

This *fp* is called an elementary pinning force.

IV. PINNING CENTERS IN HIGH TEMPERATURE SUPERCONDUCTORS

There are several candidates for pinning centers in high temperature superconductors. Here I just focus on Y_2BaCuO_5 inclusions in the $YBa_2Cu_3O_7$ superconductors[4].

In the Y-Ba-Cu-O system, when solidified from molten region, the superconducting $YBa_2Cu_3O_x$ phase is produced by the following peritectic reaction

$$Y_2BaCuO_5 + L\,(3BaCuO_2 + 2CuO) \rightarrow 2YBa_2Cu_3O_x$$

As a result of incomplete peritectic reaction, Y_2BaCuO_5 (211) phase is occasionally trapped in Y123 matrix. It is known that nonsuperconducing particles work as pinning centers in conventional superconductors. Therefore, they are also expected to be pinning centers in Y-Ba-Cu-O. Definitely their effectiveness strongly depends on their size and density.

Suppose that the size and density of 211 inclusions are d and N_p, respectively. Here, d should be much larger than x_{GL}, since d cannot be reduced to the order of ξ_{GL} by metallurgical control. The pinning energy per interaction is given by

$$U_p = \pi\xi_{GL}^2 d\,(B_c^2/2\mu_0).$$

Therefore

$$f_p = U_p/\xi_{GL} = \pi\xi_{GL} d\,(B_c^2/2\mu_0).$$

In the case of large inclusions, only the interface is effective in flux pinning, since there is no energy difference for flux motion inside the inclusions so that the number of fluxoids which can be pinned by one inclusion is d/a_f, where a_f is the flux line lattice (FLL) spacing and given by $a_f = 1.075(\phi_0/B)^{1/2}$. It is also known that FLL of oxide superconductors is very soft due to large k (=l/x), and therefore a direct summation seems to be reasonably good approximation to obtain bulk pinning force. Then

$$F_p = N_p f_p$$

$$F_p = \pi B_c^2 \xi_{GL} N_p d^2 / 4\mu_0 \phi_0^{1/2} B^{3/2}.$$

This relation indicates that F_p is proportional to $N_p d^2$ which is equal to V_f/d where V_f ($=N_p d^3$) is the volume fraction of the 211. Hence, F_p can be increased either by increasing V_f or by decreasing d of the 211.

We have succeeded in dispersing fine 211 inclusions in the 123 matrix by controlling the above peritectic reaction. Figure 5 shows an optical micrograph of melt-processed Y-Ba-Cu-O where 211 inclusions <1μm are distributed in the 123 matrix.

V. DIAMAGNETISM DUE TO PINNING EFFECT

The magnetization process of type II superconductors with pinning effect differs from that of pure superconductors. As already shown in Fig. 1(b), diamagnetism is weakened when the field exceeds H_{c1} in pure superconductors without pinning effect. This is due to the fact that the field enters the superconductor above H_{c1}. However, when the superconductor contains pinning centers, the penetrated fluxoids cannot easily move into the superconductor. As shown in Fig. 6, the fluxoids will be trapped by pinning centers as soon as they enter the superconductor. In order to move the fluxoids deeper, we need to increase magnetic pressure to overcome the pinning force. In general, the pinning effect depends on the kind and

Fig. 5. An optical micrograph of melt processed Y-Ba-Cu-O superconductors. Black small particles distributed are Y_2BaCuO_5 inclusions and grey matrix is $YBa_2Cu_3O_7$.

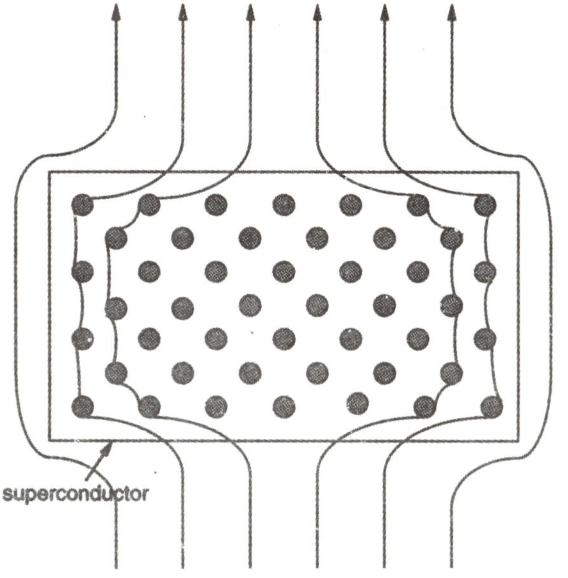

Fig. 6. Schematic illustration of flux exclusion by a pinned superconductor.

Fig. 7. Levitation of a person using a repulsive force of melt processed Y-Ba-Cu-O against Fe-Nd-B magnet.

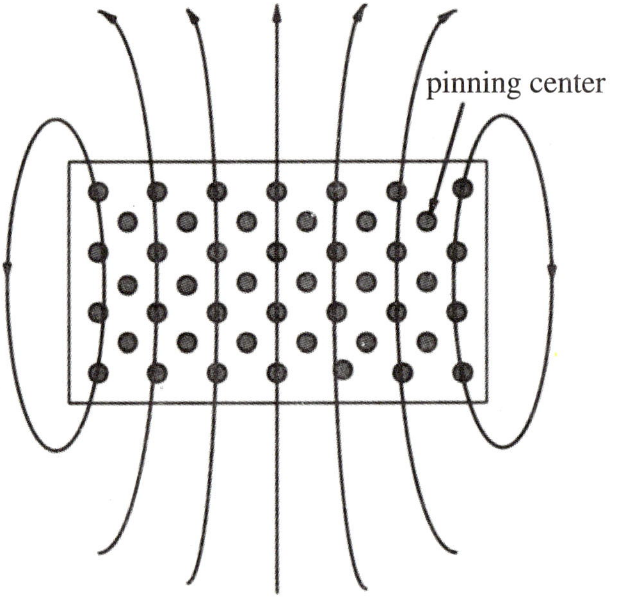

Fig. 8. Schematic illustration of flux trapping by a pinned superconductor.

distribution of pinning centers. The control of microstructure of superconductors enables us to obtain a very large pinning effect which is even effective in very high fields exceeding several teslas. When the pinning force is large, diamagnetic properties will be observed up to very high fields, and thereby a large repulsive force can be obtained.

Using such diamagnetism due to flux pinning, we succeeded in levitating a person at 5cm gap, as shown in Fig. 7. Here we used melt processed Y-Ba-Cu-O superconductors containing fine Y_2BaCuO_5 inclusions cooled by liquid nitrogen. This clearly demonstrate that flux pinning can produce a large repulsive force, although it is not complete flux exclusion.

Several applications of such levitation are now under consideration. Those are magnetic bearing, flywheel energy storage system, and load transport.

VI. FERROMAGNETISM DUE TO PINNING EFFECT

It is also important to note that strong pinning superconductors can attract a large magnetic field due to flux pinning, which bring about a large attractive force. Figure 8 shows schematically how fluxoids are trapped by pinning centers.

When we apply magnetic field before cooling, the field easily penetrates the superconductor since it is paramagnetic. If the superconductor is cooled under this condition, fluxoids will be trapped by pinning centers, and the field remains trapped even when the external field is removed, which indicates that pinned superconductors become strong permanent magnets. We have found that a small piece of melt processed Y-Ba-Cu-O with dimensions 1x1x3mm^3 exhibited BH_{max} of 80MGOe at 5K[5]. It is also interesting to note that the trapped field can be increased by increasing either J_c values or its size. At present, a superconducting bulk magnet which can generate 1T at 77K has already been fabricated using melt-processed Y-Ba-Cu-O sample 4cm in diameter and 2cm in thickness.

Recently we have developed the process which enables us to fabricate Nd-Ba-Cu-O with high T_c and large J_c values. When Nd-Ba-Cu-O is melt processed in air, the sample exhibits low T_c and broad transition due to the presence of Nd-Ba solid solution. In contrast, when they are melt processed in a reduced oxygen atmosphere, they exhibit high T_c and large J_c values, which are accompanied by the secondary peak effect. We believe that the substitution of Nd on Ba site is largely suppresed when processed in low Po2, which is the source of high T_c, as yet partially substituted regions remain in the sample with lower T_c and H_{c2}. Such regions will be driven normal when the magnetic field is applied, and thus can act as field-induced pinning centers. According to our estimate, a field >3T can be trapped at 77K in Nd-Ba-Cu-O sample 4cm in diamter. Such a superconductor pellet will serve as a new function materials, that is a superstrong permanent magnet.

VII. SUMMARY

Magnetic properties of type II superconductors with pinning effect exhibits paramagnetic, diamagnetic, and ferromagnetic properties depending on the temperature and the magnetization process. Particularly it is interesting that the same materials can exhibit both strong diamagnetic and ferromagnetic properties. At present, various applications are now under consideration using such properties.

REFERENCES

1. M. Murakami, IEEE Trans. J. Mag. Jpn. **8**, 405 (1993).
2. A.C. Rose-Innes and E.H. Rhoderick, *Introduction to Superconductivity*, Oxford (1978).
3. A.M. Campbell and J.E. Evetts, *Critical Currents in Superconductors*, Taylor and Frances, London (1972).
4. M. Murakami, *Melt Processed High Temperature Superconductors*, World Scientific, Singapore (1991).
5. S. Gotoh, M. Murakami, N. Koshizuka and S. Tanaka, Physica B **165**, 1379 (1990).

SIZE-EFFECTS IN FLUX-LINE SOLIDS AND LIQUIDS

M.H. Theunissen[1], E. Van der Drift[2] and P.H.Kes[1]

[1] *Kamerlingh Onnes Laboratory, Leiden University, 2300 RA Leiden, The Netherlands.*
[2] *Delft University of Technology, DIMES/S, P.O. Box 5046, 2600 GA Delft, The Netherlands.*

Flow of 2D flux lines confined to narrow channels has been studied. In the solid, (in)commensurability between lattice and channel width cause mobility-oscillations from which the actual channel width follows. In the liquid, the oscillations disappear and flow is dictated by the shear viscosity η. By repeated experiments for different channel widths, η is determined for the first time. It diverges when the melting line is approached from above according to $\eta \propto \xi_+^2(T,B)$, with ξ_+ the translational correlation length. An indication for the hexatic-to-liquid transition is also observed.

I. INTRODUCTION

The appearance of a flux liquid just below the upper critical field B_{c2} has been predicted to be a general property of the mixed state[1]. This prediction was largely ignored until recently Nelson suggested that the absence of zero resistance in a magnetic field observed in high temperature superconductors (HTS), is caused by a transition to the flux liquid state[2]. At present, a variety of experiments provided evidence for melting (even in presence of weak disorder), both for a three dimensional (3D) vortex lattice[3] and for a 2D vortex lattice[4]. Now the existence of a flux liquid is well established, one may focus on its properties, particularly, the appearance of a dynamic shear viscosity $\eta(T,B)$ in place of a shear modulus c_{66}. The shear modulus should discontinuously disappear at melting when the shear fields vary slowly both in space and time[5,6]. The behavior of η close to freezing is specially interesting, since the (hexatic) liquid then may be considered as a solid with a low density of free dislocations, (i.e. dislocation loops in 3D[7] and edge dislocations in 2D[8]). Concentrating on the 2D case being the actual situation in our experiments, we follow Nelson's suggestion[9] that η is inversely proportional to the density of free dislocations n_f which itself is determined by the translational correlation length ξ_+ in the liquid given by

$$\eta(T,B) \sim \frac{1}{a_0^2 \tilde{\mu} n_f} \sim \xi_+^2(T,B) \tag{1}$$

with a_0 the lattice spacing and $\tilde{\mu}$ a typical dislocation mobility. Approaching the melting line $T_m(B)$ from above, η mimics the strong divergence of ξ_+, i.e.

$$\xi_+ = \xi_{+0}\exp[b'(\frac{T_m}{T-T_m})^{\nu}] = \xi_{+0}\exp[b''(\frac{B_m}{B-B_m})^{\nu}] \tag{2}$$

with $\nu = 0.36963...$, b' and b'' nonuniversal constants of order unity and $\xi_{+0} \approx a_0$, being the smallest length scale of the system. In order to probe this behavior we extended the experiment of Pruijmboom *et al.*[10] who studied the flow of a vortex solid through narrow channels in a double layer device. We designed similar devices consisting of an amorphous Nb_3Ge bottom layer covered with a NbN layer, see Fig. 1. Small channels are tched through the NbN layer into the a-Nb_3Ge layer, using nanolithographic techniques. Different samples with channel widths ranging between 238 nm $< w <$ 1185 nm were made in a four-point terminal configuration with the voltage probes next to the channel array and the current flowing perpendicular to the channels. In perpendicular magnetic field the flux lines experience a Lorentz force in the direction of the channels. In case of the solid the condition for flow is determined by the shear-modulus c_{66} of the lattice and in case of the liquid this condition for flow is controlled by the viscosity η.

II. THEORY

In the solid, the flux lines outside the channels are pinned in the strong pinning NbN layer (with j_c in NbN $\geq 10^4 j_c$ in NbGe), while those inside the channels will start to move as soon as the Lorentz force surpasses the shear force of the flux-line lattice (FLL) in the channels with the pinned FLL in the NbN. The condition for plastic shear flow in the continuum approximation can be written as $F_s = j_s B = 2\tau/w$, where $\tau = Ac_{66}$ is the flow stress. The isotropic shear-modulus in the high κ limit[11] is

$$c_{66}(b) = \frac{B_c^2(t)}{4\mu_0} b(1-b)^2(1-0.58b+0.29b^2) \tag{3}$$

with $b = B/B_{c2}$ the reduced field and $B_c(t)$ the thermodynamical critical field. The parameter A depends on the flux line lattice configuration inside the channels, e.g. when the lattice parameter, $a_0 = 1.075\sqrt{\phi_0/B}$, is commensurate to the channel width $A \approx 0.05$, but at incommensurability misfit dislocations develop along the edges which reduce A.

In the liquid phase, $c_{66} = 0$ and the oscillations in $F_s(B)$ should disappear. The flow through the channels (defined by $y = \pm w/2$) is now controlled by the Bardeen-Stephen drag parameter γ[12] and the shear viscosity η. The velocity profile follows from the (simplified) hydrodynamic equation[13]

$$-\gamma\vec{v} + \eta\nabla_\perp^2\vec{v} + \vec{f_T} = 0 \tag{4}$$

and the boundary condition $\vec{v}(\pm w/2) = 0$. The driving force $\vec{f_T}$ is in the x-direction. The velocity profile is maximum at $y = 0$ and decays exponentially at the edges in a length scale $\delta = \sqrt{\eta/\gamma}$. The integrated velocity determines the net flow through the channel and the flux flow resistivity ρ_f to be measured :

$$\rho_f = \rho_f^0[1 - \frac{2\delta}{w}\tanh(\frac{w}{2\delta})] \tag{5}$$

where $\rho_f^0 = B^2/\gamma$ is the Bardeen-Stephen value in which small corrections for weak pinning are supposed to be absorbed.

III. PREPARATION AND CHARACTERIZATION

The double layer is fabricated by rf-sputter deposition in a Leybold Z400 sputtering system, see also ref.14. The films are Ar-sputtered (10^{-2} mbar), with a base pressure of 10^{-6} mbar, on a Si (100) wafer, which is attached to a water-cooled substrate-table. The double layer consists of a 50 nm-thick layer of polycrystalline NbN deposited on top of a 550 nm layer of amorphous-$Nb_{0.64}Ge_{0.34}$ (denoted a-Nb_3Ge). The amorphousness was confirmed by X-ray diffraction, and the average stoichiometry and thicknesses were determined by electron microprobe analysis (EPMA) and Rutherford backscattering (RBS).

The nanometer channel pattern is realized by the tri-level-stencil technique developed by Jackel[15], which is successfully applied to the double layers[10,16]. Pattern transfer into the NbN layer is performed by reactive ion etching in a mixed $SiCl_4$-Cl_2-He plasma (3:1:4) at 3 Pa pressure, a total gas flow of 100 sccm and a power density of about 0.48 W/cm^2. Direct-write electron beam exposure was performed using an electron-beam pattern generator operating at 50 kV. The beam spot size was roughly 50 nm for the samples with the smallest widths and 100 nm for the samples with larger channel widths. The transfer of the four-terminal pattern is etched in the same reactive Cl_2 plasma. In each sample 300 identical, parallel channels were etched through the NbN layer into the NbGe, one channel in each 10 μm, down to a depth of 240 nm from the top surface leaving a layer of 310 nm a-Nb_3Ge. The length of each channel, and thereby the width of the sample strip, is 300 μm. For reference we fabricated one sample, with the NbN toplayer etched off, this sample is denoted as the reference sample.

We carried out a series of four-probe current-voltage (IV) measurements on all samples, both in a *dc* and an *ac* (120 Hz) mode, as a function of field at T = 1.70, 2.00, 2.20 and 2.40 K and as a function of temperature at B = 1.00, 0.70, 0.50 and 0.30 T. Characteristic parameters of our two layer devices are : T_c = 11.4 K and $B_{c2}(0) \approx 16$ T for the NbN top layer, and T_c = 2.68 K, ΔT_c = 10 mK and $B_{c2}(0)$ = 3.8 T, residual resistivity $\rho_0 = 2.17\mu\Omega$m and $dB_{c2}/dT|_{T=T_c} = -2.1$ T/K for the NbGe reference sample[5]. The upper critical field was determined by linear extrapolation of $\rho_{ac}(T,B)$ to the normal state resistivity ρ_n[4]. The superconducting parameters determined from these data are : $\xi(0)$ = 7.7 nm, $\lambda(0) = 0.94\mu$m, κ = 75, and $B_c(0)$ = 30.0 mT.

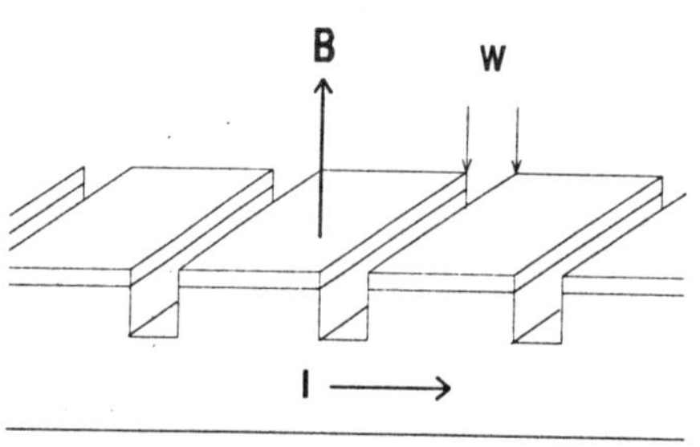

Fig.1 A schematic view of the channel configuration in our experiment.

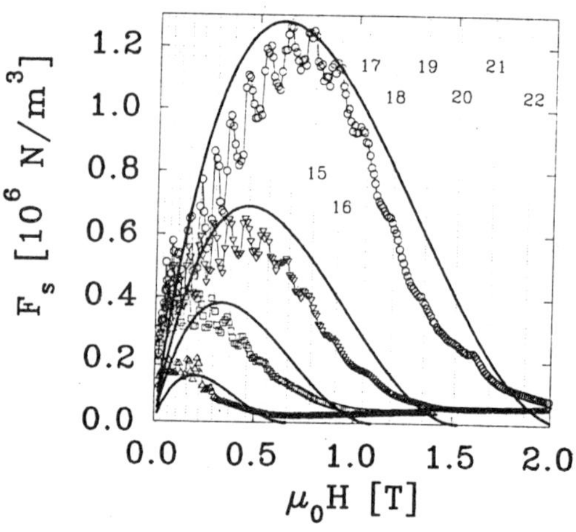

Fig.2 F_s as a function of field for four different temperatures, T=1.7 K (○), 2 K (▽), 2.2 K (□) and 2.4 K (△). The solid lines represent the continuum approximation model, using A=0.048.

IV. RESULTS AND DISCUSSION OF THE *SOLID*-PHASE

In the solid, the flux lines outside the channels are pinned in the strong pinning NbN layer (with j_c in NbN $\geq 10^4 j_c$ in NbGe), while those inside the channels will start to move as soon as the Lorentz force surpasses the shear interaction with the pinned vortices in the NbN. The condition for plastic shear flow in the continuum approximation can be written as

$$F_s(B) = J_s B = \frac{2\tau}{w} \tag{6}$$

where $\tau = Ac_{66}$ is the flow stress. The isotropic shear modulus c_{66} is known, see Eq.(3) and the parameter A depends on the flux line lattice configuration inside the channels, e.g. when the lattice parameter is commensurate to the channel width $A \approx 0.05$, but at incommensurability misfit dislocations will develop along the edges which reduces A. Measured as a function of field, F_s thus globally follows c_{66}, but also oscillates with changing commensurability being maximum for perfect matching.

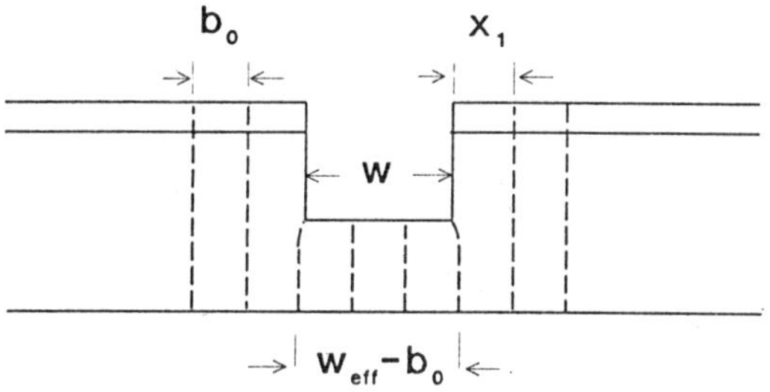

Fig.3 The schematic view of the crossection of one of the channels, where the flux-lines in the NbN near the edge are shifted to a new equilibrium position x_1, due to the interaction with the channel edge. The effective width w_{eff} is explained in text. The bend flux-lines terminate at the corners because of the step in thickness, see text. The reason for not drawing more flux-lines in the NbN is beause the flux lines in the NbN are strongly pinned and therefore the periodicity of the lattice will disapper.

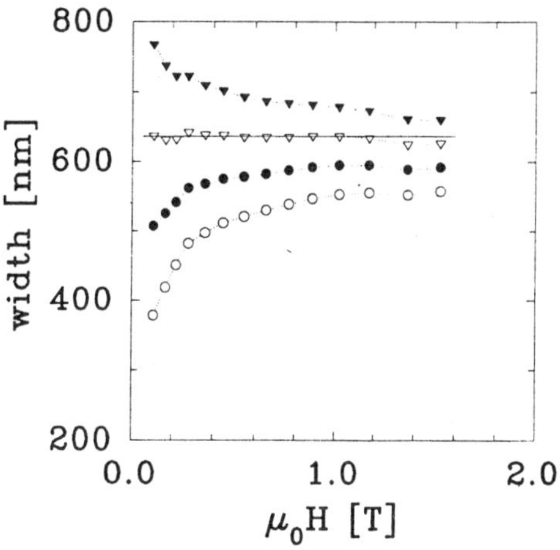

Fig.4 The channel width from the peak fields. From top to bottom results for N_0=8,7,6,5. For N_0=7 a horizontal line is obtained at w = 637 nm.

In Fig. 2 the deshearing force $F_s(B)$ at different temperatures is plotted for decreasing field for S637. The peaks in $F_s(B)$ reflecting commensurability of the FLL in the channel are clearly present. In our experiment we will see that flow of the closed packed planes parallel to the channels is the only stable orientation. Before we are going to discuss the commensurability effects, we first compare our data with the continuum approximation of the flux-line solid, Eq.(6). The continuum approximation is placed as an envelope curve over the data with a value of $A = 0.048$, giving a good description for the sample S637. Nice agreement is seen for the temperature dependence in $F_s(B)$, which is determined by the shear-modulus $c_{66}(t) \sim B_c^2(t) \sim (1-t^2)^2$ for $t \geq 0.5$. A substantial contribution to F_s ($\geq 10^5$ N/m^3) is related to the flux flow ($B < B_{c2}$) or the normal state resistivity

($B > B_{c2}$). Obviously F_s should be zero in the latter case. The $F_s(B)$ data below 0.2 T lies above the prediction of the continuum approximation. This is because the depinning current density J_c of the NbGe becomes of the order of the de*shearing* current density J_s. For the reference sample we determined using a similar criterion, a critical current density $J_c(0.2\text{T}) \simeq 5 \times 10^5$ A/m^2 (not shown here). This confirms the behavior of F_s for these low fields; for this regime a combination of the de*shearing* force F_s and depinning F_p has to be considered.

Focussing on the peaks in $F_s(B)$ it is shown in Fig. 2 that the peaks-positions are clearly temperature independent, which is expected, because the lattice spacing a_0 is temperature independent. (In)commensurability of the FLL (lattice spacing a_0) with the channel-width w produces the periodicity in the $F_s(B)$ data curves. In principle, it should be possible to recalculate the channel-width w from the peak fields B_{max}, assuming that the FLL commensurates with the channel-width at these fields, i.e. $w = N\times$ *periodicity*. However, these calculations resulted in widths which slightly deviated from the measured widths, although the periodicity was found much closer to $\frac{1}{2}\sqrt{3}a_0$ than to $\frac{1}{2}a_0$, the two possible orientations observed by Pruymboom *et al.* [10]. From the latter, we could conclude that the closed packed planes were moving parallel to the channels at the vortex velocities in our experiment.

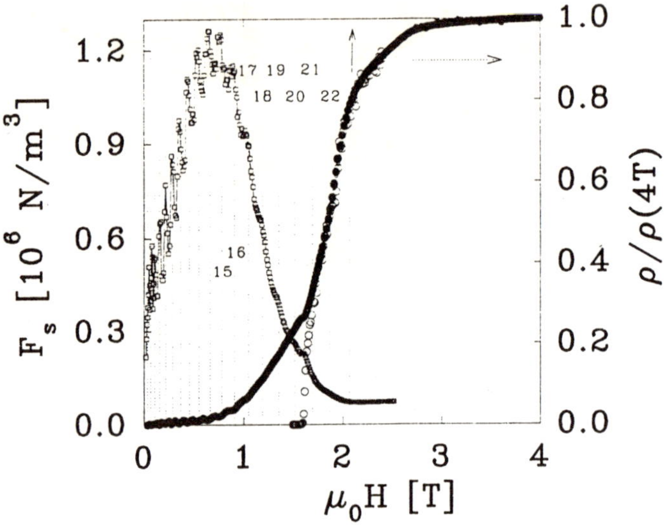

Fig.5 Shear force F_s(□) and ρ_{ac}(○) and ρ_{dc}(•) as a function of field at T = 1.7 K for the sample with w = 637 nm. Dashed lines are predictions for peak-fields. The transition from ohmic to non-ohmic behavior coincides with the "unexpected" peak between N = 20 and 21. The arrow pointing upwards, denotes B_{c2}.

To understand the details of the $F_s(S)$ curve a more sophisticated picture is needed. The first flux-line row in the NbN is close to the edge of the channel. The supercurrents of these vortices will be influenced and this will result in a new equilibrium positions. The mechanism describing a vortex close to a surface edge is proposed by Ternovskii and Shekhata[18] and by Clem[19] and in a more tractable way by Koshelev[20]. This latter description is used in our experiment to describe the positions of vortices at the edge. We therefore introduce an effective channel width w_{eff}. This parameter is essential for the interpretation of the data in the liquid regime, see below. In Fig. 3 we have schematically drawn a cross section of the channel with the configuration of the FLL near the channel edge. We assume that w_{eff} is the distance between the first rows of strongly pinned vortices in the NbN layer near the channel edges, i.e. $w_{eff} = w + 2x_1 - b_0$, where x_1 is the distance of the first row to the channel edge. Since the depth of the channel from the top is 40 percent of the total length of the vortex, we take for x_1 the equilibrium value[18,20]: $x_1 = (b_0/2)[1 + f_e(B/B_{c2})]$, where $f_e(B/B_{c2})$ expresses the contribution due to screening currents at the surface. Using an interpolation formula given in Ref. 21, x_1 can be computed at any reduced field b with

$$f_e(b) = \sqrt{\frac{1}{3} + \frac{2}{\pi\sqrt{3}}((1 + 2b - 1.08b^2)\ln\left(\frac{0.35}{b}\right) + 4b - 1.98b^2)} \tag{7}$$

As noted above, the vortex lattice at B_{max} is commensurate with the effective channel width, which gives

$$w_{eff} = w + 2x_1(B_{max}) - b_0 = (N - 1)b_0(B_{max}) \tag{8}$$

with integer $N \geq 3$. In order to find both N and w we computed w from Eq.(8) for different N_0 in $N = N_0 + n$, where $n = 0, 1, 2, 3, ...$ numbers the maxima. For the correct N_0 this procedure should give a horizontal line in a plot of w versus B_{max}, as illustrated in Fig. 4. We thus obtain $N_0 = 7$ and $w = 637 \pm 3$ nm. The resulting values of N are given in Fig. 2.

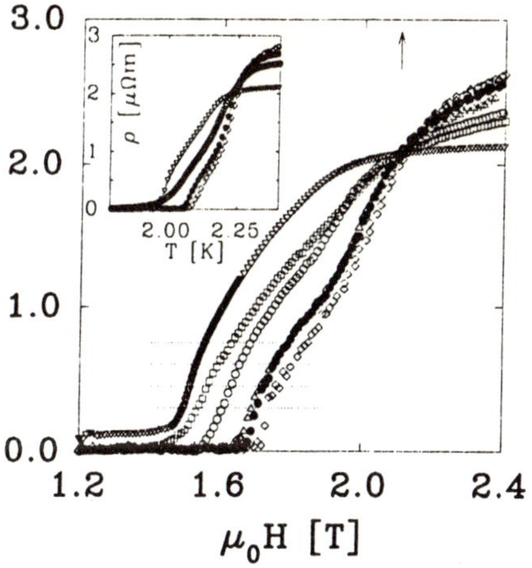

Fig.6 Size effect of the vortex lattice melting as shown by $\rho_{ac}(B)$ data at T=1.7 K for several channel widths: the reference sample (▽), w=1185 nm (□), 637 nm (○), 373 nm (△), 306 nm (•) and 245 nm (◇). Inset shows the data as a function of temperature at B=1 T.

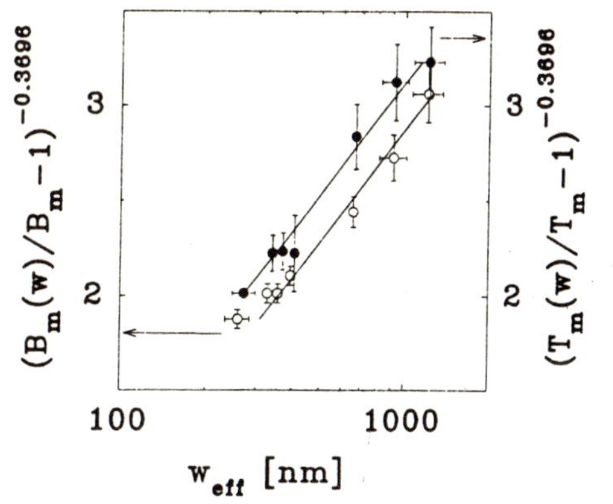

Fig.7 Finite size effect of the melting temperature $T_m(w_{eff})$ at B=1.0 T (•) and the melting field $B_m(w_{eff})$ at T=1.7 K (○). Identifying ξ_+ with the effective channel width w_{eff}, linear fits determine the nonuniversal constants b', b" and the typical length scale ξ_{+0}.

The value of w differs very little (4 percent) from the result one would obtain by simply taking $w = Nb_0$ following a scenario[22] according to Mkrtchyan *et al.*[23]. In this picture, a step in the penetration depth (which would occur at a step in thickness) gives rise to a potential well trapping the vortex at a position ξ from the step edge in the region with the smallest λ, thus with the largest film thickness. This additional mechanism is indicated in Fig. 3 by the flux-lines, which are positioned in the corners of the channels. These rows of flux-lines will terminate at the corner of bottom of the channel, because in this case the line energy ($\phi_0^2/4\pi\mu_0\lambda^2$) will be much lower than in case the flux-lines are straight. Correcting for ξ, i.e. $w = Nb_0 - 2\xi$, we get excellent agreement between both scenarios. This means that the vortex configuration at commensurability is very stable, irrespective the static (and due to pinning probably disordered) vortex configuration in the regions between the channels. The values of w obtained from the analysis are in good agreement with the values obtained from AFM. From the nice agreement of the calculated peak-positions with the measured ones, we conclude that the only stable orientation is the one with the closed packed planes parallel to the channels. We can conclude that flow experiments of the solid are explained in an adequate way by using the continuum approximation (isotropic shear modulus c_{66}) for the solid. Deviations are found for fields $B \leq 0.2$ T and for the two samples with largest widths. (In)commensurability of the solid with the channel, as observed in peaks in the de*shearing* force, is described by a simple model, which assumes flow of the closed packed planes parallel to the channel edge and an equilibrium distance for the flux-lines from the channel edges.

V. RESULTS AND DISCUSSION OF THE *LIQUID*-PHASE

Size effects in the flux line liquid

In Fig. 5 we denote the expected positions of the peaks by means of dotted lines. It shows that there is an unidentified peak at $B \approx 1.6$ T and no further peaks at $N = 21$ and 22. The origin of the extra peak is clarified by adding in Fig. 5 plots of the *dc* resistivity ($j_{dc} = 1.1 \times 10^6$ A/m^2) and *ac* resistivity ($j_{dc} = 0$ and $j_{ac} = 2.8 \times 10^3$ A/m^2). A dip in ρ_{dc} is seen at the position of the extra peak at 1.6 T. Just below this field ρ_{ac} begins to increase steeply and merges with ρ_{dc} at a field which lies slightly above the dip in ρ_{dc}. The transition from ohmic to non-ohmic behavior is known to be the best indication for the melting transition[3,4] in a weakly disordered vortex lattice. We interpret the dip in ρ_{dc} as

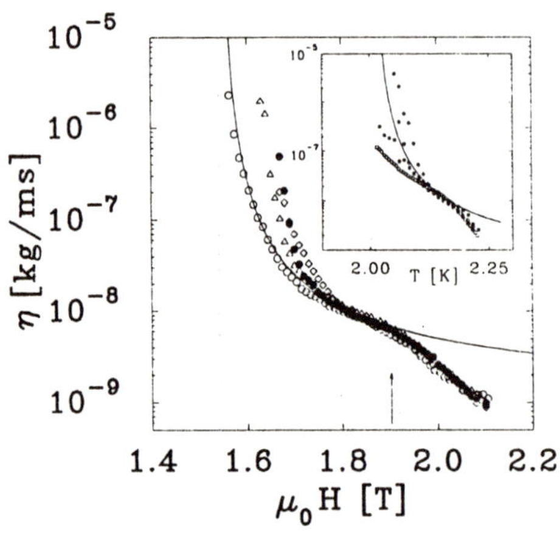

Fig.8 The shear viscosity $\eta(B)$ of the flux-line liquid at $T = 1.7$ K calculated from the resistivity data for $w = 1185$ nm (□), 637 nm (○), 373 nm (△), 306 nm (•) and 245 nm (◇). The dashed line indicates the thermodynamic melting field B_m. The curve represents $\eta = \eta\xi_+^2/b_0^2$ with $\eta_0 = 2.5 \times 10^{-10} kgm^{-1}s^{-1}$. The arrow is the possible location of the hexatic-to-(isotropic) liquid transition. The inset shows $\eta(T)$ results for the same samples at 1 T.

an indication that just before melting the shear modulus decreases due to plasticity[4,24], but remains finite in congruence with the dynamic picture of melting at finite wavelengths related to weak disorder[9]. We conclude that about 1.6 T we observe a distinct, but gradual transition from a vortex solid to a liquid.

Up to B_{c2} we probe the liquid phase which explains the absence of peaks in F_s. A kink in $\rho(B)$ is seen at $\sim$ 1.9 T which is also the field where the peak at $N = 22$ would be expected. This feature seems to indicate a change in the viscosity of the liquid, which will be further discussed below. The unexpected increase of $\rho(B)$ above B_{c2} till about 3.0 T (not observed in the reference sample) is due to a combination of surface superconductivity at the NbGe channel edges and a proximity effect at the NbN/NbGe interface.

For analyzing the melting phenomena in the samples with channels we take a constant ρ_{ac} value, which is close to the non-ohmic to ohmic transition as the criterion for melting. In the further analysis the melting transition of the reference layer with a smallest size of 300μm, is taken as the thermodynamic melting transition $B_m(\infty)$. Using Eq. 2 it can be shown that the difference between size dependent melting fields for the case of $\xi_+ = 1$ m and $\xi_+ = 300\mu$m is very small and is in fact indistinguishable within our experimental resolution. The length scale dependence of the vortex liquid properties is demonstrated in Fig. 6, where a series of ρ_{ac} measurements for different channel widths at $T = 1.70$ K is plotted.The resistivity ρ is defined as follows

$$\rho = \frac{R_{meas}A}{nw} \tag{9}$$

with R_{meas} the measured resistance in Ω, $n = 300$ is the number of channels for each sample and $A = 300\ \mu\text{m} \times 3$ mm is the crossectional area. This does not take into account the surface effects of the channel edges.

Noting that the transition from ohmic to non-ohmic behavior in Fig. 5 occurs at about $\rho = 0.75\mu\Omega$m, we use this value as a criterion to determine the finite size melting field $B_m(w)$. Similar results of $\rho_{ac}(T)$ at $B = 1.0$ T are shown in the inset. The free dislocations in the liquid near melting will in our configuration line up along the channel edges which allows us to identify ξ_+ with w_{eff} and check Eq. 2. We show the results in Fig. 7 where we compare the finite size effect of melting both as a function of temperature and field. The thermodynamic values ($B_m = 1.54$ T and $T_m = 2.01$ K) are taken from the data of the reference sample for which $\xi_+ \approx 300\mu$m. The lines represent fits of Eq.(2) providing the values for $b' = 1.2 \pm 0.1$, $\xi_{+0} = 26 \pm 5$ nm and $b'' = 1.2 \pm 0.1$, $\xi_{+0} = 34 \pm 5$ nm. As should be, the nonuniversal constants are of order one and ξ_{+0} is of order b_0, the latter being 31-35 nm between 1.5 and 1.9 T and 42 nm at 1.0 T. This confirms quite nicely the predictions of the Nelson-Halperin theory[8]. We should note, however, that similar plots with $\nu = 0.5$ gave equally reasonable results. So, we can not discriminate here between a solid-to-hexatic or a hexatic-to-isotropic liquid transition, although the former seems the more likely one, because the channel will promote orientational order. It should be noted that the criterion of $\rho = 0.75\ \mu\Omega$m is rather arbitrary. In fact, Fig. 6 shows that below this value ρ_{ac} drops rapidly but not in a sharp step which means that the shear viscosity of the liquid grows quickly, impeding the flow velocity in the channel. The choice of a lower criterion does not change the plots qualitatively, but gives slightly different values of the resulting parameters. Similar results are obtained from the experiments at other temperatures and fields, see Table 1. The general behavior of the data is the same. However, the most accurate results for the fit parameters are obtained for $T = 1.70$ K and $B = 1.00$ T, because there we had by far the highest resolution of the experiment.

T [K]	b''	ξ_{+0} [nm]	ξ_{+0}/b_0	B [T]	b'	ξ_{+0} [nm]	ξ_{+0}/b_0
1.70	1.2 ± .1	34 ± 5	1.2 ± .1	1.00	1.2 ± .1	26 ± 5	.6 ± .2
2.00	1.1 ± 0.2	42 ± 5	1.0 ± .2	0.70	1.0 ± .1	18 ± 5	.4 ± .1
2.20	2.0 ± .4	16 ± 9	.5 ± .3	0.50	1.4 ± .4	17 ± 8	.3 ± .2
2.40	2.3 ± .8	14 ± 19	.2 ± .2	0.30	.8 ± .2	21 ± 22	.3 ± .3

Table 1. Obtained parameters b'', b' and ξ_{+0} for the flux-line liquid as a function of temperature and field, using the 0.75 $\mu\Omega m$ criterion.

Probing the shear viscosity η of the flux-line liquid

Finally we can, by using Eq. (5) with w_{eff}, determine the shear viscosity from the different ρ_{ac} curves in Fig. 6. For ρ_f^0 we use the ρ_{ac} of the reference sample, instead of the Bardeen-Stephen value. The data should result in one curve of η versus B (or T) that follows the functional dependence of ξ_+^2/b_0^2 versus B (or T). This behavior is indeed observed as can be seen in Fig. 8 for $\eta(B)$ at T = 1.70 K and for $\eta(T)$ at 1.0 T (inset). The lines display $\eta = \eta_0 \xi_+^2(B)/b_0^2$ with $\eta_0 = 2.5 \times 10^{-10}$ kgm^{-1}s^{-1}. Assuming that we are dealing with a 2D liquid the diffusion constant far in the liquid phase would follow from $\eta_0 = k_B T/D_0 d$[15] and we obtain $D_0 \approx 0.3$ mm^2/s and a typical diffusion time $\tau_0 = a_0^2/D_0$ of 2 nsec.

A remarkable feature is the shoulder in $\eta(B)$ at 1.9 T. Such a feature is unexpected unless it denotes the transition from a hexatic to isotropic liquid. Below 1.9 T, $\eta(B)$ would then rather be the effective viscosity η_{eff} of the hexatic phase which at the transition reduces to η, the viscosity of the isotropic liquid. Both viscosities are related by $(\eta_{eff} - \eta) \propto K_A$, where the Frank constant K_A measures the stiffness against bond angle deformations in the hexatic phase[9]. Near the melting line it diverges like $K_A \sim \xi_+^2$, i.e. making the behavior of η_{eff} indistinguishable from that of η. At the hexatic-liquid transition it drops to zero. This second transition is somewhat smeared, but qualitatively agrees with the theoretical prediction[8]. The increase of η_{eff} at the shoulder agrees reasonably well with the increase by a factor of 2 at the disclination unbinding temperature observed in simulations by Frenkel and McTague[25]. It is remarkable that in our relatively thick layers (310 nm) the characteristics of a 2D system are observed. This indicates that the tilt deformations and entanglement of the flux lines play a minor role.

VI. CONCLUSIONS

Flow experiments were carried out on flux line solids and liquids confined to narrow channels of various widths w. Flow-velocity oscillations are observed related to commensurate-incommensurate transitions of the solid with respect to w. From the period we conclude that the lattice flows along the direction of closed packing and we could determine the effective channel width. An extra peak in the resistivity could be identified as the precursor of the melting transition. The field and temperature of this transition depends on w. We propose that free dislocations in the liquid line up along the channel edges so that the correlation length $\xi_+ = w_{eff}$. The temperature and field dependence of ξ_+ agrees well with the Nelson-Halperin theory. The shear viscosity was determined as well. It follows the expectation $\eta = \eta_0 \xi_+^2/b_0^2$ with $\eta_0 = 2.5 \times 10^{-9}$ Poise. A sudden decrease of η in the liquid may indicate a second transition from the hexatic-to-isotropic flux liquid.

VII. ACKNOWLEDGMENTS

This work is supported by the Dutch Foundation for Fundamental Research on Matter (FOM).

REFERENCES

1. G.Eilenberger, Phys.Rev. **164**, 628 (1967).
2. D.J.Bishop *et al.*, Bull.Am.Phys.Soc. **33**, 606 (1988); D.R.Nelson, Phys.Rev.Lett. **66**, 1973 (1988).
3. H.Safar *et al.*, Phys.Rev.Lett. **69**, 824 (1992); W.K.Kwok *et al.*, Phys.Rev.Lett. **69**, 3370 (1992); W.K.Kwok *et al.*, Phys.Rev.Lett. **72**,1092 (1994); S.L.Lee *et al.*, Phys.Rev.Lett. **71**,3862 (1993); Cubitt *et al.* Nature **365**, 407 (1993); E.Zeldov *et al.*, Nature **375** , 373 (1995); H.Pastoriza *et al.*, Phys.Rev.Lett. **75**, 3525 (1995).
4. P.Berghuis *et al.*, Phys.Rev.Lett. **65**, 2583 (1990); P.Berghuis *et al.*, Phys.Rev.**B** **47**, 262 (1993); A.Yazdani *et al.*, Phys.Rev.Lett. **70**, 505 (1993).
5. G.Blatter, B.Ivlev, Yu. Kagan, M.H.Theunissen, Y.Volokitin, and P.H.Kes, Phys.Rev. **B 50** 13013 (1994).
6. A.Zippelius, B.I.Halperin, and D.R.Nelson, Phys.Rev.B. **22** 2514 (1980).
7. W.Shockley, in *L'Etat Solide*, Proceedings of the *Neuvième Conseil de Physique*, ed. R.Stoops , Institut International de Physique Solvay, Brussels (1952); M.C.Marchetti and D.R.Nelson, Phys.Rev.B **41**, 1910 (1990).
8. D.R.Nelson and B.I.Halperin, Phys.Rev.B **19**, 2457 (1979).
9. D.R.Nelson, *Phase Transitions and Critical Phenomena*, ed. C.Domb and J.L.Lebowitz, Academic Press, London, vol. **7** (1983).
10. A.Pruijmboom, P.H.Kes, E.van der Drift, and S.Radelaar, Phys.Rev.Lett. **60**, 1430 (1988).
11. E.H.Brandt, Rep.Prog.Phys. **58** 1465 (1995).
12. J.Bardeen and M.J.Stephen, Phys.Rev. **140**, A1197 (1965).
13. M.C.Marchetti and D.R.Nelson, Phys.Rev.B **42**, 9938 (1990).
14. A.Pruijmboom, Ph.D. thesis (1988).
15. L.D.Jackel, R.E.Howard, E.L. Hu, D.M.Tennant and P.Grabbe, Apll.Phys.Lett. **39**, 268 (1981).
16. E.van der Drift, S.Radelaar, A.Pruijmboom, and P.H.Kes, J. Vac. Sci.Technol. B **6**, 297 (1988).
17. P.H.Kes and C.C. Tsuei, Phys.Rev.Lett. **47**, 1930 (1981).
18. F.F.Ternovskii and L.N.Shekhata, Zh. Eksp. Teor. Fiz. **62**, 2296 (1972).
19. J.R.Clem, Low Temperature Physics-LT 13, Vol. **3**, eds. K.D.Timmerhaus, W.J.O'Sullivan and E.F.Hammel, Plenum, New York, p. 102 (1974).
20. A.E.Koshelev, Physica C. **223**, 276 (1994).
21. A.E.Koshelev, Phys.Rev.B **50**, 506 (1994); Phys.Rev.B **51**, 12063 (1995).
22. A.E.Koshelev, private communication.
23. G.S.Mkrtchyan, F.R.Shakirzyanova, E.A.Shapoval, and V.V. Shmidt, Zh. Eksp. Teor. Fiz. **63**, 667 (1972).
24. S.Bhattacharya and M.J.Higgins, Phys.Rev.Lett. **70**, 2617 (1993); S.Bhattacharya and M.J.Higgins, Phys.Rev.B **49**, 10005 (1994).
25. D.Frenkel and J.P.McTague, Phys.Rev.Lett. **42**, 1632 (1979).

FLUX PINNING BY ARTIFICIAL DEFECTS IN HIGH TEMPERATURE SUPERCONDUCTORS

Harald W. Weber
Atominstitut der Österreichischen Universitäten, A-1020 Wien, Austria

Many attempts to improve flux pinning in high temperature superconductors have been reported on, but only a few were found to be effective in all families of the cuprate superconductors. This paper reviews the progress achieved by introducing artificial defects by all kinds of radiation, but mainly by fast neutrons, into single crystals, textured bulk materials as well as thin films of Y-, Bi-, and Tl-based cuprates. Particular emphasis will be placed on the nature and the distribution of the defects produced by fast neutrons, high energy heavy ions and fission products. Because of their different extensions and orientational arrangement, their interaction with the magnetic microstructures is largely different and also depends on the "dimensionality" of the superconductors, i.e. the presence of (3D) flux lines or of (2D) pancake vortices with strong or marginal coupling along the field direction. The generally beneficial effects of these defects will be illustrated by results on selected classes and various forms of high temperature superconductors. The data demonstrate that large enhancements of the critical current densities as well as strong shifts of the irreversibility lines occur especially at higher fields and temperatures, where the "as-grown" defect structures become less effective or entirely lose their flux pinning capability. Since these effects are of significant interest from an applied point of view, the paper is concluded by some remarks on the feasibility of radiation techniques for practical applications.

I. INTRODUCTION

During the past few years tremendous progress has been made in the preparation of high temperature superconducting materials. However, very little could be achieved with regard to a definitive control of the systems on an atomic level ("cation disorder") or, alternatively, with regard to a controlled modification of the microstructural properties by preparational or metallurgical means, in order to improve the flux pinning characteristics. Basically the defect - flux line interaction, which determines the critical current densities, is still decided by some unknown steps in the preparation process. Remarkable exceptions are the deliberately introduced second phase precipitates in melt-textured bulk 123-superconductors[1], a Ti-induced defect in Bi-2212 single crystals[2], and stacking faults in specially prepared PMP-processed bulk 123-materials[3]. All of these defects show characteristic strong pinning features, i.e. they interact with the flux lines or pancakes directly via the core interaction rather than collectively, such as with weakly pinning point defects. An example of a model system containing these weak pinning centers has also been reported recently (O-defects in Bi-2212 single crystals[4]).

As a conclusion of the last paragraph we are left with the fact that the most essential parameter for most applications of high temperature superconductors, i.e. the critical current density, cannot be readily influenced or optimized by preparational means. Fortunately, however, various techniques have emerged during the past few years which enable us to modify the metallurgical defect microstructure in these materials by radiation-induced defects. Whereas early experiments employing light ions of moderate energy, e.g. 5 or 10 MeV protons, have demonstrated significant improvements of the critical current densities[5], it soon became apparent, that these enhancements were restricted to low temperatures and / or low magnetic fields, i.e. that the radiation-induced defects were suitable flux pinning centers in a temperature and field range where the "as-grown" defects were also able to collectively pin the magnetic microstructure. In other words, the irreversibility or depinning line in the magnetic phase diagram [(H,T)-plane] remained unchanged, thus indicating that the nature of the new defects was similar to that of the original defect structure. This view was substantiated by TRIM calculations[6, 7] which proved that the energy transferred by these particles to the primary knock-on atoms was too small to produce extended defects with strong pinning characteristics. Hence, other irradiation conditions have to be employed.

II. RADIATION-INDUCED EXTENDED DEFECTS

Fast neutron irradiation proved to be a good choice. Clear shifts of the irreversibility line to higher fields and temperatures were reported for neutron-irradiated Bi-2212 single crystals for the first time[8]. Meanwhile, various aspects of fast neutron irradiation effects on high temperature superconductors have been explored in much detail[9-11]. It was shown in particular, that enhancements of J_c occured in every member of the cuprate families investigated so far, and that the degree of improvement in J_c and the magnitude of the shifts in the irreversibility lines depended significantly on the pre-irradiation defect microstructures and the strengths of their pinning potentials. Taking Y-123 as an example to illustrate the last remark, we find J_c-enhancements at 77 K by factors of ≈ 100 in single crystals[11], by factors of ≈ 10 in melt-textured bulk samples[12], and none in thin films[13] due to their optimized defect structures[14]. Apart from the fact that tremendous effects were found in single crystalline Tl-2223, Tl-1223 or Bi-2212[15,16] a direct transport measurement on Bi-2223[17] tapes also proved that flux pinning plays an important role, even under extreme "two-dimensional" conditions of the magnetic microstructure. The interplay of this dimensionality aspect with various flux pinning centers of different nature is currently being investigated by SQUID magnetometry on a whole set of Y-123 single crystals in a systematic way[18].

Of course, an important piece of information needed to further our understanding of the relevant flux pinning mechanisms is related to the nature, size and density of the neutron-induced defects. In view of the experimental difficulties encountered with such a program, thorough studies have been made only for Y-123 single crystals so far[19,20]. Transmission electron microscopy was employed in two stages, firstly by simulating the defect production by fast neutrons through in-situ irradiations with suitable ions (e.g. Kr or Xe with an energy of 50 or 85 keV), and secondly by directly exposing TEM samples to various reactor neutron spectra and fluences. The results may be summarized as follows. Collision cascades were identified to be the major defect, with sizes ranging from below 1 nm (the resolution limit of conventional TEM) up to 5 nm, the average defect diameter being 2.5 nm. They are surrounded by an inwardly directed (vacancy) strain field, which renders the total defect size to an approximately spherical volume with a diameter of 6 nm on average. The density of cascades was found to be $5x10^{22}$ m^{-3} at a fast neutron fluence of 10^{22} m^{-2} (E>0.1 MeV), they remained stable at temperatures up to at least 400 °C. High resolution pictures of simulated[7,19] (Xe-irradiation) or neutron-induced defects[21] showed mostly amorphous, but in some cases re-crystallized defect structures, thus confirming their suitability for flux pinning.

The salient features of the above discussion are illustrated in Figs. 1 and 2. Firstly, the response of the irreversibility line to sequential neutron irradiation is shown in Fig. 1 for Tl-2223 and Tl-1223. Both were investigated in single crystalline form by SQUID magnetometry with the field applied parallel to the c-axis. The irreversibility line itself was deduced from the appearance of distortions in the SQUID reponse[22], a technique that provides us with a very low J_c-criterion. In the most two-dimensional case, i.e. for Tl-2223, the irreversible regime moves up by more than 30 K at 1T, further enhancements being impeded by the onset of severe material degradation. At the highest fluence, the transition temperature has degraded by more than 10 K, mainly due to oxygen displacements. Annealing experiments to restore T_c, but keeping the cascades at the same time, are currently under way. Tl-1223 also shows remarkable shifts of the irreversibility line, but the effects are smaller due to its less 2D character. Still, the irreversibility line in the irradiated state exceeds that of Y-123 and demonstrates the high application potential of this material. In all cases, the critical current densities show a similar systematic dependence on neutron fluence, i.e. they increase continuously, less at low temperatures, where the "as-grown" defects pin significantly, but remarkably at temperatures of practical interest. Of course, in many cases, such as in Tl-2223, critical currents appear at temperatures and fields, where complete reversibility prevailed prior to irradiation, due to the shift of the irreversibility lines. As an example, Fig. 2 shows the J_c-enhancement in another material of practical interest, namely the newly developed melt-processed bulk Nd-123 superconductors[23]. In this case, the original "fishtail" behaviour of J_c disappears after the first irradiation step and J_c increases up to nearly 200.000 A/cm² at low fields (77 K). At the same time, the irreversibility line moves further up and reaches 11 T at 77 K (H||c).

Another (and equally or even more) effective irradiation technique was introduced by several groups[24-27] and is based on a completely different interaction mechanism with matter, i.e. the electronic energy loss of heavy ions of very high energy, e.g. 5 GeV Pb or 1 GeV Au ions. In this case, the damage process leads to complete disorder (amorphization) of the material along the direction of the incident particles, i.e. to the formation of "columnar tracks". Hence, the defect production is completely different from that of fast neutrons, in particular with regard to their arrangement, i.e. we find perfect alignment of the tracks along the particle direction as opposed to

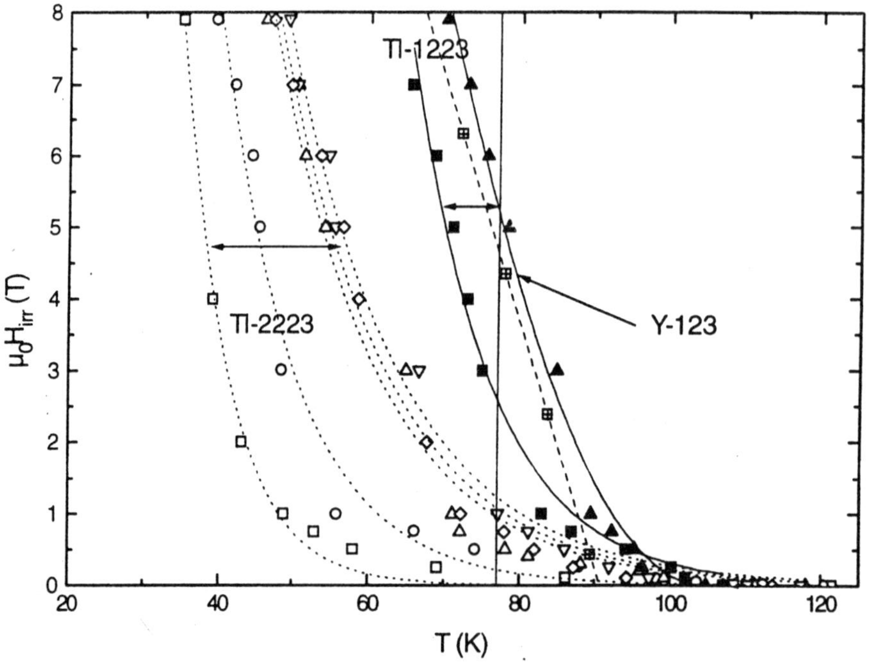

Fig. 1. Irreversibility lines in Tl-2223 and Tl-1223 single crystals (H||c) at various neutron fluences.

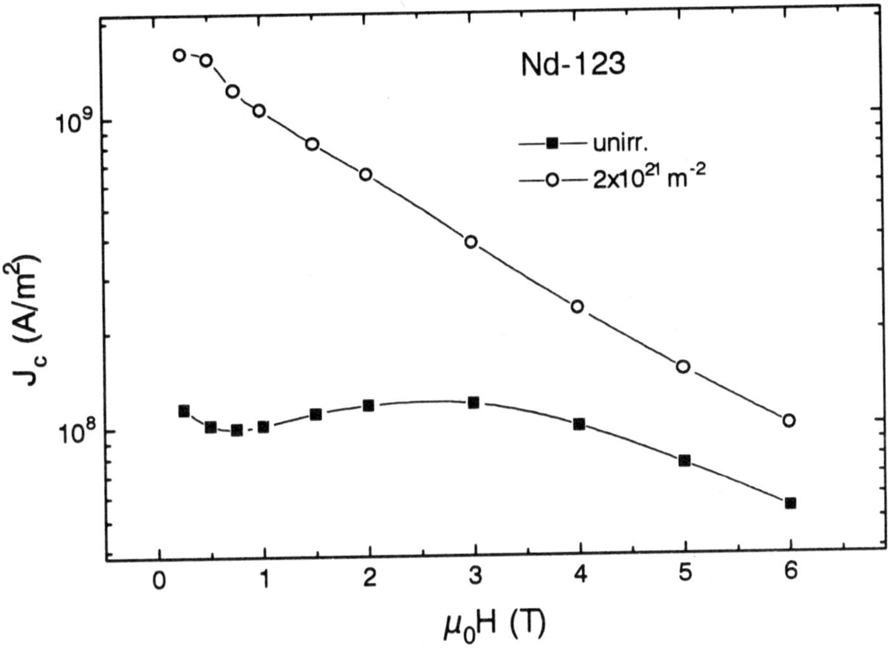

Fig. 2. Critical current density in a Nd-123 bulk superconductor prior to and following neutron irradiation.

a statistical distribution in the case of neutrons. As a consequence, the pinning action of the tracks is, in general, strongly anisotropic and optimal for the tracks being oriented parallel to the magnetic field direction. In this case, vast improvements of J_c and shifts of the irreversibility line were reported[25], which go far beyond those achievable with collision cascades. Further fundamental work that allows one to probe details of the magnetic microstructures, i.e. that is related to the question of flux lines versus pancakes, should be mentioned[28].

Again, the defects were investigated carefully by TEM, e.g.[29,30]. It turns out that each particle leads to the creation of a track, which is of roughly cylindrical shape with a diameter of the order of 5-10 nm. Its interior is amorphous. If the energy of the incident particles is below optimum, i.e. the stopping power falls below a certain material specific value, the shape of the defect along the trajectory becomes inhomogeneous, i.e. deviates from its cylindrical form and looks more like a series or string of elongated bubbles. In all cases and even with the highest energies employed so far, the thickness accessible for these types of radiation-induced defects is limited to a few tens μm. This is certainly the most serious disadvantage of the technique. The other drawback, viz. the preferential orientation of the tracks, can be at least partially removed by forming "splayed" tracks[31], but the distribution will of course never be statistical and, therefore, never effective for arbitrary orientations of the magnetic field.

A very promising version of introducing similar kinds of defects into high temperature superconducting materials, in particular in view of practical applications, has been demonstrated recently by Krusin-Elbaum et al.[32]. They employed 800 MeV protons to induce fission of the ^{209}Bi nuclei in Bi-2212[32] and -2223[33] tapes. The corresponding fission products, such as Xe or Kr, have sufficient energy to produce columnar defects in much the same way as discussed above. However, due to the statistical nature of the fission locations, the directions of the columnar tracks are random, i.e. the desired splay of these extended defects is automatically achieved. TEM on such an irradiated Bi-2212 tape showed columnar defects with diameters of 7 nm and lengths in excess of 6 μm, which is found to be consistent with the expected total range of columnar defect production by the fission tracks of 7 μm for 80 MeV Xe and of 12 μm for 100 MeV Kr. The results reported in the above two references confirm the beneficial effects of these defects on the critical current densities, the irreversibility lines as well as the creep behavior of the tapes. In addition, they confirm our previous result on neutron irradiated Bi-2223 tapes[17], namely that the *inter*grain properties do not degrade too dramatically to override the positive *intra*grain pinning effects. Obviously, more work is still needed to assess the potential of this kind of radiation treatment. However, in view of the enormous penetration range of such high energy protons into the superconductors, the most serious limitation of the track formation as discussed in the previous paragraph is clearly removed.

The final possibility of introducing extended defects into high temperature superconductors by radiation is the least well established. It was proposed and tested on ceramics quite early on[34], but investigated in a somewhat more systematic way only very recently[35]. The idea consists of adding small amounts of fissionable nuclei, such as ^{235}U, to the superconductor prior to processing and to expose the final product to a beam of thermal neutrons, thus inducing fission of the added uranium and creating fission tracks in much the same way as discussed in the last paragraph. The fission products are well known in this case, they have total energies of $\approx$ 200 MeV and create columnar defects of isotropic distribution, again with lengths of the order of $\approx$ 20 μm. Thermal neutrons have a very large penetration range in all superconductors of interest. In addition, thermal neutron beams are much more readily available and of sizes even suitable for processing of components, if desired. It should be pointed out that alternative previous suggestions[36] of adding Li to the material and relying on the Li(n,α)-reaction do not fall into the category of radiation treatments discussed in this context, since the resulting defects are too small to act as effective strong pinning centers.

III. CONCLUSIONS

In summary, various possibilities of introducing artificial defect structures into all families of high temperature superconducting materials have been explored during the past few years, only a few of them being suitable for effectively enhancing their flux pinning properties at high temperatures and magnetic fields. Among them, some techniques are more suitable for fundamental studies of the flux line - defect interactions (fast neutrons, columnar tracks), others seem to be promising also from an applied point of view (fast neutrons because of the statistical defect distribution and the large penetration range, fission tracks for the same reasons, but only if the preparational aspects, i.e. successful admixing of the doping elements without property degradations, can be solved, as has been demonstrated for Y-123[35]). Both fast neutron and thermal neutron irradiation do not seem to represent a major problem for those materials and applications, which do not require a silver sheath, as far the induced radioactivity of the superconductors is concerned. Silver, however, activates badly under a thermal neutron environment, which might make other radiation treatments, such as proton-induced fission of the Bi-nuclei, more attractive in this case.

ACKNOWLEDGEMENTS

I am pleased to acknowledge the cooperation and numerous fruitful discussions with the members of my group, Dr. F.M. Sauerzopf, Dr. G. Brandstätter, Dr. G. Samadi Hosseinali, Dipl.Ing. G.W. Schulz, and Dipl.Ing. M. Werner as well as with Dr. M. Murakami, ISTEC Tokyo, Prof. S.X. Dou, Wollongong, and Prof. R. Weinstein, Houston. This work was supported in part by the Austrian Science Foundation, Vienna, under contracts no. 6837, 7098, 7970, 8849 and 9194, and by the EU Brite Euram program under contract no. BRE2-CT93-0531.

REFERENCES

1. M. Murakami, S. Gotoh, H. Fujimoto, K. Yamaguchi, N. Koshizuka and S. Tanaka Supercond. Sci. Technol. **4,** 43 (1991).
2. T.W. Li, Ph.D. Thesis, Kamerlingh Onnes Laboratory, Leiden University (Dec. 1995).
3. P.X. Zhang, H.W. Weber and L. Zhou, Supercond. Sci.Technol. **8,** 701(1995).
4. T.W. Li, P.H. Kes, A.A. Menovsky and J.J.M. Franse, *Applied Superconductivity 1995*, Ed. D. Dew-Hughes, Institute of Physics Conference Series Number 148, IOPP, Bristol, Vol. 1, 283 (1995).
5. L. Civale, A. Marwick, M.W. McElfresh, T.K. Worthington, A.P. Malozemoff, F.H. Holtzberg, J.R. Thompson and M.A. Kirk, Phys. Rev. Lett. **65,** 1164 (1990).
6. J.P. Biersack and L.G. Haggmark, Nucl. Instr. Meth. **174,** 257 (1980).
7. M.A. Kirk and H.W. Weber, *Studies of High Temperature Superconductors*, Ed. A.V. Narlikar, Nova Science Publishers, New York, Vol. 10, 243 (1992).
8. W. Kritscha, F.M. Sauerzopf, H.W. Weber, G.W. Crabtree, Y.C. Chang, and P.Z. Jiang, Europhys. Lett. **12,** 179 (1990).
9. H.W. Weber and G.W. Crabtree, cf. Ref. 7, Vol. 9, 37 (1992).
10. H.W. Weber, Physica C **185,** 309 (1991).
11. F.M. Sauerzopf, H.P. Wiesinger, H.W. Weber and G.W. Crabtree, Phys. Rev. **B 51,** 6002 (1995).
12. C. Czurda, M. Wacenovsky, H.W. Weber and M. Murakami, Adv. Cryog. Eng. **40,** 1015 (1994).
13. R.M. Schalk, Ph.D. Thesis, Technical University of Vienna (1992).
14. R.M. Schalk, K. Kundzins, H.W. Weber, E. Stangl, S. Proyer and D. Bäuerle, Physica **C 256,** 341 (1996).
15. G. Brandstätter, G. Samadi Hosseinali, G.W. Schulz, W. Straif, X. Yang, H.W. Weber, F. Ladenberger, E. Schwarzmann, S.L. Yan, M. Manzel, W.Y. Liang and R. Flükiger, *Fourth Euro Ceramics*, Eds. A. Barone, D. Fiorani and A. Tampieri, Gruppo Editoriale Faenza Editrice S.p.A. Vol. 7, 57 (1995).
16. W. Kritscha, F.M. Sauerzopf, H.W. Weber, G.W. Crabtree, Y.C. Chang and P.Z. Jiang, Physica **C 179,** 59 (1991).
17. Q.Y. Hu, H.W. Weber, F.M. Sauerzopf, G.W. Schulz, R.M. Schalk, H.W. Neumüller and S.X. Dou, Appl. Phys. Lett. **65,** 3008 (1994).
18. G. Brandstätter, F.M. Sauerzopf, H.W. Weber and B.D. Veal, unpublished.
19. M.C. Frischherz, M.A. Kirk, J.P. Zhang and H.W. Weber, Phil. Mag. **A 67,** 1347 (1993).
20. M.C. Frischherz, M.A. Kirk, J. Farmer, L.R. Greenwood and H.W. Weber, Physica **C 232,** 309 (1994).
21. O. Eibl and H.W. Weber, unpublished.
22. F.M. Sauerzopf, H.P. Wiesinger, H.W. Weber, G.W. Crabtree, M.C. Frischherz and M.A. Kirk, Supercond. Sci. Technol. **5,** S 105 (1992).
23. S.I. Yoo, N. Sakai, H. Takaichi, T. Iguchi and M. Murakami, Appl. Phys. Lett. **65,** 633 (1994).
24. L. Civale, A.D. Marwick, T.K. Worthington, M.A. Kirk, J.R. Thompson, L. Krusin-Elbaum, Y. Sun, J.R. Clem and F. Holtzberg, Phys. Rev. Lett. **67,** 648 (1991).

25. V. Hardy, D. Groult, J. Provost and B. Raveau, Physica **C 190,** 289 (1992).
26. M. Konczykowski, F. Rullier-Albenque, E.R. Jacoby, A. Shaulov, Y. Yeshurun and P. Lejay, Phys. Rev. **B 44,** 7167 (1991).
27. W. Gerhäuser, G. Ries, H.W. Neumüller, W. Schmidt, O. Eibl, G. Saemann-Ischenko and S. Klaumünzer, Phys. Rev. Lett. **68,** 879 (1992).
28. B. Holzapfel, G. Kreiselmeyer, M. Kraus, G. Saemann-Ischenko, S. Bouffard, S. Klaumünzer and L. Schultz, Phys. Rev. **B 48,** 600 (1993); M. Kraus, M. Leghissa and G. Saemann-Ischenko, Phys. Bl. **50,** 333 (1994).
29. V. Hardy, D. Groult, M. Hervieu, J. Provost, B. Raveau and S. Bouffard, Nucl. Instr. Meth. **B 54,** 472 (1991).
30. Y. Zhu, Z.X. Cai, R.C. Budhani, M. Suenaga and D.O. Welch, Phys. Rev. **B 48,** 6436 (1993).
31. L. Civale, L. Krusin-Elbaum, J.R. Thompson, R. Wheeler, A.D. Marwick, M.A. Kirk, Y. Sun, F. Holtzberg and C. Feild, Phys. Rev. **B 50,** 4102 (1994).
32. L. Krusin-Elbaum, J.R. Thompson, R. Wheeler, A.D. Marwick, C. Li, S. Patel, D.T. Shaw, P. Lisowski and J. Ullmann, Appl. Phys. Lett. **64,** 3331 (1994).
33. H. Safar, J.H. Cho, S. Fleshler, M.P. Maley, J.O. Willis, J.Y. Coulter, J.L. Ullmann, P.W. Lisowski, G.N. Riley, M.W. Rupich, J.R. Thompson and L. Krusin-Elbaum, Appl. Phys. Lett. **67,** 130 (1995).
34. R.L.Fleischer, J.R.Hart, jr., K.W.Lay, and F.E.Luborsky, Phys. Rev. **B 40,** 2163 (1989).
35. R. Weinstein, J. Liu, Y. Ren, I.G. Chen, V. Obot, R.P. Sawh, C. Foster and A. Crapo, in Proc. of *The 1994 International Workshop on Superconductivity*, Kyoto, Japan, 39 (1994).
36. Y.J. Zhao, J.R. Liu, R.L. Meng and W.K. Chu, Physica **C 198,** 256 (1992).

SURFACE BARRIER AND IRREVERSIBILITY IN NIOBIUM AND $Bi_2Sr_2CaCu_2O_8$ SINGLE CRYSTAL

P.K. Mishra[a], G. Ravikumar[a], V.C. Sahni[a], M.R. Koblischka[b] and A.K. Grover[c]

[a]*TPPED, Bhabha Atomic Research Centre, Mumbai 400 085, India*
[b]*University of Geneva, 20, rue d'Ecole de Medicine, CH-1211 Geneva 4, Switzerland*
[c]*Tata Institute of Fundamental Research, Mumbai 400 005, India*

In hard superconductors the flux motion and the associated hysteresis is mostly controlled by bulk pinning centres. At the same time surface barriers are known to play an important role in the vortex dynamics of different superconductors with a significant contribution to the irreversible properties. Generally the effect is better manifested in systems with weak bulk pinning. In these cases, main impediments to the flux entry into the sample arise from surface barriers. In many cases, both in conventional and high temperature superconductors, an appropriate temperature and magnetic field regime can be selected where the bulk pinning strength is very much reduced so that the effect of surface barrier can be seen prominently.

For the Meissner state, $B = 0$, the role played by specimen surface influencing the magnetic behavior of type II superconductors was shown by Bean and Livingston in their pioneering paper[1]. The BL barrier provided by an ideal surface of a superconductor inhibits the entry of a vortex due to two competitive effects: (i) the repulsion of a vortex from the surface due to the interaction of the flux line with the induced shielding current, and (ii) the attraction of the entering flux line towards the surface due to its mirror image which is an antivortex. They together give rise to a barrier to vortex motion into the superconductor. For a mixed state, $B > 0$, this was considered by Clem[2] using a simple approximate approach. This leads to a perticular shape of the irreversible $M - H$ curve which is an important signature of the surface barrier. There are also some other useful early studies on the effect of surface barrier on vortex nucleation[3,4] and magnetization loop shape in the literature.

After the discovery of high T_c materials there have been many reports on the observation of surface barrier effects in these new superconductors[5,6,7]. Here, we present results of isothermal magnetization measurements on annealed niobium sample and good quality $Bi_2Sr_2CaCu_2O_8(Bi2212)$ single crystal with a view to investigate surface barrier effects through different signatures discussed below. We bring out a striking similarity between observations made on niobium and $Bi2212$ and show that under suitable conditions surface barrier effects play significant role in bestowing an irreversible magnetic behavior in these materials. We show that $M - H$ hysteresis curves in case of niobium can be scaled in the forward cycle indicating the nature of flux entry is same at different temperatures.

The *BSCCO* single crystal used in our experiment has a T_c^{onset} of 85K and a transition width of 0.2K and The sample dimensions are 2mm x2mm x.3mm for *Bi*2212 single crystal, and 2.3mm x 1.5mm x.4mm for niobium strip.

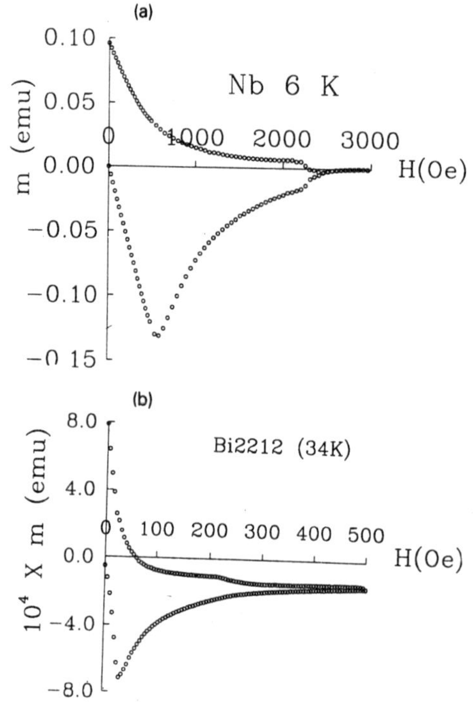

Fig.1: Hysteresis curve for (a) niobium (b) *BSCCO*

All measurements were carried out using a commercial (Quantum Design Inc.,U.S.A) SQUID magnetometer keeping the superconducting magnet in persistent mode. In case of *Bi*2212, the field was applied parallel to the c-axis. Isothermal magnetization hysteresis loops were recorded upto a field beyond H_{c2} value in the case of niobium and 500 Oe in the case of *Bi*2212.

A close look at the reverse branch of the magnetization hysteresis curves for niobium (Fig.1a) shows a flat feature akin to the hall mark of surface barrier. It had been pointed out by Campbell and Evetts[3], on the basis of Clem's model[2], that surface barrier effect manifests as a flat portion in the reverse branch of the magnetization loop. In other words, the magnetization M is small and constant while decreasing the external field H. This makes the magnetization loop very asymmetric which is in contrast to the calculations based on Bean's model. Similar features were also observed in the hysteresis curves of *Bi*2212 crystal (Fig.1b).

Clem's model also provides some additional signatures of surface barrier effect. In that model, while considering the interaction of fluxoids with surfaces of superconductors, Clem derived in the presence of only surface barrier, the magnetization is given by while increasing field,

$$M = (H^2 - H_p^2)^{1/2} - H \; for \; H > H_p$$

and the width of the hysteresis loop, for $H >> H_p$, (also see ref.8)

$$\Delta M \sim \left[\frac{H_p^2(T)}{2H}\right]$$

In other words the model predicts, for the fields $H >> H_p$, the width of the hysteresis curve ΔM to bear an inverse relationship with the applied magnetic field. When we plotted the same as a function of $1/H$ we could confirm the linear relationship over a significant range of field (Fig.2). In case of $Bi2212$ this relationship is better established in the temperature range between 30K to 40K.

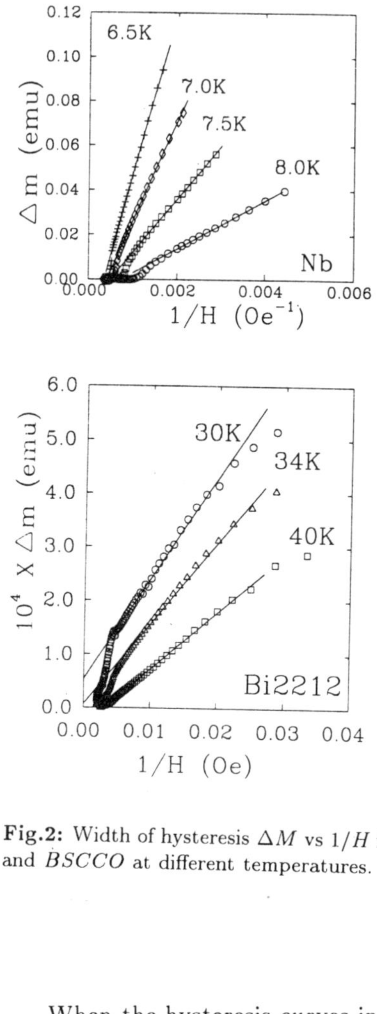

Fig.2: Width of hysteresis ΔM vs $1/H$ for niobium and $BSCCO$ at different temperatures.

The linear relationship between the width of the hysteresis, ΔM, and $1/H$, appears consistent with the claims that in $Bi2212$ system, for $T \geq 40K$, the irreversibilty originates only from surface pinning. In temperature range between 30K and 40K, the nature of pinning is a mixture of surface effect and bulk pinning, with the bulk contribution decreasing in strength with the rise in temperature. This causes a deviation from the $1/H$ behavior at higher fields where surface pinning vanishes and the bulk pinning dominates. Unlike in $Bi2212$, in the case of niobium, this high field deviation feature persists, to some extent, at all temperatures as the bulk pinning survives almost upto T_c. A comparison between the two systems indicates that in case of highT_c $Bi2212$, at 40K, there is a complete crossover to surface pinning regime.

Another interesting feature in the hysteresis curves of both niobium and $Bi2212$ is the observation of a blip or 'shoulder-like' structure in the forward as well as reverse branch of the loops. The structure is more prominent at lower temperatures. This is also explained on the basis of surface barrier effect[9].

When the hysteresis curves in case of niobium were replotted with normalized M and H, a scaling behavior is observed for forward branches of the hysteresis loops (Fig.3). This is in confirmation with our observation that the barrier to flux-entry is predominantly from the surface in this case. However we also mention that the above scaling does not appear to hold good in flux-exit case.

To summarize, we have shown how the surface barrier influences the hysteretic behavior of both conventional and high T_c superconductors. To observe the effect one has to choose a proper temperature and magnetic field regime where the bulk pinning strength is substantially reduced. This is clearly established in case of high T_c superconductor $Bi2212$ in the temperature window discussed. From a close similarity between observations made in niobium and a $Bi2212$ crystal, we have attempted to bring out the important role of surface barrier in high T_c materials.

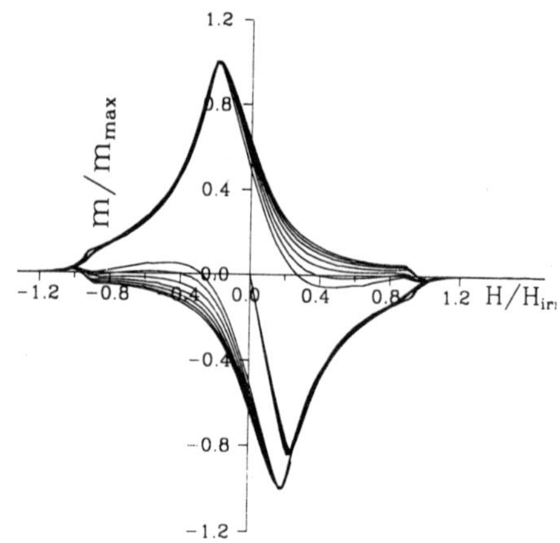

Fig.3: Scaled hysteresis curves for niobium

REFERENCES

1. C.P. Bean and J.D. Livingston, Phys. Rev. Lett. **12,** 14 (1964).
2. J.R. Clem, *Proceedings of Low Temperature Physics LT 13*, ed. K.D. Timmerhaus et al., Plenum, New York, Vol. 3, 102 (1974).
3. A.M. Campbell and J.E. Evetts, *Critical Currents in Superconductors*, Taylor and Francis, London, 140 (1972).
4. P.G. de Gennes, Solid State Commun. **3**, 127 (1965).
5. M. Konczykowski, L.I. Burlachkov, Y. Yeshurun and F. Holtzberg, Phys. Rev. B **43**, 13707 (1991).
6. Ming Xu, D.K. Finnemore, G.W. Crabtree, V.M. Vinokur, B. Dabrowski, D.G. Hinks and K. Zhang, Phys. Rev. B **48**, 10630 (1993).
7. N. Chikumoto, M. Konczykowski, N. Motohira and A.P. Malozemoff, Phys. Rev. Lett. **69**, 1260 (1992).
8. L. Burlachkov, V.B. Geshkenbein, A.E. Koshelev, A.I. Larkin and V.M. Vinokur, Phys. Rev. B **50**, 16770 (1994).
9. P.K. Mishra, G. Ravikumar, V.C. Sahni, M.R. Koblischka and A.K. Grover, Physica C 269, 71 (1996).

MULTIPLE PEAK STRUCTURE IN THE PEAK EFFECT REGION OF THE VORTEX LATTICE IN THE CUBIC SUPERCONDUCTOR $CeRu_2$

N. G. Patil[1], Subir Saha[1], S. S. Banerjee[1], K. Ghosh[1], S. Ramakrishnan[1], A. K. Grover[1], T.V. Chandrasekhar Rao[2], G. Ravikumar[2], P.K. Mishra[2], V.C. Sahni[2], E. Yamamoto[3], Y. Haga[3], M. Hedo[4], Y. Inada [4], Y. Onuki [4]

[1]Tata Institute of Fundamental Research, Mumbai - 400005, India
[2]TPPED, Bhabha Atomic Research Centre, Mumbai - 400085, India
[3]Advanced Science Research Center, Japan Atomic Energy Research Institute, Tokai, Ibaraki 319-11, Japan
[4]Faculty of Science, Osaka University, Toyonaka 560, Japan

We present new magnetization results on a clean single crystal of cubic $CeRu_2$ compound which reveal that the magnetization hysteresis anomaly below $H_{c2}(T)$ involves two distinct steps. We may identify the first step with the onset of GFFLO transition and attribute the second step to the eventual collapse of the shear modulus of the vortex lattice.

I. INTRODUCTION

The mixed state of superconducting cubic Laves (C15) phase compound $CeRu_2$ ($T_c \approx 6.3$ K) is currently in focus[1-10] because of the sharp transition of its flux line lattice associated with weak to the strongly pinned state while approaching the normal/superconducting phase boundary ($T_c(H)/H_{c2}(T)$ line) in very clean single crystals. The said transition bears strong resemblance to the ubiquitous Peak Effect (PE) phenomenon (*i.e.*, anomalous maximum in critical current density $J_c(H,T)$ just below the $T_c(H)$ line) in weakly pinned Vortex Lattices (VL) (see, for instance, Ref[11-16]. However, dilatometric and magnetization experiments of Modler *et al* [4] reveal that the magnetic phase diagram of $CeRu_2$ is similar to that of the antiferromagnetic Heavy Fermion Superconductor (HFS) UPd_2Al_3 ($T_N \approx 14$ K, $T_c \approx 2$K) and the said transition in these two compounds (and other HFS systems) reflects the emergence of "generalized Fulde-Ferrell-Larkin-Ovchinnikov (GFFLO)" state[5,17,18]. This state is characterized by *spatially modulated order parameter* giving rise to periodic array of nodal planes perpendicular to the vortices. In type II superconductors with large normal state electronic susceptibility, the FFLO state[17,18] is considered to be brought about by adequate compensation of loss of superconducting condensation energy by the gain in spin-polarization energy of conduction electrons in the core of the vortex. A later study of FFLO state [19] suggested that such a state is stable for $T_c/T_c(0) \leq 0.56$. However, Tachiki *et al* [5] and Modler *et al* [4] have recently surmised that GFFLO state could exist over extended field- temperature region.

II. EXPERIMENTAL RESULTS

We report here a new result, viz., the detection of a *two peak* structure (identified as α and β in Figs.2 and 3) in the PE region in the real part of ac susceptibility (χ') data in a single crystal of $CeRu_2$. While the peak α is located near sharp onset of PE region, the peak β is located at the edge of irreversible to quasi-reversible response. A narrow reversible region precedes the arrival of $T_c(H)$ after the occurrence of PE. This region of the mixed state is also associated with a change in sign of χ' and appearance of yet another (i.e., third) peak (identified as γ in Fig.3) in χ' data, which may be identified with the so called *Differential Paramagnetic Effect* (DPE)[20-23]. Differential magnetization (dM/dH) signal (with infinitesimal driving force)of the reversible mixed state should be paramagnetic as it implies the decrease of diamagnetism with increase of H. We will describe several notable features of our data related to the peaks α, β and γ, and based on these we examine various possible explanations involving PE phenomenon suggested in the literature, viz., (i) that the PE is associated with the onset of GFFLO state[4,5,7], (ii) the PE is linked to the inherent elastic softening of the VL via Pippard's scenario [13,16,24,25] and (iii) this phenomenon precedes the incipient unpinned state of the VL just before its destruction at H_{c2}[14].

The high quality single crystal of $CeRu_2$ ($2.75 \times 1.45 \times 0.9\ mm^3$) is the same as the one used for a de-Haas van Alphen effect study[26]. The ac magnetization measurements in dc fields have been performed using a well shielded home built ac susceptometer[27]. The dc magnetization data was recorded on a commercial SQUID magnetometer (Quantum Design Inc., USA).

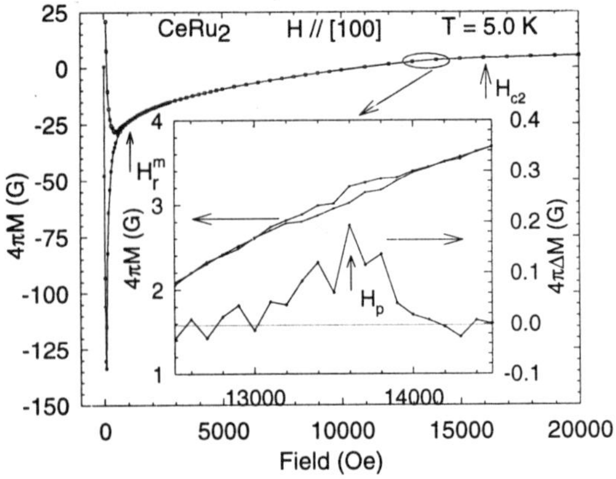

Figure 1: M-H loop in a single crystal of $CeRu_2$ at 5~K.The inset panel reveals the second anomalous opening of the hysteresis loop centered around 13.5 kOe (=H_p) and shows a peak in the hysteresis anomaly($4\pi\ \Delta M$) at H_p.

Fig.1 shows isothermal magnetization data at 5K in which the forward and reverse hysteresis curves nearly overlap above 1 kOe (=H_r^m(5 K)). However, its inset panel shows the existence of measurable hysteresis between 13 kOe and 14 kOe when data are recorded in small steps. The inset also depicts $4\pi\Delta M$ (= $4\pi(M(H\downarrow) - M(H\uparrow))$ vs H. The tiny hysteresis loop in the inset is a manifestation of PE phenomenon and the ordinates provide a measure of $J_c(H)$ ($4\pi\Delta M \propto J_c$)[11]). Note, also, that the structure of hysteresis anomaly is not fully resolved. In conformity with earlier reports[1-4,7], the hysteresis bubble in the present sample significantly expands and its center of gravity also shifts to higher fields (data not being shown here) as the temperature is lowered below 5~K. Such an anomaly is harder to observe at T > 5 K in dc magnetization data in typical samples; however, it can be tracked using ac susceptibility measurements over the entire temperature range.

Fig.2(a) shows the temperature variation of χ' under a superposed dc field of 13.5 kOe for two values of the ac field (h_{ac}) at a frequency of 211 Hz, namely, 0.5 Oe (r.m.s) and 2.0 Oe (r.m.s). Usually for a strongly pinned VL (large J_c), the screening response is nearly perfect (*i.e.*, $\chi' \approx -1$) at small values of h_{ac}. Also, for a given H, χ'(T) monotonically varies as $T \rightarrow T_c$ reflecting the $J_c(T) \rightarrow 0$ (see, for instance, χ'(T) plot in zero field in the inset of Fig.2(a)). However, in both the χ'(T) curves at 13.5 kOe in Fig.2(a), one can distinctly see around 5 K the existence of two negative peaks marked by arrows, which we designate as *precursor* PE peak (α) and *main* PE peak (β), respectively. These two peaks, we believe, are the fingerprints of the processes related to the presence of anomalous variations in J_c in the PE region identified by the appearance of the hysteresis loop around 13.5 kOe at 5 K in the inset panel of Fig.1. The height of each of the two negative peaks is a sensitive function of h_{ac}; while the *peak temperatures appear independent of it.* The h_{ac} dependence can be understood in terms of the relative strengths of the anomalous increases in J_c (ΔJ_c) represented by the two peaks. If h_{ac} is too small or too large as compared to a given ΔJ_c, its fingerprint may not be very conspicuous in χ'(T) data. Thus, for a specific value of h_{ac} in a given situation, the two peak structure can coalesce into an unresolved composite peak (*cf.* χ'(T) curves at 8 kOe for different h_{ac} in Fig.2(b)). At 8 kOe, the peaks α and β are, however, clearly resolved with $h_{ac} \leq 0.5$ Oe. In Fig.2(b), the peak α at 8 kOe is extremely sharp and the χ'(T) behaviour at 19.5 kOe in h_{ac} of 0.5 Oe shows (see the inset panel of Fig.2(b)) that the onset of peak α occurs abruptly(within 3 mK) implying a phase transition in the mixed state of $CeRu_2$ from weak to a strongly pinned state. Peaks α and β reside on a sloping χ′(T) response. A closer examination of data at different H in Fig.2 reveals that α weakens faster than β as H decreases. In fact, we could not distinctly observe peak α for H ≤ 4 kOe, whereas, a residual fingerprint of β could be easily ascertained for H down to 1 kOe at $T/T_c(0)$ of 0.98 (data not shown here).

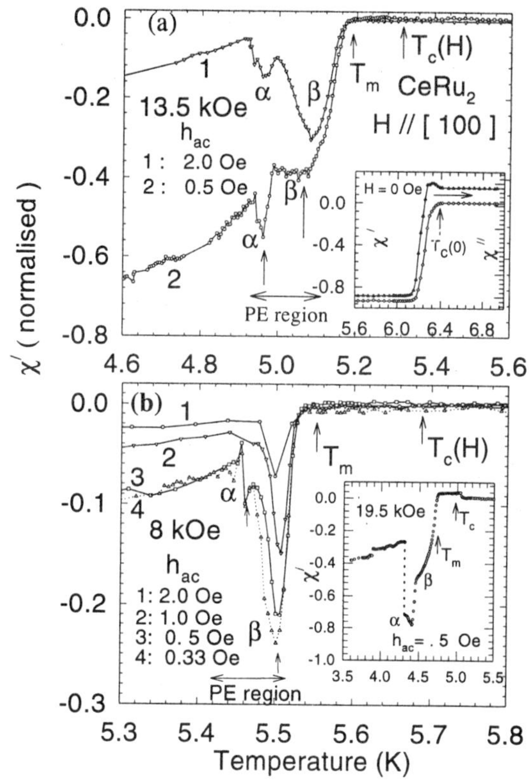

Figure 2: χ′(T) for different h_{ac} in $CeRu_2$ at (a) H= 13.5 kOe, (b) 8 kOe and 19.5 kOe (inset panel). α and β identify the precursor and the main peak respectively. $T_c(H)$ and $T_m(H)$ have been determined from the dc magnetization data (see text). The DPE peaks ($\approx 10^{-3}$) are not clearly visible in these sets of data. The inset panel in Fig.2(a) shows χ′(T) and χ″(T) in H=0.

Fig.3 shows the distinct two peak structure in the PE region in χ′data at 5 kOe. The peak α is broader than the peak β; however, in absolute terms both of them are narrow and extremely tiny. For instance, normalized χ′ has a value ≈ -0.008 at the peak position β. Further, as soon as the peak β collapses, χ′(T) undergoes a sharp change (ΔT≈10 mK) to positive values and another broad peak structure (γ) emerges that lasts up to the T_c (H) value, which has been independently determined from dc magnetization data shown in the inset panel of Fig.3. The curves (a) and (b) in the inset show 4πM vs T at H = 5 kOe on different Y-scales. The curve (a) identifies $T_c(H)$ via the onset of diamagnetic response and the curve (b) elucidates the determination of the extrapolated temperature T_m by fitting the (lower temperature) magnetization data to a mean field like linear variation in temperature. The main panel of Fig.3 shows that T_m lies just above the peak β and it closely identifies the end of PE region. We believe that the paramagnetic χ′(T) signal sandwiched between $T_m(H)$ and $T_c(H)$ is the orthodox *Differential Paramagnetic Effect*[20]. This *Effect* has in recent years[21-23] been linked to the existence of irreversibility line in the mixed state of high T_c and conventional low T_c alloy superconductors. Near H_{c2}, dM/dH is of the order $1/(2\kappa^2-1)$[20]. In $CeRu_2$ with $\kappa \approx 16$ - 21[3], DPE response is expected to be of the order 10^{-3}, consistent with the experimental data in Fig.3. A careful search for DPE response at different H revealed its presence from 19.5 kOe to 2 kOe between the respective $T_m(H)$ and $T_c(H)$ values.

III. SUMMARY AND CONCLUSIONS

To sum up, Figs.2 and 3 highlight the new findings, viz., the occurrence of two negative peaks (α and β) and one positive DPE peak (γ) in χ′(T) data. Two distinct minima, precursor and main PE peaks, have been identified in isothermal χ′(H) data as well. The negative χ′ response at temperatures prior to three peak structure implies that the dc magnetization is not truly reversible in that region ($J_c \approx 1$ A/cm^2, but not zero). Just as the location of α and β are independent of h_{ac}, the onset temperatures of DPE response as signalled by collapse of $\chi'(T)$ after β, are also independent of h_{ac}. In view of the observation that the DPE region immediately follows PE region and the belief[13] that PE phenomenon is either a fingerprint or a close precursor[14,16] of vanishing shear modulus of VL, we infer that DPE region in the present data closely identifies the reversible state below the nominally ascertained $T_c(H)$ value. Interestingly, the mean field $T_m(H)$ values also nearly coincide with the onset temperatures of the DPE response.

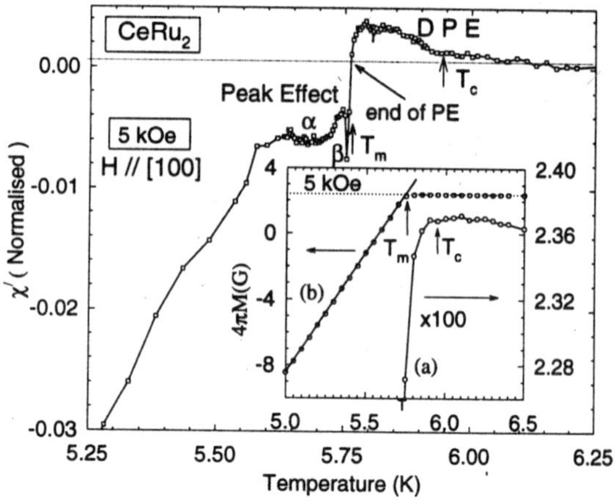

Figure 3 : $\chi'(T)$ at H=5 kOe and in h_{ac} of 2 Oe(r.m.s.) in $CeRu_2$. The PE peaks (α, β) and DPE peak (γ)have been identified and $T_m(H)$ and $T_c(H)$ are marked. The inset panel reveals the location of $T_c(H)$as the onset temperature of nominal diamagnetic response. It also shows identification of extrapolated temperature $T_m(H)$ by a linearfit of M vs H data. The main panel elucidates thatγ is sandwiched between $T_m(H)$ and $T_c(H)$.

Fig.4 elaborates the magnetic phase diagram of $CeRu_2$ in the light of our ac and dc magnetization measurements. In the main and inset panels of Fig.1, we can note the fields corresponding to five features, viz., (a) the first closing of the hysteresis loop at H_r^m (b) the second (anomalous) opening of the loop, (c) the center of gravity of that opening at H_p, (d) its closing and (e) finally, the arrival of H_{c2}. The finger prints of these features are also identified in other isofield and isothermal ac and dc magnetization data on the same sample (all data are not shown here). The phase diagram in Fig.4, therefore, comprises five lines, viz., $H_r^m(T)$, $H_i(T)$, $H_{peak}(T)$, $H_f(T)$ and $H_{c2}(T)$; these identifications incorporate the nomenclature of Modler *et al*[4].The nominal irreversibility line (T_r^m (H) or H_r^m (T)) determinable from merger of temperature dependent zero field cooled and field cooled magnetization data or the first collapse of hysteresis in M-H loops lies far below the PE region. This line usually[28] indicates the boundary across at which $J_c(H,T)$ in a typical sample falls below ≈ 10 A/cm^2. The separation between the precursor and the main PE peaks starts to grow as field (temperature) increases (decreases) above (below) about 5 kOe (5.75 K). The locus of temperatures identifying precursor peak α smoothly joins to the $H_i(T)$ line defining the second anomalous opening of the hysteresis loop in dc magnetization data. The locus of temperatures identifying peak β remains nearly coincident with the $H_p(T)$ line. The $H_f(T)$ line defining the offset of the hysteresis anomaly appears to join to the temperature/field data identifying the end of PE region in χ' data. To a first approximation, $H_f(T)$ line passes through the temperatures $T_m(H)$, which have been determined by linear fits of dc magnetization data; in this spirit, $H_f(T)$ line could be taken to identify a quasi-mean field phase boundary below which the vortex lines exist as a pinned lattice.

Through a recent field and temperature dependent ultrasonic propagation study in single crystal of $CeRu_2$, Goshima *et al*[7] have found that at H_i there is an anomalous increase in the transverse elastic modulus ($[C_{11}-C_{12}]/2$) due to an enhancement in the stiffness of the VL, and that such a feature manifests for H ≥ 5 kOe. Also, they surmise that the stiffening of the VL occurs due to (first order) GFFLO transition. It has been argued [5]that as the GFFLO state sets in, vortex lines segment into coupled pieces having lengths comparable to the wavelength of the order parameter modulation and when that happens, the segments conform more easily to the inevitably present disorder, thereby enhancing total pinning. Such a pinned lattice has to eventually get depinned on or before H_{c2}. The orthodox PE describes softening of pinned VL before H_{c2}.We are, therefore, inclined to treat $H_i(T)$ and $H_f(T)$ lines as phase boundaries marking the sharp onset of GFFLO[4] state and the collapse of shear modulus of VL, respectively. Interestingly, both these features involve phenomena related to underlying pure Abrikosov VL; while their fingerprints emerge through the presence of unavoidable disorder (pinning) in any real superconducting sample. However, entirely different physical arguments define the limiting temperature/field values upto which phenomena corresponding to lines H_i and H_f may be present. We have found that a sharp transition (peak α, i.e., H_i line) above 5 kOe and below a reduced temperature ($T/T_c(0)$ of 0.91. We interpret the absence of well defined β peak at $H \leq 1$ kOe (and at $T/T_c > 0.98$) as follows. At 1kOe, the separation between the near neighbour vortex lines ($a_o \approx 5 \times 10^4 / [B(Oe)]^{1/2}$ Å) exceeds the penetration depth κ (≈ 1100 -1400 Å [3]) of $CeRu_2$ and the interaction amongst flux lines becomes weak. In such a circumstance each vortex line is perhaps individually pinned by random disorder and the concept of collective phase transition, such as, collapse of the shear modulus at the incipient melting of VL is not applicable[13,29-31]

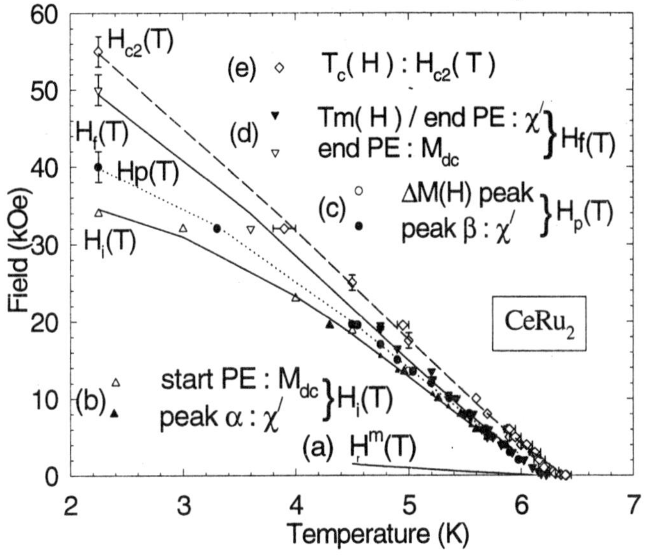

Figure 4 : H,T diagram of $CeRu_2$ summarizing information on five features (see text). The data on peaks α and β and the end of PE region have been collated in three groups, (b), (c) and (d). $H_i(T)$, $H_p(T)$ and $H_f(T)$ lines correspond to second opening of loop (see the inset of Fig.1), its center of gravity and its eventual closing, respectively.

In summary, we have reported the first observation of distinct multi-peak structure including the transition from a PE peak to DPE peak, below $T_c(H)$ phase boundary in a clean single crystal of $CeRu_2$.

REFERENCES

1 K. Yagasaki *et al*, J. Phys. Soc. Jpn. 62, 3825 (1993).
2 A. D. Huxley *et al*, J. Phys. Condens. Matter 5, 7709 (1993) and references therein; A. Huxley *et al*, Physica B (in press).
3 H. Sugawara *et al*, J. Phys. Soc. Jpn. 64, 4849 (1995).
4 R. Modler *et al*, Phys. Rev. Lett. 76, 1292 (1996) and references therein.
5 M. Tachiki, *et al*, Z. Phys. B 100, 369 (1996)
6 L. E. De Long, *et al* ,Physica B 223 and 224, 22 (1996)

7 H. Goshima *et al*, *ibid*, p. 172.
8 H. Goshima *et al*, Physica B 206-207, 193 (1995).
9 K. Matsuda *et al*, J. Phys. Soc. Jpn. 64, 2750 (1995).
10 S. B. Roy *et al*, Phys. Rev. B 52, 9847 (1995).
11 A. M. Campbell and J. E. Evetts, Adv. Phys. 21, 327 (1982) and references therein.
12 M. Isino *et al*, Phys. Rev. B 38, 4457 (1988) and references therein.
13 S. Ramakrishnan *et al*, Physica C 256, 119, (1996);K. Ghosh *et al.* Phys. Rev. Lett. 76, 4600 (1996).
14 W. K. Kwok *et al*, Phys. Rev. Lett. 73, 2614 (1994); 76, 4596 (1996).
15 G. D' Anna *et al*, Physica C 218, 238 (1993); G. D' Anna *et al.* Eur. Phys. Lett. 25, 539 (1994).
16 S. Bhattacharya and M. Higgens, Phys. Rev. Lett. 70, 2617 (1993); Physica C 257, 232 (1996) and the references therein.
17 P. Fulde and R. A. Ferrell, Phys. Rev. 135, A550, (1964).
18 A. I. Larkin and Y. N. Ovchinnikov, Sov. Phys. JETP 20, 762 (1965).
19 L. Gunther and L.W. Gruenberg, Phys. Rev. Lett. 16, 1966 (1965).
20 R. A. Hein and R. L. Falge Jr., Phys. Rev. 123, 407 (1961); see, also, R. A. Hein in : Magnetic Susceptibility of Superconductors and other Spin Systems, eds. R. A. Hein *et al* (Plenum press, New York, 1991) p. xi.
21 M. Couach and A. F. Khoder, ibid, p. 25.
22 S. Ramakrishnan *et al*, ibid., p. 389.
23 T. V. Chandrasekhar Rao *et al*, Physica C 249, 271 (1995).
24 A. P. Pippard, Philos. Mag. 19, 217 (1969).
25 A. I. Larkin *et al* Phys. Rev. Lett. 75, 2992 (1995).
26 M. Hedo *et al*, J. Phys. Soc. Jpn. 64, 4535 (1995); our present sample is very clean with electronic mean free path estimated as 1300Å, much larger than ξ of 67Å.
27 S. Ramakrishnan *et al*, J. Phys. E 18, 650 (1985).
28 R. Liang *et al*, Phys. Rev. Lett. 76, 835 (1996).
29 D. R. Nelson and P. Le Doussal, Phys. Rev. B 42, 10113 (1990).
30 G. Blatter *et al*, Rev. Mod. Phys. 66, 1125 (1994) and references therein.
31 M. I. J. Probert and A. I. M. Rae, Phys. Rev. Lett., 75, 1835 (1995).

EFFECT OF FIELD INHOMOGENEITY ON THE MAGNETISATION MEASUREMENTS IN THE PEAK EFFECT REGION OF $CeRu_2$ SUPERCONDUCTOR

G. Ravikumar[1], T.V. Chandrasekhar Rao[1], P.K. Mishra[1], V.C. Sahni[1], Subir Saha[2], S.S. Banerjee[2], N.G. Patil[2], A.K. Grover[2], S. Ramakrishnan[2], S. Bhattacharya[3], E. Yamamoto[4], Y. Haga[4], M. Hedo[5], Y. Inada[5] and Y. Onuki[5]

[1]*TPPED, BARC, Mumbai 400 085, India*
[2]*Tata Institute of Fundamental Research, Mumbai 400 005, India*
[3]*NEC Research Institute, Princeton, NJ 08540, USA*
[4]*ASRC, Japan Atomic Energy Research Institute, Tokai, Ibaraki, Japan*
[5]*Faculty of Science, Osaka University, Toyonaka 560, Japan*

Magnetisation measurements are presented for 'cubic Laves phase' superconducting single crystal $CeRu_2$. These measurements have been performed in the peak effect regime using Quantum Design SQUID Magnetic Property Measurement System (MPMS). To understand the influence of magnetic field inhomogeneity of this set up on the measured magnetisation, all the data is recorded for different scan lengths. We clearly demonstrate that some of the anomalous magnetisation results reported in literature are artefacts of field inhomogeneity. We also present a model based on Bean's critical state model to simulate the effect of field inhomogeneity on the measured magnetisation.

Using Quantum Design (**QD**) SQUID MPMS (Magnetic Property Measurement System) various studies such as isothermal magnetisation hysteresis and temperature dependent magnetisation measurements are commonly reported in literature. A magnetisation measurement in **QD** MPMS involves moving a sample across a 'second derivative' pick up coil. The pick up signal as a function of sample position is then used to obtain the magnetic moment. This is done by fitting the measured signal with the theoretical response of a magnetic point dipole. For the second derivative pick up coil geometry this theoretical response of a unit dipole moment for this system is given by

$$V(x) = c(\mu_0 R^2/2)[-[R^2+(x+Z)^2]^{-3/2} + 2[R^2+x^2]^{-3/2} - [R^2+(x-Z)^2]^{-3/2}] \quad (1)$$

where R (= 0.97 cm) is the radius of the pickup coil, $2Z$ (=3.038 cm) is the distance between the two end coils and c is a calibration factor of the instrument. However, experimentally one always observes a constant offset (a) and a linear drift (bx) in the SQUID response along the scan length. MPMS software also provides a facility to compensate for a possible off centering of the sample. Taking these factors into consideration the modified

SQUID response for a magnetic moment m can be written as

$$V^{mod}(x) = a + bx + mV(x - x_0) \tag{2}$$

where x_0 is the correction for mispositioning of the sample. The detected SQUID response is fitted to eqn (2), to obtain the magnetic moment m. This analysis implicitly assumes that the magnetizing field is perfectly homogeneous so that the induced moment m does not vary along the scan length. However, due to field inhomogeneity, m can become position dependent along the scan distance. Especially, measurements on superconducting samples, which display large magnetic hysteresis, can be affected dramatically under small magnetic history due to inhomogeneous magnetic field[1,2].

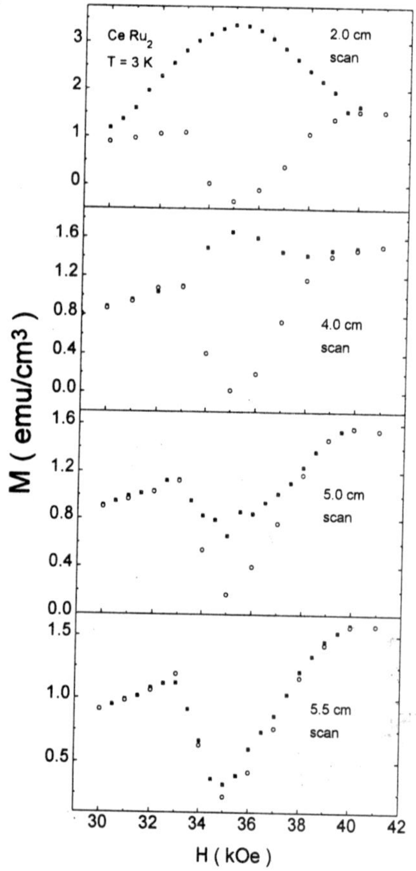

Fig.1: Magnetisation hysteresis of single crystal CeRu$_2$ recorded for different scan lengths at 3K in the peak effect regime. Open circles denote the magnetisation in increasing field (*forward*) and solid squares, the magnetisation in decreasing fields (*reverse*).

In this paper we focus on the magnetisation measurement in the peak effect regime of a superconductor. Peak effect is characterised by a sharp maximum in the critical current density vs field relation just below the second critical field H_{c2}. Thus a magnetisation hysteresis loop is expected before the superconductor becomes normal. However, there are controversial results reported in literature. Goshima et al[3] have observed a significant magnetisation hysteresis in $CeRu_2$ below H_{c2}. On the contrary, Roy et al[4] and Huxley et al[5] have reported a reversible magnetisation curve. An independent measurement using VSM confirms the hystereis in magnetisation as a function of field[6]. Huxley et al [5] have indicated that an apparent reversible magnetisation can arise due to the magnetic field inhomogeneity in the magnet which causes the magnetic moment to vary strongly along the scan length. However, so far no detailed quantitative calculations have been reported in the literarture pertaining to this problem, and this will be our objective here.

Magnetisation measurements were performed on a single crystal $CeRu_2$ (3mm X 1.5mm X 0.8mm) as a function of both magnetic field H and temperature T. The sample has a transition temperature of 6.3K at zero field. To minimise the effect of demagnetisation, the sample was mounted on the sample holder such that the magnetic field is parallel to the longest dimension. To understand the effect of field inhomogeneity on the measured magnetic moment, all the magnetisation studies were done with different scan lengths ranging from 2 to 5.5 cm.

In figure 1 we plot magnetisation hysteresis loops at $T = 3$K in the peak effect region for 2,4,5 and 5.5 cm scan lengths. The magnetisation hysteresis measured with 2 cm scan length has a peak centred around 35 kOe. The magnetisation hysteresis gradually reduces for increasing scan length and finally for 5.5 cm scan length, magnetisation becomes reversible. Associated with this peak effect, we observe a drop in regression value below unity. Regression value is a measure of deviation of the measured SQUID response from the fitted theoretical response. We find that the drop in regression is more for longer scan length suggesting the influence of field inhomogeneity on the measurements in the peak effect regime.

We also present in figure 2a, the temperature dependent magnetisation measurements with 4 cm scan length, in FC and ZFC cases at a constant applied field of 35 kOe. Both FC and ZFC magnetisation show anomalous negative peaks just below the superconducting transition. However, a similar measurement with

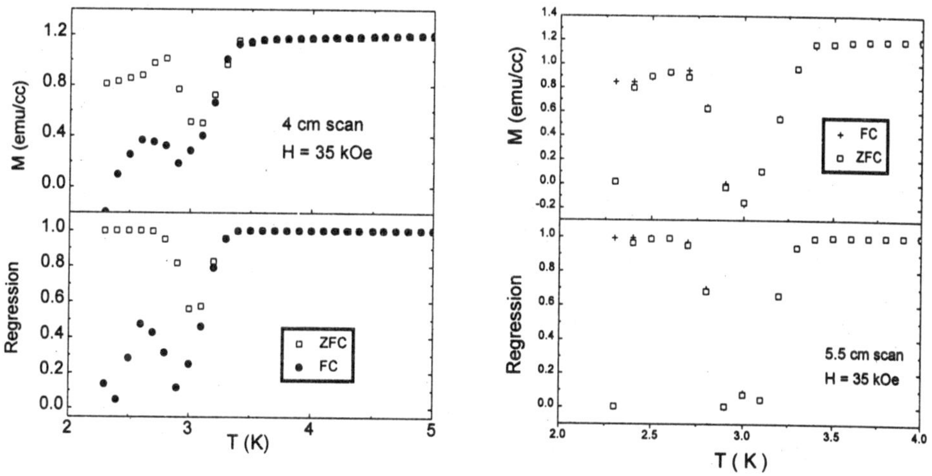

Fig.2: 'M vs T' and 'regression vs T' plots in **FC** and **ZFC** cases with an applied field of 35 kOe. (a) 4 cm scan and (b) 5.5 cm scan.

5.5 cm scan length gives a completely reversible negative peak as shown in figure 2b. The anomalous negative peaks in magnetisation are again accompanied by a sharp drop in regression value.

To simulate the effect of field inhomogeneity on the measured moment, it is necessary to know the field profile along the scan. In a technical advisory, **QD** provides the calculated field profile of the solenoid. It approximately obeys the form

$$H(x)/H_0 = 1 - \alpha x^4 \tag{3}$$

where H_0 is the field at $x = 0$, the centre of the scan. In this paper we assume that the field profile is governed only by the geometry of the superconducting coil and is fixed for a given applied field. Further, in the peak effect regime, we assume, the critical current density to be of the form

$$J_c(H) = J_{c0}\ exp[-(H - H_p)^2/2\sigma^2] \tag{4}$$

Experimentally we found that σ (measure of the peak width) is 1900 Oe which is much larger than the total field change along the scan length. Therefore J_c can be treated as field independent for a given scan.

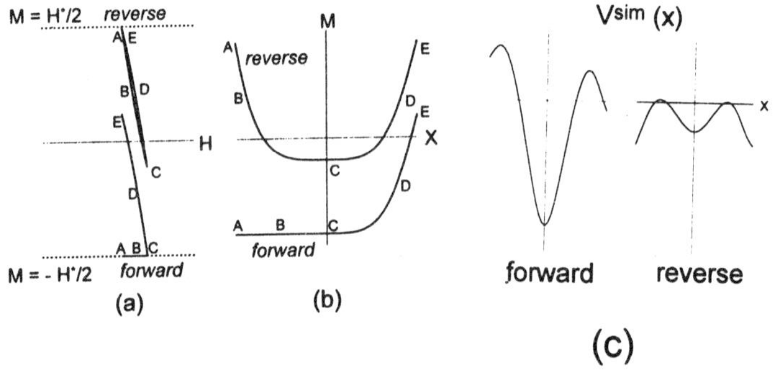

Fig.3: Schematic plot showing the magnetisation variation as a function of (a) H and (b) x for *forward* and *reverse* cases, during a scan. Points A, C and E denote the position of the sample $x = -l, 0 and +l$ respectively. (c) Schematics of the simulated SQUID response in *forward* and *reverse* cases.

During a measurement when a superconducting sample is moved along the scan length from $x = -l$ to $x = l$, it encounters an increasing field for $-l < x < 0$ and a decreasing field when $0 < x < l$. Using Bean's critical state model, variation of the sample magnetisation $M(x)$ along the scan length can be calculated. Moreover $M(x)$ depends on the magnetic history the superconductor is subjected to, while sample is at $x = -l$ (before the measurement scan begins). For the initial history of the sample we consider two cases viz., *forward* and *reverse*. In the *forward (reverse)* case, the sample is subjected to increasing (decreasing) field while at $x = -l$, till the field reaches a value $H_0(1 - \alpha l^4)$. In the *forward* case sample magnetisation is unaffected in the first half of the scan i.e., for $-l < x < 0$ (A to C), as shown in the lower part of figure 3a. However M changes sharply for $0 < x < l$ (C to E). On the other hand in the *reverse* case magnetisation changes strongly both between A to C as well as C to E. The corresponding $M\ vs\ x$ curves are shown in figure 3b. The '*simulated*' response is calculated as

$$V^{sim}(x) = M(x)V(x) \tag{5}$$

Schematics of $V^{sim}(x)$ are depicted in figure 3c for *forward* and *reverse* cases. These curves are fitted to equation (2) to obtain the parameters a, b, M^* and x_0. M^* will henceforth be termed as '*simulated*' (or apparent) magnetisation. Using this method we calculate the M^* vs H loops for different scan lengths in the peak effect regime. The results of this calculation are presented in figure 4 and the simulated hysteresis loops are in good qualitative agreement with the experimental hysteresis data.

Following our model we argue that the SQUID response would be the same on both *forward* and *reverse* magnetisation curves when the field change in a given scan is much greater than the characteristic penetration field H^*. Our data suggests that, this condition is satisfield all over the peak effect regime for a scan length of 5.5 cm. Moreover, deviation of the SQUID response would be much greater if the magnetisation in question is closer to zero, because of the large relative variation in the magnetisation induced by inhomogeneous field. In such a case the error in magnetic moment value inferred by MPMS software would be much larger.

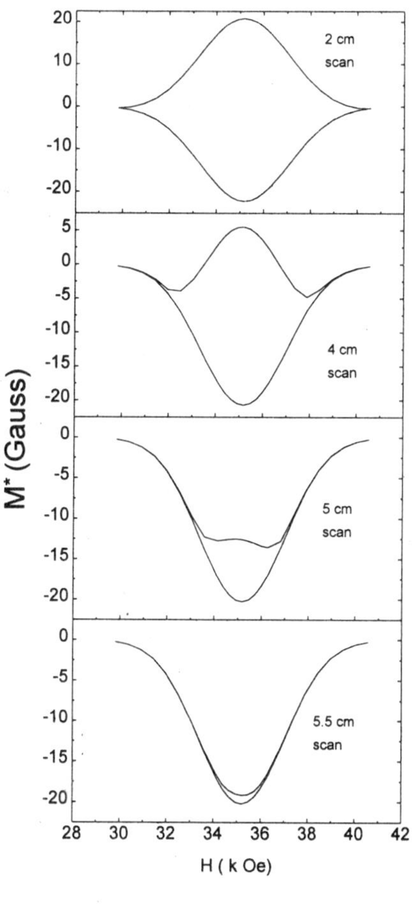

Fig.4: Simulated magnetisation hysteresis loops (M^* vs H) for different scan lengths with H_p = 35.2 kOe, σ = 1.9 kOe and H^* = 44.4 Oe. Field inhomogeneity parameter α = $4.0X10^{-5}/cm^4$.

The arguments presented above have implications on the irreversibility field estimates obtained using **QD** MPMS software. By extending the arguments of our model, it can be shown that irreversibility field would decrease with increasing scan length. The field inhomogeneity can also affect the magnetisation relaxation measurements, as the sample is subjected to a sequence of increasing and decreasing magnetic fields at specified time intervals, which is not desirable.

Temperature dependent magnetisation in both **ZFC** and **FC** cases is expected to be monotonic even in the peak effect regime[8]. However this argument fails in the presence of field inhomogeneity, because during a measurement scan at each temperature, the superconducting sample is subjected to an increasing and decreasing magnetic fields. Therefore, it can be concluded that the negative peaks in magnetisation vs temperature data shown in figure 2, is only a signature of the peak effect.

In summary, the inhomogeneity in magnetic field can affect the magnetisation obtained from the **QD** MPMS SQUID magnetometer, especially when the sample being studied is a superconductor. Particularly, in the peak effect regime, the measurements in large scan lengths can yield spurious reversible magnetisation data as observed by Roy et al[4]. The results obtained by using small scan lengths may be more reliable.

REFERENCES

1. F.J. Blunt, A.J. Perry, A.M. Campbell and R.S. Liu, Physica C **175,** 539 (1991).
2. S. Libbrecht, E. Osquiguil and Y. Bruynseraede, Physica C **225,** 337 (1994).
3. H. Goshima et al., Physica B **206,** 193 (1995).
4. S.B. Roy, A.K. Pradhan and P. Chaddah, Phys. Rev. B **52,** 984 (1995).
5. A.D. Huxley et al., J. Phys. Condens. Matter **5**, 7709 (1993).
6. K. Yagasaki et al., J. Phys. Soc. Jpn. **62,** 3825 (1993).
7. H. Sugawara et al., Physica B **206**, 196 (1995).
8. G. Ravikumar and P. Chaddah, Pramana **31**, L141 (1988).

NON-RESONANT rf ABSORPTION EVIDENCE FOR VORTEX DECOUPLING IN Bi-2212 SINGLE CRYSTALS

Y.S. Sudershan[1], Amit Rastogi[1], S.V. Bhat[1], A.K. Grover[2], Y.Yamaguchi[3], K. Oka[3] and Y. Nishihara[3]

[1]*Department of Physics, Indian Institute of Science, Bangalore 560 012, India*
[2]*Tata Institute of Fundamental Research, Colaba, Mumbai 400 005, India*
[3]*Electrotechnical Laboratory, Tsukuba, Ibaraki 305, Japan*

The magnetic field dependent rf losses in Bi2212 single crystals show two distinct features. 1) Unlike the usual behavior, the loss is a minimum at a non-zero field $H_M(T)$ and 2) dP/dH exhibits a maximum at $H_D(T)>H_M(T)$ with an accompanying minimum in the frequency of the tank circuit of the rf oscillator. We interpret feature 1 as due to the enhanced net coupling (Josephson + magnetic) between the CuO bilayers at $H_M(T)$. Feature 2 is a signature of the decoupling of the CuO bilayers and we argue that $H_D(T)$ is also the field at which flux lines decompose into 2D vortices in the planes. The temperature behavior of the decoupling field is in agreement with the predictions of Glazman and Koshelev [Phys. Rev. **B43**, 2835 (1991)].

The results of a variety of experiments investigating the vortex structure as function of field and temperature in $Bi_2Sr_2CaCu_2O_8$ (Bi2212) have been interpreted either as melting of the vortex lattice or decoupling of the 3D vortex system into 2D vortices. The discontinuous jump in the local field measurements by Zeldov et al.[1] using miniature hall probes is shown to be the consequence of a first order vortex lattice melting transition. However they find that their data fits the (1/t - 1) [2] (where $t=T/T_C$) behavior, conventionally used for the decoupling, better. The same temperature dependence has been unambiguously used to describe the decoupling transition by Doyle et al [3] and Revaz et al. [4].

The technique of non-resonant electromagnetic absorption studies the magnetic field dependent high frequency dissipation in the superconducting state. To study the rf losses we have used a home made CW NMR spectrometer equipped with a low rf level Robinson oscillator operating in the 10-22 MHz range[5] and an Oxford instruments CF200 continuous He flow cryostat. Use of magnetic field modulation and lock-in-detection results in the derivative dP/dH of the absorption being recorded as a function of field.

The field dependent losses at high frequencies in HTSC arise due to either or both of the two processes: 1) the flux flow losses when the fluxons are driven[6] by rf/microwave currents; 2) the weak superconducting order parameter in the junctions being suppressed by applied magnetic field and becoming resistive[7, 8] . In single crystals, the junctions are usually stacking faults, twins, inter-growths etc. at the inter-unit cell level[9]. But in single crystals of Bi2212, due to the layered nature, additionally the BiO-SrO layers sandwiched between the CuO bilayers also behave as intrinsic Josephson junctions[10]. The order parameter in these junctions is weaker than that in the CuO bilayer and it can be suppressed at lower fields. When this occurs, flux lines decompose into 2D vortices residing in the bilayers. With H_{rf} // a, the induced rf currents are along the c and b axes. With H // c , the losses due to c- axis rf currents are affected by the alignment of the pancakes in the CuO bilayers while the in-plane dynamics of the pancakes is determined by the b-axis component of the rf current. In general, it is difficult to distinguish between the losses due to the two components in this orientation, but by comparing the losses with those observed for H_{rf} // a, H // c where the induced rf currents are in the CuO bilayers we believe the losses observed for H_{rf} // H // c are predominantly c-axis losses.

In figures 1a and 1b we show the dP/dH *vs.* H signals at various temperatures for an air annealed crystal with a T_C of 83 K. Note that at any temperature dP/dH initially falls to a minimum, then increases to cross the field axis at H_M beyond which it shows a maximum at H_D. Such a behavior corresponds to a

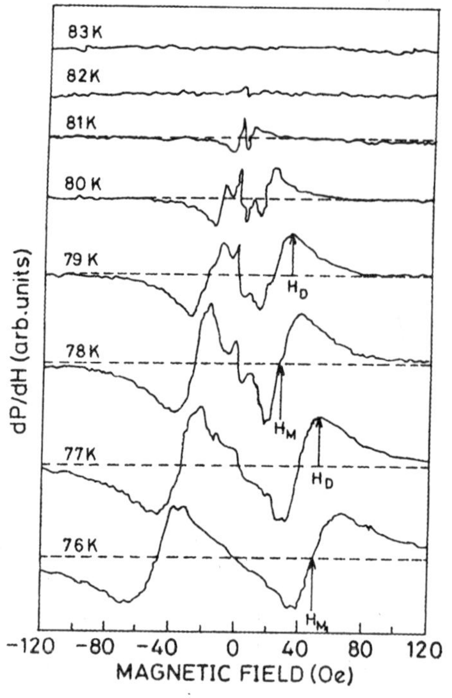

Figure 1a. dP/dH *vs.* H between T_C and 76 K.

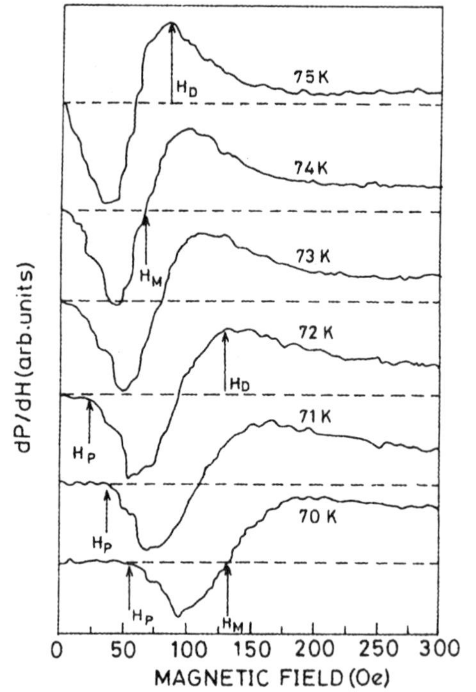

Figure 1b. dP/dH *vs.* H between 75 K and 70 K

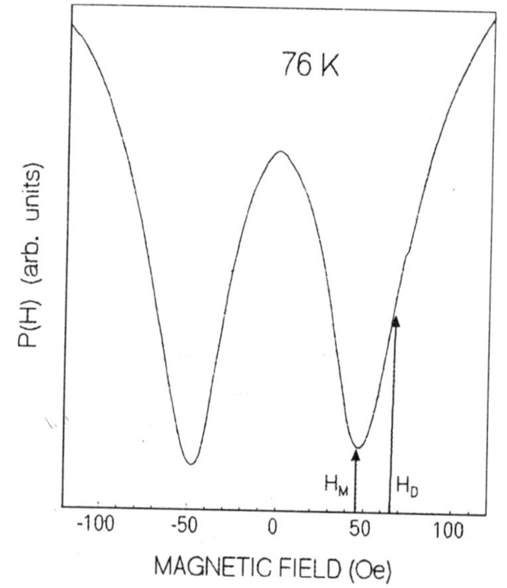

Figure 2. P(H) vs. H at 76 K obtained by numerically integrating dP/dH.

P(H) shown in figure 2, which is obtained by numerically integrating the signal at 76K. It is seen that P(H) decreases to a minimum at H_M and increases again as function of field. The arrows in fig. 2 also indicate the field position of the features marked in fig. 1. At lower temperatures (see fig.1b), dP/dH shows an additional feature in the form of its field independence till H_P. This again corresponds to a P(H) that is independent of field till H_P and we identify H_P with the H_{C1} of the sample uncorrected for geometrical barrier and demagnetization effects.

In the vicinity of H_{C1}, the flux nucleation occurs in the form of 3D flux line liquid [2]. These lines comprise of weakly coupled 2D vortex currents in the planes. As the field is increased, the density of the flux lines increases and the inter-vortex separation decreases causing the in-plane repulsive forces between them to grow stronger. For instance, the vortex diameter $\Lambda \approx 2\lambda_{ab}$ at 76 K is 7340 Å where $\lambda_{ab}(T) = \lambda_{ab}(0)/(1-t^4)^{1/2}$, $\lambda_{ab}(0) = 2000$ Å, $t = T/T_C$. At this temperature, the minimum occurs at at $H_M = 45$ Oe and the corresponding inter-vortex separation $R \approx (2\phi_0/\sqrt{3}\,H_M)^{1/2}$ is 7270 Å. Therefore $R \approx \Lambda$ at H_M and in this regime of separation, the inter-vortex repulsive forces are significant and these suppress the thermal fluctuations and the motion of the vortex currents in the planes. Further on increasing the field continuously from zero onwards, a crossover occurs from R>Λ for $H<H_M$ to R<Λ for $H>H_M$. When R≅Λ, attractive electromagnetic forces between two pancakes in adjacent layers become effective[11]. This in cooperation with the in-plane repulsive forces, confines the 2D vortex currents to the stack thus enhancing the stiffness of the vortex line. Therefore, it becomes increasingly difficult for the rf currents to drive the flux lines. The decrease in loss between H_P and H_M indicates this gradual stiffening of the 3D flux lines. At H_M, the lines are the stiffest and represent a transition into the vortex lattice. A detailed quantitative analysis of this behavior will be presented elsewhere. In this report, we concentrate on the increase in P(H) from H_M onwards.

The superconducting order parameter in the BiO-SrO layers between the CuO bilayers is weaker than that in the bilayers. In addition to the above in-plane and out-of-plane interactions that come into play as the field is increased, the weak order parameter in the intervening medium is also continuously suppressed. This effect becomes more severe for $H>H_M$ slowly turning the BiO-SrO regions resistive, thus causing the observed loss to increase beyond H_M. We identify H_D, at which the maximum in dP/dH occurs, with the decoupling field. At this field the flux lines decompose into 2D vortices residing in the planes because of the complete suppression of the superconducting order parameter in the BiO-SrO layers.

Our conclusion is supported by observed behavior of the frequency $f(H,T) = 1/2\pi(LC)^{1/2}$ of the tank circuit of the oscillator, where $L=L_C+L_S$ is the total inductance, L_C is coil inductance that is practically independent of field, L_S the sample inductance and C the capacitance. Figure 3a and 3b show f *vs* H at various temperatures. It is seen that f decreases with increasing field and at $H \approx H_D$ it exhibits a minim-

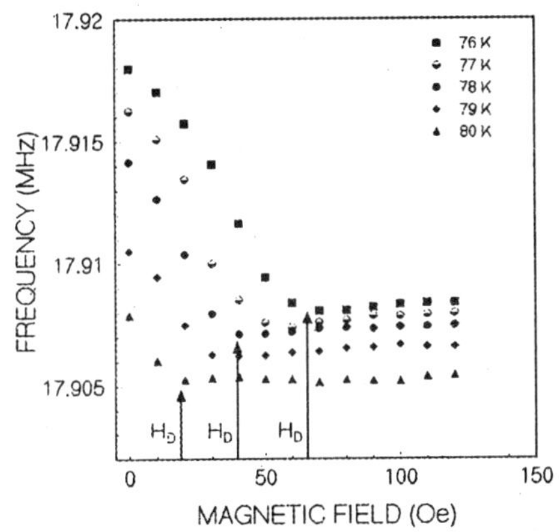

Figure 3a. Frequency of the tank circuit f *vs*. H between 80 K and 76 K

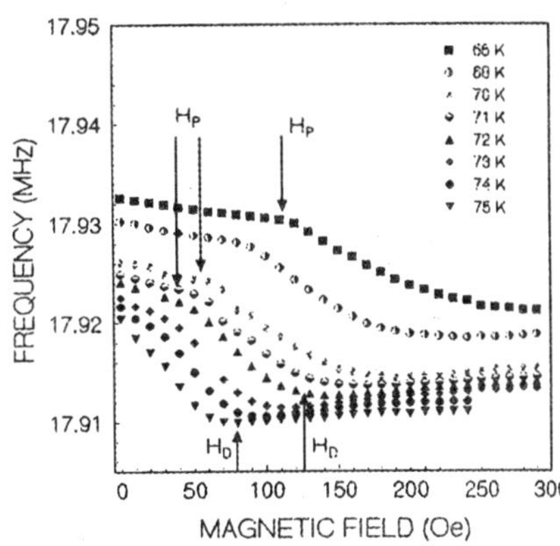

Figure 3b. Frequency of the tank circuit f *vs H*. between 75 K and 70 K

um. The minimum occurs when L_s is maximum. This is an indication of the complete loss of rf screening at H_D and hence the maximum rf penetration into the sample, which is a consequence of the suppression of the superconducting order parameter in the BiO-SrO layers. Therefore at H_D, the 3D vortex lines decouple in discrete 2D vortices residing in the CuO bilayers. It also marks the field at which the bilayers are decoupled.

In figure 4 we show the $H_D(T)$ *vs*. t for the air-annealed crystal (labeled A) and the two as-grown crystals (labeled B & C). To further prove the validity of our identification that H_D is the decoupling field, we have fitted our experimental data to the equation [ref. 2] $H_D(T) = H_0(1/t - 1)$, where $H_0 = \alpha_m \phi^3/(4\pi \gamma \lambda_{ab}(0))^2 s T_C k_B$, $\alpha_m = 0.1$, $\lambda_{ab}(0) = 2000$ Å, s = 15 Å, γ is the anisotropy ratio (λ_c/λ_{ab}), and $k_B=1.38\times10^{-16}$ ergs/K. The continuous lines indicate the fits obtained with H_0 as the fitting parameter. It is seen that the data fit very well with the predicted behavior supporting our identification of $H_D(T)$ with the decoupling field. From the values of H_0 we obtained the γ values of 95, 119 and 166 for the crystals A, B & C respectively. These values are in agreement with those obtained by other techniques[1, 3, 4].

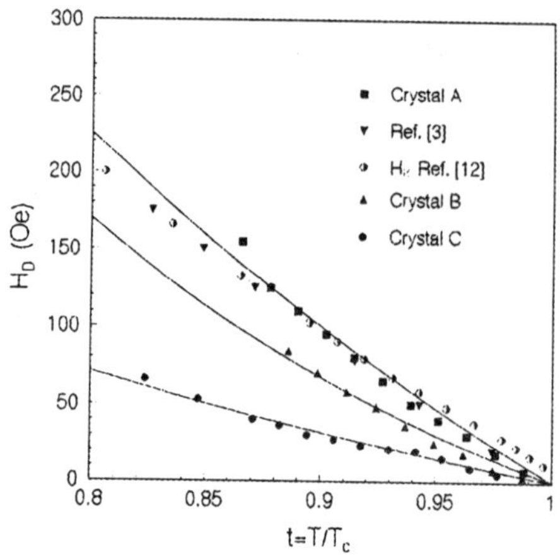

Figure 4. The closed symbols show $H_D(T)$ vs. T for crystals A, B & C. The lines are the fits to the (1/t - 1) behavior. The data taken from refs. [3 & 12] is also shown.

To further corroborate our results, we show the decoupling data taken from the local mutual inductance method of Doyle et al [3] and the magnetization measurements of Yamaguchi et al [12] in fig. 4. Given the differences of the crystals and the techniques of measurement the agreement is excellent thus supporting our conclusion.

In summary, we have observed an unusual decrease in rf losses as the field is increased. We have explained the behavior as due to the gradual stiffening of the vortex lines and the minimum at H_M indicates a transformation of the line liquid near H_P into the vortex lattice. The maximum in dP/dH observed at H_D is identified with the decoupling field. This is supported by th e observed minimum in the frequency of the tank circuit and also by the excellent fits obtained to the (1/t - 1) behavior. The anisotropy values extracted from our experiments are in agreemment with the values quoted in the literature.

ACKNOWLEDGMENTS

Y.S. Sudershan and Amit Rastogi would like to thank UGC and CSIR, India for financial support.

REFERENCES

1. E. Zeldov et al., Nature **365**, 373 (1995).
2. L.I. Glazman and A.E. Koshlev, Phys. Rev. **B43**, 2835 (1991).
3. R.A. Doyle et al., Phys. Rev. Lett. **75**, 4520 (1995).
4. B. Revaz et al., Europhys. Lett. **33**, 701 (1996).
5. S.V. Bhat et al., J. Phys. **C20**, L559 (1987).
6. K.W. Blazey in *Earlier & Recent Aspects of Superconductivity* edited by J.G. Bednorz & K.A. M. Muller, Springer series in Solid State Sciences, vol. 90 (Springer Verlag, Berlin, 1990), p. 262; For recent reviews see *studies of High Temperature Superconductors,* ed. A. Narlikar (Nova Science Pub., New York, 1996), vol. **18**.
7. A. Dulcic et al., Europhys. Lett. **10**, 593 (1989).
8. M. Pozek et al., Physica **C197**, 175 (1992).
9. F. Iga et al., Phys. Rev. **B51**, 8521 (1995).
10. R. Kleiner and P. Muller, Phys. Rev. **B49**, 1327 (1994).
11. G. Blatter et. al., Rev. Mod. Phys. **66**, 1125 (1994).
12. Y. Yamaguchi et al., *Advances in Superconductivity VIII,* 209, eds. H. Hayakawa and Y. Enomoto, Springer Verlag, Tokyo 1996.

ORIGIN AND DEVICE IMPLICATIONS OF STRONG GAP ANISOTROPY IN THE HIGH T_c CUPRATES

Guy Deutscher

School of Physics and Astronomy, Tel Aviv University, Tel Aviv 69978, Israel

A strong gap anisotropy, including the extreme case of a gap with lines of nodes, is now recognized as a general property of the High T_c cuprates. Irrespective of the mechanism of superconductivity - for instance whether it is through anti-ferromagnetic fluctuations or through the electron-phonon interaction - a strong gap anisotropy is a general consequence of weak electronic screening, itself a result in the case of the cuprates of their quasi-two dimensionality. In conjunction with a short coherence length, the gap anisotropy leads to a gap reduction near surfaces and interfaces. Grain and twin boundaries can behave as weak links. In discussing the properties of High T_c boundaries, it is important to distinguish between the case of tunneling barriers, and that of Sharvin contacts whose conductance is limited by the number of independent conducting channels. Some device implications are discussed.

I. INTRODUCTION

Amongst recent advances in the unveiling of the fundamental properties of the High Tc cuprates, the strong anisotropy in the superconducting gap is certainly one of the most important ones. While there is still a lack of agreement as to the symmetry of the gap - is it anisotropic s-wave, d-wave, or some mixture of s and d-wave - all the available data points out to a spread of gap values and, where available, to a strong angular dependence (1).

This strong anisotropy is in contrast with the case of low Tc superconductors, for which the anisotropy is at most of a few 10 per-cent. The existence of a strong gap anisotropy is not by itself a proof that the mechanism for High Tc is necessarily very different from that for low Tc, but it places certain restrictions on a description of the High Tc cuprates in their normal state. If one assumes an itinerant electron model with a van Hove singularity in the Density of States (DOS) (2), due to the quasi-two dimensional character of the cuprates, and furthermore that this singularity is close to the Fermi level, one can obtain a large gap anisotropy under the condition of a weak electronic screening. This was pointed out by Abrikosov et al.(3), and by Bok et al.(4). We summarize this argument in the first Section, as it is important for further discussion. In the quoted works, it has been assumed that the coupling matrix elements are isotropic; in that case the origin of the gap anisotropy lies in the van Hove singularity. If one assumes that anisotropic matrix elements are in fact at the origin of the gap anisotropy, weak electronic screening is also required to obtain this anisotopy. Therefore, xweak screening can be regarded as a fundamental property of the cuprates, whatever the exact mechanism for superconductivity, and whatever the exact nature of the normal state. Weak screening can by itself lead to a change in gap symmetry (5)

In the second Section, we describe some recent Point Contact measurements on $La_{2-x}Sr_xCuO_4$ (LSCO) single crystals near optimum doping, that give access to the value of the gap anisotropy. We find that for this cuprate, the gap anisotropy is finite; the data seem to exclude a d-wave symmetry. For the model reviewed in Section 1, the measured value of the anisotropy can be used to obtain a value for the screening length. It turns out that for the studied LSCO crystal, this value is significantly larger that the inter-atomic distance, in stong contrast with the case of usual metals where it is shorter than this distance. Some arguments presented in Section 1 can explain such a large screening length based on the quasi-two dimensional character of the cuprates. However, it must be recognized that a screening length larger than the inter-atomic distance puts the cuprates in the category of semiconductors rather than in the category of metals. This has necessarily some profound implicatons for their normal state properties, which are beyond the scope of this article.

In the third Section, we discuss some of the consequences of a strong gap anisotropy, combined with a short coherence length, on device properties. More specifically we consider the properties of grain boundaries, usually called grain boundary junctions (GBJ). We include in this discussion the extreme case of a d-wave symmetry (as may apply to the case of YBCO). Some experiments are suggested that could clarify the symmetry question, taking into account the possibility that GBJs could in fact be Andreev contacts rather than tunnel junctions, as has been usually assumed in previous theoretical analysis.

II. WEAK SCREENING

In the Jellium model, the matrix elements that enter into the BCS theory can be written in the form:

$$V_{kk'} = \frac{4\pi e^2}{(qa)^2 + (q_o a)^2} \, . \, f(\omega) \tag{1}$$

where $q = (k-k')$, a is the lattice parameter and q_o is the inverse screening length.

These matrix elements enter into the BCS gap equation:

$$\Delta_k = \Sigma \, V_{kk'} \frac{\Delta_{k'}}{(\xi_{k'}^2 + \Delta_{k'}^2)^{1/2}} \, . \tag{2}$$

where $\xi_{k'}$ is the energy measured from the Fermi level in the normal state for wave vector k'.

In usual metals ans alloys, $(q_o a) > 1$ with a typical value of 3. In that case, carrying the sum in Eq.2 involves all points on the Fermi surface with almost equal weight, since the maximum value for q is $(2\pi/a)$. In that case, any factor that could in principle induce a gap anisotropy (such as for instance an anisotropic DOS, or an anisotropic $V_{kk'}$), would be averaged out, leaving only a negligible anisotropy. This is the case in usual superconducting metals and alloys.

Suppose however that $(q_o a) < 1$. In that case, when computing Δ_k from Eq.2, only values of k' close to k will play a role. This will allow for gap anisotropy.

Now, it is well known that for a three dimensional metal the screening length is rather unsensitive to the carrier concentration:

$$q_o^2 = \frac{4 \, n_o^{1/3}}{a_o} \tag{3}$$

where n_o is the carrier concentration, and a_o is the Bohr radius. Even for the rather low carrier concentration characteristic of the cuprates (about $5.10^{21}/cm^3$), $(q_o a) > 1$.

The situation is different if we consider a two dimensional metal sheet, embedded in a three dimensional insulator having a dielelectric constant κ . In that case, due to electrostatic interaction through the third dimension, the screening potential falls off as a power law instead of an exponential, and the screening length is proportional to the dielctric constant. The q Fourier component of the screening potential is given by (6):

$$A_q = \frac{Ze}{\kappa \, (q + q_s)} \tag{4}$$

where $q_s = (\kappa a_0)^{-1}$, for a screened charge Ze located in the plane.

The actual situation is in fact more complex, i.e. we do not have isolated two-dimensional sheets but a stack of sheets with a finite electronic coupling between them. This finite coupling is reflected in the finite electronic conductivity in the direction perpendicular to the planes. In that case we expect a cross-over between a two-dimensional behavior at short distances, and a three dimensional behavior at large distances. To describe this three-dimensional behavior, we must use a screening length based on some effectve electronic density. A possibke guess is to use for this density a value that takes into account the intersheet coupling, as reflected through the resistivity anisotropy ratio.:

$$q_0^{*2} = \frac{4\, n_0^{*1/3}}{a_0^*} \qquad (5)$$

where $n_0^* = n_0\,(\rho_{ab}/\rho_c)$, and $a_0^* = \kappa\, a_0$. Here ρ_{ab} and ρ_c are the in-plane and out-of-plane resistivity, respectively. Using a typical resistivity ratio of 100, and a dielectric value of 10, we obtain a screening length of the order of 10A, which corresponds to a value of $(q_0 a)$ of the order of three.

Another consequence of the quasi-two- dimensional character is the existence of a logartihmic divergence of the DOS at the van Hove singularity. The two effects combined - a weak screening and an anisotropic DOS - lead to gap anisotropy (3,4). Explicit calculations of gap anisotropy values have been performed (7); when the van Hove singularity is close enough to the Fermi level, and for screening parameters similar to our estimates. The gap anisortopy then falls in the range of 3 to 5. As we shall now see in the next section, such an anisotropy value is indeed observed in at least one of the cuprates.

III. POINT CONTACT MEASUREMENTS ON LSCO SINGLE CRYSTALS

The characteristics of Point Contacts between a normal metal N and a superconductor S having an isotropic s-wave gap Δ have been calculated by Blonder et al. (BTK, 8). An important parameter in this calculation is the transparency of the contact. For an ideal contact, having no dielectric barrier and a perfect matching of the Fermi velocities, the conductance of the contact $\sigma(V) = (dI/dV)$ is enhanced by a factor of 2 at biases smaller than (Δ/e), as compared to its value at larger biases (normal state conductance σ_N). This enhancement is due to the Andreev reflection mechanism, in which an incoming electron from N is reflected as a hole, while a Cooper pair flows in the S side. For a non-ideal contact, the conductance ratio $(\sigma(0)/\sigma_N)$ is smaller than 2, and in fact goes to zero for very weak contacts (tunnel barriers). In all cases of non-ideal contacts, $\sigma(V)$ is non-monotonic: it first rises, reaches a maximum value σ_{max} at $V = (\Delta/e)$, and then decreases down to σ_N. The ratio $(\sigma_{max}/\sigma_N) > 2$, and diverges for a tunnel barrier.

Point Contacts between a gold point and LSCO single crystals have been found to follow this behavior (9), but to differ quantitatively from the BTK predictions in two respects (10): for a given value of $\sigma(0)$, the value of σ_{max} is smaller than predicted, and the return to the normal state value σ_N is slower than predicted. The experimental characteristics can in fact be fitted to a generalized BTK theory, taking an anisortopic gap of the form (anisotropic s-wave):

$$\Delta = \Delta_0 + \Delta_1 \cos(4\theta) \qquad (6)$$

with $(\Delta_0 - \Delta_1) = 2$ to 5 meV, $(\Delta_0 + \Delta_1) = 10$ t0 15 meV, giving an anisotropy ratio between 3 and 5 (Fig.1). As we have seen in the previous section, such an anisotropy ratio can be explained by the combination of a van Hove singularity in the DOS near the Fermi level, and reasonable weak screening parameters. Further measurements of the gap anisotropy on other cuprates of various doping levels, will show if there exists a correlation between Tc and the screening parameter, as has been suggested recently (11).

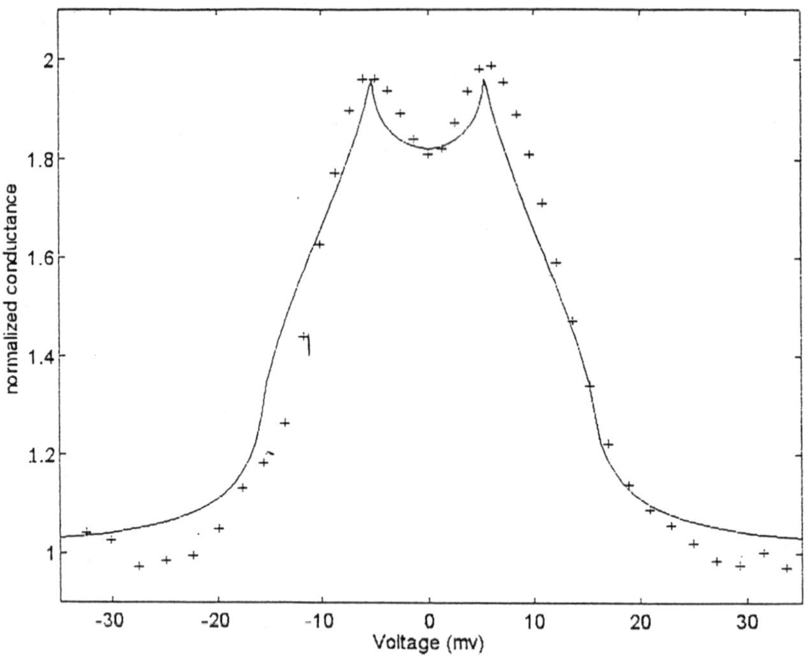

Fig.1 The initial increase of this Au-LSCO Point contact is inconsistent with a d-wave symmetry of the gap. Continous line: fit to thedata points with $\Delta_{max} = 15$ meV, $\Delta_{min} = 5$ meV.

Not less importantly, these Point Contact data are not compatible with a d-wave gap symmetry. This is because in that case, as shown by Hu (12), surface bound states of zero energy (at the Fermi level) are created, due to the interference between Andreev reflections along directions corrsponding to opposite gap signs. As shown in the detailed calculations of Tanaka (13), these zero energy bound states increase the zero bias conductance beyond the Andreev enhancement factor of 2, over a narrow energy range: irrespective of the barrier transparency. $\sigma(V)$ is then a **decreasing function** of the bias at small bias, in contrast with the BTK predictions for a generalized s-wave gap, and our observations on LSCO. The data is therefore inconsistent with LSCO being a d-wave superconductor. A small d-wave component cannot be excluded, but if present it is certainly not dominant.

IV. GAP ANISOTROPY, SYMMETRY AND JUNCTION PROPERTIES

One of the striking (and damaging) properties of the cuprates is the fact that the critical current density j_c of a GBJ is strongly affected by a small misorientation. This effect strongly reduces the critical current density of polycrystalline materials, and as such it is of great concern for high current applications. On the other hand, this same property has been used advantageously to create Josephson junctions. In any case, a full understanding of this high sensitivity to grain disorientation remains of great interest to the community.

In general terms, it is understood that the strong reduction of j_c at grain boundaries is a result of the extremely short coherence length of the cuprates (14). In a low temperature, long coherence length superconductor, the effect of any local perturbation on the order parameter, due to an inhomogeneity of the material on the atomic scale (such as impurities, surfaces etc...), is averaged out over the coherence length, and has a negligible impact on the value of the order parameter. One of the best examples of the validity of this statement is the case of a free surface. In a low Tc material, it has no effect on the order parameter, whose gradient can be taken as zero at the surface (15).

This situation is changed radically in the high Tc cuprates. A local change of composition (for instance, a change in oxygen concentration, as can well occur at a grain boundary), will induce a substantial reduction of the order parameter at and in the immediate vicinity of the defected region. It has been argued that even a free, ideal surface, will have a similar effect (14), in contrast with the case of low Tc, long coherence length superconductors. In the general case, it is extremely difficult to determine composition and structure with an atomic resolution. But the general argument is quite solid, and therefore there is really no reason to be surprised by a critical current reduction at grain boundaries. It is also quite reasonable that such reduction will become stronger with the misorientation, as the concentration of defects can be expected to increase with it (16). As the detailed defect structure may well vary from one cuprate to another, we may also expect such reduction in j_C to be general, but non-universal in its details. This seems indeed to be the case. For instance, it is known empirically to be stronger in LSCO than in YBCO (17).

The influence of the depression of the order parameter at a grain boundary on j_C, measured accross the boundary, is twofold. First, the depairing current will be reduced. Second, in the presence of vortices (that can for instance be induced by the self-field effect), these will flow preferentially along the grain boundary. This is because the pinning energy is small for a vortex moving *along* the boundary, since in this region the gap is small. Both effects will reduce j_C, the second one being usually the dominant one if no precautions are taken to avoid the presence of vortices, and if j_C is defined by a small dissipation level.

This "channeling" effect has been observed very recently on uniaxially twinned YBCO films (18), for which it has been found that the critical current is smaller when it flows in the direction perpendicular to the twins (the Lorentz force then moves the vortices along the twins). Theory predicts that the depression of the order parameter should become larger, and in fact diverge, as Tc is approached (14). Thus, the critical current anisotropy should diverge in this limit. This is exactly what is observed experimentally (19). This divergence is a subtle effect: the value of the order parameter at the boundary is proportional to $(b/\xi(T))$, where b is a fixed length and $\xi(T)$ is the temperature dependent coherence length which diverges at Tc as $(T\text{-}Tc)^{-1/2}$. Because the critical current is proportional to the square of the order parameter, the anisotropy should increase as $(Tc\text{-}T)^{-1}$, which it does.

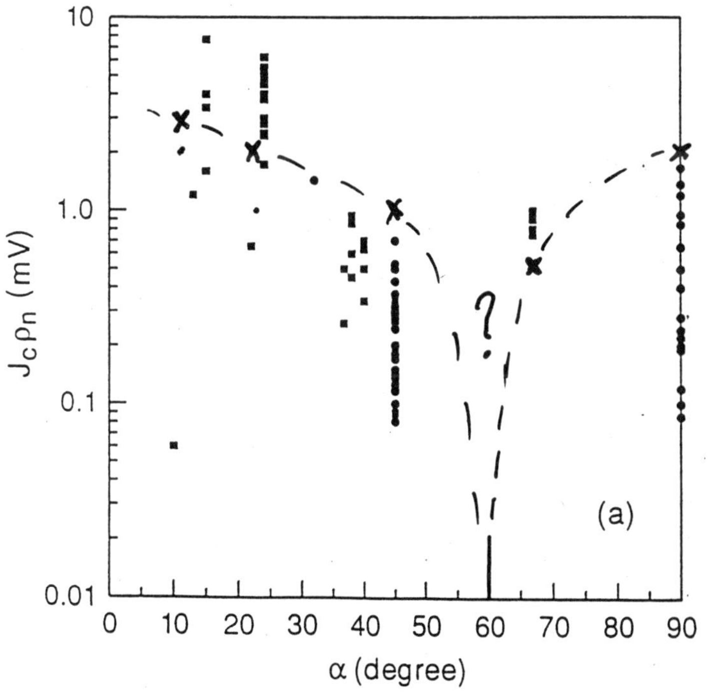

Fig.2 Grain boundary critical current dependence on the disorientation. The line is the theory prediction for a Sharvin contact. For a tunneling contact, the critical current should be zero for a 45^0 disorientation, in contradiction with experimental data (see Ref.20).

A strong gap anisotropy will make these short coherence length effects even stronger. Even in the absence of any local change in composition, right at the boundary, the order parameter "sees" two different crystallographic orientations. Because it cannot adjust to any one of them, the corresponding cost in condensation energy will result in a depressed order parameter.

A case of special interest is of course that of a d-wave superconductor *having a short coherence length.* Suppose for instance that a boundary separates two grains, having respectively their [100] and [110] axis perpendicular to it. For the reason explained above, we can certainly expect a depression of the order parameter at the boundary. This effect has not yet been considered in detail theoretically.

In fact, for reasons of symmetry, the critical current should cancel out anyhow for this orientation. A more subtle case arises if we consider a boundary between two grains having their [110] axis perpendicular to it. If we assume that the boundary is a tunnel junction, its critical current should be strictly zero since the direction of the tunneling current corresponds to nodes in the gap. This is not however in agreement with the existing data (20). A possible way to reconcile the data with a d-wave symmetry is to assume that the boundary is in fact a Sharvin contact, whose conductance is limited by the number of independent conducting channels, and not by the tunneling matrix elements. In that case, all channels corresponding to points on the Fermi surface contribute to the current; and since gap lobes of equal signs face each other, the critical current is finite and in fact not much smaller than for a zero degree boundary. This does corresponds to the data. This Sharvin model of a grain boundary also predicts that j_c is zero for a misorientaion close to 30°. This prediction has not yet been tested experimentally, but it is not in contradiction with the available data (Fig.2). It would certainly be of use to obtain more detailed data in the angular range of interest.

Grain boundary junctions are definitely worth further studies. For instance, Hilgenkamp et al. have observed $j_c(B)$ patterns on 45° assymmetric YBCO boundaries, that they have ascribed to facetting of the boundaries in conjonction with a d-wave symmetry. If their interpretation is correct, and if as we have claimed, LSCO is *not* a d-wave superconductor, such patterns should not be seen for LSCO 45° boundaries.

V. CONCLUSIONS

A strong gap anisotropy seems to be a universal property of the cuprates, but the symmetry of the gap may not be universal. While it is widely believed that it is d-wave in YBCO, our Point Contact measurements on LSCO indicate that in this material it is anisotropic s-wave. We have proposed that the basic property of the cuprates is weak electronic screening rather than a d-wave gap. Further studies are necessary to establish whether there exists a correlation between Tc and the weak screening parameter. The grain boundary properties special to the cuprates are a result of the short coherence length, combined with this anisotropy.

ACKNOWLEDGMENTS

This work was supported by the Israel National Science Foundation, by the Heinrich Hertz Center for High Temperature Superconductivity and by the Oren family Chair of Experimental Solid State Physics.

REFERENCES

1. Z.X.Shen e al., Phys.Rev.Lett. **70**, 1553 (1993)
2. J.Labbe and J.Bok, Europhys.Lett. **3,** 1225 (1987)
3. A.A.Abrikosov, Physica **C222,** 191 (1994)
4. J.Bouvier and J.Bok, Physica **C249,** 117 (1995)
5, G.Santi et al., Physica **C259,** 253 (1996)
6. T.Ando, Rev.Mod.Phys. **54,** 437 (1982)
7. J.Bouvier, J.Bok and G.Deutscher, to be published
8. G.E.Blonder, M.Tinkham and T.M.Klapwijk, Phys.Rev. **B25,** 4515 (1982)

9. N.Achsaf et al. , in "Coherence in High Tc Superconductors" Eds. G.Deutscher and A.Revcolevschi, World Scientific, p.428 (1996)
10. N.Achsaf et al., to appear in J.of Superconductivity (Proceedings of MOS 96)
11. G.Deutscher, Phys.Rev.Lett. **76,** 1555 (1996)
12. C.R.Hu, Phys.Rev.Lett. **72,** 1526 (1994)
13. Y.Tanaka, in "Coherence in High Tc Superconductors", Eds.G.Deutscher and A.Revcolevschi, World Scientific. p.393 (1996)
14. G.Deutscher and A. Muller, Phys.Rev.Lett. **59,** 1745 (1987)
15. P.G.de Gennes, "Superconductivity in Metals and Alloys", Benjamin, New York (1966)
16. D.Dimos et al., Phys.Rev.Lett. **61,** 210 (1988)
17. H.Hilgenkamp, private communication
18. C.Villard et al., Phys.Rev.Lett.**77,** 3913 (1996)
19. G.Koren and C.Villard, private communication
20. See H.Hilgenkamp, these Proceedings
21 G.Deutscher and R,Maynard, Europhys.Lett. **30,** 361 (1995)

IMPLICATIONS OF A $d_{x^2-y^2}$ SYMMETRY COMPONENT OF THE ORDER PARAMETER ON POTENTIAL HIGH-T_c DEVICES

H. Hilgenkamp[1], J. Mannhart[1], B. Mayer[1], Ch. Gerber[1], J. R. Kirtley[2], and K. A. Moler[2]
[1]*IBM Research Division, Zurich Research Laboratory,*
8803 Rüschlikon, Switzerland
[2]*IBM Research Division, Thomas J. Watson Research Center,*
P.O. Box 218, Yorktown Heights, NY 10598, USA

Various experiments have indicated that the order parameter in the high-T_c cuprates has a predominant $d_{x^2-y^2}$ symmetry. This has important implications for devices based on Josephson junctions. The $d_{x^2-y^2}$ symmetry leads for example to intrinsic weak link formation, inhomogeneous current distributions in the junctions and reduced I_cR_n products. This article points out the implications a $d_{x^2-y^2}$ symmetry component has for Josephson junction devices, with particular emphasis on devices based on grain boundaries in high-T_c superconducting thin films.

In recent years, it has been debated extensively whether the symmetry of the order parameter in high-T_c superconductors is best described by an s or a $d_{x^2-y^2}$ function [1, 2] (see Fig. 1). This symmetry is of interest in order to identify the underlying mechanisms of high-T_c superconductivity, but is also important for device applications. A better knowledge of the order parameter symmetry can help provide better control over the properties of devices such as Josephson junctions.

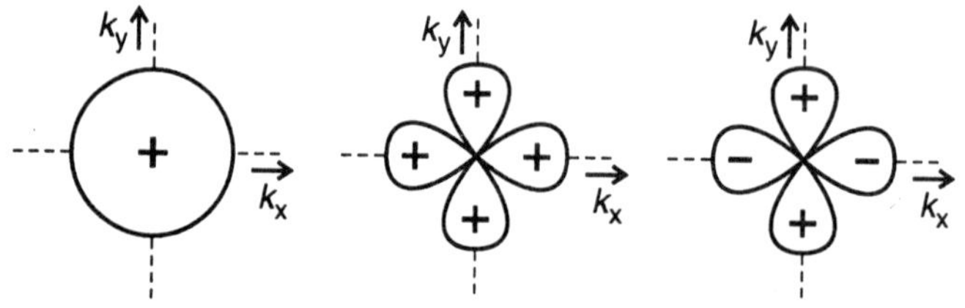

Figure 1: Schematic representation of an isotropic s, an anisotropic s, and a $d_{x^2-y^2}$ symmetry of the order parameter, respectively. In the case of $d_{x^2-y^2}$ symmetry, a π phase difference occurs between the orthogonal directions in momentum ($\vec{k}$) space.

Various experiments have provided strong evidence of a predominant $d_{x^2-y^2}$ symmetry [1, 3, 4], possibly in combination with a small (perhaps anisotropic) s component [5]. On the other hand, experiments on biepitaxial polygons [6] and results of c-axis tunneling studies [7] are considered evidence of s symmetry. It is noted, however, that the results of the former experiments are consistent with $d_{x^2-y^2}$ symmetry if spontaneous generation of magnetic flux in the biepitaxial boundaries is taken into consideration [8-10] and that the latter experiments do not rule out the possibility of a predominant $d_{x^2-y^2}$ symmetry component in the a, b-plane.

Here, we discuss the influence of a $d_{x^2-y^2}$ symmetry component of the order parameter on the electrical transport properties of high-T_c Josephson devices. We will focus mainly on grain boundary junctions (GBJs). Owing to the misalignment of the superconducting grains, these junctions are particularly influenced by the anisotropy of the order parameter. For a schematic representation of the most commonly used grain boundary configuration, the [001] tilt boundary, see Fig. 2.

Considering various types of junctions, a number of groups have discussed the influence of a $d_{x^2-y^2}$ symmetry component of the order parameter on the transport properties of contacts between high-T_c superconductors [11-16]. The dependence of the Josephson current on the respective orientations of two $d_{x^2-y^2}$ superconducting electrodes, separated by an insulating barrier layer, has

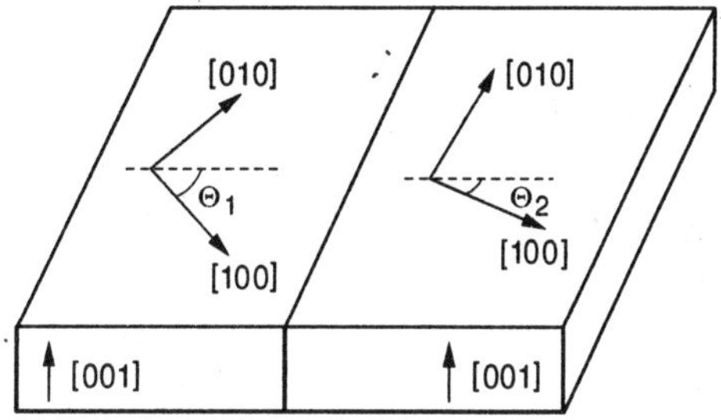

Figure 2: [001] tilt grain boundary, schematic. From Ref. [17], © 1996 IEEE.

been described by Sigrist and Rice [11]. In their work, the implications of the angular dependent magnitude of the gap in the direction perpendicular to the barrier plane are considered. Subsequently, it was shown by Barash et al. [12] that, in addition to these density-of-states effects, also the $d_{x^2-y^2}$ symmetry can give rise to a depression of the order parameter in the superconducting electrodes abutting an insulating or normal-conducting barrier layer. The magnitude of the order parameter depression depends on the orientations of the superconducting electrodes with respect to the boundary plane and the transparency of the barrier. Furthermore, Deutscher and Maynard [13] discussed the influence of the misorientation of the Fermi surfaces on the angular dependence of the critical current density of Sharvin point contacts between two misaligned superconductors.

We recently proposed that a boundary region with a depressed order parameter is created intrinsically at the interface between two $d_{x^2-y^2}$ wave superconductors that form a contact at an angle $\Theta = \Theta_1 - \Theta_2$ (see Fig. 2) [18]. Typically this boundary region extends over distances of the order of the coherence length from the boundary. The depression arises solely from the frustration caused by the different crystallographic orientations of the superconductors on either side of the boundary and by the finite value of the coherence length. No insulating or normal-conducting barrier layer has to be present *a priori*, but in the case of a complete depression of the order parameter at the boundary a normal-conducting barrier layer will arise intrinsically. It is expected that the depression of the order parameter gives rise to quasi-particle states, bound by Andreev reflections, in the boundary region, in a way comparable to the quasi-particle states occurring in the cores of Abrikosov vortices [19]. The magnitude of the intrinsic order parameter depression depends, among other things, on Θ_1, Θ_2, temperature, and on the materials involved. It is strongest for 45° grain boundaries, in which the maxima of the order parameter of one superconductor coincide with the nodes in the gap function of the other.

In addition to the effects arising from an anisotropy of the order parameter, the angle-dependent grain boundary properties are also influenced by other mechanisms [20], such as a depressed order parameter at the boundary owing to dislocations [21, 22], structural disorder [23, 24], and deviations from bulk stoichiometry [25].

Independent of the details of the current transport across the junction (such as e.g. the degree of directionality of the charge-transfer process) the order parameter orientations can, in specific configurations, give rise to an additional π phase shift over the junction (π junction). This is, for example, the basis of spontaneous generation of half flux-quanta in specially designed tricrystal rings [4]. Intrinsic π phase shifts also play an important role in the properties of single grain boundaries, because thin-film grain boundaries are not straight, but generally consist of facets of various orientations and with typical dimensions <100 nm (see e.g. [26] and Fig. 3). This faceting in combination with a $d_{x^2-y^2}$ symmetry component of the order parameter leads to an intrinsically inhomogeneous distribution of the Josephson current in the GBJs [27], including regions for which an additional π phase shift arises from the order parameter orientation. For a constant phase of the order parameters in the grains, the Josephson current across these facets flows in the opposite

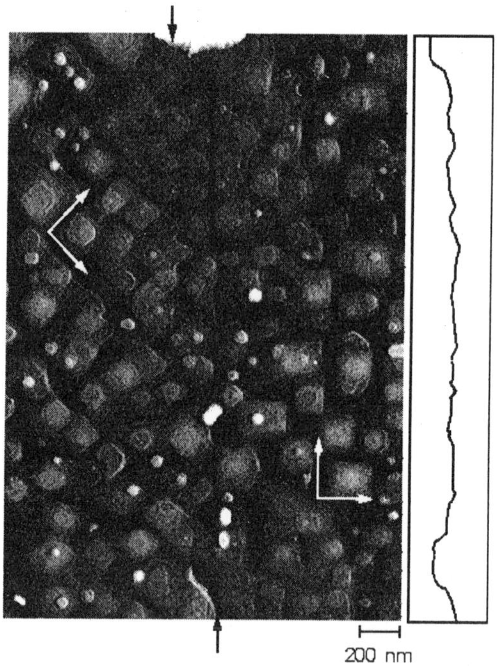

Figure 3: Atomic force microscopy image of the surface of a 150-nm-thick $YBa_2Cu_3O_{7-\delta}$ film with an asymmetric 45° [001] tilt boundary. The location of the grain boundary is indicated by arrows. The meandering path of the boundary has been replotted at the right. From [10].

direction ('negative' critical current density) from the current across the other facets.

The inhomogeneity is most prominent for asymmetric 45° [001] tilt GBJs ($\Theta_1 = 0$, $\Theta_2 = 45°$), which is the configuration typically obtained with the template biepitaxial process [28]. Using the expression proposed by Sigrist and Rice for the Josephson critical current density between two $d_{x^2-y^2}$ superconductors,

$$J_c \propto \prod_{m=1,2} ((\sin\Theta_m)^2 - (\cos\Theta_m)^2) \tag{1}$$

the spatial distribution of the critical current density was calculated for the boundary depicted in Fig. 3 and in a similar way as was described for other grain boundary angles in Ref. [27]. The result is shown in Fig. 4.

For such asymmetric 45° [001] tilt boundaries, the inhomogeneity results in anomalous magnetic field dependences of the critical current, I_c, in remarkable contrast to a Fraunhofer pattern [27, 29, 30]. A striking characteristic is the absence of a global maximum of I_c at $H_a = 0$. Figure 5 shows an example of such an anomalous magnetic field dependence. The highly symmetric characteristics show that the anomaly is not caused by magnetic flux quanta trapped in, or close to, the boundary. Also displayed is the $I_c(H_a)$ dependence, recorded after the critical temperature of the sample (i.e., of the grains) had been reduced from 90 to 78 K by annealing the sample in vacuum. This procedure reduced the T_c of the grain boundary to below 30 K. Except for some minor differences, no significant change in the magnetic field dependence was observed. Interestingly, also for asymmetric 24° samples that display less distorted Fraunhofer-like $I_c(H_a)$ dependences, no fundamental change in the magnetic field dependence was observed as T_c was reduced stepwise to 60 K using the same procedure. Furthermore it is noted that, except for a change in the absolute value of I_c, these characteristics do not change significantly as a function of temperature for temperatures well below T_c.

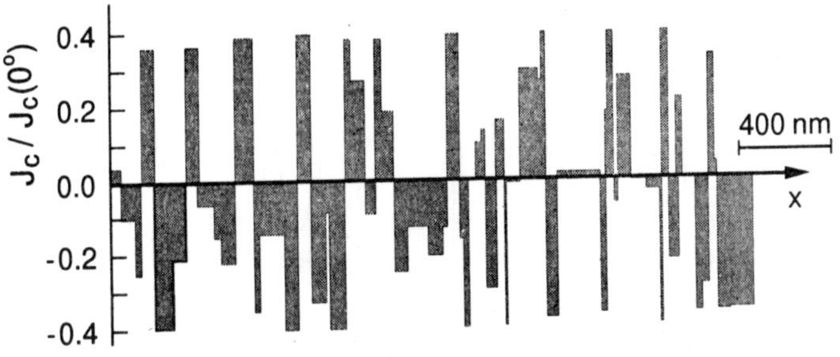

Figure 4: Spatial distribution of J_c according to Eq. (1) for the asymmetric 45° [001] tilt grain boundary depicted in Fig. 3.

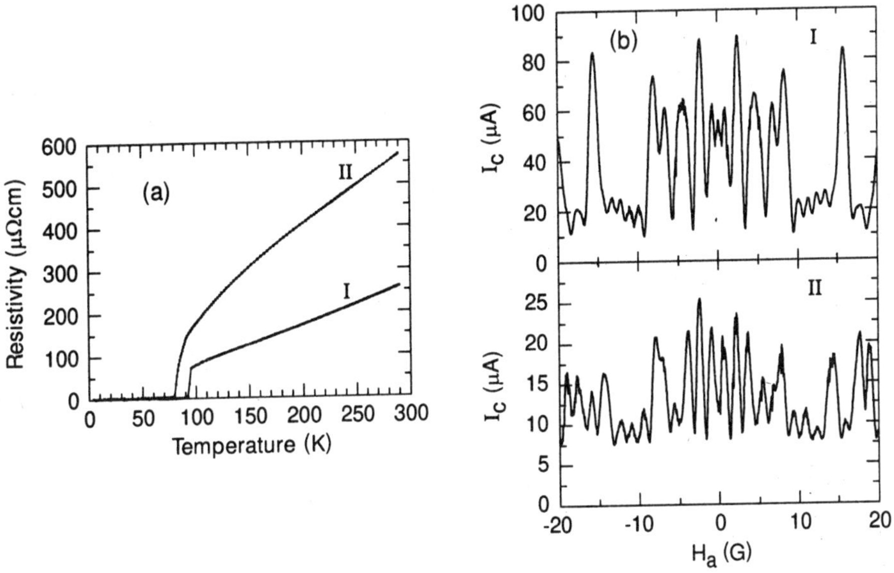

Figure 5: (a) Resistivity as a function of temperature for a bridge structure containing an asymmetric 45° [001] tilt grain boundary in an $YBa_2Cu_3O_{7-\delta}$ film before (I) and after (II) annealing in vacuum. (b) Critical current as a function of applied magnetic field at $T = 4.2$ K for the same sample before and after annealing (thickness: 180 nm, width 10 μm). With the annealing procedure the normal-state resistivity of the junction has increased from 1.9×10^{-8} Ωcm^2 (I) to 3.0×10^{-8} Ωcm^2 (II). From Ref. [17], © 1996 IEEE.

Figure 6 shows the $I_c(H_a)$ dependence calculated from the inhomogeneous J_c distribution of Fig. 4. The characteristic features observed in the experiments, such as the distorted periodicity and the absence of a global maximum of I_c at $H_a = 0$, are also found in the calculated $I_c(H_a)$ dependences. It is noted that the difference in the magnetic field scale result from the fact that the boundary of Fig. 3 is about an order of magnitude smaller than the one used for the experiments of Fig. 5 and because flux focusing has not been taken into consideration.

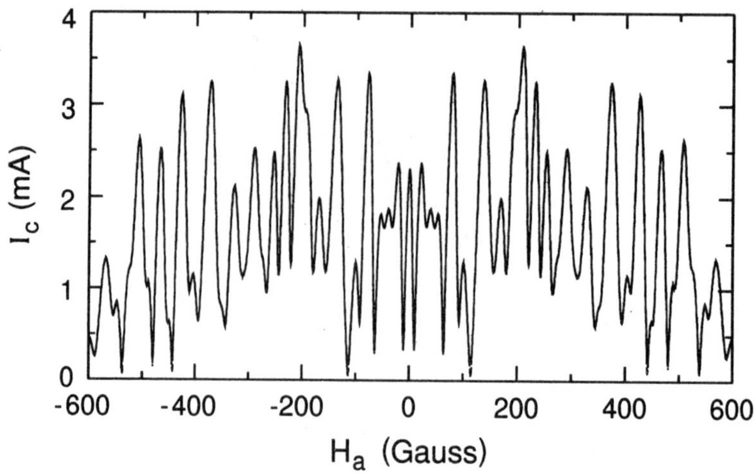

Figure 6: Magnetic field dependence of the critical current calculated for the grain boundary of Fig. 3, using the spatial distribution of the critical current density shown in Fig. 4.

The lowest energy state of an inhomogeneous grain boundary can include a spontaneously generated phase profile in the junction. The Josephson coupling energy E_J of a GBJ can be expressed as

$$E_J = \frac{\phi_0}{2\pi}\int_0^t\int_0^w J_c(x,y)(1-\cos(\varphi(x,y)-\alpha(x,y)))dxdy \quad , \tag{2}$$

where $J_c(x,y)$ is the (positive) local critical current density, $\varphi(x,y)$ the local phase difference, t the film thickness, and w the width of the junction. The additional phase shift arising from the order parameter orientation is denoted by α, which is equal to 0 or to π. The phase difference $\varphi(x,y)$ tends to adjust to the variations of $\alpha(x,y)$ in order to minimize the total free energy of the junction [10]. This energy is given by the sum of E_J and the magnetic energy E_M associated with a nonconstant $\varphi(x,y)$. In the steady superconducting state, $\varphi(x,y)$ follows the generalized sine-Gordon equation:

$$\Delta\varphi(x) - \lambda_J^{-2}(x,y)\sin(\varphi(x,y)-\alpha(x,y)) = 0 \quad , \tag{3}$$

where $\lambda_J(x,y) \propto J_c(x,y)^{-1/2}$ is the local Josephson penetration depth. If one assumes a local $J_c(x,y) \lesssim 10^5$ A/cm^2, then $\lambda_J(x,y) \gtrsim 1$ μm. As $t \leq 200$ nm, $\varphi(x,y)$ will not vary strongly over the thickness of the film (i.e., in the y-direction). In the x-direction, $\varphi(x)$ will vary to best exploit $\alpha(x,y)$. The variations of $\varphi(x)$ may be promoted by self-field effects and, in contrast to homogeneous junctions, the transition from electrically small to wide junctions may lead to an increase of the critical current for very inhomogeneous junctions.

The spontaneous generation of magnetic flux, corresponding to the formation of a phase profile in the grain boundary, has been visualized by scanning SQUID microscopy (SSM) [10]. In the instrument used, an 8 μm$\times$8 μm superconducting pickup loop, an integral part of a low-T_c SQUID, is scanned a few microns above the sample surface to image the local magnetic fields. Figure 7 shows an SSM image of a 400 μm$\times$400 μm area of an asymmetric 45° [001] tilt grain boundary, cooled to 4.2 K in a nominally zero applied magnetic field. Under the conditions used for this experiment, $1\phi_0$ in the sample corresponds to $0.44\phi_0$ coupled into the pickup loop. The line scan along the boundary shows randomly distributed flux, with the smallest variations limited by the 8-μm resolution of the pickup loop.

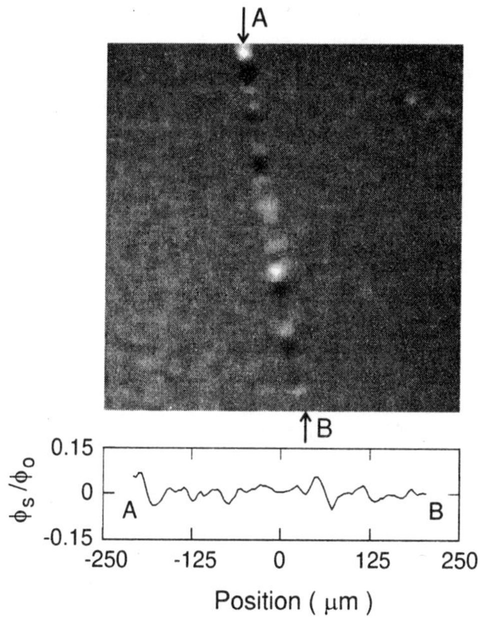

Figure 7: Scanning SQUID microscope image of a 400 μm$\times$400 μm area including an asymmetric 45° [001] tilt $YBa_2Cu_3O_{7-\delta}$ bicrystal grain boundary taken with a square pickup loop with a side length of 8 μm. The image shows spontaneously generated magnetic flux along the grain boundary. The bottom section shows a cross section through the data along the grain boundary, measured in units of ϕ_0 penetrating the pickup loop. From Ref. [10].

We have shown above that owing to the combination of faceting and a $d_{x^2-y^2}$ symmetry component of the order parameter, the grain boundaries are intrinsically inhomogeneous. The inhomogeneity is most prominent for asymmetric 45° [001] tilt grain boundaries, typically the configuration obtained with the template biepitaxial process. Transport properties measured in practical GBJ experiments reflect the averaged behavior of these inhomogeneous junctions, and therefore are no direct measure of intrinsic grain boundary properties. The dependence of J_c on grain boundary misorientation, the $I(V)$ characteristics, and the I_cR_n product are not known for individual facets. The reduction of the junction critical current resulting from the occurrence of regions with negative Josephson current leads to reduced I_cR_n products. Furthermore, the local Josephson penetration depth λ_J can be substantially smaller than the value derived from the junction critical current density, according to $\lambda_J \propto J_c^{1/2}$.

The $d_{x^2-y^2}$ symmetry component does not only affect the characteristics of grain boundary junctions, but also influences those of other high-T_c Josephson junction configurations, such as ramp-type junctions, weakened microbridges, and SNS step edge junctions. For these junctions the orientation of the barrier with respect to the a,b-lattice of the superconductors will largely determine their characteristics. Depending on the orientation of the crystal with respect to the boundary plane, the order parameter is depressed near the barrier [12]. The poor reproducibility of these junctions, compared to bicrystal GBJs, may in part be due to the fact that often in-plane alignment is not adequately taken into consideration.

The presence of self-generated flux is suggested to explain the contradictory results of the symmetry experiments performed on tricrystal rings and on biepitaxial polygons. The magnetic flux is a potential source of noise for GBJs and for their applications, for example in SQUID magnetometers. Furthermore the spontaneous magnetization can modify the current-phase relation of the junction and enhance critical current fluctuations. In addition, the flux will cause a paramagnetic component to the grain boundary susceptibility.

Acknowledgments

The authors thank J. G. Bednorz, A. A. Golubov and R. P. Hübener for useful discussions. Part of this work was done in the framework of the Esprit Project 8132 "WELITTD-HTS" and supported by grants from the Swiss Federal Office of Education and Science.

References

[1] D. J. Scalapino, Phys. Rep. **250**, 329 (1995).

[2] J. Annett, N. Goldenfeld, and A. J. Leggett, in *Physical Properties of High Temperature Superconductors*, Vol. 5, D. M. Ginsberg, Ed., World Scientific: Singapore, 1996.

[3] D. J. Van Harlingen, Rev. Mod. Phys. **67**, 515 (1995).

[4] C. C. Tsuei, J. R. Kirtley, C. C. Chi, L. S. Yu-Jahnes, A. Gupta, T. Shaw, J. Z. Sun, and M. B. Ketchen, Phys. Rev. Lett. **73** 593 (1994).

[5] K. A. Müller, Nature **377**, 133 (1995).

[6] P. Chaudhari and S.-Y. Lin, Phys. Rev. Lett. **72**, 1084 (1994).

[7] A. G. Sun, D. A. Gajewski, M. B. Maple, and R. C. Dynes, Phys. Rev. Lett. **72**, 2267 (1994).

[8] A. J. Millis, Phys. Rev. B **49**, 15408 (1994).

[9] J. R. Kirtley, P. Chaudhari, M. B. Ketchen, N. Share, S.-Y. Lin, and T. Shaw, Phys. Rev. B **51**, 12057 (1995).

[10] J. Mannhart, H. Hilgenkamp, B. Mayer, Ch. Gerber, J. R. Kirtley, K. A. Moler, and M. Sigrist, Phys. Rev. Lett. **77**, 2782 (1996).

[11] M. Sigrist and T. M. Rice, Rev. Mod. Phys. **67**, 503 (1995).

[12] Yu. S. Barash, A. V. Galaktionov, and A. D. Zaikin, Phys. Rev. B **52**, 665 (1995).

[13] G. Deutscher and R. Maynard, Europhys. Lett. **30**, 361 (1995).

[14] S. Kashiwaya, Y. Tanaka, M. Koyanagi, H. Takashima, and K. Kajimura, Phys. Rev. B **51**, 1350 (1995).

[15] C. Bruder, A. van Otterlo, and G. T. Zimanyi, Phys. Rev. B **51**, 12904 (1995).

[16] Y. Tanaka and S. Kashiwaya, Phys. Rev. B **53**, R11957 (1996).

[17] H. Hilgenkamp, J. Mannhart, B. Mayer, Ch. Gerber, J. R. Kirtley, and K. A. Moler, in: Proc. 1996 Applied Superconductivity Conf. (1996 ACS), Pittsburgh, PA, Aug. 25-30, 1996. IEEE: New York, in press.

[18] H. Hilgenkamp and J. Mannhart, submitted to Appl. Phys. A.

[19] H. F. Hess, R. B. Robinson, R. C. Dynes, J. M. Valles, Jr., and J. V. Waszczak, Phys. Rev. Lett. **62**, 214 (1989); J. D. Shore, M. Huang, A. T. Dorsey, and J. P. Sethna, *ibid.* **62**, 3089 (1989).

[20] P. Chaudhari, D. Dimos, and J. Mannhart, in *Superconductivity*, J. G. Bednorz and K. A. Müller, Eds., Springer: Berlin, 1990, p. 174.

[21] D. Dimos, P. Chaudhari, J. Mannhart, and F. K. LeGoues, Phys. Rev. Lett. **61**, 219 (1988).

[22] J. A. Alarco and E. Olsson, Phys. Rev. B **52**, 13625 (1995).

[23] S. E. Babcock and D. C. Larbalestier, J. Mater. Res. **5**, 919 (1990).

[24] M. F. Chisholm and S. J. Pennycook, Nature **351**, (1991).

[25] D. M. Kroeger, A. Choudhury, J. Brynestad, R. K. Williams, R. A. Padgett, and W. A. Coghlan, J. Appl. Phys. **64**, 331 (1988).

[26] J. A. Alarco, E. Olsson, Z. G. Ivanov, P. Å. Nilsson, D. Winkler, E. A. Stepantsov, and A. Ya. Tzalenchuk, Ultramicroscopy **51**, 239 (1993).

[27] H. Hilgenkamp, J. Mannhart, and B. Mayer, Phys. Rev. B **53**, 14586 (1996).

[28] K. Char, M. S. Colclough, S. M. Garrison, N. Newman, and G. Zaharchuk, Appl. Phys. Lett. **59**, 733 (1991).

[29] C. A. Copetti, F. Rüders, B. Oelze, Ch. Buchal, B. Kabius, and J. W. Seo, Physica C **253**, 63 (1995).

[30] J. Mannhart, B. Mayer, and H. Hilgenkamp, Z. Phys. B **101**, 175 (1996).

EFFECT OF ELECTRIC FIELDS ON THE NORMAL STATE OF SUPERCONDUCTORS

Lydia S. Lingam and Keshav N. Shrivastava
School of Physics, University of Hyderabad, Hyderabad 500 046, India

Two mechanisms have been proposed for the variation of resistance of a superconducting film in a normal state as a function of electric field. Both the theoretical approximations are in agreement with the experimental measurements of the variation of normal state resistivity of $YBa_2Cu_3O_{7-x}$ with applied electric field.

We propose two mechanisms for the variation of resistance of a film as a function of applied electric field. The resistivity calculated for the field-driven movement of the mobile charge carriers in the normal state of a superconductor in the static approximation varies as the square root of the electric field, $\rho \propto E^{1/2}$ only as long as the dielectric constant is independent of the field. We also predict that the resistivity of a system with conduction electrons localized along a one-dimensional chain varies as the square root of the applied electric field. We have found that for the electric field larger than about half of the break down field, E > $\frac{1}{2}$ E_{BD}, the measured resistivity varies as the square root of the electric field,$\rho \propto E^{1/2}$. Thus both of the theoretical approximations are in agreement with the experimental measurements of the normal state resistivity of $YBa_2Cu_3O_{7-x}$.

The current along the z direction due to polarization caused by the gate voltage is given by

$$nev = \sigma E \tag{1}$$

where v is the velocity of the electrons and σ the change in conductivity due to polarization, $\sigma = \frac{1}{R}$, which is the inverse of the change in resistivity along the z direction. We can write the electron concentration as

$$n_0 = \frac{1}{3\pi^2}\left(\frac{4\pi P k^2}{\vec{E}}\right)^{3/2}, \tag{2}$$

so that from (1), the resistance is found to become,

$$R = \frac{3\pi^2 E^{5/2}}{ev(4\pi P k^2)^{3/2}}. \tag{3}$$

The energy imparted by the electric field to the electrons is $e\epsilon_o Ed$ where d is the thickness of the film along the z direction. Therefore, the velocity of the electrons along the z direction may be determined from the relation $\frac{1}{2}mv^2 = e\epsilon_o Ed$ and substituted in (3) so that the resistance becomes

$$R = \frac{3\pi^2 m^{1/2} E^2}{\sqrt{2}(4\pi e P)^{3/2} k^3 \epsilon_o^{1/2} d^{1/2}} \tag{4}$$

which varies as E^2. The maximum value of k is π/d so that the change in resistance becomes,

$$R = \frac{3\pi^2 m^{1/2} E^2 d^{5/2}}{\sqrt{2}(4\pi e P)^{3/2} \epsilon_o^{1/2} \pi^3}. \tag{5}$$

From which the change in resistance due to the applied field is found to be proportional to the 5/2 th power of the depth of the film. A static approximation of (5) also exists.Since $\epsilon_o \simeq 25$ may be treated as a constant we can ignore the dynamics of the system and write,$4\pi P = \epsilon_o E$ which substituted in (5) gives

$$R = \frac{3\pi^2 m^{1/2} E^{1/2} d^{5/2}}{\sqrt{2} e^{3/2} \epsilon_o^2 \pi^3} \tag{6}$$

which shows that the resistance varies as the square root of the applied electric field. In a previous paper[1] we found that in the energy of the system, one term is linear in E and another term is a small contribution proportional to E^2. In the normal state, this energy gives rise to a current determined by eRJ so that from eqn(6) of ref.1, we can write

$$eRJ = |E|[< p > + \sum_n \frac{< 0|E.p|n >< n|p|0 >}{E_n - E_o}] \tag{7}$$

where $< p >$ is the average dipole moment. Since the voltage,$V = |E|d$,the first term gives the Ohm's law type correction to the resistance and the second term gives the resistance proportional to $|E|$.

We consider the electrons in a Fermi sphere of volume $\frac{4}{3}\pi k_F^3$. The volume element in the k-space is $(2\pi/L)^3$. The total number of orbitals is $N = k_F^3 L^3/(3\pi^2)$. Therefore, the Fermi wave vector is $k_F = \left(\frac{3\pi^2 N}{L^3}\right)^{1/3}$ and the Fermi energy is $\epsilon_F = \frac{\hbar^2}{2m}\left(\frac{3\pi^2 N}{L^3}\right)^{2/3}$. We thus find that the number of carriers is given by

$$N = \frac{L^3}{3\pi^2}\left(\frac{2m\epsilon}{\hbar^2}\right)^{3/2} \tag{8}$$

and hence the density of states for a three dimensional solid is given by

$$\frac{dN}{d\epsilon} = \frac{L^3}{2\pi^2}\left(\frac{2m}{\hbar^2}\right)^{3/2} \epsilon^{1/2} \tag{9}$$

Similarly for a two dimensional solid, $N = k_F^2 L^2/2\pi$ so that the Fermi wave vector becomes $k_F = (2\pi N/L^2)^{1/2}$. The energy of the carriers in a plane is then given by

$$\epsilon = \frac{\hbar^2}{2m}\frac{2\pi N}{L^2} \tag{10}$$

so that the number of orbitals in a plane of area L^2 becomes

$$N = \frac{L^2}{2\pi}\frac{2m\epsilon}{\hbar^2} \tag{11}$$

and hence the density of states is given by,

$$\frac{dN}{d\epsilon} = \frac{L^2}{2\pi}\frac{2m}{\hbar^2} \tag{12}$$

which is a constant independent of the carrier energy, ϵ.

In one dimension, $k_F = \pi N/L$ so that $\epsilon = \frac{\hbar^2}{2m}(\frac{\pi N}{L})^2$. The number of carriers in a line becomes

$$N = \frac{L}{\pi}\left(\frac{2m\epsilon}{\hbar^2}\right)^{1/2} \tag{13}$$

and the density of states becomes

$$\frac{dN}{d\epsilon} = \frac{L}{2\pi}\left(\frac{2m}{\hbar^2}\right)^{1/2}\epsilon^{-1/2}. \tag{14}$$

Thus for d= 3, 2 and 1 we find that the density of states varies with energy as $\epsilon^{1/2}, \epsilon^0$ and $\epsilon^{-1/2}$, respectively. It has been reported by Klein et al[2] that the conductivity varies as the relative change in the density of states,δ_N/N_o according to ,

$$\frac{\delta G}{G} = \int \frac{\delta N}{N_o}\frac{d}{d\epsilon}[-f(\epsilon + eV)]d\epsilon \tag{15}$$

where δG is the change in the conductivity, G, upon the application of the electric voltage, V and f(ϵ) is the Fermi distribution. Thus the conductivity is predicted to vary as $V^{1/2}, V^0$ and $V^{-1/2}$ for the dimensionality d = 3, 2 and 1, respectively. For conduction along a chain the conductivity is thus expected to vary as $V^{-1/2}$ and hence the resistivity,

$$\rho \propto E^{1/2} \tag{16}$$

where E is the applied electric field. For a d-dimensional system, the resistivity is proportional to $E^{(1-d/2)}$ so that for one dimensional conduction in layered systems, where the conduction is along the c-axis, the resistivity varies as the square root of the voltage or the applied electric field.

The resistivity of $YBa_2Cu_3O_{7-x}$ as a function of electric field has been measured by Frey et al[3] . For small values of the electric field, E, the curve is linear as expected from eq(16). For the fields larger than about half of the breakdown voltage the variation of resistivity as a function of the electric field is shown in Fig.1. We find that $R \propto E^{1/2}$ which is in agreement with that predicted approximately from the field driven movement of the mobile charge carriers in the normal state of the superconductor as long as the dielectric constant is independent of the electric field, and it agrees also with the resistivity of a system of electrons, the density of states of which is localized in one dimension along a chain as in (16) due to conduction along z direction.

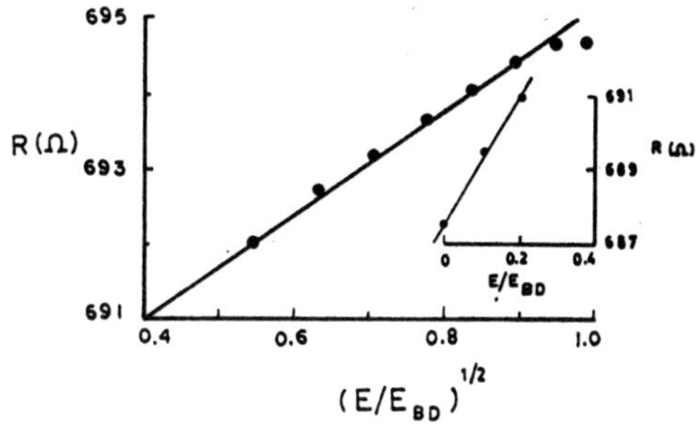

Fig.1. Plot of normal state resistivity vs square root of electric field at 100K.

Fig . 1:We have plotted the resistivity as a function of the square root of the electric field at a temperature of 100 K . The experimental measurements in $YBa_2Cu_3O_{7-x}$ are taken from Frey et . al[3]. We find that resistivity varies as $E^{1/2}$. For small fields, $E/E_{BD} \leq 0.2$, the resistivity shown in the inset is linear, $\rho \propto E$.

Frey et al[3] have also made an independent direct measurement of charge from which they obtain the resistance as a function of charge. This measurement also shows that atleast two mechanisms are involved out of which one of them agrees with the variation of resistivity as the square root of the electric field.

In conclusion, we find that the experimental variation of resistivity as a function of increasing electric field shows two separate mechanisms. (a) For small electric fields, the resistivity is a linear function of electric field, and (b) for $E > \frac{1}{2}E_{BD}$, the resistivity varies as $E^{1/2}$. These findings are in agreement with the theory. Although d.c. fields are discussed in the present work, a.c.fields have also[4] been published and hence our work is of interest for the understanding of a wide variety of phenomena.

REFERENCES

1. K . N . Shrivastava , J . Phys . Condens . Matter **5**, L 597, (1993).

2. T . Klein , O . G . Symko ,D . N . Davydov and A . G . M . Jansen , Phys . Rev . Lett . **74**,3656, (1995) .

3. T . Frey , J . Mannhart ,J . G . Bednorz and E . J . Williams, Phys . Rev . B**51**, 3257,(1995) .

4. Y . Gim, C . Doughty, X . X . Xi, A . Amar, T . Venkatesan and F . C . Wellstood, Appl . Phys . Lett,**62**,3198 (1993).; S . N . Mao, X . X . Xi,Q . Li, I . Takeuchi, S . Bhattacharya, C . Kwon, C . Doughty, A . Walkenhorst, T . Venkatesan, C . B . Whan, J . L . Peng and R . L . Greene, Appl . Phys . Lett . **62**,2425 (1993) .

THEORY OF HIGH TEMPERATURE SUPERCONDUCTIVITY AND ELECTRONIC PROPERTIES

S. Maekawa[a,b]

[a]*Department of Applied Phyiscs, Nagoya University, Nagoya 464-01, Japan*

[b]*Institute for Materials Research, Tohoku University, Sendai 980-77, Japan*

The relation among the crystal structure, electronic structure and superconductivity in high temperature superconducting cuprates is reviewed.

1. Introduction

Numerous high temperature suprconductors have been found since the first one, La-Ba-Cu-O, was discovered in 1986,[1)] and the transition temperature is steadily rising. This trend in rising transition temperatures is likely to continue, but selecting materials for study that are likely to exhibit high transition temperatures from amongst the infinite number of elemental combinations is very diffucult. Some form of guiding principle is necessary.

Fortunately, the high temperature superconductors that have been discovered so far have very distinctive features. Therefore, we may categorize these features properly and understand the role they play in the emergence of the supercondcutivity. High temperature superconductors have the following features: (i) elemental substitution and carrier doping occur in different layers of the layered perovskite structure, which contains the basic elements of copper and oxygen and has a strong two-dimensionality; (ii) when carriers are introduced into the parent material, which is an antiferromagnetic insulator, the CuO_2 plane becomes metallic and superconductivity emerges; and (iii) the concentration of carriers required for the emergence of sueperconductivity is about 0.15 ~ 0.2 per copper ion, and the antiferromagnetic effect remains quite strong. Here, we would like to discuss the relationship between these features and the conditions necessary for the emergence of high temperature superconductivity, in terms of the electronic structure.

2. Electronic Structure

The parent materials of high temperature superconductors are ionic crystals. Let us consider La_2CuO_4, where one Cu^{2+} ion and its four O^{2-} ligands form a CuO_2 plane. This CuO_2 plane is alternately stacked with an $(LaO)_2$ layer (called the block layer)[2] containing La^{3+} and O^{2-} ions. Evidently, the CuO_2 plane has a negative charge and the $(LaO)_2$ layer has a positive charge. Therefore, the positive and negative charges are stacked alternately in a direction perpendicular to the layers (called the c direction).

If a Sr^{2+} ion is substituted for part of the La^{3+} ion, a surplus of one negative charge ($-e$) per Sr^{2+} ion occurs. Accordingly, the electrical charge of the CuO_2 plane must decrease by one in order to keep the entire crystal electrically neutral. In other words, one hole is added to the CuO_2 plane. We note that the crystal structure of the CuO_2 planes, into which the carriers were introduced, has not changed. Unlike conventional semiconductors, carrier doping can be performed in the superconductors without disturbing the structure of the conductive CuO_2 plane. This is an important feature of high temperature superconductors.

The concentration of carriers required for the emergence of high temperature superconductivity is 0.15 ~ 0.2 carriers per copper ion. Within this range of concentration, the CuO_2 plane exhibits the metallic properties. But, the electrical resistivity is considerably higher in the c direction. In the superconducting state, therefore, the ionic character existing in the c dirction are quite strong. This duality of the ionic crystal structure and the metallic property of the electonic structure is important for carrier doping and effectively gives the electronic structure its two-dimensionality, which is also important for high temperature sueprconductivity.

The CuO_2 plane in the parent material is an ionic crystal. The Cu^{2+} ions have a $3d^9$ electonic structure, and only one electron exists in the $3d(x^2-y^2)$ orbit. On the other hand, the 2p orbits of the O^{2-} ions surrounding the copper ion are entirely occupied by electrons. Figure 1 shows the energy levels of these orbits.[3] Figures 1 shows that the energy of the $3d(x^2-y^2)$ orbit is lower than that of the $2p_\sigma$ orbit. However, if a Cu^{1+} ion is produced through the addition of another electron into the $3d(x^2-y^2)$ orbit, the energy level of the orbit is elevated due to Coulomb repulsion between the electrons. This increase in energy (U) is about 8eV, and the effect of this repulsion is called the electron correlation. Threfore, even though the energy of

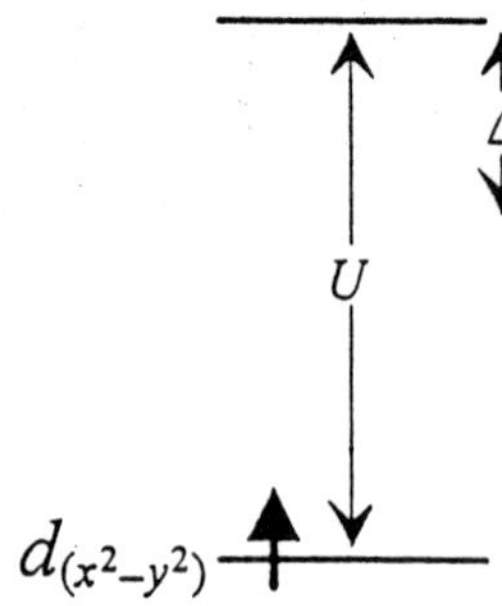

Fig. 1 Energy levels of copper and oxygen ions. The energy of the copper ion's $3d(x^2-y^2)$ orbit is lower than the $2p_\sigma$ orbit of the oxygen ion. However, when the $3d(x^2-y^2)$ orbit contains two electrons, the energy is elevated due to the inter-electron Coulomb interaction (U). Therefore, a vacancy remains in the $3d(x^2-y^2)$ orbit of the Cu^{2+} ion.

the $3d(x^2-y^2)$ orbit is lower than that of the $2p_\sigma$ orbit, the $3d(x^2-y^2)$ orbit always has a vacancy of one electron or hole. Thus the copper ion is magnetic and has a spin of S=1/2.

The energy required for an electron to move from the oxygen ion to the copper ion (expressed by t_{pd}) is estimated to be about 1 eV. As seen in Fig. 1, when an electron jumps from the oxygen ion to the copper ion, the energy of this electron increases by Δ. This energy is called the charge transfer energy and is equal to about 2 ~ 3 eV. Therefore, although relocation of the electron is possible, in reality the electron can not move about in the crystal and is effectively localized around the ion, becoming an insulator. However, a small amount of quantum mechanical seepage is possible, and as a result, an antiferromagnetic exchange interaciton begins to act between the spins of neighboring copper ions. The magnitude of this interaction (J) is estimated to be

$$J \approx t_{pd}^4 / \Delta^3, \tag{1}$$

which is about 0.1 eV.

If a Sr^{2+} ion is substituted for a La^{3+} ion and carriers are introduced into the CuO_2 plane, an electron is removed from the oxygen ion and a hole is introduced, as expected in Fig. 1. When this hole moves in the crystal, electrical conductivity emerges. Since the oxygen ion's hole has a spin (S=1/2), an antiferromagnetic exchange interaction occures between the hole and the adjacent copper ion. Since, the distance between the oxygen and copper ions is shorter than the distance between two copper ion, the interaction is very strong and is estimated to be

$$J_K \approx t_{pd}^2 / \Delta, \qquad (2)$$

which is about 1 eV. Therefore, the hole captures the spin of one of the adjacent copper ions and produces an antiferromagnetic bond, resulting in the formaton of the spin singlet state. Moreover, since the four surrounding oxygen ions are equivalent to the copper ion, the energy of the singlet state is lowered when the hole is distributed equally to the four oxygen ions, forming a singlet.[4)] Thus, by substituting Sr^{2+} for La^{3+} and introducing a hole into the oxygen ions, the hole is coupled to the spin of the adjacent copper ion and erases the spin, after forming a singlet. This creates the image of positively charge "holes" emerging and moving about on the CuO_2 plane. This state is shown in Fig. 2. The basis of high temperature superocnductivity is thus the positively charged holes that "float" on the CuO_2 plane with a spin of S=1/2, upon which the antiferromagnetic exchange interactions act. The kinetic energy of the positve charges is expressed by using the parameter, t. Also, the magnitude of the antiferromagnetic exchange interactions that act between the spins is expressed by J. Therefore, this theoretical model is called the t-J model.

3. Superconductivity

As previously stated, the neighboring spins (S=1/2) of the copper ions on the CuO_2 plane exert an antiferromagnetic interaction (J) upon each other. If holes are

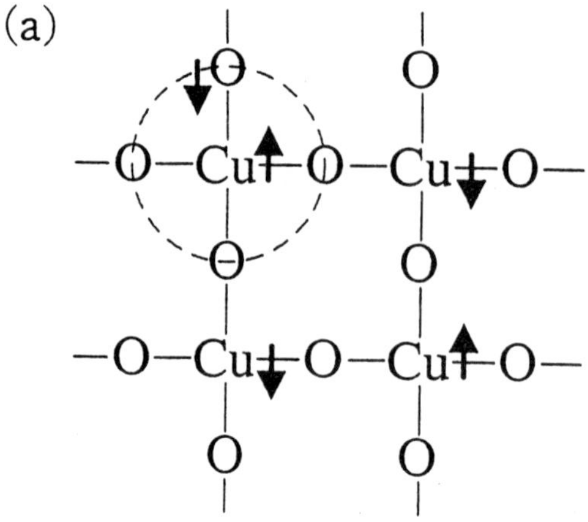

Fig. 2(a) CuO_2 plane. A hole introduced into the oxygen ion forms a singlet state with the spin in a copper ion.

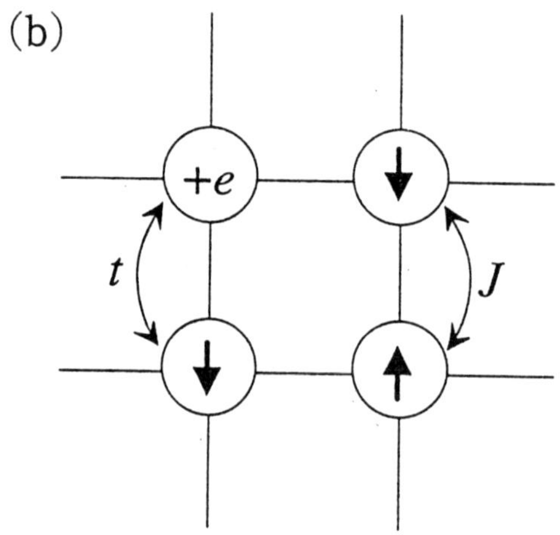

Fig. 2(b) t-J model. The antiferromagnetic exchange interactions between spins are expressed by J and the energy required for the hole and spin to exchange positions is expressed by t.

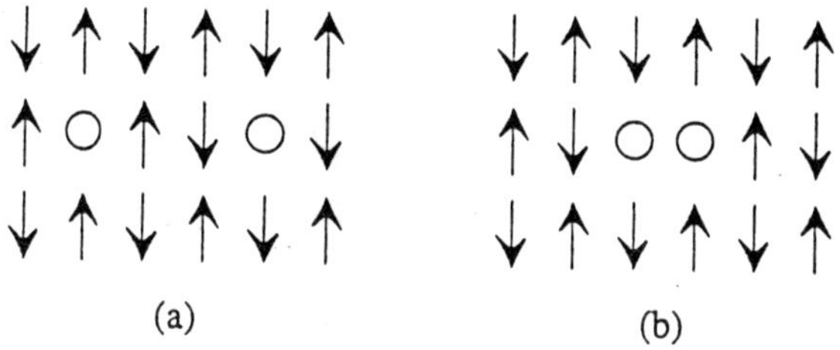

Fig. 3 Holes in a two-dimensional antiferromagnetic body. The arrows show spin and the circles show the holes. (a) Two holes separated. (b) Two holes adjoined.

introduced, they move about with a positive charge in the antiferromagnetic body. The parameter expressing this motion is t.

Let us assume, for simplicity, that t is very small when compared with J. In this case, each hole introduced into the antiferromagnetic body destroys four antiferromagnetic bonds, as shown in Fig. 3. Therefore, the energy is increased by $4J$ per hole. On the contrary, when two holes are adjoined, the erengy is increased by $7J$. Therefore, the difference J=$8J$-$7J$ is the attractive interaction of holes. However, if the kinetic energy of the holes is small, a phase separation with hole domains and spin domains will occur due to this attractive interaction. Superconductivity emerges when the attractive interaction and the kinetic energy are well balanced. Figure 4 is a schematic phase diagram of the two-dimensional t-J model obtained after various physical quantitites were obtained by the exact diagonalization method.[6,7] The abscissa shows the amount of doped holes (n_h). The shadowed domain shows the superconducting state in which Cooper pairs have a $d(x^2-y^2)$ symmetry and form a ground state.

A large J value is thus considered to be a necessary condition for high temperature superconductivity. The value of high-temperature superconductors is extremely large ($J \approx 0.1$eV) when compared to those of other antiferromagnetic materials. This observation can be explained by the small difference in energy levels (Δ) between copper and oxygen, as seen in eq. (1), and the condsiderable overlapping

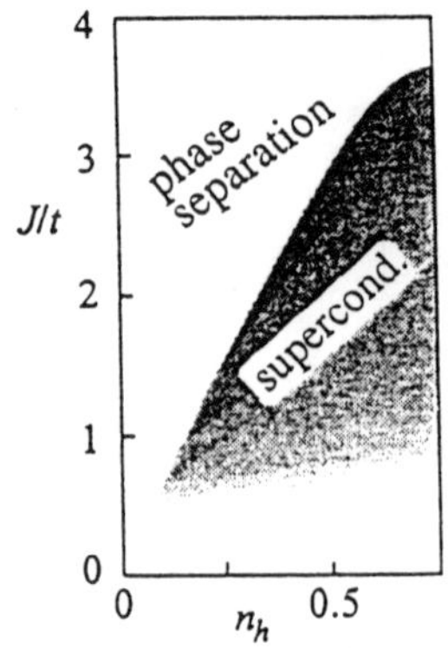

Fig. 4 Schematic phase diagram of the two-dimensional *t-J* model. Abscissa shows the concentration of holes (n_h).

(t_{pd}) of the Cu3$d(x^2-y^2)$ orbit and the O2p orbit. The suppressed emergence of the antiferromagnetic ordered state, despite such a large J value, is attributable to the strong two-dimensionality of the high temperature superconductors and the S=1/2 value of the copper ion's spin, both of which cause the large quantum fluctuatrion.

Note that as the number of holes increases, the number of spins decreases and the antiferromagnetic fluctuation is weakened. Therefore, since antiferromagnetism plays its role on the holes, the amount of doping should not be increased much. The fact that the optimum hole concentration for superconductivity is about $0.15 \sim 0.2$ per copper ion seems to reflect this restriction.

This work was supported by the Grant-in Aid of the Ministry of Education, Science and Culture of Japan and NEDO. This paper is based on my article published in ISTEC Journal Vol.8, No.3 (1995) 20. For a more comprehensive and detailed discussion, we direct readers to this article and references 3 and 7.

References

1. J. Bednorz and K. A. Muller: Z. Phys. B**64**(1986)189.
2. Y. Tokura: *Physics of High-temperarure Superconductors*, eds. S. Maekawa and M. Sato. Vol.**106**, (Springer Series in Solid State Sciecnes, 1991) 191.

3. S. Maekawa, Y. Ohta and T. Tohyama: *Physics of High-temperarure Superconductors*, eds. S. Maekawa and M. Sato, Vol.**106**, (Springer Series in Solid State Sciecnes, 1991) 20.

4. F. C. Zhang and T. M. Rice: Phys. Rev B**37**(1988)3759.

5. T. Tohyama and S. Maekawa: J. Phys. Soc. Jpn. **59**(1990) 1760; H. Eskes, G. A. Sawatzky and L. F. Feiner: Physica C**160**(1989) 424; M. S. Hybertsen, E. B. Stechel, M. Schluter and D. R. Jenison: Phys. Rev. B**41**(1990)11068; H. Fukuyama and H. Matsukawa: *Strong Correlatioon and Superconductivity*, eds. H. Fukuyama, S. Maekawa and A. P. Malozermoff, Vol.**89** (Springer Series in Solid State Sciences, 1989) 45.

6. E. Dagotto and J. Riera: Phys. Rev. Lett. **70**(1993)682.

7. Y. Ohta, T. Shimozato, R. Eder and S. Maekawa: Phys. Rev. Lett. **73** (1994)324.

SYMMETRY OF ORDER-PARAMETERS IN HIGH-T_c LAYERED SUPERCONDUCTORS

A.K. Rajagopal[1] and Sudhanshu S. Jha[2]

[1] *Naval Research Laboratory, Washington D.C. 20375-5320, USA*
[2] *Tata Institute of Fundamental Research, Bombay 400 005, India*

It is well known that the anisotropy and wave-vector dependence of the energy-gap function determine many important properties of a superconductor which are relevant for device applications. Apart from a weak dependence on the wave-vector $\vec{k}$ in the direction perpendicular to the reciprocal layer-plane of a high-T_c layered superconductor, it is shown that anisotropic superconducting order parameters for intra-layer pairing in the class of such materials with orthorhombic crystal structures, can have either pure s-wave like symmetry or mixed d-wave and anisotropic extended s-wave like symmetries in the reciprocal layer-plane. However, in such materials with tetragonal crystal structures, it is possible to have a pure s-wave like symmetry, which may be either isotropic or anisotropic in the layer $\vec{k}$-space, or a pure d-wave like symmetry, as far the $\vec{k}$-dependence in the reciprocal layer plane is concerned. In view of this, some suggestions for analysing experimental data will also be presented.

I. INTRODUCTION

The symmetry and the wave vector dependence of the order parameter, $\Delta(\vec{k})$, of the superconducting state in high-T_c copper oxides have attracted great attention recently[1,2]. From these studies, it is apparent that the $\vec{k}$-dependence of the energy gap function and the corresponding quasiparticle excitation energies are quite complex unlike the conventional low-T_c superconductors. Since most of the physical properties of these materials depend on the detailed form of $\Delta(\vec{k})$, determination of their exact behavior crucially depends on the knowledge of $\Delta(\vec{k})$.

The high-T_c superconductors are layered materials with very weak interlayer coupling of electrons in conducting copper oxide layers. It is often assumed that the superconducting pairing takes place within these quasi-two-dimensional copper oxide layers. The crystal structure of these systems is either tetragonal or orthorhombic. The experimental observations determining the symmetry of $\Delta(\vec{k})$

is still confusing[2]. The recent Josephson-coupling experiments[1,3] **have provided** evidence for a planar d-wave-like symmetry $\left(\left(k_x^2 - k_y^2\right) - \text{type}\right)$. **However, it** has been argued by many workers that there are several **experiments[1] which are** consistent with either an anisotropic extended s-wave-like **symmetry[4] or a mixed** anisotropic s-wave and d-wave-like symmetry.

In this talk, we examine carefully various possible forms of the order **parameters** that may arise from purely crystal structure symmetry of these materials. **In** layered materials with N-layers ($J = 1, 2, \cdots N$) per unit cell, although **there is a** possibility of both intra-layer ($\Delta_{JJ}(\vec{k})$) and inter-layer ($\Delta_{JJ'}(\vec{k})$, $J' \neq J$) pairings, here we will focus our attention only on the intra-layer pairing function, $\Delta_{JJ}(\vec{k})$, for each copper oxide layer in the unit cell.

II. EQUATIONS FOR ORDER PARAMETERS

In the general three-dimensional case for spin singlet Cooper pairing, the generalized BCS equation for the order parameter $\Delta(\vec{k})$ is

$$\Delta(\vec{k}) = -\sum_{\vec{k}'} V(\vec{k}, \vec{k}')\, \Delta(\vec{k}')\, \frac{\tanh(E(\vec{k}')/2k_B T)}{2E(\vec{k}')} \tag{1}$$

where the quasi-particle energy is given by

$$E(\vec{k}) = \left[\xi^2(\vec{k}) + |\Delta(\vec{k})|^2\right]^{1/2}, \quad \xi(\vec{k}) = (\varepsilon(\vec{k}) - \mu), \tag{2}$$

in terms of the single particle band energy, $\varepsilon(\vec{k})$, and chemical potential μ. Near $T = T_c$, the equation for $\Delta(\vec{k})$ is given by the linearized version of Eq. (1) in which $E(\vec{k})$ is replaced by $\xi(\vec{k})$. Here, $V(\vec{k}, \vec{k}')$ represents the effective electron-electron interaction describing the low energy scattering of two electrons of opposite spins in the wave vectors $\vec{k}$ and $-\vec{k}$ to two electrons of wave vectors $\vec{k}'$ and $-\vec{k}'$. For a given crystal structure, the symmetry of $\Delta(\vec{k})$ for a general wave vector $\vec{k}$, is determined by the corresponding point group symmetry of the unit cell. It satisfies the relation

$$\Delta(\vec{k} + \vec{G}) = \Delta(\vec{k}) \tag{3}$$

where $\vec{G}$ is any reciprocal lattice vector. $\Delta(\vec{k})$ can be decomposed [3,5] in terms of the basis functions of the corresponding irreducible representations (IR) of the point group. For spin singlet pairing, only those basis functions which are symmetric under the transformation $\vec{k} \to -\vec{k}$ enter in this expansion. These basis functions for $\Delta(\vec{k})$ can be constructed by using the expansion in terms of the direct lattice vectors, $\vec{R}$,

$$\Delta(\vec{k}) = \sum_{\{\vec{R}\}} \Delta(\vec{R})\, e^{i\vec{k}\cdot\vec{R}} = \sum_{\Gamma} \Delta_\Gamma\, \psi_\Gamma(\vec{k}) \tag{4}$$

where one rearranges the plane waves, $e^{i\vec{k}\cdot\vec{R}}$, for each irreducible representation Γ of the point group. For the effective interactions, $V(\vec{k},\vec{k}')$, which is symmetric with respect to the interchange $\vec{k} \rightleftharpoons \vec{k}'$, one has the general expansion

$$V(\vec{k},\vec{k}') = \sum_{\Gamma,\Gamma'} V_{\Gamma,\Gamma'}\ \psi_\Gamma(\vec{k})\ \psi_{\Gamma'}(\vec{k}'). \tag{5}$$

Near the transition temperature $T \simeq T_c$, where the linearized form of Eq. (1) is applicable, if one assumes that there is only one IR in the expansion for $\Delta(\vec{k})$ in Eq. (4) which gives the maximum T_c, then only the diagonal terms with $\Gamma' = \Gamma$ may be retained in Eq. (5). In such a situation, there is no mixing of the representations for $\Delta(\vec{k})$. However, as T goes below T_c one has the full nonlinear equation, Eq. (1), which reduces the symmetry for the solution $\Delta(\vec{k})$, and may mix the representations.

In layered systems, with only weak coupling between the layers, one may classify the symmetry of the crystal structure and the pairing differently. For each layer, we can use the symmetry of the corresponding unit cell in the x-y plane. In the z-direction for each of the N layers in the unit cell there is a translational symmetry w.r.t transformation $z \to z + pd$, $(p = 0, \pm 1, \pm 2, \cdots)$ where d is the periodicity of the unit cell in the z-direction. The order parameters which determine the quasi-particle spectrum are now labelled by $\Delta_{JJ'}(\vec{k}_t, k_z)$. Here, $\Delta_{JJ}(\vec{k}_t, k_z)$ corresponds to intra-layer pairing in the Jth copper oxide conducting layer in the unit cell, and $\Delta_{JJ'}, J' \neq J$, describes the inter-layer pairing, where $\vec{k}_t$ is the wave vector in the reciprocal layer plane. In this case, the energy gap E_g for quasi-particle excitations at the Fermi surface, e.g., for 2 inequivalent layers per unit cell, is given by

$$E_g = (\Delta_{11} + \Delta_{22}) \pm \left[(\Delta_{11} - \Delta_{22})^2 + 4\Delta_{12}\Delta_{21}\right]^{1/2} \tag{6}$$

For only intra-layer pairings to which we will restrict our present discussion, in the case of N equivalent layers per unit cell, the quasi-particle energy is, however, given by a simple decoupled equation

$$E_J(\vec{k}_t, k_z) = \left[\xi_J^2(\vec{k}_t, k_z) + |\Delta_{JJ}(\vec{k}_t, k_z)|^2\right]^{1/2} \tag{7a}$$

with the energy gap

$$E_g = 2\Delta_{JJ} \tag{7b}$$

where,

$$\xi_J(\vec{k}_t, k_z) \simeq \varepsilon_J(\vec{k}_t) + 2t_3(1 - \cos k_z d) - \mu. \tag{7c}$$

In what follows, we discuss the symmetry and $\vec{k}$-dependence of Δ_{JJ} which we denote by Δ_J.

III. BASIS FUNCTIONS FOR LAYERS IN TETRAGONAL AND ORTHORHOMBIC CRYSTALS

Since high-T_c systems occur as tetragonal or orthorhombic crystals, we now focus our discussion only to corresponding layered systems.

Using[5] the expansion procedure described earlier, one has the following leading order basis functions for different IR's in these two cases.

(A) <u>2-D layer for Tetragonal system (a: square unit cell length in x, y directions)</u>

$$
\begin{aligned}
A_{1g}(\sim 1, k_x^2 + k_y^2) &: 1, (\cos k_x a + \cos k_y a),\ \cos k_x a \cos k_y a, \\
&\quad (\cos 2k_x a + \cos 2k_y a), \cdots \quad (\text{s} - \text{wave like}) \\
B_{1g}(\sim (k_x^2 - k_y^2)) &: (\cos k_x a - \cos k_y a), \\
&\quad (\cos 2k_x a - \cos 2k_y a), \cdots \quad (d_{x^2-y^2}-\text{wave like}) \\
B_{2g}(\sim k_x k_y, (k_x^3 k_y + k_x k_y^3)) &: \sin k_x a \sin k_y a, \\
&\quad (\sin 2k_x a \sin k_y a + \sin k_x a \sin 2k_y a), \cdots \\
&\quad (d_{xy}-\text{wave like}) \\
A_{2g}(\sim k_x k_y (k_x^2 - k_y^2)) &: (\sin 2k_x a \sin k_y a - \sin k_x a \sin 2k_y a), \cdots
\end{aligned}
$$

(B) <u>2-D layer for orthorhombic system (a,b: unit cell dimensions in x, y directions)</u>

$$
\begin{aligned}
\Gamma_{1g}(\sim 1, k_x^2, k_y^2) &: 1, \cos k_x a,\ \cos k_y b,\ \cos k_x a \cos k_y b, \\
&\quad \cos 2k_x a, \cos 2k_y b, \cdots \rightarrow (A_{1g} + B_{1g}) \\
\Gamma_{2g}(\sim k_x k_y, k_x^3 k_y, k_x k_y^3) &: \sin k_x a \sin k_y b,\ \sin 2k_x a \sin k_y b, \\
&\quad \sin k_x a \sin 2k_y b, \cdots \rightarrow (A_{2g} + B_{2g})
\end{aligned}
$$

Note that for the tetragonal case, where one has $\pi/2$ rotation symmetry in the plane, for small $\vec{k}_t$, $B_{1g} \sim (k_x^2 - k_y^2)$ which is characteristic of a $d_{x^2-y^2}$-wave-like symmetry of a spherical system. In this case, A_{1g} is s-wave-like and B_{2g} transforms like $d_{k_x k_y}$. However, for the orthorhombic case, there is no $\pi/2$ rotation symmetry, and so their representations cannot be classified in general in terms of pure s-wave and d-wave-like symmetries. In fact, the basis functions for Γ_{1g} are linear combinations of s- and d-wave-like symmetries (A_{1g} and B_{1g} of case (A)), and Γ_{2g} are linear combinations of B_{2g} and A_{2g} of case (A). Except for the basis function 1 in Γ_{1g} which transforms like an isotropic s-wave, other basis functions have always mixed character when classified in terms of the symmetry of the corresponding tetragonal or spherical systems. It should be noted that even the s-wave-like A_{1g} functions, $\cos k_x a + \cos k_y a$, $\cos k_x a \cos k_y a$, etc., have nodes in the $\vec{k}_t$-plane of the Brillouin zone. For a suitable large size of the Fermi surface, such extended s-wave-like gap functions will exhibit nodes at the Fermi surface.

IV. GAP EQUATION FOR $\Delta_J(\vec{k}_t, k_z)$

Confining our attention to intralayer spin singlet pairing, T_c is determined by the equation

$$\Delta_J(\vec{k}_t, k_z) = -\sum_{J', \vec{k}'_t, k'_z} V_{JJ'}(\vec{k}_t k_z;\ \vec{k}'_t k'_z) \Delta_{J'}(\vec{k}'_t, k'_z) \frac{\tanh \xi_{J'}(\vec{k}'_t, k'_z)/2k_B T_c}{2\xi_{J'}(\vec{k}'_t, k'_z)}. \tag{8}$$

The solutions of this equation have been analyzed by us earlier[6] using the usual approximation in which the interactions are assumed to be relevant in a narrow energy band near the Fermi surface (FS) so that $\vec{k}, \vec{k}'$ lie near the FS. By going out to the variables ξ_J, ϕ, k_z, where $k_x = k_t \cos\phi$, $k_y = k_t \sin\phi$, $\xi_J = \xi_J(k_t, \phi, k_z)$, and performing the energy integral at FS, one then obtains a simplified equation for the gap function, $\tilde{\Delta}_J(\phi, k_z)$, which now depends only on ϕ and k_z. In this case, the effective interaction enters in the form of coupling constants, $\lambda_{JJ'}(\phi k_z; \phi' k'_z) = [N_J(0)]^{1/2}\ V_{JJ'}(\phi k_z, \phi' k'_z)[N_{J'}(0)]^{1/2}$, where $N_J(0)$ is the FS density of states in layer J. To leading orders, our analysis shows that

$$\begin{aligned} \lambda_{JJ'}(\phi k_z; \phi' k'_z) &= \lambda^{(0)}_{JJ'}(\phi; \phi') + \lambda^{(1)}_{JJ'}(\phi; \phi')\ \cos k_z d \cos k'_z d \\ &\quad + \lambda^{(2)}_{JJ'}(\phi; \phi')(\cos k_z d + \cos k'_z d) \end{aligned} \tag{9a}$$

where for $\alpha = 0, 1, 2$,

$$\begin{aligned} \lambda^{(\alpha)}_{JJ'}(\phi, \phi') &= \lambda^{(\alpha)0s}_{JJ'} + \lambda^{(\alpha)1s}_{JJ'}\ \cos 4\phi\ \cos 4\phi' \\ &\quad + \lambda^{(\alpha)2s}_{JJ'}(\cos 4\phi + \cos 4\phi') + \lambda^{(\alpha)1d}_{JJ'}\ \cos 2\phi\ \cos 2\phi' \\ &\quad + \lambda^{(\alpha)1d'}_{JJ'} \sin 2\phi \sin 2\phi' + \lambda^{(\alpha)2d}_{JJ'}\ (\cos 2\phi + \cos 2\phi') \\ &\quad + \lambda^{(\alpha)2d'}_{JJ'}(\sin 2\phi + \sin 2\phi') \end{aligned} \tag{9b}$$

for the orthorhombic unit cell symmetry. For the tetragonal case, one has invariance under simultaneous $\pi/2$ rotations of ϕ, ϕ', which implies, $\lambda^{(\alpha)2d}_{JJ'} = \lambda^{(\alpha)2d'}_{JJ'} = 0$. In the tetragonal case, the first three terms lead to s-wave-like order parameters, the fourth term with $\cos 2\phi \cos 2\phi'$ term leads to $d_{x^2-y^2}$-type symmetry, and the fifth term with $\sin 2\phi \sin 2\phi'$ term leads to d_{xy}-type symmetry. However, in the orthorhombic system, because of the presence of the last two terms in Eq. (9), which break the invariance under simultaneous $\pi/2$-rotation of ϕ, ϕ', one always obtains either pure s-wave like symmetry or mixed d-wave and s-wave like symmetries.

V. MAIN RESULTS AND DISCUSSION

When all the terms in Eq. (9b) are present, the general solution of the equation for the order parameter has the form

$$\begin{aligned}\tilde{\Delta}_J(\phi, k_z) &= C_1 + C_2 \cos 4\phi + C_3 \cos 2\phi \\ &\quad + C_4 \cos k_z d + C_5 \cos 4\phi \cos k_z d + C_6 \cos 2\phi \cos k_z d \end{aligned} \tag{10a}$$

or

$$\begin{aligned}\tilde{\Delta}_J(\phi, k_z) &= C_1' + C_2' \cos 4\phi + C_3' \sin 2\phi \\ &\quad + C_4' \cos k_z d + C_5' \cos 4\phi \cos k_z d + C_6' \sin 2\phi \cos k_z d \end{aligned} \tag{10b}$$

where C's and C''s are constants, independent of k_z and ϕ. This shows that the general solutions in the orthorhombic case have mixed s- and d-wave type planar symmetries with terms which are either multiplied by $\cos k_z d$ or by 1. Thus, there are terms which have anisotropy in reciprocal lattice (k_x, k_y) plane together with anisotropy containing the k_z-direction.

For the tetragonal case, it is found that planar d-wave solutions containing terms $\cos 2\phi, \cos 2\phi \cos k_z d$, or $\sin 2\phi, \sin 2\phi \cos k_z d$ do not mix with the planar s-wave type (first four terms in Eq. (10a) or (10b)). In such a case, because of the orthogonality of $\sin 2\phi$ and $\cos 2\phi$ terms, it is possible to have pure d-wave-type of superconductivity of the form which is either like $d_{x^2-y^2}(\sim \cos 2\phi)$ or like $d_{xy}(\sim \sin 2\phi)$. In many experiments on high-T_c superconductors, the evidence seems to be for the form $d_{x^2-y^2}$-like symmetry. It is clear that a variety of scenarios can be constructed[6] depending on the choice of the coefficients of interaction constants in Eq. (9). Note that the extended s-wave part of the order parameter proportional to $\cos 4\phi$ will exhibit nodes at the FS. The confusion in interpreting experimental results may perhaps be understood by appreciating the difficulties involved in distinguishing nodes arising from anisotropic s-wave or from $d_{x^2-y^2}$-wave type symmetries.

For orthorhombic systems, it of course makes no sense in talking about exact $d_{x^2-y^2}$-wave-like symmetry. In such a case, without any bias, experimental results should be fitted with the form

$$\Delta_J(\vec{k}) = \Delta_J^{(0)}(\vec{k}_t) + \Delta_J^{(1)}(\vec{k}_t)\cos k_z d \tag{11a}$$

$$\Delta_J^{(0),(1)}(\vec{k}_t) = C_s\ \psi_s(k_x, k_y) + C_a\ \psi_d(k_x, k_y) \tag{11b}$$

where ψ_s and ψ_d are suitably normalized basis functions, such that ψ_s does not change sign under the transformation, $k_x \to k_y$, $k_y \to -k_x$ (i.e., under $\pi/2$ rotation), whereas ψ_d changes sign under this transformation. Both ψ_s and ψ_d are symmetric under the transformation $\vec{k}_t \to -\vec{k}_t$. For layered systems corresponding to orthorhombic unit cell (rectangular 2-d unit cell), one expects that either $C_s \neq 0,\ C_a = 0$ or $C_s \neq 0,\ C_a \neq 0$. For tetragonal systems, at least near T $\sim$ T_c, one may have $C_s \neq 0,\ C_a = 0$ or $C_s = 0,\ C_a \neq 0$. However, in a more general analysis which takes into account interlayer pairing and/or considers nonlinear equation for order-parameters below T_c, the situation may be more complicated even in tetragonal system.

In conclusion, it may be emphasized that d-wave like symmetry claimed in some high-T_c orthorhombic materials can at best be approximate. The symmetry behavior in all high-T_c materials, in fact, need not be universal.

ACKNOWLEDGEMENT

AKR is supported in part by the office of Naval Research. The support of this collaboration by TIFR is appreciated. More importantly, the warm hospitality of Professor K.G. Prasad and Sudha Prasad is deeply appreciated.

REFERENCES

1. J.R. Kirtley, C.C. Tsuei, J.Z. Sun, C.C. Chi, L.S. Yu-Jhanes, A. Gupta, M. Rupp and M.B. Ketchen, *Nature* **373**, 225 (1995); J.H. Miller Jr., Q.Y. Ying, Z.G. Zon, N.Q. Fan, J.H. Xu, M.F. Davis and J.C. Wolfe, *Phys. Rev. Lett.* **74**, 2347 (1995). For other experimental papers, see *J. Phys. Chem. Solids* **56**, (1995) and references in B. Goss Levy, *Phys. Today* **49** (1), 19 (1996). See also H. Hilgenkamp, in this volume.

2. D. Pines and P. Monthoux, *J. Phys. Chem. Solids* **56**, 1651 (1995); C.M. Varma, *ibid*, 1685 (1995).

3. D.J. Scalapino, *Phys. Rep.* **250**, 329 (1995). See also, M. Sigrist and K. Ueda, *Rev. Mod. Phys.* **63**, 239 (1991).

4. See, e.g., G. Deutscher, in this volume.

5. S.S. Jha and A.K. Rajagopal (to be published).

6. A.K. Rajagopal and S.S. Jha, *Phys. Rev. B* **54**, 4331 (1996).

INTERPLAY OF STRUCTURAL TRANSITION AND SUPERCONDUCTIVITY IN CUPRATES

Haranath Ghosh[1], Manidipa Mitra[1], S.N. Behera[1] and S.K. Ghatak[2]
[1]Institute of Physics, Sachivalaya Marg, Bhubaneswar - 751 005, INDIA
[2]Department of Physics and Meterology, I.I.T. Kharagpur- 721302, INDIA

The presence of lattice distortion is known to suppress the superconducting (SC) transition in the cuprates [1]. It is now accepted that electron correlation plays a dominant role in shaping the properties of these undoped and doped systems. Furthermore, since the Fermi level in these systems lies in a degenerate band of $Cu : d$ and $O : p$ orbitals the structural transition can be modeled as a band Jahn-Teller effect. We study the coexistence of superconductivity and band Jahn-Teller (J-T) distortion [2], taking into account the electron correlation within the slave boson formalism. It is shown that with increasing dopant concentration (δ), the structural transition temperature (T_s) remains constant upto a certain value and then vanishes, while the SC transition temperature (T_c) increases to a maximum value. The highest value of T_c corresponds to that value of δ where T_s vanishes. Besides with increasing lattice distortion superconductivity is suppressed. These findings are in qualitative agreement with the experimental results.

Keywords : Electron-correlation, J-T distortion, degenerate band, Cuprates

The dependence of T_c on structural properties is less understood and had been a crucial question since the discovery of high-T_c superconductivity [1]. On the other hand, effect of pronounced electronic correlation in the normal state properties of the high-T_c cuprates have been realized [2]. The structural transition associated with the lowering of crystallographic symmetry has been observed in many intermetallic compounds [3,4] as well as in high T_c cuprates [5,6] and fullerenes [7,8]. A common feature to all these systems is the existence of Fermi level (FL) lying within a orbitally degenerate band, hence are prone to undergo band Jahn-Teller interaction. It is expected that the band Jahn-Teller (BJT) distortion would strongly influence the superconductivity in such a system because with such change there will be an associated change in the electronic density of states (DOS) around the Fermi level (FL). If the FL lies at the centre of the degenerate band then the DOS at the FL will be reduced drastically and the superconducting (SC) transition temperature (T_c) will be reduced. On the other hand, T_c can be higher as phonon softening due to BJT effect may increase the coupling constant which is inversely proportional to mean square phonon energy.

All high-T_c cuprates have perovskite type crystal structure, where oxygen octahedra is centred around the J-T active Cu^{++} ions. In fact, Bednorz and Müller [1] in their historic paper conjectured that a J-T type of polaron formation may lead to high T_c. Indeed experiments indicate that in $(La_{1-x}A_x)_2CuO_4$ ($A = Ca, Ba, Sr$) with onset of T_c around 38K, each CuO_6 octahedron is J-T distorted with interlayer $Cu-O$ distance equal to 1.9A^0 and on elongated $Cu-O$ distance of 2.4A^0. A detailed neutron scattering measurement of the temperature dependence of the spontaneous strain has been reported by McK. Paul et al. [9] for $La_{1.85}Ba_{0.15}CuO_4$ ($T_c \approx 38$ K). The system exhibits a structural transition at 180K ; the magnitude of the strain rises on lowering temperature and shows an anamolous suppression below 75K. Although the suppression of the lattice distortion appears at a temperature higher than T_c, it is expected that this anamoly is associated with the appearence of superconductivity. Such structural transition has also been observed in other high T_c compounds and its strong influence being correlated with the SC state by many authors [10]. The evidence for such interplay between lattice distortion and superconductivity in $La_{1.85}Sr_{0.15}CuO_4$, $YBa_2Cu_3O_{6.5}$ and even the electron doped cuprate system $Nd_{1.85}Ce_{0.15}CuO_{4-\delta}$ has been reported by Lang et al., [11] from thermal expansion measurements. These measurements, while demonstrating the insensitivity of structural distortion to charge carriers doping clearly exhibit that the highest T_c is limited by lattice instability. Furthermore, the suppression of lattice distortion in the SC state has also been seen from extended absorption of X-ray fine structure edge measurements in the other families of cuprates e.g., the '2212' and '2223' Bi-superconductors [12]. Therefore, the well known structural transition in cuprates from the tetragonal to orthorhombic phase at a higher temperature and appearence of superconductivity at a lower temperature in the orthorhombic phase is likely to have a close impact with the BJT distortion and its interplay with the occurence of superconductivity.

The electron correlation is known to play a significant role in shaping the physical properties of the normal and SC properties of the cuprates. It is believed that it is the single band Hubbard model that contains the necessary physical ingredients to describe the $Cu-O$ systems. One of the recently contrived methods to solve the Hubbard model for any filling and any value of correlation (U) is due to slave - boson technique. The slave - boson formulation of Kotliar and Ruckenstein (KR) [13] introduces four auxilliary boson fields corresponding to the occupancy of a site, namely empty (e_i), doubly occupied (d_i), singly occupied ($p_{i\sigma}/\ p_{i-\sigma}$) with spin $\pm\sigma$ respectively. In terms of the slave boson field operators the single band Hubbard model takes the form

$$H = \sum_{ij,\sigma} t_{ij} z_{i\sigma}^{\dagger} c_{i\sigma}^{\dagger} c_{j\sigma} z_{j\sigma} + U\sum_{i} d_i^{\dagger} d_i - \mu \sum_{i\sigma} c_{i\sigma}^{\dagger} c_{i\sigma} + \sum_{i} \lambda_i (1 - e_i^{\dagger} e_i$$

$$-d_i^{\dagger} d_i - \sum_{\sigma} p_{i\sigma}^{\dagger} p_{i\sigma}) + \sum_{i\sigma} \lambda'_{i\sigma} (c_{i\sigma}^{\dagger} c_{i\sigma} - d_i^{\dagger} d_i - p_{i\sigma}^{\dagger} p_{i\sigma}) \tag{1}$$

where $z_{i\sigma} = (1-d_i^{\dagger} d_i - p_{i\sigma}^{\dagger} p_{i\sigma})^{-1/2} (e_i^{\dagger} p_{i\sigma} + p_{i-\sigma}^{\dagger} d_i)(1 - e_i^{\dagger} e_i - p_{i-\sigma}^{\dagger} p_{i-\sigma})^{-1/2}$ and the Boson field operators are not independent of each other but are constrained by the requirements of completeness and local charge conservation at a site, hence the fourth and fifth terms are added in the Hamiltonian with the help of Lagrange multipliers (λ_i and λ'_i). The values of the boson field operators and Lagrange multipliers are determined by minimizing the free energy in the saddle point approximation. This approximation diagonalizes the U term whereas renormalizes the hopping term as $\tilde{q}t_{ij}$ with $\tilde{q} = z^{\dagger}z$; $\tilde{q}$ usually being complicated function of u $(= \frac{U}{U_c})$, dopant concentration (δ) and $x(= e + d)$ carries the necessary information of electron correlation. In this approach solutions are obtained for the paramagnatic state for all values of u and band fillings [14] that reproduces the correct Brinkman-Rice result for metal-insulator transition (U_c) at half-filling. Approximately, in the weak correlation limit $\tilde{q} = 1 - u^2$ and in case of strong correlation and small values of δ, $\tilde{q} = \frac{2\delta}{\sqrt{1-u^{-1}}}$ (for details please see ref. [13,14]).

We consider a model system in which the electrons are in a two-fold degenerate e_g band and interact with the lattice as well as between themselves, the later via a BCS kind of pairing interaction mediated by some boson exchange. The Hamiltonian for such a system may be written as,

$$H = \sum_{k\sigma\alpha=1,2} [\{\tilde{q}\epsilon_k - \mu + 2U' < n_\beta >\} n_{k\sigma\alpha} - (-1)^{\alpha} Gen_{k\sigma\alpha}] + \frac{1}{2}ce^2 + \sum_{k,\alpha=1,2} (\Delta c_{k,\alpha,\uparrow}^{\dagger} c_{-k,\alpha,\downarrow}^{\dagger} + h.c.)$$

$$+bosonic\ terms \tag{2}$$

The first term corresponds to band-energy of the correlated electrons in a two-fold orbitally degenerate e_g band ; where U' is the strength of the inter-orbital Coulomb correlation treated within the mean field approximation. The second term corresponds to the symmetry breaking electron - lattice interaction that lifts the degeneracy by creating population difference and hence is responsible for strain (e). The cost in lattice energy due to such strain is given by the third term whereas the fourth term represents the usual BCS term for cooper pairs in which electron pairing is assumed to be restricted only within each subband and the strength of pairing interaction being of equal magnitudes, for simplicity [15]. The bosonic terms in the Hamiltonian does not contribute anything to the dynamics of the system. In the normal undistorted phase, it is possible to show that the strain is exactly proportional to the population difference between the shifted subbands and the selfconsistent gap equations for strain (e), SC-order parameter(Δ) are given below,

$$e = -\sum_{k,\sigma,\alpha} (-1)^{\alpha} n_{k\sigma\alpha} = -\frac{G}{c} \sum_{k,\sigma} (-1)^{\alpha} \frac{\varepsilon_\alpha(k)}{E_\alpha(k)} \tanh \beta \frac{E_\alpha(k)}{2} \tag{3}$$

$$\Delta = -V\tilde{q}^2 \sum_{k,\alpha} < c_{k\alpha\uparrow} c_{-k\alpha\downarrow} > = \sum_{k,\alpha} V\tilde{q}^2 \frac{\Delta}{2E_\alpha(k)} \tanh \frac{\beta E_\alpha(k)}{2} \tag{4}$$

where $\varepsilon_\alpha(k) = \tilde{q}\epsilon_k - \mu + 2U' < n_\beta > -(-1)^\alpha Ge$ is the band energy of the correlated normal state in the distorted phase where $E_\alpha(k) = \sqrt{\varepsilon_\alpha^2(k) + \Delta^2}$ is the SC quasiparticle energy in the subband α.

Now, inorder to study the mutual influence of lattice distortion and superconductivity as a function of band filling one has to solve the above two coupled equations (3-4) self-consistently together with the equation for dopant concentration which can be obtained as, $(1-\delta) = \frac{1}{N}\sum_{k\sigma\alpha} < n_{k\sigma\alpha} >$, the number conservation equation.

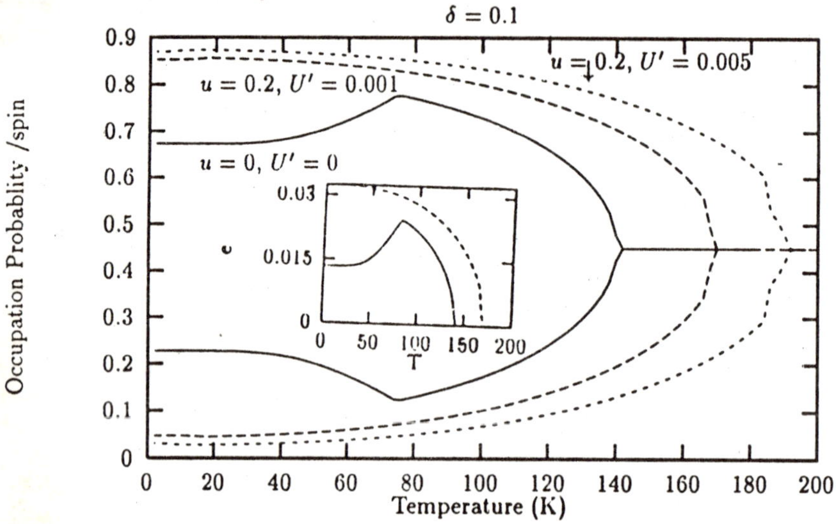

Fig. 1 Temperature variation of the occupation probablity per spin for different electronic correlations. The inset figure shows the temperature variation of the spontaneous strain for $u = 0$ (solid line) and $u = 0.2, U' = 0.001$ (dashed curve). The strain is higher in the later case.

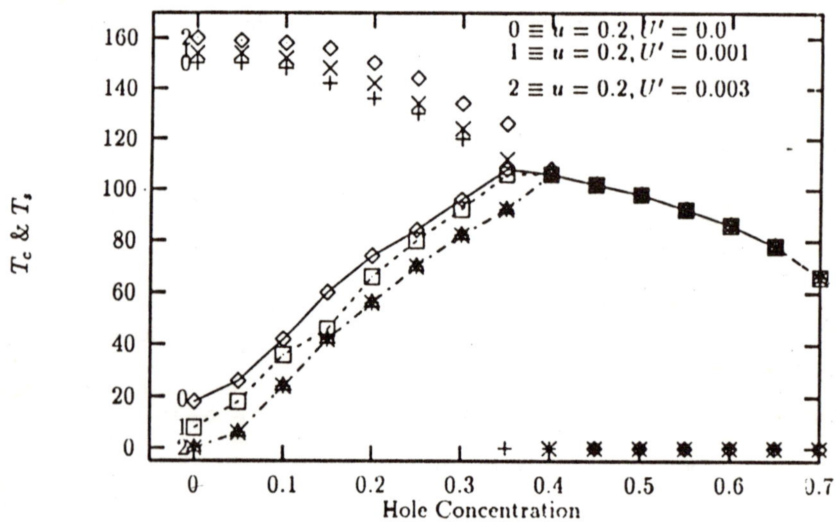

Fig. 2 The transition temperatures for the SC and the structural transitions as a function of hole concentration (δ) for different electronic correlations (the solid line represents T_c curves).

In figure 1, the occupation probablity/per spin for different orbitals are plotted as a function of Temperature. To be noted, the occupation probablity of the two orbitals is same above the structural transition temperature T_s. Below T_s, the occupation probablities being different creates a population diference between the orbitals and hence a net strain builds up. With lowering in temperature, the system undergoes SC transition and hence the orbital density is suppressed below the T_c. Needless to say, this transition resembles to the cubic to tetragonal transition in A-15 as well as in high-T_c systems. However, as a result of reduction in the orbital population density below T_c, the growth of the spontaneous strain gets arrested which is also seen in experiments [9] (cf inset fig. 1).

In figure 2, the detail phase diagram comprising of T_c & T_s as a function of hole concentration is shown for different values of intra and inter orbital correlation strength. The phase diagram (solid lines represent T_c curves) demonstrates that the T_c is largely suppresed in presence of structural distortion and the highest T_c is obtainable for an optimum doping (δ_c) where T_s vanishes. These results are in close agreement with experimental observations [11,17].

Finally, the elctron correlation within the present treatment have the following impacts : (i) band narrowing effect similar to the Gutzwiller approximation (ii) such band narrowing (enhances the density of states) enhances the Jahn - Teller splitting (iii) the pairing amplitude will suffer suppression due to correlation by a factor $\tilde{q}^2$ and also due to the presence of Jahn-Teller interaction. (iv) Inclusion of inter orbital coulomb correlation makes the J-T spliting asymmetric with respect to the centre of mass of the original band and hence enhances the lattice strain but suppresses SC further. Therefore, the present model study describes an extremely interesting phenomena regarding the interplay of structural transition with superconductivity in correlated systems like the cuprates. However, how the antiferromagnetic spin fluctuation leading to d-wave pairing superconductivity (which seems to be present in cuprates) is influenced by structural distortion still remains to be seen.

REFERENCES

[1] Bednorz, J. G., and Müller, K. A., Z. Phys. **B 64** 189(1986).

[2] Anderson, P. W., Science **235**,1196(1987).

[3] Testardi, L. R. et al., Rev. Mod. Phys. **43**, 657(1975).

[4] Suites, J. C., Sol. State. Commn., **18**, 423(1976).

[5] Axe, J. D., et al., Phys. Rev. Lett. **62**, 27551(1989) ; Takagi, T., et al., Phys. Rev. Lett. **68**, 3777(1992).

[6] Sato, M., in "*Physics of high temperature superconductivity*"; (Springer Verlag, Berlin Heidelberg) p. 239 (1992).

[7] Lannoo, M., Baraff, G. A., Schluter, M. and Tomanek, D., Phys. Rev. **B 44**, 12106(1991).

[8] Weaver, J. H., et al., Phys. Rev. Lett., **66**, 1741(1991).

[9] Paul, D. Mck, et al., Phys. Rev Lett. **58**, 1976(1987).

[10] Jargensen et al., Phys. Rev. **B 36**, 573(1987) ; Chen Wei et al., Solid State Commun. **63**, 997(1987).

[11] Lang, M., et al., Phys. Rev. Lett. **69**, 482 (1992).

[12] Rohler, J. and Larsich, A., in "*Electronic Properties of high-T_c superconductors and related compounds*" ; (Springer Verlag, Berlin)vol. **99**, p. 152, Editors : Kuzmamy, H., Ching, M. and Fink, J.

[13] Kotliar, G. and Ruckenstein, A. E., Phys. Rev. Lett. **57**, 1362(1986) ; Lavagna, M. Phys. Rev. **B 41**, 142(1990) and Entel, P., Behera. S. N., Zielinski, J. and Kaufmann, E., Int. Jr. Mod. Phys. **B 5**, 271(1991).

[14] Mitra, M., and Behera., S. N., Cond. Matt. Mat. Commun. **1**, 247(1994).

[15] Behera., S. N., Ghosh Haranath, Ghatak., S. K. and Ray., D. K., Ind. Jr. Phys. (1996) to appear. Also see for details, Ghosh Haranath et al., IP/BBSR/95-68 (and also communicated to Physica C (1996)) ; Haranath Ghosh, Ph. D. Thesis.

[16] Ghosh Haranath, Mitra Manidipa, Behera, S. N., Ghatak, S. K., and Ray, D. K., Unpublished calculation.

[17] Dabrowski, B., et al., Phys. Rev. Lett. **76**(8),1348 (1996).

ASMCCD '96 : SUMMARY

R. Srinivasan
Raman Research Institute, Bangalore 560 080, India

Mr. Chairman and friends,

Let me first thank Professors Malik and Pinto of the TIFR for inviting me to participate in this ASMCCD Symposium. I enjoyed attending all the lectures which were very informative.

It is a matter of pride for all the Indian participants that the TIFR is completing its fiftieth year of dedicated service to science. As a premier scientific institution in the country it has served as a role model for other institutions in India. On my behalf and on behalf of all the participants let me convey our good wishes to the scientists of TIFR for many more years of distinguished service in the cause of science.

I have been given the formidable task of summarising the Symposium. There were 37 invited talks, 8 oral presentations and 88 poster papers. I attempted to listen to all the lectures and look at some of the posters. Considering the range of topics covered and my limited knowledge of the subject, I have misgivings about rendering justice to the task devolved on me. If I fail in my task I may kindly be excused.

On new materials and their properties there were six talks and thirty two posters. Prof. L.C. Gupta gave a brief survey of the preparation, structure and properties of the quaternary borocarbides. The first report of superconductivity, in these materials came from the TIFR. There are several interesting questions that need to be answered in these materials. Does nickel have a magnetic moment? A recent neutron scattering study appears to indicate a moment of about 0.3 Bohr magneton on nickel. This needs to be confirmed. If a magnetic moment indeed is present on nickel, one may have to review all our ideas about the origin of superconductivity in these materials. In $DyNi_2B_2C$ the Neel temperature is 11K while the superconducting transition temperature is around 6K. This implies that superconductivity develops in an ordered antiferomgnetic structure. When superconductivity sets in will the internal field due to the Dy moments decrease due to the superconducting shielding currents? Doping with a non-magnetic material like Lutetium causes a rapid drop in T_c. Is this drop due to disorder induced by doping? The rather large (5 K) interval between T_N and T_c in this material provides a convenient window to study many interesting properties with sufficient temperature resolution. The double re-entrant superconducting transition in the holmium nickel boride occurs in a very narrow range of compositions. What is the reason for such behaviour? It appears to me that a lot of new physics may come out of an intensive study of these quaternary borocarbides and I believe that more groups in this country should be involved in such studies.

When superconductivity was first discovered in YBCO there were hopes that new materials with higher and higher T_c values will be discovered soon. This hope has been realised and we have now a profusion of cuprate superconductors with transition temperatures ranging up to 133 K under ambient pressure and up to 160 K under high pressure. I often get confused with the chemical formulae for the various compounds. In all these compounds there are blocks of charge reservoirs and blocks of superconducting layers. Perovskite-like and rock salt-like layers are organised in a certain sequence. Dr. Yamauchi explained the nomenclature which can be conveniently used to classify the structures of these materials and to clarify the role of the various blocks. He also laid down certain empirical rules for achieving the high T_c and reported on the preparation of several carbonate cuprates with transition temperature in excess of 100 K.

Dr. Tholence described the preparation and properties of the various homologous mercury compounds both as sintered pellets and as single crystals and reported on their properties. There is an optimal oxygen doping to achieve maximum T_c in these compounds. But an application of pressure enhances the transition temperature of even the optimally doped compound. Is it possible to obtain a higher transition temperature in these compounds by chemical pressure resulting from doping with suitable dopants? In Hg 1223 compound a critical current density of 10^7 amp/cm^2 at 77 K has been achieved. Can one realise high current applications using this material? I hope that within the next couple of years we find new materials with a T_C in excess of 200 K.

The role of Praseodymium in reducing the transition temperature of YBCO appeared to me to have been well studied, though a clear understanding of the mechanism of T_c reduction still appears to elude us. The new results presented by Dr. Suryanarayanan on the difference between oxygen annealed and argon-oxygen annealed samples of YBCO and YSrBaCuO doped with Pr made the issue more intriguing.

In the field of materials processing the techniques of quenched melt growth and melt texturing have led to YBCO with large critical current densities. Dr. Murakami showed a Japanese girl being levitated using the material prepared by him. Oxygen controlled melt growth appears to be a new repertoire in the techniques developed by Dr. Murakami and his colleagues. With this technique they have prepared Nd 123 compounds with a high J_c at high magnetic fields. Some of the underdoped Nd 123 material turns normal on the application of the field and provides new pinning centres in the rest of the material increasing J_c at high fields.

It is difficult to prepare long polycrystalline specimens of YBCO to carry large critical current densities. This is due in part to the presence of liquid phase at the grain boundaries during growth. By using pellets of 211 material to absorb the liquid phase Dr. Salama has been able to grow 10 cm long polycrystalline rods of YBCO carrying current upwards of a 1000 Amp in a cross section of 32 mm^2.

Dr. Pawar described deposition of thallium barium and copper layers sequentially on a silver substrate by electrochemical deposition technique and oxidising the material at ambient temperature to prepare superconducting compounds. Electrochemical deposition techniques offer a convenient cheap way of preparing large lengths/ areas of superconducting material. I believe this technique should receive more attention in this country.

Prof. Pinto showed that the addition of silver and gold on laser ablated films of YBCO had certain beneficial effects. Silver addition produces bigger grains and increases J_c. On the other hand addition of gold appears to stabilise the films.

Using a single target on axis biased RF sputtering technique Dr. K.V. Rao has deposited multilayers of YBCO/$Y_4B_3O_9$ on lanthanum aluminate substrate. The critical currents in these multilayers varied quadratically with temperature and decreased exponentially as the thickness of the oxide layer was increased. He has also prepared heterojunctions of YBCO with ferroelectric materials using pulsed laser ablation. But for want of time he was not able to describe the properties of such heterostructures in any detail.

In a delightful evening talk, Prof. Venkatesan described commercial applications of HTS films in cellular and mobile telephone applications and in scanning SQUID magnetic microscopy. He described work in his laboratory on growing heterostructures of YBCO and dielectric materials for microwave filters and heterostructures of YBCO and ferrites for microwave circulators.

In India there are a few laboratories in which high quality thin films of YBCO have been made. I think that these laboratories should grow such heterostructures and study their properties.

Let us now come to theory. NMR, optical conductivity and specific heat experiments appear to point to the opening of a pseudo gap in the low energy excitation spectrum of the material *in the normal state.* Dr. Randeria described a quantum Monte Carlo calculation on the attractive Hubbard model which gave a spin susceptibility decreasing with decreasing temperature. Such a behaviour is a signature of the pseudo-gap. He showed the results of recent angle resolved photo-emission experiments on single crystals of YBCO with different levels of doping. These experiments seem to suggest that as the material becomes more under-doped, the higher the temperature at which the pseudo-gap appears. I wonder whether tunneling experiments will provide a more direct evidence for such a pseudo gap in these materials.

In layered cuprates superconducting copper oxygen planes are Josephson coupled along the c-axis. In a micron size epitaxial film, there will be about 500-1000 such junctions. Convincing experimental evidence from DC and AC Josephson effects was presented by Dr. Muller to show that these junctions are phase locked one to the other. From the Shapiro steps in Josephson tunneling in the presence of microwaves, it was concluded that the order parameter has a substantial s wave component. This brings us to the question of the symmetry of the order parameter in these materials. There are some experiments to indicate that the order parameter has $d_{x^2-y^2}$ symmetry while there are other experiments which are interpreted on an admixture of s wave and d wave symmetry. Dr. Rajagopal pointed out that in an orthorhombic structure such an admixture is allowed while in a tetragonal structure the order parameter can be either s wave or $d_{x^2-y^2}$. Since these results have been obtained on group theoretical arguments they should have general validity and should be taken note of in interpreting experimental data.

Irrespective of the symmetry of the order parameter, the anisotropy of the superconducting gap parameter is a consequence of the weak screening of electrons in these quasi two dimensional materials according to Dr. Deutscher. In the gap equation only gap parameters in the immediate vicinity of a point in k space are coupled to one another and this will lead to a large gap anisotropy. One consequence of a large gap anisotropy is the suppression of the superconducting order parameter at an interface. A weak link is produced even when there is no physical barrier at the interface.

If the order parameter has indeed $d_{x^2-y^2}$ symmetry and the grain boundary is faceted, then along the length of the grain boundary there will be a distribution of zero and pi junctions. The Josephson currents at the two junctions will be oppositely directed causing an overall reduction in the Josephson current through the grain boundaries. Circulating currents at neighbouring zero and pi junctions will lead to a trapped magnetic flux even in the absence of an externally applied current or magnetic field. In addition, the suppression of the order parameter at the interface will play a part. The implications of all these effects for device applications were highlighted by Dr. Hilgenkamp. I think the question of the symmetry of the order parameter in the different cuprate superconductors is not completely settled in an unambiguous fashion.

We now come to flux pinning and flux flow in the presence of a current. In a type II superconductor, when the applied field exceeds H_{c1}, magnetic flux penetrates the material in the form of quantized vortices. The HTS materials consist of superconducting CuO_2 planes which are Josephson coupled to one another. In such a situation in each layer there are two dimensional pancake vortices and the vortices in different planes are connected by Josephson strings. If the Josephson coupling is absent then the vortices interact among themselves only through magnetic interaction both within a plane and between the planes. The intra-plane interaction is repulsive while the inter-plane interaction is attractive. The latter aligns the pancakes one above the other along the c-axis at T = 0. As the temperature increases the pancake vortices in a vertical stack are subjected to random thermal displacements. When the mean square displacement diverges the vertical stack of vortices break up. The behaviour of such pancake vortices in a finite set of layers when the top and bottom layers carry current was discussed by Dr. Clem.

The vortices are normally weakly pinned. The pinning strength can be increased by various techniques such as precipitating impurity phases or creating defects by neutron or heavy ion irradiation. At low temperatures the vortices form an ordered array called the vortex lattice. As the temperature approaches T_c, the lattice will melt to form a vortex liquid. When melting takes place the irreversibility in magnetisation as a function of the applied field, which is a characteristic of the pinned vortex lattice, will disappear at a critical magnetic field, H_{irr}. Just as H_{c1} as a function of temperature marks a boundary between the Meissner shielded phase and the flux lattice phase, a plot of the irreversibility magnetic field, H_{irr}, as a function of temperature marks a phase boundary between the vortex lattice and the vortex liquid states. The phase boundary between the vortex liquid and the normal state is marked by H_{c2} as a function of temperature.

In the presence of a transverse current, the vortices experience a force. When the force exceeds the pinning force the vortices start moving perpendicular to the direction of the vortex and direction of current flow. One can form a constricted channel through which vortices can move by having a weakly pinned region sandwiched between two strongly pinned regions. At low temperature the vortex lattice moves as a whole and one can measure the mobility of the vortex lattice. This mobility shows a multi-peaked structure, the peaks occurring whenever the effective width of the channel is commensurate with the lattice parameter of the flux lattice. In the disordered regime, the mobility does not show such peaked structure. By combining these measurements with transport measurements one can study the transition from the lattice to a liquid state through an intermediate plastic flow regime. The viscosity of the flowing vortices can be determined as a function of temperature. There is a prediction that from the vortex lattice phase one first observes a transition to a hexatic phase and subsequently to the vortex liquid phase. A small hump in the viscosity versus temperature is taken to be an indication of such a transition. These results were discussed by Dr. Kes.

In an extremely anisotropic superconductor like 2H-$NbSe_2$ the pinning is very weak. The rigidity of the lattice depends on the applied magnetic field. This is used to explore the transition from the elastic flow to plastic flow to fluid flow as the lattice softens with increase in temperature. The plastic flow regime has a characteristic finger print in the differential resistance as a function of current. In the plastic flow region large chunks of flux lines get depinned and different stages of depinning are shown up in the differential resistance as a function of current. At high current values disorder due to flux pinning is less important and a nearly defect free coherent motion is observed. Dr. Bhattacharya described the dynamics of flow in the three regimes.

Dr. Gammel reported on the studies of pinned and flowing flux lines in single crystals of $NbSe_2$ and Nb. These studies were carried out using small angle neutron scattering combined with decoration techniques, transport and thermodynamic measurements. For low fields in which the flux lines are pinned in $NbSe_2$, the relation between correlation lengths and transport current density is found to be in excellent agreement with the Larkin-Ovechennikov collective pinning theory. Near H_{c2}, the theory breaks down. A peak in the critical current density is seen here. In $NbSe_2$ it was not possible to see any connection between flux lattice melting and the peak effect. But in Nb SANS technique was used to image the flux lattice. At the peak effect a drastic disordering of the flux lattice was seen. While the flux lines remain correlated along their length, the positional order, followed by orientational order transverse to the flux line, disappears. The transport current density remains finite and shows history effects. This indicates a pinned glue of flux lines near H_{c2} rather than a vortex liquid state. Near the critical current two stages of depinning are seen first to a plastic flow regime and later to a flowing state.

Dr. Kunchur reported his studies on the freely flowing flux lattice and the current induced modification of the order parameter. To get the free flux flow one has to use currents far in excess of the critical currents. To avoid heating effects at such high currents one uses a pulsed mode of measurement. With such pulsed modes one can go up to power densities of 10^8 watts/cm^3. He showed that the flux flow resistivity normalised to the normal state resistivity was a linear function of the applied magnetic field in accordance with the Bardeen-Stephen equation. The Magnus effect on the vortices makes the Hall angle to be different from that in the normal state. The difference in Hall angle increases with temperature and this curve is independent of the applied field. Measurements were made of the flux flow resistivity as a function of temperature at a constant magnetic field and different current densities. The onset temperature of the resistive transition is not changed much when the current density is increased. However, the mid-point of the transition changes with current density. The change in the mid-point temperature varies as the two-thirds power of the current density. This is the first experimental evidence for the depairing of electrons due to current.

In all the above experiments the current is at right angles to the applied field. When the current is parallel to the applied field no force is experienced by the vortices. This case was discussed by Dr. Campbell who concluded that this problem is not still completely solved.

Mention has already been made of the peak effect. In YBCO the peak effect occurs due to clusters of point defects. Dr. Kupfer discussed the case of peak effect in oxygen deficient YBCO. Here the field B_P at which the peak in current density is seen as well as B_O the field at which current density starts increasing towards the peak, move to lower fields as the oxygen deficiency increases. Below B_O, elastic interaction between bundles of vortices is predominant, while, above B_p, plastic flow of the flux lines dominates. Between B_o and B_P the transverse correlation length decreases, ultimately becoming less than the vortex spacing.

Dr. Nishihara dealt with magnetisation measurements in single crystals of LSCO as a function of doping. In underdoped sample a peak in current density is seen around 0.2 T while in over-doped samples this peak is seen around 1 T. In each region the peak position does not depend on the carrier density. The results indicate two types of pinning centres, a weak pinning centre in under-doped regime, and a strong pinning centre in the over-doped regime.

$NbSe_2$ is archetypal of a highly anisotropic layered weakly pinned superconductor. In such a superconductor the melting line is expected to show a re-entrant behaviour. If at a constant temperature field is increased it will intersect the melting line twice. Since the peak effect is associated with a transition from a vortex lattice to a vortex fluid regime, one must get two peak effects in this case. Dr. Grover presented AC susceptibility results to indicate precisely this behaviour. The results were supported by DC magnetisation measurements. In BSCCO non-resonant rf absorption studies carried out by Dr. Bhat also show evidence for a re-entrant melting line. In cubic $CeRu_2$ multiple peaks are seen around 1 T. The reason for multiple peaks is not clear.

For practical applications one should get as large a critical current density as possible. This is achieved by increasing the strength of pinning by creating defects having a size comparable to the coherence length. In the HTS the coherence length is a few nm. Irradiation by neutrons produces a conglomeration of defects of the correct size to cause strong core pinning. Dr. Weber presented results on many HTS to show that neutron irradiation (a) increases the current density at self field by about an order of magnitude, (b) increases the current density at high fields and (c) increases current density at high temperatures. Energetic heavy ions in the range of 500 MeV to 1 GeV produce columnar defects which also serve to provide a back bone for the c-axis. Such columnar defects lead to a current density which depends on the angle between the direction of the heavy ion beam and the magnetic field. Such columnar defects can also be produced by fission fragments. Since these fragments produce splayed columnar defects the current density is less dependent on the direction of the field.

Dr. Budhani presented his results on critical current density measurements on the layered HTS materials subjected to heavy ion irradiation. The anisotropy of the flux flow resistivity with respect to applied field is clearly demonstrated in these experiments.

We have so far discussed about weak and strong pinning effects in HTS. It would be nice if we can produce these defects in a regular geometric array to study their behaviour. This can be done by producing an artificial array of flux pinning centres using electron beam lithography and lift off technique. One can create a regular pattern of holes of nm diameter in a film of superconductor. For a given diameter of the hole there will be a maximum number of vortices which can be accommodated. Additional vortices introduced will have to go into interstitial sites where flux pinning will be weak. Using such patterned holes in WGe amorphous superconducting film Dr. DeLong showed that the existence of different vortex phases (the perfectly pinned lattice, weak interstitial vortices and the vortex fluid) can be distinguished by the angular, field, frequency and drive dependencies.

Now we come to applications. When HTS were discovered it was expected, overoptimistically, that high current applications of these materials at liquid nitrogen temperature will be realised soon. However, the short coherence length and granular nature of the materials pose many problems for processing the materials. In the quenched melt growth technique one preecipitates finely divided 211 impurity phase in 123 material to provide effective flux pinning. One can also texture the material to align the grains. Dr. Murakami has utilised such a processing technique to make 123 material for application in magnetic bearings. A photograph of a fly wheel for energy storage (50 Whr) operating at 2000 rpm was shown by Dr. Murayama. Dr. Balachandran reported on the development of several kilometer lengths of BSCCO tapes with high current densities. A 200 HP motor operating at 27 K has been demonstrated successfully. A 2.4 kV and 2.2 kA fault current limiter has also been tested successfully. It looks like HTS materials have come of age in high current applications.

The efforts in Bhabha Atomic Research Centre to develop NbTi wires and to use them in some applications were highlighted by Dr. Sahni. There have been other attempts, notably at the National Physical Laboratory, Delhi, which did not find mention in the Symposium. Two superconducting high magnetic field gradient separators using NbTi wires have been built, one by the Department of Atomic Energy institutions and one by Bharath Heavy Electricals in collaboration with NPL, Delhi. A 5 MVA superconducting generator has been built by BHEL with collaboration from NPL, Delhi, the Indian Institute of Science, Bangalore, the Indian Institutes of Technology at Madras and Kharagpur. The generator has been tested at Hyderabad. The NPL, Delhi, has built a high homogeneity SC magnet using NbTi wires for NMR applications. All these applications were given a push when HT superconductivity was discovered. There have been some efforts at developing BSCCO tapes at NPL. But these are still in the rudimentary stage. It is a pity that major funding for superconductivity research was suddenly cut in the country. A long term policy with far sighted goals is lacking in this country.

One of the major applications of HTS materials is in the development of microwave filters. Dr. Pinto of TIFR described measurements of microwave surface resistance in films of YBCO. Dr. Sarkar of SAMEER, Mumbai, has elaborated on the design aspects of some microwave devices. Stripline resonators operating at S, C and X bands with YBCO films on

MgO and $LaAlO_3$ substrates have been developed. HTS filters at C band are under development. As mentioned previously there are a few laboratories in India which are capable of producing high quality HTS films. But however with the exception of TIFR, no serious attempts have been made in other laboratories to develop applications of these thin films.

While on microwave applications of HTSC material, mention may be made of the talk of Dr. Pedersen who used two techniques to measure the noise in different types of junctions. He found that the variation of noise temperature with junction temperature is different for different types of junctions. The reason for such differences is not known.

Dr. A.K. Gupta described his work on setting up a Josephson voltage standard in NPL. He has also studied different types of junctions in thick and thin films of HTS materials and has developed a RF SQUID out of these materials operating at 77K. A RF SQUID device for application purposes has been built.

Dr. T.S. Radhakrishnan has set up facilities for preparing high quality $Nb\text{-}Al_2O_3\text{-}Nb$ Josephson junctions. He has made a DC SQUID and a number of applications for the DC SQUID are being planned.

I have come to the end of my summary. I am sure there were other important results reported in the posters and I hope the participants had discussions on these aspects with the authors. To present this review has been a daunting task. But I ventured on this to present a summary of all the important talks for the benefit of the young Indian scientists who have just started to work in this area. I hope they find this summary helpful.

This was a highly successful conference and gave an opportunity for the Indian participants to meet distinguished scientists from abroad and to discuss with them their research results. To relieve the tedium of long hours spent in the lecture hall, a cultural program was organised on one of the days of the symposium. The dance program was colourful and vibrant and was enjoyed by all. All the arrangements were made with meticulous care.

On my behalf and on behalf of all the participants, I would like to convey our thanks to Professors Malik, Pinto and all their colleagues for sparing no efforts to make this Symposium a success.

Thank you.

AUTHOR INDEX

Participants ASMCCD '96
A Golden Jubilee event of T.I.F.R., September 17-20, 1996, Mumbai, India.